TOXICOLOGY OF GLUTATHIONE TRANSFERASES

T0172728

TOXICOLOGY OF GLUTATHIONE TRANSFERASES

Edited by

Yogesh C. Awasthi

University of Texas Medical Branch
Galveston, Texas, U.S.A.

CRC Press
Taylor & Francis Group
Boca Raton London New York

CRC Press is an imprint of the
Taylor & Francis Group, an **informa** business

CRC Press
Taylor & Francis Group
6000 Broken Sound Parkway NW, Suite 300
Boca Raton, FL 33487-2742

First issued in paperback 2019

© 2010 by Taylor & Francis Group, LLC
CRC Press is an imprint of Taylor & Francis Group, an Informa business

No claim to original U.S. Government works

ISBN-13: 978-0-8493-2983-8 (hbk)
ISBN-13: 978-0-367-39053-2 (pbk)

A CIP record for this book is available from the British Library.

Library of Congress Cataloging-in-Publication Data available on application

Visit the Informa Web site at
www.informa.com

and the Informa Healthcare Web site at
www.informahealthcare.com

Preface

Ever since the discovery of an enzyme catalyzing the conjugation of electrophilic xenobiotics to cellular nucleophilic glutathione, voluminous studies have probed important questions about the pharmacological and physiologic significance of this enzyme referred to as glutathione S-transferase (GST), or glutathione transferase. Several aspects of this rather unusual enzyme have fascinated investigators. Its rigid requirement for GSH on one hand, and the promiscuity enabling it to handle structurally diverse electrophilic compounds as the second substrate on the other hand, have attracted the attention of enzymologists and toxicologists alike. Because of its ability to catalyze the conjugation of various xenobiotics, the enzyme system was correctly characterized as heterogeneous in earlier studies, and the enzymes were assigned different names based on substrate preferences. Studies leading to the identification of GST with ligandin correctly predicted its two-pronged role in providing defense against toxic electrophiles through conjugation to GSH and through sequestration by nonenzymatic binding. The discovery of induction of GSTs by antioxidants and subsequent protection against chemical carcinogens *in vitro* and *in vivo* shown in later studies established the role of this enzyme as a major player in the defense against electrophilic xenobiotics. Among phase II enzymes, GST became the favorite enzyme of toxicologists.

Earlier studies on GST focused on identifying the multitude of its substrates and its relationship with ligandin. During this period, the structural elucidation of these enzymes through classical protein chemistry techniques led to major advances in this field. The discovery of immunologically distinct forms of GSTs from erythrocytes, placentas, and livers during the period from the later 1970s to the early 1980s showed that this family of enzymes is encoded by several genes and that GSTs are composed of at least three immunologically distinct subunits. This provided the basis of the classification of GSTs into three distinct subfamilies (Alpha, Mu, and Pi), members of which had overlapping catalytic and structural properties. The existence of a gene superfamily of these enzymes was thus realized, and the advent of techniques for molecular cloning led not only to the discovery of several new members of the Alpha, Mu, and Pi families, but also to the identification of several new gene families within the superfamily of GSTs. Genomic cloning of these enzymes provided information on the structure of genes, regulation of their expression, and phylogenetic interrelationships among these genes. Perhaps even more important, these studies led to the discovery of polymorphisms within members of the GST gene family, and an important area of research emerged that defined the clinical and toxicological significance of these polymorphisms in human populations.

In parallel with studies on the identification of genes and on structural and kinetic properties of gene products, another area of GST research has focused on the role of these enzymes in chemical carcinogenesis and multidrug resistance. The ability of GSTs to neutralize potent carcinogens, such as benzo[*a*]pyrene (B[*a*]P) metabolites including (+) B[*a*]PDE, led to a surge of studies to identify generally regarded as safe (GRAS) micronutrients that induce GSTs and could be used as effective chemoprotective agents. These studies have resulted in the discovery of potentially promising chemoprotective agents, of natural as well as of synthetic origin. In addition, investigators dealing with this aspect of GSTs have immensely contributed to our current knowledge of GST gene expression, of reaction mechanisms of GST-mediated catalysis, and also of the mechanisms of chemical carcinogenesis. The ability of GSTs to detoxify electrophilic alkylating drugs, such as melphalan and chlorambucil, and their overexpression in drug-resistant cell lines and certain tumors provided evidence for a role of GSTs in the mechanisms of the drug resistance of cancer cells, which continues to be the major impediment in treating cancer. This encouraged studies to develop efficient GST inhibitors and accelerated the pace of efforts to determine the three-dimensional structures of GSTs. Consequently, at present, crystal structures of GSTs of almost all families are known.

In the last two decades of the previous century, studies on GSTs heavily focused on cancer, which was the major theme of international workshops and conferences on GSTs organized in Europe and the United States. While it may be argued that studies on the role of GSTs in drug resistance have as yet not yielded results that could be clinically applicable, these studies have contributed greatly toward the understanding of the mechanisms of multidrug resistance that, as we understand now, are pleotropic in nature. In particular, these studies have been catalysts in understanding the significance of the transport of GSH conjugates of electrophiles (GS-E), and have led to discoveries that suggest that the rate of GS-E transport could be a determinant of not only drug resistance of cells, but also a regulator of stress-mediated signaling. These rather exciting offshoots of GST research are covered in this volume along with other newly discovered physiological roles of GSTs.

Even though a remarkably high level of GST protein is constitutively present in the liver and is further induced upon xenobiotic exposure, GSTs were primarily thought to be of pharmacological relevance. Only after two decades of the discovery of GSTs, the Se-independent glutathione peroxidase (GP_x) activity and the antioxidant potential of these enzymes are being realized. Recent studies demonstrating a prominent role of the Alpha-class GSTs in negating the toxicity of oxidants through their GP_x activity open a new area of GST research centered on their role as antioxidant enzymes. Perhaps even more important, the physiological role of GSTs in the regulation of stress-mediated signaling is the focus of current work. GSTs seem to affect signaling processes through their multiple activities. Recent studies suggest an important role of the electrophilic end products of lipid peroxidation such as 4-hydroxynonenal (4-HNE) as "signaling molecules." Two of the activities of GSTs — the GP_x and transferase activities — can regulate the intracellular concentration of these signaling molecules and thereby modulate signaling. In addition, the ligandin-type binding activity of GSTs also seems to function in modulation of signaling, and there is compelling evidence that through binding with specific kinases, GSTs regulate stress signaling.

The endogenous electrophiles formed during stress conditions are conjugated to GSH to form the conjugate (GS-E) whose intracellular levels are determined through the coordinated functions of GSTs and of transporters extruding GS-E from cells. Recent studies show that GS-E is also involved in signaling mechanisms. There is mounting evidence that the intracellular concentration of GS-E is linked to one of the signal-terminating mechanisms involving endocytosis of receptors. This opens up a new exciting field on the physiological significance of the transferase activity of GSTs, particularly toward the endogenously generated electrophiles under stress conditions.

While excellent reviews and volumes covering specific aspects of GSTs are available, there is no recent comprehensive account of GST research that would cover new directions in the context of previous results. Through rather limited scope, this volume attempts to fill that void. The chapters are written by experts who have been involved in all aspects of GST research and have actively contributed to the progress in this field. Starting with a historical perspective, the book provides updated information on gene families, structure and regulation of gene expression, reaction mechanisms and substrates, three-dimensional structures of GSTs, design of proteins with GST activity, and various approaches to develop inhibitors for clinical use to overcome drug resistance. The physiological significance and the mechanisms of activation of microsomal GSTs are also covered. These topics, along with the role of GSTs in detoxification of chemical carcinogens and possible implications of GST polymorphism in human health and disease, should make this volume useful to researchers in the field of pharmacology and toxicology.

A possible role of GSTs in defense mechanisms against chronic oxidative stress-linked disorders such as atherosclerosis is also covered. The newly emerging area of GST-mediated modulation of stress signaling through its catalytic as well as binding activities is also extensively covered. Interestingly, most of the contributors have at least casually mentioned this role of GSTs in their chapters. Recognition of this fast emerging physiological role of GSTs by all the experts underscores the physiological and pharmacological significance of this novel function of GSTs. An intriguing hypothesis covered in the book strongly suggests an important role of the factors that regulate the GSH-electrophile conjugate (GS-E) homeostasis in the regulation of cellular processes. These studies point to a novel role of disturbances in GST-linked processes in the etiology of disorders including diabetes and athero-sclerosis. This book should therefore be of interest to clinicians involved in basic as well as translational research, particularly those studying oxidative stress-related degenerative diseases. Chapters on known and potential endogenous electrophilic substrates and a major role of GSTs in suppressing stress-induced lipid peroxidation reinforce this newly emerging role of GSTs.

In order to make this volume useful to new investigators, a chapter on techniques for purification, quantitation, and kinetic studies with various substrates has been included so that the volume could serve as a reference book for bench workers. Likewise, a chapter on the composition of GST isozymes in human tissues has been included for the benefits of investigators using animal models in toxicological research. The latter chapters cover common and widely known aspects of GSTs familiar to the experts; these chapters should be helpful to investigators less familiar with this enzyme system.

Even though each chapter focuses on a particular aspect, readers may find some overlap, particularly in chapters on relatively new aspects such as signaling, polymorphisms, physiological functions, and so on. While this was completely unavoidable because most areas of GST research are intricately interconnected, some overlap was, in fact, encouraged, with the expectation that considering a topic in different contexts will foster its deeper understanding. It is hoped that the volume will attract interest from a wide readership and will contribute not only to a toxicological understanding of GSTs but will be useful to biochemists, physiologists, and clinicians alike. While glutathione transferase is the scientifically correct name, use of the name glutathione S-transferase and its abbreviation GST is highly prevalent in scientific literature. Therefore, the names glutathione transferase, glutathione S-transferase, and the abbreviation GST have been used interchangeably throughout this volume.

I sincerely thank all of the contributors to this book. I am also thankful to colleagues in my laboratory who have helped me in many ways during this venture. In particular, I thank Mrs. Rose Byrdlon-Griggs for her help in typing and Dr. Shaheen Dhanani for her help in proofreading, formatting, and checking the references. This book was supported in part by NIH Grants EY04396 and ES012171.

Contributors

Yoko Aniya
Laboratory of Molecular
  Pharmacology
Graduate School of Medicine
University of the Ryukyus
Japan

G.A. Shakeel Ansari
Department of Biochemistry and
  Molecular Biology
University of Texas Medical Branch
Galveston, Texas

Irwin M. Arias
Departments of Physiology and
  Medicine
Tufts University School of Medicine
Boston, Massachusetts
and
Cell Biology and Metabolism Branch
National Institutes of Child Health and
  Development
National Institutes of Health
Bethesda, Maryland

Narayan G. Avadhani
Department of Animal Biology
School of Veterinary Medicine
University of Pennsylvania
Philadelphia, Pennsylvania

Yogesh C. Awasthi
Department of Biochemistry and
  Molecular Biology
The University of Texas Medical
  Branch
Galveston, Texas

Sanjay Awasthi
Department of Chemistry and
  Biochemistry
University of Texas at Arlington
Arlington, Texas

Paul Boor
Department of Pathology
The University of Texas Medical
  Branch
Galveston, Texas

William Carroll
Human Genomics and Disease Group
Institute for Science and Technology in
  Medicine
Keele University Medical School
Hartshill Campus
Staffordshire, England

Shaheen Dhanani
Department of Biochemstry and
  Molecular Biology
Galveston, Texas

Kenneth Drake
Department of Chemistry and
  Biochemistry
University of Texas at Arlington
Arlington, Texas

Victoria J. Findlay
Department of Cell and Molecular
  Pharmacology and Experimental
  Therapeutics
Medical University of South Carolina
Charleston, South Carolina

Anthony Fryer
Human Genomics and Disease Group
Institute for Science and Technology in
  Medicine
Keele University Medical School
Hartshill Campus
Staffordshire, England

John D. Hayes
Biomedical Research Centre
University of Dundee
Ninewells Hospital and Medical School
Dundee, Scotland

Paul Hoban
Human Genomics and Disease Group
Institute for Science and Technology in
  Medicine
Keele University Medical School
Hartshill Campus
Staffordshire, England

Sarah L. Holley
Human Genomics and Disease Group
Institute for Science and Technology in
  Medicine
Keele University Medical School
Hartshill Campus
Staffordshire, England

Ylva Ivarsson
Department of Bicohemistry and
  Organic Chemistry
Uppsala University
Biomedical Center
Uppsala, Sweden

Ian R. Jowsey
Unilever Safety and Environmental
  Assurance Centre
Bedfordshire, England

Irving Listowsky
Department of Biochemistry
Albert Einstein College of Medicine
Bronx, New York

Bengt Mannervik
Department of Biochemistry
Uppsala University
Biomedical Center
Uppsala, Sweden

Ralf Morgenstern
Institute of Environmental Medicine
Division of Biochemical Toxicology
Karolinska Institutet
Stockholm, Sweden

Haider Raza
Department of Biochemistry *and*
Department of Medicine and Health
  Sciences
United Arab Emerates University
Al Ain, United Arab Emirates

Rajendra Sharma
Department of Biochemistry and
  Molecular Biology
University of Texas Medical Branch
Galveston, Texas

Mikuyi Shimoji
Division of Biochemical Toxicology
Institute of Environmental Medicine
Karolinska Institutet
Stockholm, Sweden

Sharda P. Singh
University of Arkansas for Medical
  Sciences
Department of Pharmacology and
  Toxicology
Little Rock, Arkansas

Shivendra V. Singh
Department of Pharmacology and
  Urology
Biochemoprevention Program
Hillman Cancer Centre
Pittsburgh, Pennsylvania

Jyotsana Singhal
Department of Chemistry and
  Biochemistry
University of Texas at Arlington
Arlington, Texas

Sharad Singhal
Department of Chemistry and
  Biochemistry
University of Texas at Arlington
Arlington, Texas

Richard C. Strange
Clinical Biochemistry and Research
  Laboratory
Keele University School of Medicine
Hartshill Campus
University of Hospital of North
  Staffordshire
Staffordshire, England

Kenneth Tew
Department of Cell and Molecular
  Pharmacology and Experimental
  Therapeutics
University of South Carolina
Charleston, South Carolina

Danyelle M. Townsend
Department of Pharmaceutical Sciences
Medical University of South Carolina
Charleston, South Carolina

Hui Xiao
Department of Pharmacology
University of Pittsburgh School of
  Medicine
Pittsburgh, Pennsylvania

Sushma Yadav
Department of Chemistry and
  Biochemistry
University of Texas at Arlington
Arlington, Texas

Yonghzen Yang
Department of Pathology
The University of Texas Medical
  Branch
Galveston, Texas

Yusong Yang
Department of Pathology
SUNY Downstate Medical Center
Brooklyn, New York

Piotr Zimniak
Department of Pharmacology and
  Toxicology
University of Arkansas for Medical
  Sciences
and
Central Arkansas Veterans Healthcare
  System
Little Rock, Arkansas

Contents

1 Ligandin and Glutathione S-Transferases: Historical Milestones

Irving Listowsky and Irwin M. Arias

CONTENTS

1.1   EARLY HISTORY

Mercapturate formation as a mechanism for the physiological metabolism and excretion of xenobiotics was known in the nineteenth century.[1,2] However, it was not until 1961 that an enzymatic reaction responsible for the first step in the conjugation of xenobiotics with glutathione was recognized. Thus, Combes, Boyland, and others[3-5] reported on the conjugating activity in liver extracts. The enzymes, named glutathione transferases (GSTs), were originally classified on the basis of presumed substrate specificities (such as aryltransferase, alkyltransferase, epoxidetransferase, and so on),[5,6] until it became clear that the enzymes actually had overlapping specificities.

1.2   LIGANDIN–GST RELATIONSHIPS

Three roads merged in 1971 largely due to advances in gel permeation chromatography and the availability of Sephadex matrices. The research interests of the groups of Ketterer et al. (azodye carcinogen mechanisms),[7,8] Litwack and Morey (cortisol metabolism),[9] and Arias and coworkers (uptake of bilirubin and other organic anions by hepatocytes)[10] coalesced about an interesting binding protein that was subsequently named ligandin.[11] Use of Sephadex columns permitted separation of cytoplasmic proteins based on size rather than on charge. In each case, chromatographic methods revealed that following administration of radiolabeled azocarcinogen, cortisol, or bilirubin to rats, the cytoplasmic fraction of liver contained three easily resolved components. The first was largely ignored because it eluted with many proteins in the void volume of the columns. A second labeled fraction coeluted with proteins of about 50 kDa and a third low-molecular-weight fraction of less than 20 kDa. The proteins responsible for the binding of specific ligands were isolated and initially identified as azocarcinogen metabolite-binding protein, cortisol metabolite-binding protein, and "Y protein" (which bound bilirubin and many other organic anions). The third peak was similarly labeled and included the "Z protein" fraction (which bound organic anions, particularly long-chain fatty acids). The three investigators interpreted the results in line with their own research interests. These findings occurred during early stages of other studies on cytochrome P450s, receptors and membrane transporters, and pathways of drug metabolism. Advances in those fields later provided much information regarding the basic mechanisms involved in the interests of the three experimentalist groups.

At about the time Litwack, Ketterer, and Arias reported their observations,[11] Jakoby et al. at NIH used ion exchange and gel permeation chromatographic methods and succeeded in purifying a family of 50-kDa liver cytoplasmic proteins having catalytic activity for conjugating glutathione with a series of electrophilic compounds;[12] these enzymes were deemed to be the glutathione transferases (GSTs) predicted by Boyland and Combes. Jakoby's work laid to rest the concept that putative GSH transferases could be characterized on the basis of their substrate specificities. The purified enzymes showed overlapping activities with different substrates. The major GST in rat liver, GST B (later shown to be among the Alpha-class GSTs), had many biochemical features similar to those of the three soluble proteins identified by the groups of Arias, Ketterer, and Litwack mentioned above. In an important fundamental study, use of cross-immunologic and immunodiffusion methods established that the Ketterer, Litwack, and Arias proteins and Jakoby's GST B were, in fact, the same protein.[13] After much discussion, the investigators agreed to call the 50-kDa protein ligandin, a functionally benign term indicating only the general binding capacity of the protein. Shortly thereafter a fourth road appeared on the scene when Talalay and coworkers identified the soluble delta 5, 3-ketosteroid isomerase activity as a glutathione transferase.[14]

Subsequently, the roads diverged. Ketterer continued studies on the relation of ligandin to chemical carcinogenesis.[15] The Arias group examined the regulation of expression of ligandin, its role in organic anion uptake by the liver, and its dual role as a GST and binding protein. The results of those studies provided data consistent with the function of ligandin as an inducible intracellular binding protein[16] that could regulate the efflux of organic anions between plasma and intracellular sites.[17] Habig et al. employed high-resolution methods to separate multiple forms of the 50-kDa protein GST fraction, and peptide maps suggested that they were derived from different gene products. The GST supergene family thus emerged.

1.3 INTRACELLULAR LIGAND BINDING

In early studies ligandin and subsequently the GSTs were shown to be abundant cytoplasmic proteins which exhibited characteristic broad specificity binding capacity for diverse nonsubstrate endogenous and exogenous ligands. This feature led to suggestions that the ligandin/GST system could function as intracellular stoichiometric ligand-binding proteins,[18] in support of the observation that hepatic net uptake of organic anions, such as bilirubin and BSP, correlated positively with the intracellular concentration of ligandin in developing livers.[16]

Wolkoff et al.[19] developed an ingenious method to quantify hepatic influx and efflux rates in perfused rat livers. Studies with radiolabeled bilirubin and BSP quantitatively related the concentration of ligandin, during development or after induction, inversely with the efflux rate of organic anions back into the plasma following hepatic uptake.

Based on the studies, it was proposed that ligandin is the major organic anion-binding protein in rat liver and regulates the net uptake by controlling their efflux into the plasma following hepatic uptake[18,20] by what we now know to be multiple membrane transporters. Ligandin permits these organic anions to remain within the liver where they subsequently interact with diverse detoxification and secretory systems, including UDP glucuronosyl transferase, sulfate transferases, and GST as well as downstream membrane transporters for moving conjugated or unconjugated ligands into the bile. Similar handling was proposed for the proximal tubule of the nephron, small intestinal mucosa, and, subsequently, the choroid plexus.

1.4 MAJOR ADVANCES

GSH affinity matrices used by Simons and Vander Jagt[21,22] facilitated purification procedures; a single step allowed for purification of GSH-binding proteins in a cell extract including the family of GSTs. This later led to the widespread notion that GST-fusion proteins can be used to purify diverse proteins based on binding affinities and specificities of GSTs for GSH. Advances in the GST field subsequently paralleled general technological advances in biology in the next two decades. Thus, in the mid-1980s, GST genes were cloned by several groups including those of Board, Daniel, Pickett, Tu, and others.[23-30] It was soon recognized that mammalian GSTs are products of gene superfamilies. The GSTs were classified into subfamilies based on sequence homologies and other common properties.[31,32] There are at least seven different cytosolic classes that have been identified (Alpha, Mu, Pi, Theta, and so on) and subclasses within each class that probably derived from gene duplication or conversion events.[33] The extent of the genetic diversity of mammalian GSTs has recently been disclosed because sequences of human, rodent, and other genomes have been determined. Unfortunately, with the emerging complexity of the system, inconsistencies in nomenclature used in the literature resulted in confusion among investigators in naming and assigning properties to individual GST forms. Currently adopted nomenclature has rectified this problem to some extent.[34]

In the early 1990s, crystal structures of proteins from the three major cytosolic GST classes (Alpha, Mu, and Pi) were solved.[35–37] Investigators were thereby able to probe fine points about ligand binding and catalytic mechanisms of the GSTs.

1.5 REGULATION OF GST GENE EXPRESSION

To follow early studies suggesting that ligandin is inducible,[16] Pearson, Benson, and Talalay[38] showed that rodent GST gene expression is enhanced by structurally diverse compounds. That disclosure opened new directions in this field. Many of the inducers were also implicated among compounds that protect animals and perhaps humans against chemical carcinogens and are considered to be chemopreventive agents. This led to the hypothesis that induction of certain GSTs protects the cell against noxious agents, such as chemical carcinogens. The work of Tew and other investigators also pointed to elevations of GSTs in the acquisition of drug resistance to chemotherapeutic agents.[39]

Studies on the observed induction mechanisms were promoted by observations in the laboratories of Pickett and Daniel.[40–45] Those workers detected sequence motifs in the 5'-flanking regions of genes for rodent GSTs that were shown to be responsible for the inducible gene expression. These motifs, called ARE (antioxidant response elements) or EpRE (electrophile response elements), have opened major areas of exploration in the GST field.

Perhaps the most significant development in this field has been the recent identification by Hayes and Yamamoto of a transcription factor Nrf2 that appears to bind to the ARE sequences and enhance transcription of GSTs and other detoxification genes.[46–54] Mechanisms of interaction of Nrf2 and a cytoskeletal-bound Keap1 protein, translocation of Nrf2 to the nucleus, or degradation mechanisms involving this system have advanced our understanding of the relevant regulatory mechanisms.[46–54]

1.6 GSTS AND SIGNALING PATHWAYS

Recent evidence has linked GSTs to signaling pathways by their effects on certain small molecules or by protein–protein interactions. The involvement of GSTs in prostaglandin and leukotriene biosynthesis and metabolism had been postulated for a long period of time.[55–59] In addition, certain toxic products of fatty acid oxidation, including lipid hydroperoxides and alkeneals, notably 4-hydroxynon-2-enal (4-HNE), are substrates for some GSTs.[60–62] More recently it has been suggested that cyclopentenone prostaglandins, such as 15-deoxy-$\Delta^{12,14}$-prostaglandin J_2 (15d-PGJ$_2$), are substrates for GSTs.[63–65] 4-HNE is thought to function in intracellular signaling molecules and 15d-PGJ$_2$ is a ligand for peroxisome proliferation activator receptors (PPARγ). Recent studies have shown that by catalyzing conjugation reactions, GSTs can modulate signaling pathways involving these compounds (for reviews, see References 66–68). GSTs have also been shown to engage in direct protein–protein interactions with important signaling proteins.[69–71] By interactions with cellular components it is likely that the abundant GSTs can influence physiological functions in ways yet to be determined.

1.7 PROSPECTS

With the advances in the molecular biological aspects, the GST field shifted entirely into the topics mentioned above and are considered in the following review articles in this treatise. Unfortunately, the transport aspects have been sidelined, although science, like history, repeats itself. Recent studies have rediscovered the ligand-binding properties of GST. Definitive testing of the functional ligandin hypothesis would involve studies of either an inheritable disorder in which the protein is defective or hepatocytes in which ligandin has been depleted. Although availability of potential mouse models with targeted disruptions of the GST genes may be suited for testing this hypothesis, no such studies have yet been described clinically or experimentally.

This brief historical review suggests that GSTs can function as intracellular binding proteins that regulate net cellular uptake and perhaps certain downstream events. This hypothesis should be revisited using up-to-date techniques including methods to probe the binding and catalytic sites. Advances in proteomics suggest that GSTs engage in protein–protein interactions with signaling and other proteins.[70–72] Moreover, the recent proposal that GSTs have the potential to interact selectively with proteins that have been S-glutathiolated under conditions of oxidative stress[72] predicts additional functions for this abundant class of proteins.

REFERENCES

1. Baumann, E. and Preusse, C., Uber bromphenyl-mercaptursaüre, *Ber. Dtsch. Chem. Ges.,* 12, 806–810, 1879.
2. Jaffe, M., Über die nach einfuhring von brombenzol und chlorbenzol im organismus entstehenden schuetelhaltigen sauren, *Ber. Dtsch. Chem. Ges.,* 12, 1092–1098, 1879.
3. Combes, B. and Stakelum, G.S., A liver enzyme that conjugates sulfobromophthalein sodium with glutathione, *J. Clin. Invest.,* 40, 981–988, 1961.
4. Boyland, E., Ramsay, G.S., and Sims, P., Metabolism of polycyclic compounds 18. The secretion of metabolites of naphthalene. 1,2-dihydronaphthalene and 1,2-epoxy-1,2,3,4-tetrahydronaphthalene in rat bile, *Biochem. J.,* 78, 376–384, 1961.
5. Boyland, E. and Williams, K., An enzyme catalysing the conjugation of epoxides with glutathione, *Biochem. J.,* 94, 190–197, 1965.
6. Booth, J., Boyland, E., and Sims, P., An enzyme from rat liver catalyzing conjugations with glutathione, *Biochem. J.,* 79, 516–524, 1961.
7. Ketterer, B. and Christodoulides, L., Two specific azodye-carcinogen-binding proteins of the rat liver. The identity of amino acid residues which bind the azodye, *Chem. Biol. Interact.,* 1 (2), 173–183, 1969.
8. Ketterer, B. et al., Interactions of azodye carcinogens and azodye carcinogen conjugates with specific proteins in the rat liver, *Chem. Biol. Interact.,* 3 (4), 285–286, 1971.
9. Litwack, G. and Morey, K.S., Cortisol metabolite binder. I. Identity with the dimethylaminoazobenzene binding protein of liver cytosol, *Biochem. Biophys. Res. Commun.,* 38 (6), 1141–1148, 1970.
10. Levi, A.J., Gatmaitan, Z., and Arias, I.M., Two hepatic cytoplasmic protein fractions, Y and Z, and their possible role in the hepatic uptake of bilirubin, sulfobromophthalein, and other anions, *J. Clin. Invest.,* 48 (11), 2156–2167, 1969.

11. Litwack, G., Ketterer, B., and Arias, I.M., Ligandin: a hepatic protein which binds steroids, bilirubin, carcinogens and a number of exogenous organic anions, *Nature,* 234 (5330), 466–467, 1971.

12. Pabst, M.J., Habig, W.H., and Jakoby, W.B., Mercapturic acid formation: the several glutathione transferases of rat liver, *Biochem. Biophys. Res. Commun.,* 52 (4), 1123–1128, 1973.

13. Habig, W.H. et al., The identity of glutathione S-transferase B with ligandin, a major binding protein of liver, *Proc. Natl. Acad. Sci. USA,* 71 (10), 3879–3882, 1974.

14. Benson, A.M. et al., Relationship between the soluble glutathione-dependent delta 5-3-ketosteroid isomerase and the glutathione S-transferases of the liver, *Proc. Natl. Acad. Sci. USA,* 74 (1), 158–162, 1977.

15. Ketterer, B. et al., Ligandin, *Biochem. Soc. Trans.,* 3 (5), 626–630, 1975.

16. Reyes, H. et al., Organic anion-binding protein in rat liver: drug induction and its physiologic consequence, *Proc. Natl. Acad. Sci. USA,* 64 (1), 168–170, 1969.

17. Kamisaka, K. et al., Interactions of bilirubin and other ligands with ligandin, *Eur. J. Biochem.,* 14 (10), 2175–2180, 1975.

18. Listowsky, I. et al., Intracellular binding and transport of hormones and xenobiotics by glutathione-S-transferases, *Drug. Metab. Rev.,* 19 (3–4), 305–318, 1988.

19. Wolkoff, A.W. et al., Role of ligandin in transfer of bilirubin from plasma into liver, *Am. J. Physiol.,* 236 (6), E638–E648, 1979.

20. Listowsky, I., Gatmaitan, Z., and Arias, I.M., Ligandin retains and albumin loses bilirubin binding capacity in liver cytosol, *Proc. Natl. Acad. Sci. USA,* 75 (3), 1213–1216, 1978.

21. Simons, P.C. and Vander Jagt, D.L., Purification of glutathione S-transferases by glutathione-affinity chromatography, *Methods Enzymol.,* 77, 235–237, 1981.

22. Simons, P.C. and Vander Jagt, D.L., Purification of glutathione S-transferases from human liver by glutathione-affinity chromatography, *Anal. Biochem.,* 82 (2), 334–341, 1977.

23. Board, P.G., Biochemical genetics of glutathione-S-transferase in man, *Am. J. Hum. Genet.,* 33 (1), 36–43, 1981.

24. Daniel, V. et al., Rat ligandin mRNA molecular cloning and sequencing, *Arch. Biochem. Biophys.,* 227 (1), 266–271, 1983.

25. Daniel, V. et al., Mouse glutathione S-transferase Ya subunit: gene structure and sequence, *DNA,* 6 (4), 317–324, 1987.

26. Lai, H.C. et al., The nucleotide sequence of a rat liver glutathione S-transferase subunit cDNA clone, *J. Biol. Chem.,* 259 (9), 5536–5542, 1984.

27. Lai, H.C., Grove, G., and Tu, C.P., Cloning and sequence analysis of a cDNA for a rat liver glutathione S-transferase Yb subunit, *Nucleic Acids Res.,* 14 (15), 6101–6114, 1986.

28. Pickett, C.B. et al., Rat liver glutathione S-transferases. Complete nucleotide sequence of a glutathione S-transferase mRNA and the regulation of the Ya, Yb, and Yc mRNAs by 3-methylcholanthrene and phenobarbital, *J. Biol. Chem.,* 259 (8), 5182–5188, 1984.

29. Tu, C.P. et al., Cloning and sequence analysis of a cDNA plasmid for one of the rat liver glutathione S-transferase subunits, *Nucleic Acids Res.,* 10 (18), 5407–5419, 1982.

30. Ding, G.J., Lu, A.Y., and Pickett, C.B., Rat liver glutathione S-transferases. Nucleotide sequence analysis of a Yb1 cDNA clone and prediction of the complete amino acid sequence of the Yb1 subunit, *J. Biol. Chem.,* 260 (24), 13268–13271, 1985.

31. Mannervik, B. et al., Identification of three classes of cytosolic glutathione transferase common to several mammalian species: correlation between structural data and enzymatic properties, *Proc. Natl. Acad. Sci. USA*, 82 (21), 7202–7206, 1985.

32. Mannervik, B., The isoenzymes of glutathione transferase, *Adv. Enzymol. Relat. Areas Mol. Biol.*, 57, 357–417, 1985.

33. Lai, H.C. et al., Gene expression of rat glutathione S-transferases. Evidence for gene conversion in the evolution of the Yb multigene family, *J. Biol. Chem.*, 263 (23), 11389–11395, 1988.

34. Mannervik, B., Awasthi, Y.C., Board, P.G., Hayes, J.D., Di Ilio, C., Ketterer, B., Listowsky, I., Morgenstern, R., Muramatsu, M., Pearson, W.R., Pickett, C.B., Sato, K., Widersten, M., and Wolf, C.R., Nomenclature for human glutathione transferases, *Biochem. J.*, 282, 305–306, 1992.

35. Sinning, I. et al., Structure determination and refinement of human alpha class glutathione transferase A1-1, and a comparison with the Mu and Pi class enzymes, *J. Mol. Biol.*, 232 (1), 192–212, 1993.

36. Reinemer, P. et al., The three-dimensional structure of class pi glutathione S-transferase in complex with glutathione sulfonate at 2.3 A resolution, *Embo. J.*, 10 (8), 1997–2005, 1991.

37. Ji, X. et al., The three-dimensional structure of a glutathione S-transferase from the mu gene class. Structural analysis of the binary complex of isoenzyme 3-3 and glutathione at 2.2-A resolution, *Biochemistry*, 31 (42), 10169–10184, 1992.

38. Pearson, W.R. et al., Increased synthesis of glutathione S-transferases in response to anticarcinogenic antioxidants. Cloning and measurement of messenger RNA, *J. Biol. Chem.*, 258 (3), 2052–2062, 1983.

39. Wang, A.L. and Tew, K.D., Increased glutathione-S-transferase activity in a cell line with acquired resistance to nitrogen mustards, *Cancer Treat. Rep.*, 69 (6), 677–682, 1985.

40. Daniel, V., Tichauer, Y., and Sharon, R., 5′-Flanking sequence of mouse glutathione S-transferase Ya gene, *Nucleic Acids Res.*, 16 (1), 351, 1988.

41. Daniel, V., Sharon, R., and Bensimon, A., Regulatory elements controlling the basal and drug-inducible expression of glutathione S-transferase Ya subunit gene, *DNA*, 8 (6), 399–408, 1989.

42. Friling, R.S. et al., Xenobiotic-inducible expression of murine glutathione S-transferase Ya subunit gene is controlled by an electrophile-responsive element, *Proc. Natl. Acad. Sci. USA*, 87 (16), 6258–6262, 1990.

43. Paulson, K.E. et al., Analysis of the upstream elements of the xenobiotic compound-inducible and positionally regulated glutathione S-transferase Ya gene, *Mol. Cell Biol.*, 10 (5), 1841–1852, 1990.

44. Rushmore, T.H., Morton, M.R., and Pickett, C.B., The antioxidant responsive element. Activation by oxidative stress and identification of the DNA consensus sequence required for functional activity, *J. Biol. Chem.*, 266 (18), 11632–11639, 1991.

45. Rushmore, T.H. and Pickett, C.B., Transcriptional regulation of the rat glutathione S-transferase Ya subunit gene. Characterization of a xenobiotic-responsive element controlling inducible expression by phenolic antioxidants, *J. Biol. Chem.*, 265 (24), 14648–14653, 1990.

46. Enomoto, A. et al., High sensitivity of Nrf2 knockout mice to acetaminophen hepatotoxicity associated with decreased expression of ARE-regulated drug metabolizing enzymes and antioxidant genes, *Toxicol. Sci.*, 59 (1), 169–177, 2001.

47. Hayes, J.D. et al., The Nrf2 transcription factor contributes both to the basal expression of glutathione S-transferases in mouse liver and to their induction by the chemopreventive synthetic antioxidants, butylated hydroxyanisole and ethoxyquin, *Biochem. Soc. Trans.*, 28 (2), 33–41, 2000.

48. Ishii, T., Itoh, K., and Yamamoto, M., Roles of Nrf2 in activation of antioxidant enzyme genes via antioxidant responsive elements, *Methods Enzymol.*, 348, 182–190, 2002.

49. Kobayashi, A., Ohta, T., and Yamamoto, M., Unique function of the Nrf2-Keap1 pathway in the inducible expression of antioxidant and detoxifying enzymes, *Methods Enzymol.*, 378, 273–286, 2004.

50. Kobayashi, M. and Yamamoto, M., Molecular mechanisms activating the Nrf2-Keap1 pathway of antioxidant gene regulation, *Antioxid. Redox. Signal*, 7 (3–4), 385–394, 2005.

51. Kwak, M.K. et al., Role of transcription factor Nrf2 in the induction of hepatic phase 2 and antioxidative enzymes *in vivo* by the cancer chemoprotective agent, 3H-1, 2-dimethiole-3-thione, *Mol. Med.*, 7 (2), 135–145, 2001.

52. McMahon, M. et al., The Cap'n'Collar basic leucine zipper transcription factor Nrf2 (NF-E2 p45-related factor 2) controls both constitutive and inducible expression of intestinal detoxification and glutathione biosynthetic enzymes, *Cancer Res.*, 61 (8), 3299–3307, 2001.

53. McMahon, M. et al., Keap1-dependent proteasomal degradation of transcription factor Nrf2 contributes to the negative regulation of antioxidant response element-driven gene expression, *J. Biol. Chem.*, 278 (24), 21592–21600, 2003.

54. Motohashi, H. and Yamamoto, M., Nrf2-Keap1 defines a physiologically important stress response mechanism, *Trends Mol. Med.*, 10 (11), 549–557, 2004.

55. Bach, M.K. et al., Leukotriene C synthetase, a special glutathione S-transferase: properties of the enzyme and inhibitor studies with special reference to the mode of action of U-60,257, a selective inhibitor of leukotriene synthesis, *J. Allergy Clin. Immunol.*, 74 (3 Pt. 2), 353–357, 1984.

56. Tsuchida, S. et al., Purification of a new acidic glutathione S-transferase, GST-Yn1Yn1, with a high leukotriene-C4 synthase activity from rat brain, *Eur. J. Biochem.*, 170 (1–2), 159–164, 1987.

57. Urade, Y. et al., The major source of endogenous prostaglandin D2 production is likely antigen-presenting cells. Localization of glutathione-requiring prostaglandin D synthetase in histiocytes, dendritic, and Kupffer cells in various rat tissues, *J. Immunol.*, 143 (9), 2982–2989, 1989.

58. Ujihara, M. et al., Prostaglandin D2 formation and characterization of its synthetases in various tissues of adult rats, *Arch. Biochem. Biophys.*, 260 (2), 521–531, 1988.

59. Beuckmann, C.T. et al., Identification of mu-class glutathione transferases M2-2 and M3-3 as cytosolic prostaglandin E synthases in the human brain, *Neurochem. Res.*, 25 (5), 733–738, 2000.

60. Alin, P., Danielson, U.H., and Mannervik, B., 4-Hydroxyalk-2-enals are substrates for glutathione transferase, *FEBS Lett.*, 179 (2), 267–270, 1985.

61. Danielson, U.H., Esterbauer, H., and Mannervik, B., Structure-activity relationships of 4-hydroxyalkenals in the conjugation catalysed by mammalian glutathione transferases, *Biochem. J.*, 247 (3), 707–713, 1987.

62. Singhal et al., Several closely related glutathione S-transferase isozymes catalyzing conjugation of 4-hydroxynonenal are differentially expressed in human tissues, *Arch. Biochem. Biophys.*, 311, 242, 1994.

63. Park, E.Y., Cho, I.J., and Kim, S.G., Transactivation of the PPAR-responsive enhancer module in chemopreventive glutathione S-transferase gene by the peroxisome prolif-erator-activated receptor-gamma and retinoid X receptor heterodimer, *Cancer Res.*, 64 (10), 3701–3713, 2004.

64. Paumi, C.M. et al., Glutathione S-transferases (GSTs) inhibit transcriptional activa-tion by the peroxisomal proliferator-activated receptor gamma (PPAR gamma) ligand, 15-deoxy-delta 12,14prostaglandin J2 (15-d-PGJ2), *Biochemistry,* 43 (8), 2345–2352, 2004.

65. Jowsey, I.R., Smith, S.A., and Hayes, J.D., Expression of the murine glutathione S-transferase alpha3 (GSTA3) subunit is markedly induced during adipocyte differen-tiation: activation of the GSTA3 gene promoter by the pro-adipogenic eicosanoid 15-deoxy-Delta12,14-prostaglandin J2, *Biochem. Biophys. Res. Commun.*, 312 (4), 1226–1235, 2003.

66. Cheng, J.Z et al., Accelerated metabolism and exclusion of 4-hydroxynonenal through induction of RLIP76 and hGST5.8 is an early adaptive response of cells to heat and oxidative stress, *J. Biol. Chem.*, 276, 41213, 2001.

67. Awasthi, Y.C. et al., Regulation of 4-hydroxynonenal mediated signaling by S-glu-tathione S-transferases, *Free Rad. Biol. Med.*, 37, 607, 2004.

68. Cho, S.G. et al., Glutathione S-transferase mu modulates the stress-activated signals by suppressing apoptosis signal-regulating kinase 1, *J. Biol. Chem.*, 276 (16), 12749–12755, 2001.

69. Adler, V. et al., Regulation of JNK signaling by GSTp, *Embo. J.*, 18 (5), 1321–1334, 1999.

70. Dorion, S., Lambert, H., and Landry, J., Activation of the p38 signaling pathway by heat shock involves the dissociation of glutathione S-transferase Mu from Ask1, *J. Biol. Chem.*, 277 (34), 30792–30797, 2002.

71. Manevich, Y., Feinstein, S.I., and Fisher, A.B., Activation of the antioxidant enzyme 1-CYS peroxiredoxin requires glutathionylation mediated by heterodimerization with pi GST, *Proc. Natl. Acad. Sci. USA*, 101 (11), 3780–3785, 2004.

72. Listowsky, I., Proposed intracellular regulatory functions of glutathione transferases by recognition and binding to S-glutathiolated proteins, *J. Pept. Res.*, 65 (1), 42–46, 2005.

2 Families of Glutathione Transferases

Piotr Zimniak and Sharda P. Singh

CONTENTS

2.1 INTRODUCTION AND PHYLOGENETIC DISTRIBUTION

A glutathione transferase (abbreviated as GST, based on the former name glutathione S-transferase) is defined as an enzyme capable of catalyzing the basic reaction in which glutathione is added to an electrophilic center of an acceptor molecule. The resulting conjugate (thioether) is typically the end product of the reaction, although in some cases (further discussed in Chapter 5 in this book) the binding of glutathione is transient, leading to reaction products other than conjugates. Overall, the major function of GSTs is the metabolism of electrophiles, of both endogenous and external origin. The ubiquitous nature of electrophiles and their potential to disrupt biological homeostasis are likely to have caused the evolution of a defense mechanism very early in the history of life. Accordingly, GSTs are found in a wide range of taxa. GSTs are present in proteobacterial subclasses Alpha, Beta, and Gamma and in cyanobacteria (see Reference 1 for bacterial taxonomy) but not in other phyla within the domain Bacteria, nor in Archaea.[2] It has been concluded[2] that prokaryotes that utilize compounds other than glutathione as their major intracellular thiol[3] lack GSTs. In eukaryotes, GSTs have been identified in all major taxa including fungi, plants, and animals. Multiple GST isoforms are expressed in any given organism. For example, analysis of the human genome yielded an estimate of 40 GSTs or GST-related proteins in human tissues.[4] The estimate is reasonable given that 24 human GSTs are listed in a current review;[5] additional GSTs may still be identified, and proteins exist that are not conventional GSTs but contain a GST-like domain (see later in this chapter). Forty GSTs would account for approximately 0.2% of all proteins encoded by the human genome. This percentage remains relatively constant across species, including insects (mosquitoes and fruit flies), worms (*Caenorhabditis elegans*), yeast, plants (*Arabidopsis thaliana*), fish (*Fugu rubripes*), and humans.[4]

2.2 OVERVIEW: CLASSIFICATION OF GSTS

Four major groups of enzymes with GST activity have been identified: the bacterial fosfomycin-resistance proteins, membrane-associated proteins in eicosanoid and glutathione metabolism (MAPEG), Kappa-class GSTs, and canonical GSTs.

The bacterial fosfomycin-resistance metalloproteins[6,7] belong to the vicinal oxygen chelate (VOC) superfamily[8] and do not appear to be related to the other groups of GSTs. These specialized enzymes will not be discussed in this chapter further. Of the remaining GST groups (Figure 2.1), two, or perhaps all three, share a common ancestor containing a thioredoxin-like fold.[9,10] The versatile thioredoxin structure has the ability to bind glutathione, which can then be used for a variety of processes including redox reactions, disulfide exchange, conjugative reactions, redox sensing, and others. Reactions of GSTs are further discussed in Chapter 5; a number of excellent recent reviews deal with processes catalyzed by other thioredoxin fold-containing proteins.[11–17]

Among the four groups of GSTs, the soluble, cytosolic ("canonical"[18]) GSTs are most numerous and most extensively studied. Historically, these enzymes have been first identified in mammalian tissues. It quickly became evident that the group is heterogeneous, and much of the early effort in the field was directed toward separation and characterization of individual GST isoforms. The introduction of several parallel naming schemes for GSTs has caused considerable confusion, which, to some extent, continues today. Following the example of naming conventions adopted for other multigene families of detoxification enzymes, most notably cytochromes P450,[19–22] a consensus systematic nomenclature for human GSTs was introduced.[23] The same convention was then widely adopted for GSTs from other vertebrate species. More recently, it was proposed to extend the naming scheme to GSTs of invertebrate origin.[24]

The mitochondrial (Kappa-class) GSTs,[10,25] reported to be also present in peroxisomes,[26] were originally thought to belong to the canonical GST superfamily.

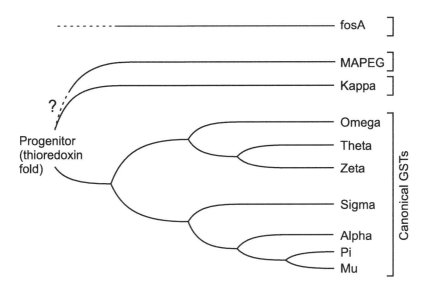

FIGURE 2.1 Phylogenetic tree of major GST classes represented in mammals. The dendrogram is based on published data[32,58,59,95,104,105,110] as well as on analyses carried out using MrBayes[31] and MEGA3[111] software. The branch lengths are arbitrary. The four groups of proteins with GST activity (fosfomycin-resistance proteins [fosA], MAPEG, Kappa-class GSTs, and canonical GSTs) are indicated by brackets. See text for details.

An analysis of their structure indicates that Kappa-class GSTs form their own superfamily, which branched off the canonical GSTs at the stage of a thioredoxin-fold precursor.[10]

Members of the MAPEG superfamily,[27] also known as microsomal GSTs (MGST), are trimeric integral membrane proteins. No X-ray crystal structure of a MAPEG protein is yet available, but a low-resolution structure has been solved on the basis of electron diffraction.[28,29] A unique conformational change in response to oxidative stress has been reported for MGST1.[30] The structure of MAPEG proteins does not bear any obvious resemblance to the other groups of GSTs. Nevertheless, GST phylogeny reconstructions using different algorithms, including Bayesian inference (our unpublished results obtained with the MrBayes program[31]) and neighbor joining,[32] show with high probability a relationship between MAPEG and GST Kappa sequences. It is not clear whether this reflects an evolutionary convergence driven by requirements imposed by the common GST function, or whether a common precursor, such as a thioredoxin domain, was recruited by two otherwise distinct proteins. The latter possibility is illustrated in Figure 2.1.

2.3 PRINCIPLES OF SYSTEMATIC NOMENCLATURE OF CANONICAL GSTS[23,24]

Mammalian canonical GSTs have been originally subdivided into classes according to substrate selectivity and immunological cross-reactivity. Later, amino acid sequence similarity has been established as the criterion for assigning an enzyme to a class.[23] Although there is no rigid threshold, it is generally accepted that enzymes belonging to the same class should show 40% or more amino acid sequence identity.[24] Since sequence similarities between classes typically do not exceed 25%, the classification is unequivocal. The following classes of canonical GSTs are recognized in humans and other mammals: Alpha, Mu, Pi, Sigma, Theta, Zeta, and Omega. Even though they are not members of the canonical GSTs, Kappa-class enzymes, but not MAPEG proteins, conform to the systematic nomenclature. The finding that gene structure (pattern of intron-exon boundaries) is well conserved within but not between classes[33] indicates that the classes are not just convenient artificial constructs but that they reflect evolutionary history.

GST protein monomers are named by a capital letter designating the class, followed by an Arabic numeral, which signifies the chronological order in which the sequence within that class was reported for the given species. In species for which the genome sequence is available but no GSTs have been previously named, the numbering reflects the position of the genes on chromosomes. Gene names are identical to those of protein subunits, except that gene names are italicized and the capitalization follows the rules established for the species in question. For example, the gene encoding the first published human Alpha-class GST sequence is named *GSTA1*, while a mouse gene would be designated as *Gsta1*. Although some effort was made to assign the same number to orthologs in various species, this was not always possible because the sequence publication chronology criterion takes precedence. Canonical GSTs are dimeric proteins. Even though dimers of subunits drawn from different classes have been generated experimentally,[34] they are rare or, most likely, nonexistent *in vivo*. However, both homodimers and heterodimers of subunits belonging to the same class are possible. Consequently, a dimer is unequivocally described by the notation GSTA1-1 or, for a heterodimer, GSTA1-2; the designations "GSTA1-A1" and "GSTA1-A2" would be redundant. Allelic forms of proteins are distinguished by lowercase letters, e.g., GSTA1a-1a; in the corresponding gene name, the subunit number is linked to the letter signifying the allele by an asterisk (*GSTA1*A* for a human gene). Finally, a one-letter prefix is used to define common species (mostly h, m, and r for human, mouse, and rat, respectively), and a two-letter prefix derived from systematic taxonomy is used for other species. Thus, a human enzyme is shown as hGSTA1-1, while a *Caenorhabditis elegans* Pi-class GST, whose sequence was the second to be reported within that class and species, is designated as CeGSTP2-2.

The systematic nomenclature summarized above fulfills its purpose very well for vertebrate GSTs and, with a few modifications,[24] for many invertebrate GSTs. The use of this nomenclature is strongly encouraged as it provides a common language valid across laboratories. This is especially important because of recent renewed interest in invertebrate GSTs, such as in insects. The unfortunate practice of introducing multiple, often laboratory-centered, naming systems mirrors the chaotic situation prevalent 15 years ago in the field of mammalian GSTs. In some cases, well-established systems of genetic nomenclature collide with the systematic GST nomenclature. For example, names of *C. elegans* genes are set by the Caenorhabditis Genetics Center (http://biosci.umn.edu/CGC/Nomenclature/nomenguid.htm) according to a set of rules.[35] GST genes are named *gst-xx*, where "xx" stands for a consecutive number assigned to the gene within a particular gene class (such as *gst*). In such situations, we propose to conform to the established gene-naming system but use the systematic GST nomenclature for the proteins (gene products).

The usefulness of the GST nomenclature declines or breaks down for enzymes derived from plants, fungi, or bacteria, for one or both of the following reasons. First, it is well established that the three-dimensional structure (folding pattern) of canonical GSTs is more highly conserved than the amino acid sequence.[36] Algorithms to identify GSTs by characteristic motifs rather than by simple sequence alignments have been proposed[2] but have not been widely used. Thus, it is easy to miss GST genes while examining genomes. On the other hand, many nonmammalian GSTs identified by bioinformatic or classical biochemical approaches have very little sequence similarity to established classes. This leads to a proliferation of new classes, which renders the classification system unwieldy.

The second major limitation of GST nomenclature derives from the fact that the ancient thioredoxin fold has been adopted in the course of evolution as a structural element of a variety of proteins. For example, similarity (and perhaps homology) to GSTs is found in the CLIC chloride channels,[37] in components of the protein synthesis machinery (translation elongation factor EF1γ[32,38] and amino acyl-tRNA synthetases[2,39]), in a zinc finger protein,[39] in the prion-like protein URE2[40] that is involved in Tor-mediated nitrogen catabolite repression in yeast,[41] and in the bacterial SspA protein that regulates RNA polymerase.[2] The function of the GST domain in these larger proteins is not known but may be related to glutathione and redox sensing. Should such proteins be included in the GST superfamily? It has been proposed to use glutathione-conjugating activity as a criterion for classifying a protein as a GST.[24] However, the translation elongation factor EF1γ has glutathione-conjugating activity,[42] while such activity may not be obvious in bona fide GSTs, especially if they fail to catalyze the conjugation of the model substrate 1-chloro-2,4-dinitrobenzene (CDNB). For example, the *Drosophila melanogaster* enzyme DmGSTS1-1 has been initially described as lacking enzymatic activity,[43] but it was later found to catalyze the conjugation 4-hydroxynonenal.[44] In light of the ancient evolutionary nature and wide phylogenetic distribution of GSTs, as well as the utilization in other proteins of structural motifs characteristic of GSTs, gray zones in the classification of these enzymes are inevitable.

2.4 BRIEF DESCRIPTION OF INDIVIDUAL CLASSES OF GSTS

In this overview, only a few selected highlights of the individual classes of GSTs will be touched on. Subsequent chapters in this book contain detailed information including substrate specificity, reaction mechanisms, physiological and pathophysiological roles, and other aspects of the proteins.

2.4.1 MAPEG

The microsomal GSTs are homotrimeric, integral membrane proteins. It is unknown whether they share a very early common ancestor with Kappa and with canonical

GSTs, as tentatively shown in Figure 2.1, or whether MAPEG enzymes and other GSTs acquired glutathione transferase activity independently (convergent evolution). Even if a common ancestor existed, MAPEG proteins must have diverged very early because they are present in such distantly related taxa as eubacterial prokaryotes, plants, insects, and vertebrates.[45] Eukaryotic MAPEG enzymes are subdivided into six groups, most of which are involved in the metabolism of eicosanoids[45] (derivatives of arachidonate, a C_{20} fatty acid). However, some of the MAPEG proteins have detoxification functions, including high glutathione peroxidase activity. In particular, the glutathione-conjugating (with CDNB) and glutathione peroxidase activities (with cumene hydroperoxide) of a fish MAPEG isolated from *Esox lucius* (pike) rivaled or exceeded those of the soluble, canonical GSTs.[45] Thus, these enzymes appear to be well suited to protect membranes from the action of lipid hydroperoxides.

2.4.2 KAPPA CLASS

GSTs of the Kappa class and canonical GSTs evolved from a common thioredoxin-like ancestor but have a distinct domain architecture.[10,25,26] Kappa-class GSTs are localized to mitochondria and to peroxisomes. Although they accept the model substrates CDNB and cumene hydroperoxide, their physiological substrates are not known.

2.4.3 ALPHA, MU, AND PI CLASSES

These are the prevalent mammalian GST classes; for example, Alpha-class GSTs account for as much as 3% of the soluble protein in human liver.[46] Alpha, Mu, and Pi GSTs are considered to be the most "highly evolved" GSTs, as judged by their functional and structural adaptation[47,48] to the prototypical activity of GSTs, namely glutathione conjugation of electrophiles. This view may, however, be in part due to the fact that they are the most studied, and thus best-understood, GSTs. The structures of proteins belonging to the three classes follow the canonical structure but differ at two functionally important sites. One is the subunit interface, which is conserved within each class but is divergent between classes. This is the structural basis for the formation of heterodimers within but not between classes. The second difference is the active site, which is relatively open in Pi-class enzymes, partially occluded by the Mu loop in proteins belonging to the Mu class, and covered by a mobile C-terminal α-helix (α9) in Alpha-class GSTs.[49] In addition, the electrophile-binding site within the active sites of Alpha and Mu enzymes is relatively hydrophobic, whereas that in Pi-class GSTs is partly hydrophilic and partly hydrophobic.[50] These differences account for the distinct substrate specificity profiles of Alpha-, Mu-, and Pi-class enzymes.

A noteworthy aspect of the Alpha-class GSTs is the existence of a subclass of these enzymes specialized in the conjugation of electrophilic lipid peroxidation products such as 4-hydroxynon-2-enal. This reaction is described in more detail in another chapter of this book. The best-studied examples of this subclass are the murine mGSTA4-4[51,52] and the human hGSTA4-4.[53] The human and mouse enzymes do not appear to be orthologs, as they are not cross-reactive immunologically, and their amino acid sequences show only a moderate similarity (59% identity/77% similarity, versus 74%/85% for hGSTA1-1 and mGSTA1-1). X-ray crystal structures are available for both mGSTA4-4[54,55] and hGSTA4-4.[56] The structures indicate that the two enzymes utilize different substrate-binding and electrophilic assistance modes (see Chapter 5). This suggests that the two enzymes independently acquired the ability to conjugate 4-hydroxynon-2-enal.

Genetic polymorphisms affecting either the coding sequence or regulatory regions have been identified in all human GSTs.[5,57] Because of the multiple roles of GSTs in detoxification, drug metabolism, and regulation, changes in the expression level or in catalytic properties of these enzymes are relevant to human health. These topics, as well as the role of GSTs (especially of the Pi class) in cancer, will be discussed elsewhere in this book.

2.4.4 SIGMA CLASS

Phylogenetic studies[58,59] indicate that Sigma-class GSTs are relatively closely related to the Alpha/Mu/Pi cluster (Figure 2.1). Sigma GST was originally isolated from squid.[60] Interestingly, in cephalopods (including squid and octopus) the eye lens crystallin is derived from Sigma-class GST, while other proteins have been recruited in other taxa to carry out this function.[61] Sigma GST is present in mammals where it has a specialized function in prostaglandin synthesis[62,63] but is more abundant in invertebrate species. For example, the *D. melanogaster* Sigma-class GST (DmGSTS1-1, formerly designated Class II insect GST) constitutes approximately 1% of all soluble proteins of the fly.[44] The protein is particularly abundant in the indirect flight muscle where it is bound to troponin H and may act as a stretch sensor or have another role in muscle function,[43] or it may provide antioxidant or anti-electrophile protection through its conjugating activity toward 4-hydroxynonenal.[44] In addition to the indirect flight muscle, DmGSTS1-1 is present in the head of the fly.[43] A null mutant of *GstS1* exacerbates the loss of dopaminergic neurons in a Drosophila model of Parkinsonism,[64] a finding consistent with a protective role of DmGSTS1-1 in neurons.

2.4.5 THETA CLASS

Theta-class GSTs are sometimes considered to be close to an ancestral protein common to all GSTs. While this notion is probably incorrect, it is evident that the Theta branch diverged early from the branch that led to the Alpha/Mu/Pi cluster (Figure 2.1). Theta-class GSTs are unusual in that most of them do not accept CDNB as a substrate. Perhaps more interesting is the very high, millimolar K_M of Theta-class GSTs for glutathione and glutathione conjugates. This property accounts for the finding, which delayed the identification of these enzymes, that Theta-class GSTs cannot be purified by glutathione affinity chromatography. It has been proposed[65] that it is the Alpha/Mu/Pi cluster, rather than Theta, that is unusual. While high, the K_M of Theta GSTs is still at, or below, the prevalent glutathione concentration in cells. In contrast, Alpha, Mu, and Pi GSTs are always saturated with glutathione or, more significantly, with their own reaction products (glutathione conjugates). With product release becoming rate-limiting, GSTs of the Alpha/Mu/Pi cluster lose catalytic efficiency but gain the ability to act as a sink for their own products, while Theta-class proteins come closer to the model of a "pure" enzyme catalyst.[65]

Structurally, Theta-class GSTs differ from Alpha, Mu, Pi, and Sigma in that the active-site tyrosine, essential for activation of glutathione, is replaced by a serine that carries out the same function.

2.4.6 ZETA CLASS

Zeta- and Theta-class GSTs are related; they share the already mentioned unusual feature of having an essential active-site serine instead of the more typical tyrosine.[66,67] Zeta-class GSTs are identical with maleylacetoacetate isomerase, an enzyme required for tyrosine catabolism.[68,69] In addition, the enzymes catalyze the metabolism of α-haloacids such as dichloroacetate.[70,71]

2.4.7 OMEGA CLASS

This class of GSTs is phylogenetically widely distributed as it is found in insects, worms, and mammals.[72] Omega-class GSTs contain an active-site cysteine, which forms a disulfide bond with glutathione. In this, Omega-class GSTs resemble thioredoxin-like proteins and could be a link between thioredoxin and GSTs. Accordingly, Omega GSTs show some activities normally associated with glutaredoxins/thioredoxins, such as thioltransferase and dehydroascorbate reductase,[72] but have low or no activity with typical GST substrates.[73]

2.4.8 BACTERIAL GSTs: BETA CLASS

Beta-class GSTs, at least those that have been studied in sufficient detail, share with Omega-class enzymes the presence in the active site of a cysteine that forms a disulfide bond with glutathione.[2,58] This indicates a relationship to glutaredoxins.[74,75] The physiological role of bacterial GSTs is thought to differ from that of eukaryotic GSTs. The latter are mostly involved in detoxification of endogenous and xenobiotic toxicants. In contrast, the major function of bacterial GSTs appears to be the catabolism of organic compounds, which permits the bacteria to grow on a variety of carbon sources,[2] including persistent environmental contaminants. In addition to Beta-class GSTs, other GSTs are likely to be present in bacteria. For example, a Theta-like GST was isolated from *Ochrobactrum anthropi*.[76]

2.4.9 INSECT GSTs: DELTA AND EPSILON CLASSES

Historically, insect GSTs have been studied because of their involvement in insecticide resistance,[77–79] although the role of these enzymes in the defense against oxidative and electrophilic stress has attracted recent attention.[44,80,81] Insects express Sigma-, Theta-, Zeta-, and Omega-class GSTs but no Alpha, Mu, or Pi enzymes.[24,59,77] In addition, there are at least two insect-specific classes. Of these, the phylogenetically related[77,80] Delta and Epsilon classes have been most extensively studied. In *D. melanogaster*, several Delta GSTs have been purified and characterized, and the regulation of their expression has been defined in a series of pioneering papers.[82–86] A gene encoding a *D. melanogaster* GST of the Epsilon class was described later.[87] Many insect GSTs, such as the first member of the Epsilon class[87] or the Delta-class enzyme from *Lucilia cuprina*,[88] were initially classified as Theta. We characterized the Epsilon GST as well as ten members of the Delta cluster by heterologous expression in *E. coli*, purification, and determination of catalytic properties.[80] Furthermore, we identified nine additional members of the Epsilon gene cluster.[80] Analysis of genome sequences of *D. melanogaster*[89] and of other insects such as *Anopheles gambiae*[4] led to the identification of further genes of these classes. Currently, 11 Delta-class and 14 Epsilon-class genes are known to exist in *D. melanogaster*.[77] Such genes are present also in mosquitoes, although their number is different because cluster expansion probably occurred after divergence of the two insect lineages.[77]

Sequence alignments and available crystal structures[88,90,91] indicate that Delta, and most likely Epsilon, GSTs utilize an active-site serine, rather than tyrosine, to ionize glutathione to the thiolate anion. In this, the insect-specific GSTs resemble Theta GSTs. The biological functions of Delta and Epsilon GSTs are only partially known. Resistance to defensive toxicants synthesized by plants as well as to man-made insecticides is certainly among these functions.[77,78,92] The findings that many Delta-class enzymes are able to conjugate lipid peroxidation products[80] and that Delta and Epsilon GSTs are induced by oxidative stress[81,93] indicate a role in antioxidant and anti-electrophile defenses. Exacerbation of the *parkin* mutant phenotype in a *D. melanogaster* model of Parkinsonism by a loss-of-function mutation of an Epsilon-class GST[94] is consistent with a protective function of the latter enzyme. More specialized functions have also been observed. For example, an olfactory-specific GST from *Manduca sexta*, which belongs to the Delta or Epsilon class (phylogenetic analysis shows a close relationship to both classes; data not shown), may protect the pheromone-sensitive sensilla of the insect from toxicants, but it is probably also involved in signal termination by metabolizing aldehyde odorants.[95] It is likely that new functions of these versatile proteins will be uncovered in the future.

2.4.10 Worm GSTs

GSTs in worms are of interest for two reasons. In parasitic worms (helminths), GSTs — which differ from those of the mammalian host — can be used as drug targets. Thus, characterization of helminth GSTs has practical significance. The soil nematode *Caenorhabditis elegans* is widely used in research as a simple and well-understood multicellular organism that can be handled by techniques similar to those used for microorganisms.[96] *C. elegans* is a convenient model organism for study of the aging process. Since detoxification processes are important in aging,[97] understanding GST function in this nematode is of considerable interest.

Several GSTs from parasitic worms such as *Fasciola hepatica* and *Schistosoma*, *Ascaris*, and *Onchocerca* species have been characterized, and crystal structures are available for several of them.[58,98] Interestingly, some of the worm GSTs are related to Sigma, Mu, and Pi classes;[58] the latter two are rarely found outside of vertebrate species. However, the majority of worm GSTs bear little resemblance to vertebrate enzymes, and a nematode-specific class has been proposed.[99] The analysis of GST types present in parasitic worms is complicated by the phylogenetic relationships of the organisms. *Fasciola* and *Schistosoma* are Platyhelminthes (flatworms), while *Ascaris* and *Onchocerca* belong to the Nematoda (roundworms). In terms of evolutionary distance, Platyhelminthes and Nematoda diverged from each other earlier than Nematoda from insects[100] and would thus be expected to express different sets of GSTs.

Bioinformatics screening of the genome of *C. elegans* predicts the existence of 57 genes encoding proteins that match the C-terminal portion of GSTs.[4] WormBase (www.wormbase.org) lists more than 40 well-annotated GST genes, several of them belonging to the Pi class. So far, few of the *C. elegans* enzymes have been characterized functionally. One of the Pi-class GSTs, CeGSTP2-2 (a product of the *gst-10* gene; formerly known as CeGST 5.7) has the ability to conjugate 4-hydroxynonenal.[101] In mammals, 4-hydroxynonenal-conjugating activity is associated mainly with Alpha-class GSTA4-4 (compare Chapter 5). This suggests that the activity evolved independently in the nematode Pi-class GST. The *gst-4* gene product has been found to confer resistance against oxidative stress without affecting the life span of the nematode.[102] The product of the *gst-5* gene was found to interact with other (non-GST) proteins in *C. elegans*,[103] suggesting the possibility of regulatory interactions similar to those exerted by mammalian Pi- and Mu-class GSTs on signaling kinases (see Chapter 5).

2.4.11 PLANT GSTs: PHI AND TAU CLASSES

Two unique plant GST classes, Phi (F) and Tau (U), have been identified.[104–106] Both classes are represented by multiple genes that formed by duplication events. In addition, plant genomes carry smaller numbers of genes encoding the ubiquitous Theta- and Zeta-class GSTs, as well as genes for proteins termed DHAR (dehydroascorbate reductase) and Lambda,[107] which contain an active-site cysteine and could be distantly related to the Omega class. DHAR and Lambda proteins may be involved in redox and thiol transfer reactions. Enzymes of the major Phi and Tau classes utilize an active-site serine to ionize glutathione, in which they resemble Theta as well as the insect-specific Delta and Epsilon proteins.

The physiological role of GSTs is somewhat puzzling in plants that do not ingest toxicant-containing food. Because of allelopathic interactions (interspecies competition),[108] plants are exposed to potentially harmful compounds that need to be metabolized. Such a detoxification function of GSTs has been demonstrated, and it is of particular importance regarding man-made herbicides. However, the concentrations of allelopathic substances in a normal habitat would be expected to be low, in striking contrast to the high abundance of GSTs, which can reach 2% of all proteins in leaves.[104] Therefore, other functions of plant GSTs have been proposed. They include defense against oxidative stress via glutathione peroxidase activity of the enzymes, signaling functions, noncatalytic binding of flavonoids, and participation in intermediary metabolism such as tyrosine catabolism.[104–106]

2.4.12 FUNGAL GSTs

Sixty-seven GST-like sequences have been identified by screening of full or partial genomes of 21 fungal species.[32] By phylogenetic analysis, these sequences could be grouped into five clusters that do not coincide with GST classes found in other organisms. Some of the identified proteins contain GST domains in the context of a larger protein with a known function, such as the translation elongation factor EF1Bγ whose GST domain structure has been confirmed by X-ray crystallography.[109] Such proteins probably evolved through "domain shuffling."[32] Fungal GSTs or GST-like proteins may participate in stress response and in redox reactions, although experimental evidence remains sparse.

ACKNOWLEDGMENT

The authors' work related to GSTs is supported by National Institutes of Health grants AG 18845 and ES 07804 (to P.Z.).

REFERENCES

1. Dworkin, M., *The Prokaryotes*, 3rd ed., Springer, New York, NY, 2001.
2. Vuilleumier, S. and Pagni, M., The elusive roles of bacterial glutathione S-transferases: new lessons from genomes, *Appl. Microbiol. Biotechnol.* 58, 138, 2002.
3. Hand, C.E. and Honek, J.F., Biological chemistry of naturally occurring thiols of microbial and marine origin, *J. Nat. Prod.* 68, 293, 2005.
4. Holt, R.A. et al., The genome sequence of the malaria mosquito *Anopheles gambiae*, *Science* 298, 129, 2002.
5. Hayes, J.D., Flanagan, J.U., and Jowsey, I.R., Glutathione transferases, *Annu. Rev. Pharm. Toxicol.* 45, 51, 2005.
6. Bernat, B.A., Laughlin, L.T., and Armstrong, R.N., Fosfomycin resistance protein (FosA) is a manganese metalloglutathione transferase related to glyoxalase I and the extradiol dioxygenases, *Biochemistry* 36, 3050, 1997.
7. Rife, C.L. et al., Crystal structure of a genomically encoded fosfomycin resistance protein (FosA) at 1.19 A resolution by MAD phasing off the L-III edge of Tl+, *J. Am. Chem. Soc.* 124, 11001, 2002.
8. Babbitt, P.C. and Gerlt, J.A., Understanding enzyme superfamilies: chemistry as the fundamental determinant in the evolution of new catalytic activities, *J. Biol. Chem.* 272, 30591, 1997.
9. Martin, J.L., Thioredoxin — a fold for all reasons, *Structure* 3, 245, 1995.
10. Ladner, J.E. et al., Parallel evolutionary pathways for glutathione transferases: structure and mechanism of the mitochondrial class kappa enzyme rGSTK1-1, *Biochemistry* 43, 352, 2004.
11. Shelton, M.D., Chock, P.B., and Mieyal, J.J., Glutaredoxin: role in reversible protein S-glutathionylation and regulation of redox signal transduction and protein translocation, *Antioxid. Redox Signal.* 7 (3–4), 348, 2005.
12. Nakamura, H., Thioredoxin as a key molecule in redox signaling, *Antioxid. Redox Signal.* 6, 15, 2004.
13. Das, K.C., Thioredoxin system in premature and newborn biology, *Antioxid. Redox Signal.* 6, 177, 2004.
14. Pekkari, K. and Holmgren, A., Truncated thioredoxin: physiological functions and mechanism, *Antioxid. Redox Signal.* 6, 53, 2004.

15. Fernandes, A.P. and Holmgren, A., Glutaredoxins: glutathione-dependent redox enzymes with functions far beyond a simple thioredoxin backup system, *Antioxid. Redox Signal.* 6, 63, 2004.

16. Gromer, S., Urig, S., and Becker, K., The thioredoxin system — from science to clinic, *Med. Res. Rev.* 24, 40, 2004.

17. Watson, W.H. et al., Thioredoxin and its role in toxicology, *Toxicol. Sci.* 78, 3, 2004.

18. Armstrong, R.N., Structure, catalytic mechanism, and evolution of the glutathione transferases, *Chem. Res. Toxicol.* 10, 2, 1997.

19. Nebert, D.W. et al., The P450 gene superfamily: recommended nomenclature, *DNA* 6, 35075, 1987.

20. Nebert, D.W. et al., The P450 superfamily: updated listing of all genes and recommended nomenclature for the chromosomal loci, *DNA* 8, 35077, 1989.

21. Nebert, D.W. et al., The P450 superfamily: update on new sequences, gene mapping, and recommended nomenclature, *DNA Cell Biol.* 10, 35078, 1991.

22. Nelson, D.R. et al., The P450 superfamily: update on new sequences, gene mapping, accession numbers, early trivial names of enzymes, and nomenclature, *DNA Cell Biol.* 12, 18629, 1993.

23. Mannervik, B. et al., Nomenclature for human glutathione transferases, *Biochem. J.* 282, 305, 1992.

24. Chelvanayagam, G., Parker, M.W., and Board, P.G., Fly fishing for GSTs: a unified nomenclature for mammalian and insect glutathione transferases, *Chem. Biol. Interact.* 133, 256, 2001.

25. Jowsey, I.R. et al., Biochemical and genetic characterization of a murine class Kappa glutathione S-transferase, *Biochem. J.* 373, 559, 2003.

26. Morel, F. et al., Gene and protein characterization of the human glutathione S-transferase kappa and evidence for a peroxisomal localization, *J. Biol. Chem.* 279 (16), 16246, 2004.

27. Morgenstern, R., Guthenberg, C., and DePierre, J.W., Microsomal glutathione S-transferase. Purification, initial characterization, and demonstration that it is not identical to the cytosolic glutathione S-transferases A, B, and C, *Eur. J. Biochem.* 128, 243, 1982.

28. Holm, P.J., Morgenstern, R., and Hebert, H., The 3-D structure of microsomal glutathione transferase 1 at 6 angstrom resolution as determined by electron crystallography of p22(1)2(1) crystals, *Biochim. Biophys. Acta* 1594, 276, 2002.

29. Thoren, S. et al., Human microsomal prostaglandin E synthase-1: purification, functional characterization, and projection structure determination, *J. Biol. Chem.* 278, 22199, 2003.

30. Busenlehner, L.S. et al., Stress sensor triggers conformational response of the integral membrane protein microsomal glutathione transferase 1, *Biochemistry* 43, 11145, 2004.

31. Ronquist, F. and Huelsenbeck, J.P., MrBayes 3: Bayesian phylogenetic inference under mixed models, *Bioinformatics* 19, 1572, 2003.

32. McGoldrick, S., O'Sullivan, S.M., and Sheehan, D., Glutathione transferase–like proteins encoded in genomes of yeasts and fungi: insights into evolution of a multifunctional protein superfamily, *FEMS Microbiol. Lett.* 242, 1, 2005.

33. Hayes, J.D. and Pulford, D.J., The glutathione S-transferase supergene family: regulation of GST and the contribution of the isoenzymes to cancer chemoprotection and drug resistance, *Crit. Rev. Biochem. Mol. Biol.* 30, 445, 1995.

34. Pettigrew, N.E. and Colman, R.F., Heterodimers of glutathione S-transferase can form between isoenzyme classes pi and mu, *Arch. Biochem. Biophys.* 396, 225, 2001.

35. Horvitz, H.R. et al., A uniform genetic nomenclature for the nematode *Caenorhabditis elegans*, *Mol. Gen. Genet.* 175, 129, 1979.

36. Wilce, M.C. and Parker, M.W., Structure and function of glutathione S-transferases, *Biochim. Biophys. Acta* 1205, 1, 1994.

37. Cromer, B.A. et al., From glutathione transferase to pore in a CLIC, *Eur. Biophys. J.* 31 (5), 356, 2002.

38. Koonin, E.V. et al., Eukaryotic translation elongation factor 1γ contains a glutathione transferase domain — study of a diverse, ancient protein superfamily using motif search and structural modeling, *Protein Sci.* 3, 2045, 1994.

39. Dai, M.S. et al., Identification and characterization of a novel *Drosophila melanogaster* glutathione S-transferase-containing FLYWCH zinc finger protein, *Gene* 342, 49, 2004.

40. Bousset, L. et al., Structure of the globular region of the prion protein Ure2 from the yeast *Saccharomyces cerevisiae*, *Structure* 9, 39, 2001.

41. Cooper, T.G., Transmitting the signal of excess nitrogen in *Saccharomyces cerevisiae* from the Tor proteins to the GATA factors: connecting the dots, *FEMS Microbiol. Rev.* 26, 223, 2002.

42. Kobayashi, S., Kidou, S., and Ejiri, S., Detection and characterization of glutathione S-transferase activity in rice EF-1ββ′γ and EF-1γ expressed in *Escherichia coli*, *Biochem. Biophys. Res. Commun.* 288, 509, 2001.

43. Clayton, J.D. et al., Interaction of troponin-H and glutathione S-transferase-2 in the indirect flight muscles of *Drosophila melanogaster*, *J. Muscle Res. Cell Motil.* 19, 117, 1998.

44. Singh, S.P. et al., Catalytic function of *Drosophila melanogaster* glutathione S-transferase DmGSTS1-1 (GST-2) in conjugation of lipid peroxidation end products, *Eur. J. Bioch.* 268, 2912, 2001.

45. Bresell, A. et al., Bioinformatic and enzymatic characterization of the MAPEG superfamily, *FEBS J.* 272, 1688, 2005.

46. van Ommen, B. et al., Quantification of human hepatic glutathione S-transferases, *Biochem. J.* 269, 609, 1990.

47. Caccuri, A.M. et al., Human glutathione transferase T2-2 discloses some evolutionary strategies for optimization of substrate binding to the active site of glutathione transferases, *J. Biol. Chem.* 276 (8), 5427, 2001.

48. Caccuri, A.M. et al., Human glutathione transferase T2-2 discloses some evolutionary strategies for optimization of the catalytic activity of glutathione transferases, *J. Biol. Chem.* 276 (8), 5432, 2001.

49. Armstrong, R.N., Glutathione S-transferases: structure and mechanism of an archetypical detoxication enzyme, *Adv. Enzymol. Rel. Areas Mol. Biol.* 69, 1, 1994.

50. Ji, X. et al., Structure and function of the xenobiotic substrate-binding site and location of a potential non-substrate-binding site in a class π glutathione S-transferase, *Biochemistry* 36, 9690, 1997.

51. Zimniak, P. et al., A subgroup of class alpha glutathione S-transferases: cloning of cDNA for mouse lung glutathione S-transferase GST 5.7, *FEBS Lett.* 313, 173, 1992.

52. Zimniak, P. et al., Estimation of genomic complexity, heterologous expression, and enzymatic characterization of mouse glutathione S-transferase mGSTA4-4 (GST 5.7), *J. Biol. Chem.* 269, 992, 1994.

53. Hubatsch, I., Ridderstrom, M., and Mannervik, B., Human glutathione transferase A4-4: an Alpha-class enzyme with high catalytic efficiency in the conjugation of 4-hydroxynonenal and other genotoxic products of lipid peroxidation, *Biochem. J.* 330, 175, 1998.

54. Krengel, U. et al., Crystal structure of a murine α-class glutathione S-transferase involved in cellular defense against oxidative stress, *FEBS Lett.* 422, 285, 1998.

55. Xiao, B. et al., Crystal structure of a murine glutathione S-transferase in complex with a glutathione conjugate of 4-hydroxynon-2-enal in one subunit and glutathione in the other: evidence of signaling across the dimer interface, *Biochemistry* 38, 11887, 1999.

56. Bruns, C.M. et al., Human glutathione transferase A4-4 crystal structures and mutagenesis reveal the basis of high catalytic efficiency with toxic lipid peroxidation products, *J. Mol. Biol.* 288, 427, 1999.

57. Hayes, J.D. and Strange, R.C., Glutathione S-transferase polymorphisms and their biological consequences, *Pharmacology* 61, 154, 2000.

58. Sheehan, D. et al., Structure, function, and evolution of glutathione transferases: implications for classification of non-mammalian members of an ancient enzyme superfamily, *Biochem. J.* 360 (Part 1), 1, 2001.

59. Ding, Y. et al., The *Anopheles gambiae* glutathione transferase supergene family: annotation, phylogeny, and expression profiles, *BMC Genomics* 4, 35, 2003.

60. Harris, J. et al., The isolation and characterization of the major glutathione S-transferase from the squid *Loligo vulgaris*, *Comp. Biochem. Physiol. Part B* 98, 511, 1991.

61. Tomarev, S.I. and Piatigorsky, J., Lens crystallins of invertebrates — diversity and recruitment from detoxification enzymes and novel proteins, *Eur. J. Biochem.* 235, 449, 1996.

62. Meyer, D.J. and Thomas, M., Characterization of rat spleen prostaglandin H D-isomerase as a sigma-class GSH transferase, *Biochem. J.* 311, 739, 1995.

63. Jowsey, I.R. et al., Mammalian class Sigma glutathione S-transferases: catalytic properties and tissue-specific expression of human and rat GSH-dependent prostaglandin D2 synthases, *Biochem. J.* 359, 507, 2001.

64. Whitworth, A.J. et al., Increased glutathione *S*-transferase activity rescues dopaminergic neuron loss in a *Drosophila* model of Parkinson's disease, *Proc. Natl. Acad. Sci. USA* 102, 8024, 2005.

65. Meyer, D.J., Significance of an unusually low Km for glutathione in glutathione transferases of the α, μ,and π classes, *Xenobiotica* 23, 823, 1993.

66. Board, P.G. et al., Clarification of the role of key active site residues of glutathione transferase zeta/maleylacetoacetate isomerase by a new spectrophotometric technique, *Biochem. J.* 374, 731, 2003.

67. Ricci, G. et al., Binding and kinetic mechanisms of the zeta-class glutathione transferase, *J. Biol. Chem.* 279, 33336, 2004.

68. Thom, R. et al., The structure of a zeta-class glutathione S-transferase from *Arabidopsis thaliana*: characterisation of a GST with novel active-site architecture and a putative role in tyrosine catabolism, *J. Mol. Biol.* 308 (5), 949, 2001.

69. Polekhina, G. et al., Crystal structure of maleylacetoacetate isomerase/glutathione transferase zeta reveals the molecular basis for its remarkable catalytic promiscuity, *Biochemistry* 40 (6), 1567, 2001.

70. Tong, Z., Board, P.G., and Anders, M.W., Glutathione transferase zeta-catalyzed biotransformation of dichloroacetic acid and other alpha-haloacids, *Chem. Res. Toxicol.* 11, 1332, 1998.

71. Tong, Z., Board, P.G., and Anders, M.W., Glutathione transferase zeta catalyses the oxygenation of the carcinogen dichloroacetic acid to glyoxylic acid, *Biochem. J.* 331, 371, 1998.

72. Whitbread, A.K. et al., Characterization of the human Omega-class glutathione transferase genes and associated polymorphisms, *Pharmacogenetics* 13 (3), 131, 2003.

73. Board, P.G. et al., Identification, characterization, and crystal structure of the omega-class glutathione transferases, *J. Biol. Chem.* 275 (32), 24798, 2000.

74. Caccuri, A.M. et al., GSTB1-1 from Proteus mirabilis: a snapshot of an enzyme in the evolutionary pathway from a redox enzyme to a conjugating enzyme, *J. Biol. Chem.* 277, 18777, 2002.

75. Caccuri, A.M. et al., Properties and utility of the peculiar mixed disulfide in the bacterial glutathione transferase B1-1, *Biochemistry* 41, 4686, 2002.

76. Favaloro, B. et al., Molecular cloning, expression, and site-directed mutagenesis of glutathione S-transferase from *Ochrobactrum anthropi*, *Biochem. J.* 335, 573, 1998.

77. Enayati, A.A., Ranson, H., and Hemingway, J., Insect glutathione transferases and insecticide resistance, *Insect Mol. Biol.* 14, 3, 2005.

78. David, J.P. et al., The *Anopheles gambiae* detoxification chip: a highly specific microarray to study metabolic-based insecticide resistance in malaria vectors, *Proc. Natl. Acad. Sci. USA* 102, 4080, 2005.

79. Prapanthadara, L. et al., Isoenzymes of glutathione S-transferase from the mosquito *Anopheles dirus* species B: the purification, partial characterization, and interaction with various insecticides, *Insect Biochem. Mol. Biol.* 30, 395, 2000.

80. Sawicki, R. et al., Cloning, expression, and biochemical characterization of one Epsilon-class (GST-3) and ten Delta-class (GST-1) glutathione S-transferases from *Drosophila melanogaster*, and identification of additional nine members of the Epsilon class, *Biochem. J.* 370, 661, 2003.

81. Ding, Y. et al., Characterization of the promoters of Epsilon glutathione transferases in the mosquito *Anopheles gambiae* and their response to oxidative stress, *Biochem. J.* 387, 879, 2005.

82. Toung, Y.-P.S., Hsieh, T., and Tu, C.-P.D., The glutathione S-transferase D genes. A divergently organized, intronless gene family in *Drosophila melanogaster*, *J. Biol. Chem.* 268, 9737, 1993.

83. Tang, A.H. and Tu, C.-P.D., Biochemical characterization of *Drosophila* glutathione S-transferases D1 and D21, *J. Biol. Chem.* 269, 27876, 1994.

84. Lee, H.C. and Tu, C.-P.D., *Drosophila* glutathione S-transferase D27: functional analysis of two consecutive tyrosines near the N-terminus, *Biochem. Biophys. Res. Commun.* 209, 327, 1995.

85. Akgul, B. and Tu, C.P.D., Evidence for a stabilizer element in the untranslated regions of *Drosophila* glutathione S-transferase D1 mRNA, *J. Biol. Chem.* 277 (38), 34700, 2002.

86. Akgul, B. and Tu, C.-P.D., Pentobarbital-mediated regulation of alternative polyadenylation in *Drosophila* glutathione S-transferase D21 mRNAs, *J. Biol. Chem.* 279, 4027, 2004.

87. Singh, M. et al., Cloning and characterization of a new theta-class glutathione-S-transferase (GST) gene, *gst-3*, from *Drosophila melanogaster*, *Gene* 247, 167, 2000.

88. Wilce, M.C.J. et al., Crystal structure of a theta-class glutathione transferase, *Embo. J.* 14, 2133, 1995.

89. Adams, M.D. et al., The genome sequence of *Drosophila melanogaster*, *Science* 287, 2185, 2000.

90. Oakley, A.J. et al., The crystal structures of glutathione S-transferases isozymes 1–3 and 1–4 from *Anopheles dirus* species B, *Protein Sci.* 10 (11), 2176, 2001.

91. Udomsinprasert, R. et al., Identification, characterization, and structure of a new Delta-class glutathione transferase isoenzyme, *Biochem. J.* 388, 763, 2005.

92. Lumjuan, N. et al., Elevated activity of an Epsilon class glutathione transferase confers DDT resistance in the dengue vector, *Aedes aegypti, Insect Biochem. Mol. Biol.* 35, 861, 2005.

93. Zou, S. et al., Genome-wide study of aging and oxidative stress response in *Drosophila melanogaster, Proc. Natl. Acad. Sci. USA* 97, 13726, 2000.

94. Greene, J.C. et al., Genetic and genomic studies of *Drosophila parkin* mutants implicate oxidative stress and innate immune responses in pathogenesis, *Hum. Mol. Genet.* 14, 799, 2005.

95. Rogers, M.E., Jani, M.K., and Vogt, R.G., An olfactory-specific glutathione-S-transferase in the sphinx moth *Manduca sexta, J. Exp. Biol.* 202, 1625, 1999.

96. Brenner, S., The genetics of *Caenorhabditis elegans, Genetics* 77, 71, 1974.

97. Gems, D. and McElwee, J.J., Broad spectrum detoxification: the major longevity assurance process regulated by insulin/IGF-1 signaling? *Mech. Aging Dev.* 126 (3), 381, 2005.

98. Perbandt, M. et al., Structure of the major cytosolic glutathione *S*-transferase from the parasitic nematode *Onchocerca volvulus, J. Biol. Chem.* 280, 12630, 2005.

99. Campbell, A.M. et al., A common class of nematode glutathione S-transferase (GST) revealed by the theoretical proteome of the model organism *Caenorhabditis elegans, Comp. Biochem. Physiol. Part B* 128, 701, 2001.

100. Tessmar-Raible, K. and Arendt, D., Emerging systems: between vertebrates and arthropods, the Lophotrochozoa, *Curr. Opin. Genet. Dev.* 13, 331, 2003.

101. Engle, M.R. et al., Invertebrate glutathione transferases conjugating 4-hydroxynonenal: CeGST 5.4 from *Caenorhabditis elegans, Chem. Biol. Interact.* 133, 244, 2001.

102. Leiers, B.R. et al., A stress-responsive glutathione S-transferase confers resistance to oxidative stress in *Caenorhabditis elegans, Free Rad. Biol. Med.* 34, 1405, 2003.

103. Greetham, D. et al., Evidence of glutathione transferase complexing and signaling in the model nematode *Caenorhabditis elegans* using a pull-down proteomic assay, *Proteomics* 4, 1989, 2004.

104. Dixon, D.P., Lapthorn, A., and Edwards, R., Plant glutathione transferases, *Genome Biol.* 3, reviews 3004.1, 2002.

105. Edwards, R., Dixon, D.P., and Walbot, V., Plant glutathione S-transferases: enzymes with multiple functions in sickness and in health, *Trends Plant Sci.* 5, 193, 2000.

106. Frova, C., The plant glutathione transferase family: genomic structure, functions, expression, and evolution, *Physiol. Plant.* 119, 469, 2003.

107. Dixon, D.P., Davis, B.G., and Edwards, R., Functional divergence in the glutathione transferase superfamily in plants. Identification of two classes with putative functions in redox homeostasis in *Arabidopsis thaliana, J. Biol. Chem.* 277, 30859, 2002.

108. Gidman, E. et al., Investigating plant–plant interference by metabolic fingerprinting, *Phytochemistry* 63, 705, 2003.

109. Jeppesen, M.G. et al., The crystal structure of the glutathione *S*-transferase-like domain of elongation factor 1Bγ from *Saccharomyces cerevisiae, J. Biol. Chem.* 278, 47190, 2003.

110. Hannon, G.J., RNA interference, *Nature* 418, 244, 2002.

111. Kumar, S., Tamura, K., and Nei, M., MEGA3: integrated software for Molecular Evolutionary Genetics Analysis and sequence alignment, *Briefings Bioinform.* 5, 150, 2004.

3 Mammalian Glutathione S-Transferase Genes: Structure and Regulation

Ian R. Jowsey and John D. Hayes

CONTENTS

3.1 INTRODUCTION

The glutathione S-transferases (GSTs: EC 2.5.1.18) represent a superfamily of detoxication enzymes. Their substrates include endogenous products of oxidative stress and electrophilic xenobiotics. In addition to detoxification, GSTs are also intimately involved in the biosynthesis of leukotrienes, prostaglandins, testosterone, and progesterone, as well as in the degradation of tyrosine. An increased understanding of the catalytic properties of the transferases has brought with it an appreciation that they may modulate signaling pathways and also the need to better understand how the genes for these enzymes are regulated. Seminal papers in this area have demonstrated the importance of the antioxidant response element and the Nrf2 (*N*uclear factor-erythroid 2 p45-*r*elated *f*actor 2) transcription factor in controlling the expression of rodent GST genes. It is presently unclear whether human GST genes are similarly regulated. Given the link between increased expression of detoxication enzymes and cancer chemoprevention, it is essential that we understand how human GST gene expression is controlled.

3.2 GST FAMILIES

Mammalian GSTs are represented by three distinct protein families: the cytosolic GST, the mitochondrial GST, and the MAPEG (*m*embrane-*a*ssociated *p*roteins in *e*icosanoid and *g*lutathione metabolism; for a recent review, see Hayes et al.).[1] This chapter will focus primarily on the cytosolic and mitochondrial families, and the reader is referred to other excellent articles for further information on the MAPEG family.[2,3]

Based on amino acid sequence similarities, the cytosolic GST family is further subdivided into seven classes, designated Alpha, Mu, Pi, Sigma, Theta, Omega, and Zeta. The mitochondrial family is synonymous with the Kappa-class GST. The cytosolic GSTs within a class typically share >40% amino acid sequence identity while those in different classes share <25% identity. At least 17 cytosolic GST subunits exist in the human, while only a single mitochondrial GST subunit has been identified. These subunits are between 199 and 244 amino acids in size. They function largely as homodimers, although heterodimerization occurs between subunits within Alpha- and Mu-class transferases,[4] further increasing the number of isoenzymes that can be generated. Heterodimerization between Mu- and Pi-class polypeptides *in vitro* has also been reported in the pig[5] but it has not been observed in other species.

3.3 CATALYTIC PROPERTIES OF GSTs

GSTs catalyze the nucleophilic attack by reduced glutathione (GSH) on nonpolar compounds that contain an electrophilic center by reducing the pKa of GSH from ~9.0 in solution to between 6 and 7 in the active site of the enzyme.[4] The reaction that best typifies the GST gene family is the conjugation of GSH with xenobiotics as the first stage in the mercapturic acid biosynthetic pathway. The role of GST in catalyzing the conjugation of GSH with endogenous products of oxidative damage to lipids (acrolein, crotonaldehyde, 4-hydroxynonenal), DNA (base propenals), and catecholamines (aminochrome, dopachrome, noradrenochrome, adrenochrome) is also well established.[6,7] GSH-conjugation reactions are catalyzed by cytosolic and mitochondrial GST. Other enzymes, such as lipoxygenase, that can generate free radicals during catalysis, also appear to be able to promote the formation of a conjugate between GSH and certain electrophiles through the indirect activation of glutathione or the second substrate. In the case of lipoxygenase, GSH conjugation with ethacrynic acid (EA) has been reported through a mechanism involving formation of either a GS• or EA• radical intermediate.[8]

Members of both cytosolic and mitochondrial GST families exhibit Se-independent GSH peroxidase activity and catalyze the reduction of organic hydroperoxides such as cumene hydroperoxide. Unlike Se-dependent GSH peroxidase, GSTs are inactive with H_2O_2 as a substrate.

The cytosolic transferases also exhibit isomerase activity. Most notable in this regard is the activity of the Sigma-class GST that catalyzes the isomerization of PGH_2 to PGD_2[9,10] and that of the Zeta-class GST that catalyzes the isomerization of maleylacetoacetate to fumarylacetoacetate.[11] These observations reveal that GSTs are intimately involved in prostaglandin biosynthesis and tyrosine catabolism and should not be regarded solely as enzymes that detoxify foreign compounds. Isomerase actvity has not yet been demonstrated for the Kappa-class GST, despite its homology with bacterial 2-hydroxychromene-2-carboxylate isomerase, a GSH-dependent oxidoreductase that catalyzes the conversion of 2-hydroxychromene-2-carboxylate to trans-O-hydroxybenzylidenepyruvate.[12] These three enzyme activities — GSH conjugation, reduction, and isomerization — extend over a diverse range of substrates, many of which are metabolized by more than one GST isoenzyme.

Besides their catalytic function, cytosolic GSTs also covalently and noncovalently bind hydrophobic nonsubstrate ligands[13] such as azo-dyes, bilirubin, heme, polycyclic aromatic hydrocarbons, steroids, and thyroid hormones. This activity contributes to intracellular transport, sequestration, and disposition of such compounds. Alpha-, Mu-, and Pi-class GSTs have multiple binding sites for xenobiotics and appear to be able to bind nonsubstrate ligands in the cleft between the two subunits.[14–16] This finding may help explain why cytosolic GSTs within these classes form heterodimers.

3.4 BIOLOGICAL FUNCTIONS OF GSTs

The biological function for which GST isoenzymes are renowned is biotransformation of foreign compounds through catalyzing the first step in the synthesis of mercapturic acids. This function encompasses exogenous substrates such as drugs, pesticides, herbicides, environmental pollutants, and carcinogens, as well as endogenous substrates including products of oxidative stress.[1] An aspect of GSTs that is often overlooked is their biosynthetic function. This includes the role of human GSTA3-3 in the synthesis of steroid hormones[17] and the ability of various transferases to catalyze the synthesis of prostaglandins. Although early studies in this area suggested that numerous GSTs catalyze the isomerization of PGH_2 to a mixture of PGD_2 and PGE_2, or reduce it to $PGF_{2\alpha}$,[18,19] it is now clear that some transferases possess remarkable specificity for certain reactions. Thus, an enzyme known as GSH-dependent prostaglandin D_2 synthase is a Sigma-class GST.[9,10] Similarly, the inducible GSH-dependent PGE_2 synthase is a member of the MAPEG family but has little activity toward prototypic transferase substrates such as 1-chloro-2,4-dinitrobenzene.[20–22]

The observation that Sigma-class GSTs are likely to contribute not only to the synthesis of PGD_2 but also to the production of its downstream metabolite, 15-deoxy-$\Delta^{12,14}$-prostaglandin J_2 (15d-PGJ_2), is interesting because this prostanoid has important biological properties (Figure 3.1). Although the effects of "classical" prostaglandins such as PGD_2, PGE_2, and $PGF_{2\alpha}$ are mediated through specific G-protein-coupled receptors, the cyclopentenone prostaglandin 15d-PGJ_2 serves as an activating ligand for the peroxisome proliferator-activated receptor γ (PPARγ).[23,24] This

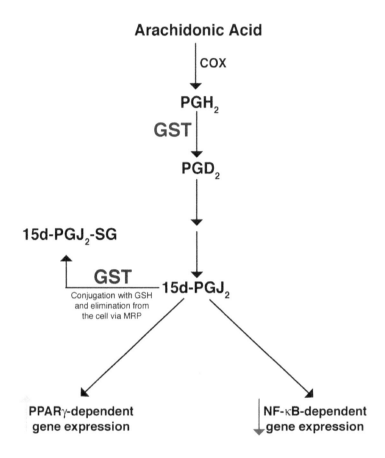

FIGURE 3.1 GST and the synthesis and elimination of prostaglandins. The biosynthetic pathway for 15d-PGJ$_2$ and the transcription factors that are activated by the prostaglandin are shown.

transcription factor is a critical regulator of adipogenesis and is the molecular target for the thiazolidinedione class of insulin-sensitizing drugs. The involvement of Sigma-class GSTs in the synthesis of an endogenous agonist for PPARγ places this type of enzyme as an important component in signaling pathways stimulated by 15d-PGJ$_2$. Furthermore, this cyclopentenone PG has also been shown to influence nuclear factor κB-dependent gene expression.[25]

In the context of PG metabolism, it is noteworthy that human GSTA1-1, A2-2, M1a-1a, and P1-1 can catalyze the conjugation of GSH with PGA$_2$ and PGJ$_2$, thereby presumably contributing to termination of their activity.[26] The PGA$_2$ and PGJ$_2$ cyclopentanone-containing prostanoids are considered to represent noncatalytically formed dehydration products of PGE$_2$ and PGD$_2$, respectively. It is thus clear that GSTs are intimately involved in both the synthesis and the elimination of prostaglandins and can therefore have both positive and negative effects on transactivation of gene expression by PPARγ and nuclear factor κB (Figure 3.1). Determining the precise role of GSTs in the context of these signaling pathways is an area that requires further study. There is evidence that cyclooxygenase (COX) 2 is coinduced with the MAPEG member PGES1.[27,28] This helps ensure that PGH_2, produced by COX-2 from arachidonic acid, is effectively converted to PGE_2. However, little is known about mechanisms through which expression of GSTs and MAPEGs that synthesize PGs, and those that help eliminate the prostanoids, are controlled in an appropriate, integrated and coordinated fashion.

3.5   REGULATION OF GST GENE EXPRESSION

In addition to the extensive biochemical studies that have helped elucidate the substrates and biological functions for GST isoenzymes, much effort has also been devoted to deciphering how the expression of the genes encoding GST subunits is regulated. The stimulus for these studies undoubtedly arose, at least in part, from the observation that hepatic GST enzyme activity toward 1-chloro-2,4-dinitroben-zene (CDNB) could be induced by a diverse range of xenobiotics, many of which contained an electrophilic center; over 100 compounds that induce GST in rats and mice have been identified.[4] In their studies of cancer chemopreventive agents, Talalay and colleagues found that induction of GST activity toward CDNB in rodents is paralleled with an increase in NAD(P)H:quinone oxidoreductase 1 (NQO1) activity.[29,30] Using NQO1 activity in mouse Hepa1c1c7 cells in a high-throughput assay to screen inducers, it has been found that as many as 11 chemical groups of compounds increase oxidoreductase activity; these include cycloalkanones, dithiolethiones, isothiocyanates, mercaptans, peroxides, phenylpropenoids, quinones, tetrapyrroles, triterpenoids, trivalent arsenicals, and flavones.[31–34] Some of these inducing agents possess cancer chemopreventive properties.[35]

The fact that many agents that induce GST and NQO1 are thiol-active,[36,37] or are metabolized to such compounds led to the attractive hypothesis that increased GST expression is part of a process of adaptation that enables the cell to survive exposure to harmful xenobiotics. This view is supported by the fact that production of GSH and expression of the glutamate cysteine ligase catalytic (GCLC) and modifier (GCLM) subunits are also induced by chemopreventive agents.[38] It is envisaged that stimulation of this adaptive response by relatively harmless cancer chemopreventive agents can be harnessed to confer cross-resistance against noxious xenobiotics and products of oxidative stress that can initiate carcinogenesis. Historically, some of the most widely studied chemopreventive compounds have been the phenolic antioxidants butylated hydroxyanisole (BHA) and ethoxyquin, the dithiolethione oltipraz, and the flavanoid β-naphthoflavone (β-NF).[39] Attention was initially focused on identifying the genetic elements within the promoters of GST genes that are responsible for induction by these compounds.

3.5.1   IDENTIFICATION OF THE ANTIOXIDANT RESPONSE ELEMENT

The first GST genes to be examined in terms of their regulatory properties were rat *GSTA2*, mouse *gsta1*, and rat *GSTP1*. The rat *GSTA2* gene was studied by Pickett's

GENE	UPSTREAM	CORE ARE	A/T BOX
Mouse GSTA3	ACTCAGGCA	TGACATTGC	ATTTT
Human NQO1	AGTCAC-AG	TGACTCAGC	AGAAT
Rat NQO1	AGTCAC-AG	TGACTTGGC	AAAAT
Mouse NQO1	AGTCAC-AG	TGAGTCGGC	AAAAT
Rat GSTP1	AGTCAC-TA	TGATTCAGC	AACAA
Mouse GSTA1	GCTAAT-GG	TGACAAAGC	AACTT
Rat GSTA2	GCTAAT-GG	TGACAAAGC	AAACT

FIGURE 3.2 Comparison of ARE enhancers. The sequences of various ARE enhancers have been aligned. Taken from published data.[41,43,46,51,56,58,59]

laboratory because of its induction by β-NF;[40] mouse *GSTA1* was studied by Daniel's group because of its induction by β-NF, 3-methylcholanthrene, trans-4-phenyl-3-buten-2-one, dimethyl fumarate, and the BHA metabolite tert-butylhydroquinone (tBHQ);[41] rat *GSTP1* was studied by Muramatsu's laboratory because it is overexpressed in hepatic preneoplastic nodules.[42] The cis-acting elements discovered by these three groups are all closely similar and are shown aligned in Figure 3.2. The enhancers were originally given distinct designations such as β-NF responsive element, EpRE (*E*lectro*p*hile *R*esponsive *E*lement) and GPEI (*G*lutathione transferase *P*i *E*nhancer *I*). However, such designations disguise the fact that they have many functional features in common. For example, the cis-elements in rat *GSTA2*, mouse *gsta1*, and rat *GSTP1* all respond to tBHQ.[41,43] Furthermore, the GSTA2 subunit is overexpressed along with GSTP1 in rat liver nodules and hepatomas initiated by dietary administration of aflatoxin B_1,[44] and both subunits are induced by ethoxyquin in rat livers.[45]

Following mutational analysis of the 5′-upstream region of rat *GSTA2* and the recognition that it is activated by catechols and hydroquinones that can redox cycle, Rushmore et al.[46] renamed the β-NF responsive element as the ARE (*A*ntioxidant *R*esponsive *E*lement). ARE should replace EpRE because they are functionally identical and have identical nucleotide sequences. The situation with GPEI is a little more complicated because it is responsible for driving overexpression of GSTP1 during preneoplasia.[42,47] This change in gene expression during hepatocarcinogenesis in the rat appears to be due to an interaction between the ARE-like sequence in GPEI with an overlapping binding consensus for C/EBP (CCAAT enhancer binding protein).[48]

Subsequent to the coining of the term ARE, this element has been shown to mediate a transcriptional response to a broad spectrum of structurally diverse xenobiotics including Michael reaction acceptors, diphenols, and quinones.[31] It was therefore thought likely to be present in the promoters of many of the rodent GST genes that are induced by phenolic antioxidants.

Mutational analysis of the rat *GSTA2* 5′-flanking region led to the identification of the minimal ARE sequence required for both basal or inducible activity as 5′-$^A/_G$GTGACNNNGC-3′,[46] though the "core" is usually referred to as 5′-

TGACNNNGC-3'. Other genes shown to be regulated through an ARE now include those encoding human, rat, and mouse NQO1,[49–51] human GCLC and GCLM subunits,[52,53] mouse heme oxygenase-1,[54,55] and mouse GSTA3.[56] An alignment of various AREs is presented in Figure 3.2. The increasing list of such genes has been accompanied by an evolution of the definition of the ARE consensus sequence. Indeed, a comprehensive study of the mouse *nqo1*-ARE indicated the sequence requirements for enhancer activity can vary markedly between different genes, suggesting that genomic context is all important for the function of the element.[51] Notwithstanding the subtleties of individual AREs, there are hallmark features supplemental to the presence of the 5'-TGACNNNGC-3' core that can be used to predict ARE functionality from primary nucleotide sequence data. For example, the nucleotides flanking the core sequence are also relatively well conserved between AREs in different genes (Figure 3.2). Particularly obvious is the presence of the A/T-rich sequence immediately 3' of the core GC dinucleotide.[57] Interestingly, despite this conservation of the A/T-rich region between AREs in different genes, comprehensive mutational analysis of the mouse *nqo1* ARE suggests that only the first 5' "A" in this "A/T box" is essential for function.[51] The region that lies immediately 5' of the core ARE also shows regions that are conserved between genes. In the case of the mouse *gsta3* ARE, the sequence 5'-TGAGT-3' is present in reverse orientation immediately upstream of the core enhancer. A closely related sequence, 5'-TGACT-3', is found in a similar position and orientation upstream of AREs in human, rat, and mouse *NQO1* and in the GPEI site within the rat *GSTP1* gene promoter. Mutational analyses of these 5'-flanking regions have shown that this upstream sequence is important for both basal or inducible expression from reporter constructs.[43,51,58,59] Although there is an additional nucleotide spacing between this pentanucleotide motif and the ARE core in mouse *gsta3* when compared to *NQO1* and rat *GSTP1*, it nevertheless seems likely that these flanking sequences are functionally important. What significance this extra nucleotide adjacent to the mouse *gsta3* ARE core makes is unknown, but it is apparent that the 5'-CATGACATTGC-3' sequence deviates from the minimal 5'-$^A/_G$GTGACNNNGC-3' enhancer reported by Rushmore et al.[46]

3.5.2 CNC bZIP TRANSCRIPTION FACTORS MEDIATE GENE INDUCTION THROUGH THE ARE

The identity of the transcription factors that mediate regulation of ARE-driven genes has stimulated much interest. The similarity between the ARE and the binding motif for NF-E2 (*nuclear factor-erythroid 2*) led Venugopal and Jaiswal[60] to propose that this transcription factor regulates the expression of ARE-containing genes. The NF-E2 factor is a dimeric protein composed of 45 kDa (p45) and 18 kDa (p18) subunits, each belonging to the basic-region leucine zipper (bZIP) transcription-factor superfamily. The restricted expression of the p45 subunit to hematopoietic tissues was not consistent with its involvement in regulating the transcription of hepatic genes. However, the related cap'n'collar (CNC) bZIP family members, Nrf1 and Nrf2 (*NF-E2 p45-related factors 1 and 2*) exhibit a more ubiquitous tissue-specific expression profile.[61] Numerous studies have since shown that these transcription factors activate target gene expression through the ARE.[62] In particular, mice lacking Nrf2 exhibit a dramatically compromised ability to activate the expression of a number of Alpha-, Mu-, and Pi-class GST genes following exposure to typical inducing agents, as well as demonstrating lower levels of GST isoenzymes under basal conditions.[61] In livers of nrf2[-/-] mice, genes showing reduced basal and inducible expression include *GSTA1, GSTA2, GSTA3, GSTM1, GSTM2, GSTM3,* and *GSTM4*.[56,63] To date, little is known about the regulation of GST by Nrf1. In nrf1[-/-] mouse embryonic fibroblasts, expression of ARE-driven genes such as *NQO1, GCLC,* and heme oxygenase 1 (*HO1*) is impaired.[64] It was therefore anticipated that many of the GSTs that are downregulated in *nrf2[-/-]* mice would also be downregulated in *nrf1[-/-]* mice. More recently it has been found that in conditional hepatic Nrf1-nulled mice expression of GSTM3 and of GSTM6 are significantly reduced;[65] contrary to expectations, the level of GSTA1 was markedly increased in these animals. Although NF-E2 p45 is not thought to regulate ARE-driven genes in nonhemopoietic cells, NF-E2 p45-nulled mice show a marked reduction in GSTA3 expression in red blood cells.[66]

As described above, mutational analysis of the sequences that lie immediately adjacent to the 5′-end of the core ARE has revealed that they are essential for induction of a reporter gene by xenobiotics. It is unclear which transcription factors bind the sequences that lie upstream of the core ARE, but their similarity to motifs recognized by AP-1 and ATF/CREB proteins suggest that members of these families may be recruited to the 5′-flanking regions of AREs.

3.5.3 NRF2 AND KEAP1

The ability of Nrf2 to activate gene expression through the ARE depends on both the intracellular levels and subcellular localization of the transcription factor. The Nrf2 protein comprises six domains that are conserved between chicken and human proteins, designated Neh1–Neh6.[67,68] The Neh4 and Neh5 domains are involved in transcriptional activation, while the Neh2 and Neh6 regions of the protein are responsible for its negative regulation.[69] Indeed, mutant forms of Nrf2 that lack the Neh2 domain are constitutively active. Of pivotal importance in the negative regulation of the Nrf2 transcription factor is Keap1 (*K*elch-like *E*CH-*a*ssociated *p*rotein *1*), which interacts with the Neh2 domain of Nrf2.[67] Two models have been proposed to explain how Keap1 negatively regulates Nrf2. In the first, Keap1 localizes to the cytoplasm through an interaction with the cytoskeleton and thus controls the sub-cellular distribution of Nrf2.[70–72] In the second model, Keap1 controls proteasomal degradation of the transcription factor.[72,73] This is achieved because Keap1 is a substrate adaptor for the Cullin 3-Roc1 E3 ubiquitin ligase.[74–76] Keap1 is thus a key protein that negatively regulates Nrf2-dependent gene expression by controlling both the intracellular levels and localization of the transcription factor. This protein–protein interaction also seems to underlie the ability of the Nrf2/ARE transcriptional network to respond to oxidative stress. Evidence indicates that under such conditions, cysteine residues Cys273 and Cys288 in Keap1 are modified, thus compromising the interaction of Keap1 with Nrf2 and, in doing so, its propensity to negatively regulate Nrf2-dependent gene expression.[77] Importantly, the ARE-gene battery is induced in *keap1[-/-]* mice and following knockdown of Keap1 by siRNA

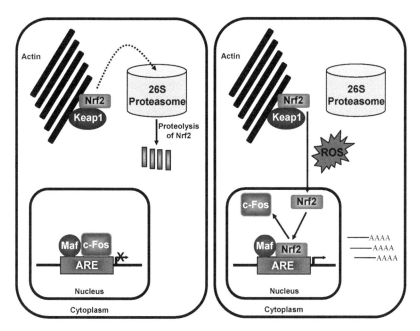

FIGURE 3.3 Regulation of ARE-dependent gene expression by Nrf2. Under basal conditions, small Maf proteins bind to the ARE along with another bZIP factor. The identity of the other binding partner for small Maf is not known and may indeed vary between target genes. Likely candidates include c-Fos (shown above), other Maf proteins, or Nrf1. Under normal homeostatic conditions, little Nrf2 is associated with the ARE, as it is targeted for proteasomal degradation through its interaction with Keap1. However, under conditions of oxidative stress (e.g., in the presence of reactive oxygen species; ROS), Nrf2 is stabilized, translocates to the nucleus, displaces repressor complexes, and activates gene expression as a heterodimer with small Maf proteins.

in human keratinocyte HaCaT cells.[78,79] These results suggest that Nrf2 does not require some redox-dependent modification, such as phosphorylation, in order for it to affect gene induction; accumulation of Nrf2 protein is sufficient to activate the ARE-gene battery. An overview of the Keap1/Nrf2/ARE transcriptional network is presented in Figure 3.3.

3.6 HUMAN GST GENES

Studies of rodents have clearly demonstrated the importance of Nrf2 and the ARE in regulating GST expression, but little is known regarding the involvement of this pathway in controlling the expression of human GST genes. This is particularly troublesome, given the efforts that have been made to identify compounds, based on their ability to induce NQO1 and GST in rodent cell lines, which could be used as cancer chemopreventive agents in humans. Several human cell lines have been reported to show an increased capacity to conjugate GSH with CDNB following treatment with xenobiotics that are known to stimulate ARE-driven gene expression. Human colon HT29 cells show increased GST activity following treatment with allyl sulfide or benzyl isothiocyanate.[80] Also, human mammary MCF-7/0 cells show increased GST activity following treatment with catechol.[81] The molecular basis for the increased GST activity is not known, however. Paucity of information in this regard may reflect in part the lack of human cell lines conducive for the required *in vitro* analyses, and possibly the fact that detection of GST induction by measuring CDNB activity is relatively insensitive and will not monitor Theta-, Omega-, or Zeta-class enzymes. Mounting epidemiological data point to an association between the consumption of foodstuffs that contain compounds know to stimulate ARE-dependent gene expression and a lower incidence of cancer.[82–84] An important challenge facing the field is that of determining whether human GST genes are regulated through an ARE and establishing the extent to which this regulatory element contributes to the apparent chemopreventive properties of such dietary components.

Although the ability to perform *in vitro* experiments in human systems is highly desirable, other routes to addressing this challenge exist in the post-genomic era. For example, the deposition of mass quantities of data through the human genome project provides a unique opportunity to identify putative regulatory elements by directly examining the promoter region of a given gene, although such analyses require information regarding the structural organization of genes of interest.

3.6.1 ORGANIZATION AND CHROMOSOMAL LOCALIZATION OF HUMAN GST GENES

In the case of human cytosolic and mitochondrial GST, genes encoding tranferases belonging to all of the GST families have been described.[11,85–90] A complete overview of human GST genes can also be obtained through bioinformatic analysis of genome sequence data. Indeed, mRNA sequences encoding human GSTA1, GSTA2, GSTA3, GSTA4, GSTA5, GSTK1, GSTM1, GSTM2, GSTM3, GSTM4, GSTM5, GSTO1, GSTO2, GSTP1, GSTT1, GSTT2, and GSTZ1 can be identified using the National Center for Biotechnology Information Web site (www.ncbi.nih.gov). The corresponding accession numbers are listed in Table 3.1. The gene loci predicted to encode each of these mRNA sequences can also be identified using the BLAT (Basic Local Alignment Tool) search facility available through the University of California at Santa Cruz Web site (www.genome.ucsc.edu). Analysis of these data provides much information, including chromosomal localization, which places the Alpha-, Kappa-, Mu-, Omega-, Pi-, Sigma-, Theta-, and Zeta-class genes on human chromosomes 6, 7, 1, 10, 11, 4, 22, and 14, respectively (Table 3.1). Predictions regarding the organization of each gene can also be made. The available human genome data demonstrate that the Alpha-, Kappa-, Mu-, Omega-, Pi-, Sigma-, Theta-, and Zeta-class genes comprise 7, 8, 8, 6, 7, 6, 5, and 9 exons, respectively (Table 3.1). These observations reveal that the gross organization of GST genes is both conserved within the different classes and between humans, rats, and mice. Information regarding the structure of human GST genes allows subsequent identification of promoter regions and detailed analysis of their sequences. Taken together with the data available concerning sequence requirements for ARE-enhancer function, this information

TABLE 3.1
Structure and Chromosomal Localization of Human GST Genes

Human GST Gene	Accession Number for mRNA	Predicted Chromosomal Localization	Predicted Gene Structure
GSTA1	NM_145740	Chromosome 6	7 Exons
GSTA2	NM_000846	Chromosome 6	7 Exons
GSTA3	NM_000847	Chromosome 6	7 Exons
GSTA4	NM_001512	Chromosome 6	7 Exons
GSTA5	NM_153699	Chromosome 6	7 Exons
GSTK1	NM_015917	Chromosome 7	8 Exons
GSTM1	NM_000561	Chromosome 1	8 Exons
GSTM2	NM_000848	Chromosome 1	8 Exons
GSTM3	NM_000849	Chromosome 1	8 Exons
GSTM4	NM_000850	Chromosome 1	8 Exons
GSTM5	NM_000851	Chromosome 1	8 Exons
GSTO1	NM_004832	Chromosome 10	6 Exons[§]
GSTO2	NM_183239	Chromosome 10	6 Exons[§]
GSTP1	NM_000852	Chromosome 11	7 Exons
GSTS1	NM_014485	Chromosome 4	6 Exons[§]
GSTT1	NM_000853	Chromosome 22	5 Exons
GSTT2	NM_000854	Chromosome 22	5 Exons
GSTZ1	NM_145870	Chromosome 14	9 Exons

Note: Accession numbers for mRNAs encoding each of the human cytosolic and mitochondrial GST subunits were identified using the National Center for Biotechnology Information Web site (www.ncbi.nih.gov). The corresponding gene loci were identified using these mRNA sequences and the BLAT (Basic Local Alignment Tool) search facility to query the May 2004 human genome assembly available through the University of California at Santa Cruz Web site (www.genome.ucsc.edu).

§Gene structure proposed based on published data.[86,90]

could form a bioinformatics-based strategy to begin to elucidate the contribution of Nrf2 and the ARE in regulating human GST gene expression.

3.6.2 POLYMORPHISMS IN HUMAN GSTs

It is pertinent to mention that numerous polymorphisms within human GST genes have been documented.[91–93] Although the precise phenotypes associated with human GST gene polymorphisms are varied, most research in this area has focused on susceptibility to tumorigenesis and response to chemotherapy. In the case of the former, early studies addressed the question of whether individuals lacking GSTM1-1 or GSTT1-1 exhibit higher incidences of bladder, breast, colorectal, head/neck, and lung cancers. Upon the discovery of allelic variants of GSTP1-encoding enzymes with compromised enzymatic activity, the possibility that combinations of polymorphisms within Mu-, Pi-, and Theta-class GSTs could contribute to disease susceptibility was explored. A general theme from studies in the field seems to be that variations within individual GST genes do not confer a major increase in cancer susceptibility, although the GSTM1-null genotype was associated with a modestly increased risk of developing lung cancer.[94] Also, loss of both GSTM1-1 and GSTT1-1 was associated with a modest effect on the incidence of head and neck cancer.[95]

Regarding the implications of genetic variation within GST genes and the efficacy of chemotherapy, *GSTP1* polymorphisms are modifiers of response to chemotherapy in patients with metastatic colorectal cancer[96] and multiple myeloma.[97] Polymorphisms within the Pi-class GST also influence risk of therapy-related acute myeloid leukemia in patients treated for breast cancer, non-Hodgkin's lymphoma, ovarian cancer, and Hodgkin's disease.[98]

In addition to the relatively limited number of studies discussed above, it is important to mention that polymorphisms have been documented for transferases belonging to other GST families[90,99,100] and that the phenotypes recorded for GST polymorphisms also include some related to inflammatory diseases such as asthma, atherosclerosis, rheumatoid arthritis, and systemic sclerosis.[101–103] Interestingly, it has recently been reported that the GSTT1-null phenotype is associated with premature morbidity and mortality in individuals with Type 2 diabetes.[104]

3.6.3 REGULATION OF HUMAN GST GENE EXPRESSION

As mentioned previously, in marked contrast to the extensive body of literature concerning the regulation of rodent GST genes, little is known regarding the role of Nrf2 and the ARE in regulating human GST gene expression. Possible reasons for this lack of data have been discussed. It is clear that this transcriptional pathway is conserved in humans, and that many of those genes known to be regulated by Nrf2 in rodents also contain functional AREs in humans. These include human *NQO1* and human *GCLC* and *GCLM*.[62] Given the conservation in GST gene structure between humans, mice, and rats, and the fact that the role of Nrf2 and the ARE in regulating the expression of other genes is similarly retained, one may well predict that human GST genes would also be regulated via this transcriptional pathway.

Despite this prediction, no human cytosolic GST genes have been shown to be regulated through an ARE. Nevertheless, there is some evidence to support this notion. Particularly relevant is the observation that compounds known to transcriptionally activate genes through an ARE have been shown to modulate the expression of GSTs in human cell lines. One such compound is sulforaphane, a naturally occurring isothiocyanate that is generated as a hydrolysis product of glucosinolates from broccoli and other cruciferous vegetables.[30] Sulforaphane has been found to induce the expression of GSTA1 in Caco2 human colon adenocarcinoma cells[105] and in LNCaP, MDA PCa 2a, and MDA PCa 2b human prostate cancer cells.[106] Sulforaphane was similarly found to increase Pi-class GST protein expression in the MCF-10F non-neoplastic human mammary cells.[107] Although not formally addressed in the studies, these observations are broadly consistent with a role for the ARE in regulating human *GSTA1* and *GSTP1* gene expression. Interestingly, both *GSTA1* and *GSTP1* are regulated through an ARE in other species. Despite the conservation in sequence between AREs in mouse- and rat-class Alpha GST genes, initial analysis of 2 kb of sequence upstream of exon 1 in the human *GSTA1* gene does not reveal the presence of an

GENE		ARE		TATA BOX
Human *GSTP1*	-72	GCGCCGTGACTCAGCACTGG--------GGCGGAGCGGGGCGGGACCACCCTTATAAGGCT		-20
Rat *GSTP1*	-67	CTGTGTTGACTCAGCATCCG------------GGGCGGGGCGCAATGCCCCTTATAAGGCT		-19
Mouse *gstp1*	-66	ACGTGTTGAGTCAGCATCCG------------GGGCGGAGCGCGATGCCCCTTATAAGGCT		-18
Mouse *gstp2*	-76	ACGTGTTGAGTCAGCATCCGGGGCGGCATCCGGGGCGGAGCGCAATGCCCCTTATAAGGCT		-16
Zebrafish *GSTP1*	-59	CGTGCATGACTCATCAAAAA--------------------CGCTGAGGCTTATAAACG		-21
Zebrafish *GSTP2*	-60	CATGAATGACTCAATAAAGT--------------------ACACATGAGGCTTTAAAAGG		-21

FIGURE 3.4 Alignment of ARE-like sequences in various gene promoters. The ARE-like sequences in the promoter region of various Pi-class GST genes have been aligned. The TATA box in the gene promoters is shown. The numbers indicate position relative to the predicted transcriptional initiation site. Data were adapted from Suzuki et al.[109]

obvious ARE. It remains to be established whether this observation excludes a role for Nrf2 in regulating the expression of this gene through an ARE, or whether it further supports the notion that there exists significant variability regarding the sequence requirements for the functionality of this enhancer. By contrast, there exists more persuasive evidence for the role of Nrf2 in regulating the expression of the human *GSTP1* gene. In the rat, much attention has focused on GPEI, which is located approximately 2.5 kb upstream of the transcription initiation site.[42] The GPEI site contains an ARE-like sequence (5′-TGATTCAGC-3′), binds an Nrf2/MafK heterodimer, and facilitates transcriptional activation of *GSTP1*-reporter constructs by ectopically expressed Nrf2.[108] However, this regulatory element appears to be unique to rat *GSTP1* and has not been identified in genes encoding other Pi-class GSTs. Instead, in other species, evidence points to a dominant role for a proximal ARE-like sequence located approximately 50 bp upstream of the transcription start site.[109] In both mouse and zebrafish *GSTP1* genes, this proximal sequence binds Nrf2 and mediates Nrf2-dependent transactivation.[109,110] This sequence is conserved in the human *GSTP1* gene (Figure 3.4), although its functionality in this genomic context is yet to be confirmed. Nevertheless, this observation is certainly suggestive of a functional role for Nrf2 and the ARE in regulating human *GSTP1* gene expression.

A recent microarray study into the influence of BRCA1 on gene expression in human prostate DU-145 cells has revealed that the BRCA1 tumor suppressor protein can induce GSTT1 and GSTZ1 mRNA.[111] Expression of the MAPEG member MGST1 was found to be particularly highly induced by BRCA1, as was MGST2.[111] Most interestingly, ectopic expression of BRCA1 increased the ratio of reduced to oxidized glutathione and activated ARE-driven reporter gene expression.[111] It is not known if Nrf2 mediates the effect of BRCA1 on expression of these transferase genes. However, previous work by Kelner et al.[112] has demonstrated that the gene promoter of human *MGST1* contains an ARE, and it therefore appears that Nrf2 mediates the effect of BRCA1 on this MAPEG member. It is possible that the promoters of human *GSTT1*, *GSTZ1*, and *MGST2* similarly contain functional AREs. This area warrants further work.

It is clear from the above discussion that much remains to be learned about the control of human GST gene expression. It seems likely that the deposition of mass data through the human genome project will expedite progress in this area over the coming years, by further embedding the bioinformatics approach described previously into this area of research.

3.7 CONCLUDING REMARKS

Historically, much work on GSTs has focused on their catalytic properties and biological functions. More recently, emphasis has been placed on the regulation of GST gene expression, in terms of both fundamental molecular mechanisms and how this relates to cancer chemoprevention. Despite the elucidation of the Nrf2/ARE transcriptional network as a key player in controlling the transcription of rodent GST (and other) genes, it is as yet unclear whether their human counterparts are similarly managed. Although some evidence suggests that this may be the case, it is clear that further study in this area is required. The contribution of GST isoenzymes to signaling pathways is also an area that is likely to receive keen interest in the future.

REFERENCES

1. Hayes, J.D., Flanagan, J.U., and Jowsey, I.R., Glutathione transferases, *Annu. Rev. Pharmacol. Toxicol.*, 45, 51, 2005.
2. Jakobsson, P.J. et al., Common structural features of MAPEG — a widespread superfamily of membrane-associated proteins with highly divergent functions in eicosanoid and glutathione metabolism, *Protein Sci.*, 8, 689, 1999.
3. Jakobsson, P.J. et al., Membrane-associated proteins in eicosanoid and glutathione metabolism (MAPEG). A widespread protein superfamily, *Am. J. Respir. Crit. Care Med.*, 161, S20, 2000.
4. Hayes, J.D. and Pulford, D.J., The glutathione S-transferase supergene family: regulation of GST and the contribution of the isoenzymes to cancer chemoprotection and drug resistance, *Crit. Rev. Biochem. Mol. Biol.*, 30, 445, 1995.
5. Pettigrew, N.E. and Colman, R.F., Heterodimers of glutathione S-transferase can form between isoenzyme classes pi and mu, *Arch. Biochem. Biophys.*, 396, 225, 2001.
6. Berhane, K. et al., Detoxication of base propenals and other alpha, beta-unsaturated aldehyde products of radical reactions and lipid peroxidation by human glutathione transferases, *Proc. Natl. Acad. Sci. USA*, 91, 1480, 1994.
7. Dagnino-Subiabre, A. et al., Glutathione transferase M2-2 catalyzes conjugation of dopamine and dopa o-quinones, *Biochem. Biophys. Res. Commun.*, 274, 32, 2000.
8. Kulkarni, A.P. and Sajan, M., Lipoxygenase — another pathway for glutathione conjugation of xenobiotics: a study with human term-placental lipoxygenase and ethacrynic acid, *Arch. Biochem. Biophys.*, 371, 220, 1999.
9. Kanaoka, Y. et al., Cloning and crystal structure of hematopoietic prostaglandin D synthase, *Cell*, 90, 1085, 1997.
10. Jowsey, I.R. et al., Mammalian-class Sigma glutathione S-transferases: catalytic properties and tissue-specific expression of human and rat GSH-dependent prostaglandin D2 synthases, *Biochem. J.*, 359, 507, 2001.
11. Blackburn, A.C. et al., Characterization and chromosome location of the gene GSTZ1 encoding the human Zeta-class glutathione transferase and maleylacetoacetate isomerase, *Cytogenet. Cell. Genet.*, 83, 109, 1998.
12. Robinson, A. et al., Modelling and bioinformatics studies of the human Kappa-class glutathione transferase predict a novel third glutathione transferase family with similarity to prokaryotic 2-hydroxychromene-2-carboxylate isomerases, *Biochem. J.*, 379, 541, 2004.

13. Listowsky, I., High capacity binding by glutathione S-transferases and glucocorticoid resistance. In *Structure and Function of Glutathione Transferases* (Tew, K.D., Pickett, C.B., Mantle, T.J., Mannervik, B., and Hayes, J.D., eds; 1993) pp. 199–209, CRC Press, Boca Raton, FL.

14. Wang, J., Bauman, S., and Colman, R.F., Photoaffinity labeling of rat liver glutathione S-transferase, 4-4, by glutathionyl S-[4-(succinimidyl)-benzophenone], *Biochemistry*, 37, 15671, 1998.

15. Vargo, M.A. and Colman, R.F., Affinity labeling of rat glutathione S-transferase isozyme 1-1 by 17beta-iodoacetoxy-estradiol-3-sulfate, *J. Biol. Chem.*, 276, 2031, 2001.

16. Ralat, L.A. and Colman, R.F., Glutathione S-transferase Pi has at least three distinguishable xenobiotic substrate sites close to its glutathione-binding site, *J. Biol. Chem.*, 279, 50204, 2004.

17. Johansson, A.S. and Mannervik, B., Human glutathione transferase A3-3, a highly efficient catalyst of double-bond isomerization in the biosynthetic pathway of steroid hormones, *J. Biol. Chem.*, 276, 33061, 2001.

18. Ujihara, M. et al., Biochemical and immunological demonstration of prostaglandin D2, E2, and F2 alpha formation from prostaglandin H2 by various rat glutathione S-transferase isozymes, *Arch. Biochem. Biophys.*, 264, 428, 1988.

19. Ujihara, M. et al., Prostaglandin D2 formation and characterization of its synthetases in various tissues of adult rats, *Arch. Biochem. Biophys.*, 260, 521, 1988.

20. Jakobsson, P.J. et al., Identification of human prostaglandin E synthase: a microsomal, glutathione-dependent, inducible enzyme, constituting a potential novel drug target, *Proc. Natl. Acad. Sci. USA*, 96, 7220, 1999.

21. Mancini, J.A. et al., Cloning, expression, and up-regulation of inducible rat prostaglandin E synthase during lipopolysaccharide-induced pyresis and adjuvant-induced arthritis, *J. Biol. Chem.*, 276, 4469, 2001.

22. Thoren, S. et al., Human microsomal prostaglandin E synthase-1: purification, functional characterization, and projection structure determination, *J. Biol. Chem.*, 278, 22199, 2003.

23. Forman, B.M. et al., 15-Deoxy-delta 12, 14-prostaglandin J2 is a ligand for the adipocyte determination factor PPAR gamma, *Cell*, 83, 803, 1995.

24. Kliewer, S.A. et al., A prostaglandin J2 metabolite binds peroxisome proliferator-activated receptor gamma and promotes adipocyte differentiation, *Cell*, 83, 813, 1995.

25. Rossi, A. et al., Anti-inflammatory cyclopentenone prostaglandins are direct inhibitors of IkappaB kinase, *Nature*, 403, 103, 2000.

26. Bogaards, J.J., Venekamp, J.C., and van Bladeren, P.J., Stereoselective conjugation of prostaglandin A2 and prostaglandin J2 with glutathione, catalyzed by the human glutathione S-transferases A1-1, A2-2, M1a-1a, and P1-1, *Chem. Res. Toxicol.*, 10, 310, 1997.

27. Han, R., Tsui, S., and Smith, T.J., Up-regulation of prostaglandin E2 synthesis by interleukin-1beta in human orbital fibroblasts involves coordinate induction of prostaglandin-endoperoxide H synthase-2 and glutathione-dependent prostaglandin E2 synthase expression, *J. Biol. Chem.*, 277, 16355, 2002.

28. Wang, X. et al., Prostaglandin E2 is a product of induced prostaglandin-endoperoxide synthase 2 and microsomal-type prostaglandin E synthase at the implantation site of the hamster, *J. Biol. Chem.*, 279, 30579, 2004.

29. Prochaska, H.J. and Talalay, P., Regulatory mechanisms of monofunctional and bifunctional anticarcinogenic enzyme inducers in murine liver, *Cancer Res.*, 48, 4776, 1988.

30. Zhang, Y. et al., A major inducer of anticarcinogenic protective enzymes from broccoli: isolation and elucidation of structure, *Proc. Natl. Acad. Sci. USA*, 89, 2399, 1992.

31. Prestera, T. et al., Chemical and molecular regulation of enzymes that detoxify carcinogens, *Proc. Natl. Acad. Sci. USA*, 90, 2965, 1993.

32. Dinkova-Kostova, A.T., Abeygunawardana, C., and Talalay, P., Chemoprotective properties of phenylpropenoids, bis(benzylidene)cycloalkanones, and related Michael reaction acceptors: correlation of potencies as phase 2 enzyme inducers and radical scavengers, *J. Med. Chem.*, 41, 5287, 1998.

33. Dinkova-Kostova, A.T. et al., Extremely potent triterpenoid inducers of the phase 2 response: correlations of protection against oxidant and inflammatory stress, *Proc. Natl. Acad. Sci. USA*, 102, 4584, 2005.

34. Fahey, J.W. et al., Chlorophyll, chlorophyllin, and related tetrapyrroles are significant inducers of mammalian phase 2 cytoprotective genes, *Carcinogenesis*, 26, 1247, 2005.

35. Hayes, J.D. et al., Cellular response to cancer chemopreventive agents: contribution of the antioxidant responsive element to the adaptive response to oxidative and chemical stress, *Biochem. Soc. Symp.*, 64, 141, 1999.

36. Spencer, S.R. et al., The potency of inducers of NAD(P)H:(quinone-acceptor) oxidoreductase parallels their efficiency as substrates for glutathione transferases. Structural and electronic correlations, *Biochem. J.*, 273, 711, 1991.

37. Dinkova-Kostova, A.T. et al., Potency of Michael reaction acceptors as inducers of enzymes that protect against carcinogenesis depends on their reactivity with sulfhydryl groups, *Proc. Natl. Acad. Sci. USA*, 98, 3404, 2001.

38. Wild, A.C. and Mulcahy, R.T., Regulation of gamma-glutamylcysteine synthetase subunit gene expression: insights into transcriptional control of antioxidant defenses, *Free Radic. Res.*, 32, 281, 2000.

39. Hayes, J.D. et al., The Nrf2 transcription factor contributes both to the basal expression of glutathione S-transferases in mouse liver and to their induction by the chemopreventive synthetic antioxidants, butylated hydroxyanisole, and ethoxyquin, *Biochem. Soc. Trans.*, 28, 33, 2000.

40. Rushmore, T.H. et al., Regulation of glutathione S-transferase Ya subunit gene expression: identification of a unique xenobiotic-responsive element controlling inducible expression by planar aromatic compounds, *Proc. Natl. Acad. Sci. USA*, 87, 3826, 1990.

41. Friling, R.S. et al., Xenobiotic-inducible expression of murine glutathione S-transferase Ya subunit gene is controlled by an electrophile-responsive element, *Proc. Natl. Acad. Sci. USA*, 87, 6258, 1990.

42. Okuda, A. et al., Structural and functional analysis of an enhancer GPEI having a phorbol 12-O-tetradecanoate 13-acetate responsive element-like sequence found in the rat glutathione transferase P gene, *J. Biol. Chem.*, 264, 16919, 1989.

43. Favreau, L.V. and Pickett, C.B., The rat quinone reductase antioxidant response element. Identification of the nucleotide sequence required for basal and inducible activity and detection of antioxidant response element-binding proteins in hepatoma and non-hepatoma cell lines, *J. Biol. Chem.*, 270, 24468, 1995.

44. Hayes, J.D. et al., Preferential over-expression of the class alpha rat Ya2 glutathione S-transferase subunit in livers bearing aflatoxin-induced pre-neoplastic nodules. Comparison of the primary structures of Ya1 and Ya2 with cloned class alpha glutathione S-transferase cDNA sequences, *Biochem. J.*, 268, 295, 1990.

45. Hayes, J.D. et al., Ethoxyquin-induced resistance to aflatoxin B1 in the rat is associated with the expression of a novel alpha-class glutathione S-transferase subunit, Yc2, which possesses high catalytic activity for aflatoxin B1-8,9-epoxide, *Biochem. J.*, 279, 385, 1991.

46. Rushmore, T.H., Morton, M.R., and Pickett, C.B., The antioxidant responsive element. Activation by oxidative stress and identification of the DNA consensus sequence required for functional activity, *J. Biol. Chem.*, 266, 11632, 1991.

47. Morimura, S. et al., Trans-activation of glutathione transferase P gene during chemical hepatocarcinogenesis of the rat, *Proc. Natl. Acad. Sci. USA*, 90, 2065, 1993.

48. Ikeda, H. et al., JBS Bio-Frontier Symposium, Tsukuba, Japan, *Poster 12*, 2003.

49. Favreau, L.V. and Pickett, C.B., Transcriptional regulation of the rat NAD(P)H:quinone reductase gene. Identification of regulatory elements controlling basal level expression and inducible expression by planar aromatic compounds and phenolic antioxidants, *J. Biol. Chem.*, 266, 4556, 1991.

50. Jaiswal, A.K., Human NAD(P)H:quinone oxidoreductase (NQO1) gene structure and induction by dioxin, *Biochemistry*, 30, 10647, 1991.

51. Nioi, P. et al., Identification of a novel Nrf2-regulated antioxidant response element (ARE) in the mouse NAD(P)H:quinone oxidoreductase 1 gene: reassessment of the ARE consensus sequence, *Biochem. J.*, 374, 337, 2003.

52. Mulcahy, R.T. et al., Constitutive and beta-naphthoflavone–induced expression of the human gamma-glutamylcysteine synthetase heavy subunit gene is regulated by a distal antioxidant response element/TRE sequence, *J. Biol. Chem.*, 272, 7445, 1997.

53. Moinova, H.R. and Mulcahy, R.T., An electrophile responsive element (EpRE) regulates beta-naphthoflavone induction of the human gamma-glutamylcysteine synthetase regulatory subunit gene. Constitutive expression is mediated by an adjacent AP-1 site, *J. Biol. Chem.*, 273, 14683, 1998.

54. Prestera, T. et al., Parallel induction of heme oxygenase-1 and chemoprotective phase 2 enzymes by electrophiles and antioxidants: regulation by upstream antioxidant-responsive elements (ARE), *Mol. Med.*, 1, 827, 1995.

55. Alam, J. et al., Mechanism of heme oxygenase-1 gene activation by cadmium in MCF-7 mammary epithelial cells. Role of p38 kinase and Nrf2 transcription factor, *J. Biol. Chem.*, 275, 27694, 2000.

56. Jowsey, I.R. et al., Expression of the aflatoxin B1-8,9-epoxide-metabolizing murine glutathione S-transferase A3 subunit is regulated by the Nrf2 transcription factor through an antioxidant response element, *Mol. Pharmacol.*, 64, 1018, 2003.

57. Wasserman, W.W. and Fahl, W.E., Functional antioxidant responsive elements, *Proc. Natl. Acad. Sci. USA*, 94, 5361, 1997.

58. Okuda, A. et al., Functional cooperativity between two TPA responsive elements in undifferentiated F9 embryonic stem cells, *Embo. J.*, 9, 1131, 1990.

59. Xie, T. et al., ARE- and TRE-mediated regulation of gene expression. Response to xenobiotics and antioxidants, *J. Biol. Chem.*, 270, 6894, 1995.

60. Venugopal, R. and Jaiswal, A.K., Nrf1 and Nrf2 positively and c-Fos and Fra1 negatively regulate the human antioxidant response element-mediated expression of NAD(P)H:quinone oxidoreductase1 gene, *Proc. Natl. Acad. Sci. USA*, 93, 14960, 1996.

61. McMahon, M. et al., The Cap'n'Collar basic leucine zipper transcription factor Nrf2 (NF-E2 p45-related factor 2) controls both constitutive and inducible expression of intestinal detoxification and glutathione biosynthetic enzymes, *Cancer Res.*, 61, 3299, 2001.

62. Nguyen, T., Sherratt, P.J., and Pickett, C.B., Regulatory mechanisms controlling gene expression mediated by the antioxidant response element, *Annu. Rev. Pharmacol. Toxicol.*, 43, 233, 2003.

63. Chanas, S.A. et al., Loss of the Nrf2 transcription factor causes a marked reduction in constitutive and inducible expression of the glutathione S-transferase Gsta1, Gsta2, Gstm1, Gstm2, Gstm3, and Gstm4 genes in the livers of male and female mice, *Biochem. J.*, 365, 405, 2002.

64. Leung, L. et al., Deficiency of the Nrf1 and Nrf2 transcription factors results in early embryonic lethality and severe oxidative stress, *J. Biol. Chem.*, 278, 48021, 2003.

65. Xu, Z. et al., Liver-specific inactivation of the Nrf1 gene in adult mouse leads to nonalcoholic steatohepatitis and hepatic neoplasia, *Proc. Natl. Acad. Sci. USA*, 102, 4120, 2005.

66. Chan, J.Y. et al., Reduced oxidative-stress response in red blood cells from p45NFE2-deficient mice, *Blood*, 97, 2151, 2001.

67. Itoh, K. et al., Keap1 represses nuclear activation of antioxidant responsive elements by Nrf2 through binding to the amino-terminal Neh2 domain, *Genes Dev.*, 13, 76, 1999.

68. Katoh, Y. et al., Two domains of Nrf2 cooperatively bind CBP, a CREB-binding protein, and synergistically activate transcription, *Genes Cells*, 6, 857, 2001.

69. McMahon, M. et al., Redox-regulated turnover of Nrf2 is determined by at least two separate protein domains, the redox-sensitive Neh2 degron and the redox-insensitive Neh6 degron, *J. Biol. Chem.*, 279, 31556, 2004.

70. Kang, M.I. et al., Scaffolding of Keap1 to the actin cytoskeleton controls the function of Nrf2 as key regulator of cytoprotective phase 2 genes, *Proc. Natl. Acad. Sci. USA*, 101, 2046, 2004.

71. Motohashi, H. and Yamamoto, M., Nrf2-Keap1 defines a physiologically important stress response mechanism, *Trends Mol. Med.*, 10, 549, 2004.

72. Itoh, K. et al., Keap1 regulates both cytoplasmic-nuclear shuttling and degradation of Nrf2 in response to electrophiles, *Genes Cells*, 8, 379, 2003.

73. McMahon, M. et al., Keap1-dependent proteasomal degradation of transcription factor Nrf2 contributes to the negative regulation of antioxidant response element-driven gene expression, *J. Biol. Chem.*, 278, 21592, 2003.

74. Kobayashi, A. et al., Oxidative stress sensor Keap1 functions as an adaptor for Cul3-based E3 ligase to regulate proteasomal degradation of Nrf2, *Mol. Cell. Biol.*, 24, 7130, 2004.

75. Zhang, D.D. et al., Keap1 is a redox-regulated substrate adaptor protein for a Cul3-dependent ubiquitin ligase complex, *Mol. Cell. Biol.*, 24, 10941, 2004.

76. Furukawa, M. and Xiong, Y., BTB protein Keap1 targets antioxidant transcription factor Nrf2 for ubiquitination by the Cullin 3-Roc1 ligase, *Mol. Cell. Biol.*, 25, 162, 2005.

77. Wakabayashi, N. et al., Protection against electrophile and oxidant stress by induction of the phase 2 response: fate of cysteines of the Keap1 sensor modified by inducers, *Proc. Natl. Acad. Sci. USA*, 101, 2040, 2004.

78. Wakabayashi, N. et al., Keap1-null mutation leads to postnatal lethality due to constitutive Nrf2 activation, *Nat. Genet.*, 35, 238, 2003.

79. Devling, T.W. et al., Utility of siRNA against Keap1 as a strategy to stimulate a cancer chemopreventive phenotype, *Proc. Natl. Acad. Sci. USA*, 102, 7280, 2005.

80. Kirlin, W.G. et al., Dietary compounds that induce cancer preventive phase 2 enzymes activate apoptosis at comparable doses in HT29 colon carcinoma cells, *J. Nutr.*, 129, 1827, 1999.

81. Sreerama, L., Rekha, G.K., and Sladek, N.E., Phenolic antioxidant-induced overexpression of class-3 aldehyde dehydrogenase and oxazaphosphorine-specific resistance, *Biochem. Pharmacol.*, 49, 669, 1995.

82. Verhoeven, D.T. et al., A review of mechanisms underlying anticarcinogenicity by brassica vegetables, *Chem. Biol. Interact.*, 103, 79, 1997.

83. International Agency for Research on Cancer, *Handbook on Cancer Prevention, Fruit and Vegetables*, IARC Press, 2003.

84. International Agency for Research on Cancer, *Handbook on Cancer Prevention, Cruciferous Vegetables, Isothiocyanates, and Indoles*, IARC Press, 2004.

85. Coggan, M. et al., Structure and organization of the human theta-class glutathione S-transferase and D-dopachrome tautomerase gene complex, *Biochem. J.*, 334, 617, 1998.

86. Kanaoka, Y. et al., Structure and chromosomal localization of human and mouse genes for hematopoietic prostaglandin D synthase. Conservation of the ancestral genomic structure of sigma-class glutathione S-transferase, *Eur. J. Biochem.*, 267, 3315, 2000.

87. Morel, F. et al., Gene and protein characterization of the human glutathione S-transferase kappa and evidence for a peroxisomal localization, *J. Biol. Chem.*, 279, 16246, 2004.

88. Reinhart, J. and Pearson, W.R., The structure of two murine class-mu glutathione transferase genes coordinately induced by butylated hydroxyanisole, *Arch. Biochem. Biophys.*, 303, 383, 1993.

89. Rohrdanz, E., Nguyen, T., and Pickett, C.B., Isolation and characterization of the human glutathione S-transferase A2 subunit gene, *Arch. Biochem. Biophys.*, 298, 747, 1992.

90. Whitbread, A.K. et al., Characterization of the human Omega class glutathione transferase genes and associated polymorphisms, *Pharmacogenetics*, 13, 131, 2003.

91. Hayes, J.D. and Strange, R.C., Glutathione S-transferase polymorphisms and their biological consequences, *Pharmacology*, 61, 154, 2000.

92. Townsend, D. and Tew, K., Cancer drugs, genetic variation, and the glutathione-S-transferase gene family, *Am. J. Pharmacogenomics*, 3, 157, 2003.

93. Coles, B.F. and Kadlubar, F.F., Detoxification of electrophilic compounds by glutathione S-transferase catalysis: determinants of individual response to chemical carcinogens and chemotherapeutic drugs? *Biofactors*, 17, 115, 2003.

94. Benhamou, S. et al., Meta- and pooled analyses of the effects of glutathione S-transferase M1 polymorphisms and smoking on lung cancer risk, *Carcinogenesis*, 23, 1343, 2002.

95. Hashibe, M. et al., Meta- and pooled analyses of GSTM1, GSTT1, GSTP1, and CYP1A1 genotypes and risk of head and neck cancer, *Cancer Epidemiol. Biomarkers Prev.*, 12, 1509, 2003.

96. Stoehlmacher, J. et al., Association between glutathione S-transferase P1, T1, and M1 genetic polymorphism and survival of patients with metastatic colorectal cancer, *J. Natl. Cancer Inst.*, 94, 936, 2002.

97. Dasgupta, R.K. et al., Polymorphic variation in GSTP1 modulates outcome following therapy for multiple myeloma, *Blood*, 102, 2345, 2003.

98. Allan, J.M. et al., Polymorphism in glutathione S-transferase P1 is associated with susceptibility to chemotherapy-induced leukemia, *Proc. Natl. Acad. Sci. USA*, 98, 11592, 2001.

99. Noguchi, E. et al., New polymorphisms of haematopoietic prostaglandin D synthase and human prostanoid DP receptor genes, *Clin. Exp. Allergy*, 32, 93, 2002.

100. Board, P.G. et al., Identification of novel glutathione transferases and polymorphic variants by expressed sequence tag database analysis, *Drug Metab. Dispos.*, 29, 544, 2001.
101. Palmer, C.N. et al., Association of common variation in glutathione S-transferase genes with premature development of cardiovascular disease in patients with systemic sclerosis, *Arthritis Rheum.*, 48, 854, 2003.
102. Gilliland, F.D. et al., Effect of glutathione-S-transferase M1 and P1 genotypes on xenobiotic enhancement of allergic responses: randomised, placebo-controlled cross-over study, *Lancet*, 363, 119, 2004.
103. Romieu, I. et al., Genetic polymorphism of GSTM1 and antioxidant supplementation influence lung function in relation to ozone exposure in asthmatic children in Mexico City, *Thorax*, 59, 8, 2004.
104. Doney, A.S. et al., Increased cardiovascular morbidity and mortality in Type 2 diabetes is associated with the glutathione S-transferase theta-null genotype. A Go-DARTS study, *Circulation*, 111, 2927, 2005 May 31.
105. Svehlikova, V. et al., Interactions between sulforaphane and apigenin in the induction of UGT1A1 and GSTA1 in CaCo-2 cells, *Carcinogenesis*, 25, 1629, 2004.
106. Brooks, J.D., Paton, V.G., and Vidanes, G., Potent induction of phase 2 enzymes in human prostate cells by sulforaphane, *Cancer Epidemiol. Biomarkers Prev.*, 10, 949, 2001.
107. Singletary, K. and MacDonald, C., Inhibition of benzo[a]pyrene- and 1,6-dinitropyrene-DNA adduct formation in human mammary epithelial cells by dibenzoylmethane and sulforaphane, *Cancer Lett.*, 155, 47, 2000.
108. Ikeda, H., Nishi, S., and Sakai, M., Transcription factor Nrf2/MafK regulates rat placental glutathione S-transferase gene during hepatocarcinogenesis, *Biochem. J.* 380, 515, 2004.
109. Suzuki, T. et al., Pi-class glutathione S-transferase genes are regulated by Nrf2 through an evolutionarily conserved regulatory element in zebrafish, *Biochem. J.*, 388, 65, 2005.
110. Ikeda, H. et al., Activation of mouse Pi-class glutathione S-transferase gene by Nrf2(NF-E2-related factor 2) and androgen, *Biochem. J.*, 364, 563, 2002.
111. Bae, I. et al., BRCA1 induces antioxidant gene expression and resistance to oxidative stress, *Cancer Res.*, 64, 7893, 2004.
112. Kelner, M.J. et al., Structural organization of the microsomal glutathione S-transferase gene (MGST1) on chromosome 12p13.1-13.2. Identification of the correct promoter region and demonstration of transcriptional regulation in response to oxidative stress, *J. Biol. Chem.*, 275, 13000, 2000.

4 Combinatorial Protein Chemistry in Three Dimensions: A Paradigm for the Evolution of Glutathione Transferases with Novel Activities

Ylva Ivarsson and Bengt Mannervik

CONTENTS

4.1   INTRODUCTION

In this chapter the design of proteins with glutathione transferase (GST) activity is considered from the vantage point of protein engineering. Covered are the scopes of reactions to be catalyzed, the tuning of substrate selectivities, and the stochastic combinations of building blocks in the construction of enzyme molecules with diverse functions. This account does not purport to describe the actual mechanisms operating in natural evolution, but it is an approach to identifying fundamental features in the molecular blueprint of GSTs. A deeper understanding of crucial principles may prove useful in the redesign of GSTs for applications in medicine and biotechnology.

4.2   CHEMISTRY

4.2.1   THE REACTIVE SULFUR OF GLUTATHIONE

The tripeptide glutathione (GSH) has been adopted by aerobic organisms for the cellular defense that counteracts the oxidative stress exerted by reactive oxygen species (ROS) and electrophilic substances produced by oxidation of cellular components or organic compounds in the surrounding environment. The unusual tripeptide structure of GSH, γ-L-glutamyl-L-cysteinylglycine, owing to the γ-glutamyl bond, is stable against peptidases that otherwise hydrolyze small peptides (Figure 4.1). The reactive sulfhydryl group of GSH has chemical properties that are essentially the same as those of most other low molecular weight thiols occurring in living systems. There is no obvious chemical reason why another biothiol could not have

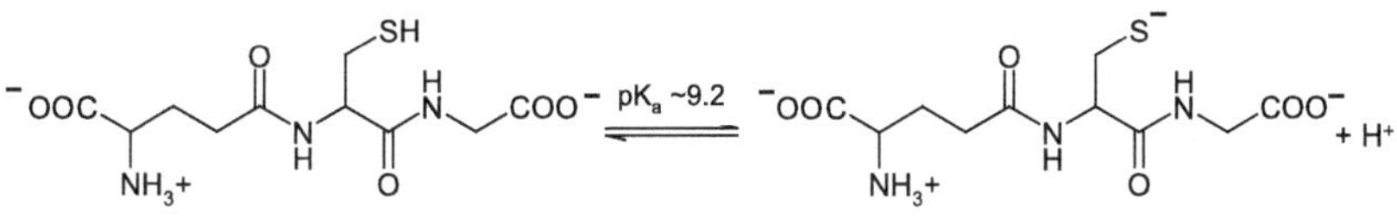

FIGURE 4.1 Structure of glutathione and the ionization of its thiol group.

been recruited in evolution for the thiol biochemistry now ascribed to GSH. In fact, many organisms depend on different thiols and appear to be devoid of GSH for their biochemical reactions.[1,2]

The majority of the toxicologically relevant reactions of GSH involve the sulfur as a nucleophile. The thiol group reacts predominantly in its ionized thiolate form, because the undissociated sulfhydryl group is a comparatively poor nucleophile. Therefore, GSH reactions are base-catalyzed when the pH of the reaction medium is lower than the pK_a of its thiol group. In general, the reactivity of a thiol depends on its basicity and its nucleophilicity. These parameters work in opposite directions, because a high pK_a (such as 9.2 for GSH) is usually associated with high nucleophilicity, but at the same time it yields only a small proportion of the thiolate at neutral pH. A balance between these contributions is reached when the pK_a is close to the pH for the reaction.

4.2.2 TWO MAJOR CATEGORIES OF DETOXICATION REACTIONS CATALYZED BY GSTS

A variety of compounds, both of biological and nonbiological origins, can be conjugated with GSH. Sulfur is a soft nucleophile and preferentially reacts with soft electrophiles. Most electrophilic organic compounds fitting this description can form GSH conjugates under suitable conditions, provided that steric hindrance or other boundary conditions are not prohibitive. Some compounds react nonenzymatically, but reaction rates vary greatly, and enzyme catalysis will often accelerate the conjugation. The nucleophilic character of the thiolate group of GSH promotes transfer of negative charge to electrophilic centers. At its extreme, electrons of the sulfur are completely shifted to the recipient molecule. This is often the case when an electronegative heteroatom is the target of the nucleophilic attack, and such reactions result in reduction of hydroperoxides, disulfides, nitrate esters, and other substances. However, in most cases the electron transfer is not complete, and the donor and acceptor molecules become a linked couple that share an electron pair. There are two main categories of such conjugative reactions: additions and substitutions.

In additions, the sulfur of GSH combines with a double bond, oxirane ring, or similar structure, resulting in a product containing all the constituents of the reacting molecules. The majority of the known physiological reactions catalyzed by GSTs are additions of this kind. An example is the conjugation of an alkenen-2-al (Figure 4.2). In substitutions, the GSH thiolate displaces a fragment (for example, an anion such as a halide) from the electrophilic substrate and replaces the leaving group. In this case two products are formed, and the net reaction usually involves release of a proton stoichiometric to the one produced in the ionization of the thiol group of GSH (Figure 4.3). The strong conjugate acid of a leaving group such as chloride is

FIGURE 4.2 Michael addition of glutathione to an alkenal.

$$GSH + H_3C\text{-}Cl \longrightarrow GS\text{-}CH_3 + Cl^- + H^+$$

FIGURE 4.3 Nucleophilic substitution of a glutathionyl group for chloride in methyl chloride.

fully dissociated, and only in reactions liberating the conjugate base of a weak acid will the proton be taken up by the displaced group.

4.3   MECHANISTIC ASPECTS OF GLUTATHIONE CONJUGATION

4.3.1   ADDITION REACTIONS

Addition reactions involving GSH and electrophiles originating in living organisms include epoxides, activated alkenes, quinones, and isothiocyanates. Common to the addition reactions is the uptake of a proton by the target substrate (formally the proton released from the thiol group of GSH). This neutralizes the negative charge transferred to the second reactant from the sulfur of GSH.

4.3.1.1   Epoxides

In the conjugation of epoxides the proton acceptor is oxygen, which transiently acquires a negative charge when the oxirane ring is attacked by GSH. Epoxides are important toxic agents found ubiquitously in the environment and are commonly formed in cellular oxygen metabolism. The stereochemical course of their reaction with GSH is sensitive to steric and electronic influences of the substituents on the oxirane carbon atoms. The diverse isomeric products that can arise are shown in the conjugation of styrene 7,8-oxide (Figure 4.4). Nonenzymatically, GSH reacts with either C7 or C8, with the benzylic C7 position favored by a factor 2 in comparison with the terminal C8 oxirane carbon.[3] Further, both conjugates occur in *S*- and *R*-enantiomeric forms, which are formed in equal amounts in the uncatalyzed reaction. In the conjugations of epoxides catalyzed by GSTs the relative amounts of the four diastereomers will be biased in favor of one or more of the different isomers, depending on the enzyme used.[4]

4.3.1.2   Alkenes

Alkenals are formed by free-radical reactions and oxidative processes in biological membranes, and they are also by-products in the biotransformation of drugs and other xenobiotics. The addition of GSH to activated alkenes is usually referred to

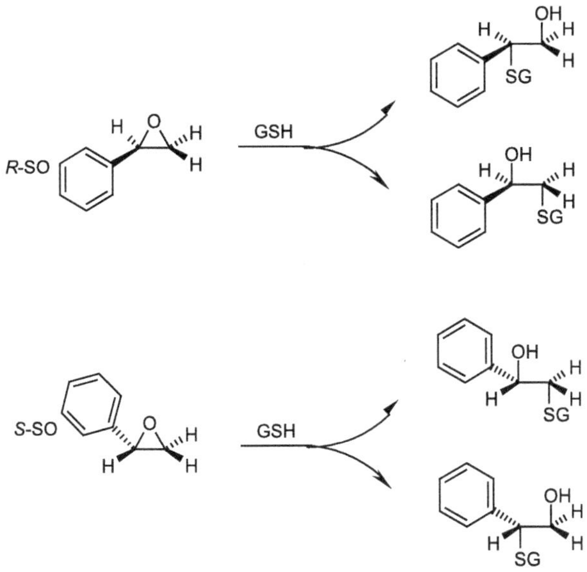

FIGURE 4.4 The four alternative stereo- and regioisomeric glutathione conjugates of styrene 7,8-oxide.

as a Michael addition and typically involves an α,β-unsaturated carbonyl compound. The stereochemistry of the addition of a nucleophile to a double bond involves an attack perpendicular to the plane of the sp^2-hybridized electrophilic atom. The approach of the attacking nucleophile may be from one side or from the other. If the substituents around the carbon atom subject to the attack are different, the alternative approaches will yield products of different steric configurations. This is the case with crotonaldehyde and other α,β-unsaturated carbonyl compounds, such as 4-hydroxynonenal, in which the β-carbon is a prochiral center and an attack from the *si* face[5] will give a chiral product with a steric configuration opposite to that arising from a *re* face attack (Figure 4.5)

In the absence of steric hindrance from the groups linked to the β-carbon, an uncatalyzed reaction in an isotropic medium will produce equal proportions of each stereoisomer but in enzyme-catalyzed reactions one of the chiral products may be formed exclusively.

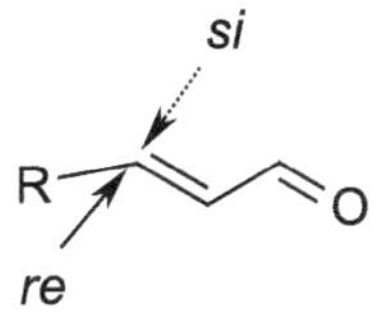

FIGURE 4.5 The two alternative stereogenic additions of glutathione to the β-carbon of an alkenal.

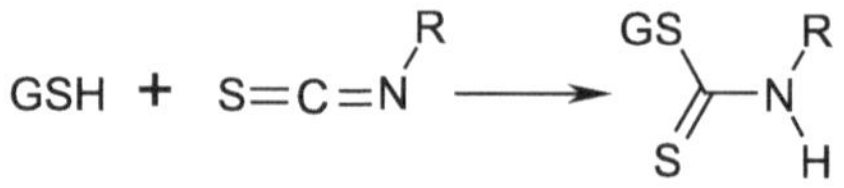

FIGURE 4.6 Formation of a dithiocarbamate by addition of glutathione to an organic isothiocyanate.

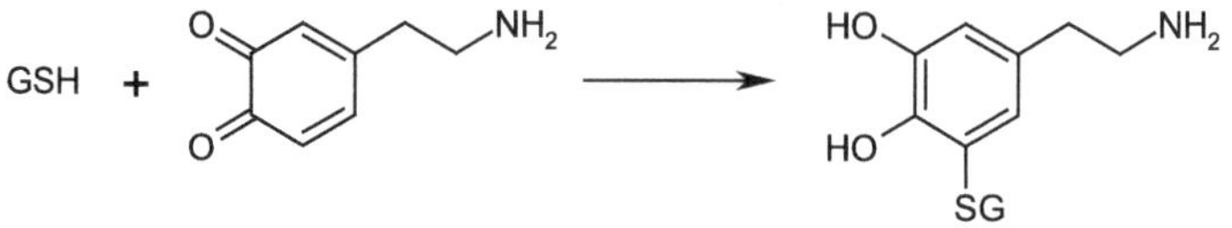

FIGURE 4.7 Formation of the 5-S-glutathionyldopamine conjugate by addition of glutathione to ortho-quinone derived from dopamine.

4.3.1.3 Organic Isothiocyanates

Another biochemical addition reaction is the formation of a dithiocarbamate from an isothiocyanate and GSH (Figure 4.6). Organic isothiocyanates occur numerously and abundantly in plants and are usually stored as unreactive glucosinolates from which they are released by the enzyme myrosinase. Even though isothiocyanates generally have been known as toxic substances, many are now recognized as chemopreventive agents. Sulforaphane, an isothiocyanate isolated from broccoli, provides protection against experimental tumors in rats.[6]

4.3.1.4 Quinones

The conjugation of GSH with quinones are additional examples of Michael additions. In these reactions, a glutathionyl hydroquinone derivative is formed (Figure 4.7), which may be regarded as a reduced quinone (even though the electron pair of the thioether bond strictly belongs to the sulfur). The GSH conjugates of many quinones are prone to oxidation and redox cycling, resulting in the production of ROS. As a consequence, they are more toxic than the parent compound.[7] However, in the case of the ortho-quinones formed from naturally occurring catecholamines such as dopa and dopamine (Figure 4.7), the GSH conjugation prevents redox cycling and appears to be an important neuroprotective mechanism.[8–10]

4.3.2 NUCLEOPHILIC SUBSTITUTIONS

4.3.2.1 Aliphatic Electrophiles

Nucleophilic substitutions (S_N2 reactions) involving GSH are often straightforward and proceed through a transition state in which the negative charge of the sulfur atom is relocated to the leaving group. For example, when a tetrahedral carbon of an alkyl halide is attacked, it is forced into a bipyramidal configuration with the approaching sulfur in one apical position and the leaving group in the other apical position (Figure 4.8). The energy barrier of the S_N2 transformation is minimized when the attacking sulfur, the electrophilic center, and the leaving ligand are all

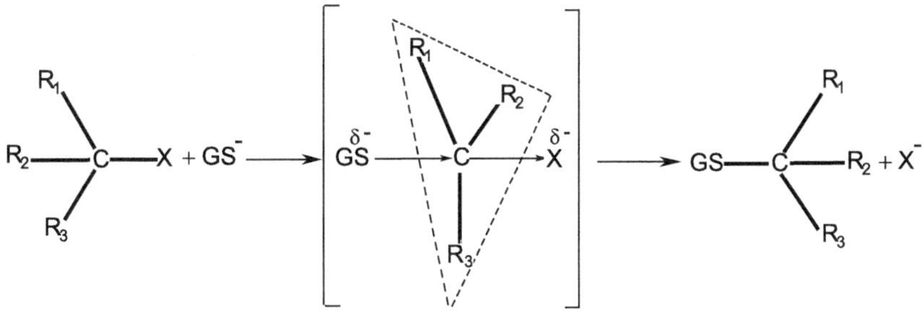

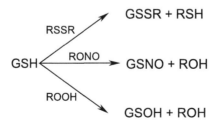

$$R_2 \overset{\displaystyle R_1}{\underset{\displaystyle R_3}{\rule{0pt}{1pt}}}\!\!\!\!-C-X + GS^- \longrightarrow \left[GS \overset{\delta^-}{\cdots} \overset{R_1}{\underset{R_3}{\rule{0pt}{1pt}}}\!\!C\!\!\cdots \overset{\delta^-}{X} \right] \longrightarrow GS \overset{\displaystyle R_1}{\underset{\displaystyle R_3}{\rule{0pt}{1pt}}}\!\!\!\!-C-R_2 + X^-$$

FIGURE 4.8 Walden inversion of sp^3-hybridized carbon of an alkyl halogenide ($CR_1R_2R_3$-X) undergoing conjugation with glutathione. In the bipyramidal intermediate a negative charge migrates from the thiolate sulfur of GSH to the leaving group (X) of the alkyl substrate.

arranged in a straight line. In aqueous media the charges of the apical groups can be stabilized by interactions with the solvent molecules. This favors the ground state and increases the activation energy of the reaction. In contrast, less polar environments, such as the active site of an enzyme, are expected to decrease the energy barrier and thereby enhance the reaction rate.

4.3.2.2 Reductive Substitutions

Reactions of GSH with organic nitrites (RONO) and hydroperoxides (ROOH) are formally similar to the more well-known thiol-disulfide interchange involving disulfides (RSSR) (Figure 4.9).

They first give rise to unstable intermediates such as sulfenic acid (GSOH) and S-nitrosoglutathione (GSNO) or mixed disulfides (GSSR), respectively, which subsequently can react with a second GSH molecule to give GSSG or undergo other chemical transformations. Each of the partial reactions is an S_N2 reaction depictable along a trajectory with attacking sulfur, electrophilic center, and leaving group along a straight line. In fact, these reductive processes may be the most important cellular substitution reactions involving GSH.

$$
\begin{array}{c}
\text{GSSR + RSH} \\
\text{RSSR} \nearrow \\
\text{GSH} \xleftarrow{\quad} \overset{\text{RONO}}{\longrightarrow} \text{GSNO + ROH} \\
\text{ROOH} \searrow \\
\text{GSOH + ROH}
\end{array}
$$

FIGURE 4.9 Reduction of electrophilic heteroatoms via substitution reactions with glutathione, followed by an additional reductive reaction with glutathione (not shown). The examples include disulfides (RSSR), nitrate esters (RONO), and organic hydroperoxides (ROOH).

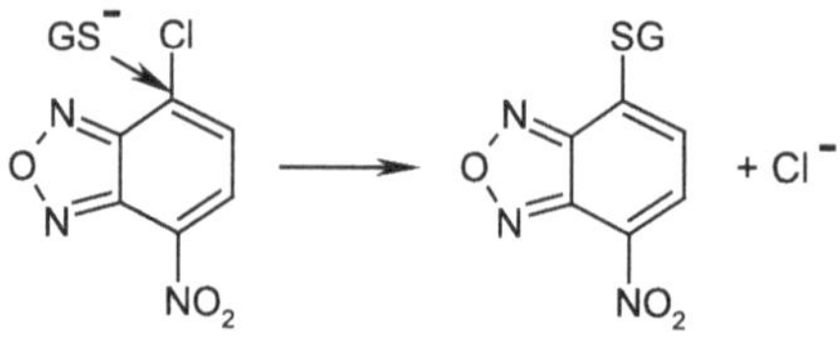

FIGURE 4.10 Aromatic substitution reaction of the fluorogenic NBD-chloride with glutathione.

4.3.2.3 Aromatic Substitutions

Aromatic nucleophilic substitutions, Ar_N2, involve an attack by the thiolate via a reaction path perpendicular to the plane of the aromatic ring. Physiologically relevant substrates undergoing aromatic substitution are scarce, but some reactions nevertheless have been instrumental in mechanistic studies of GSTs. A well-investigated example is the reaction between 1-chloro-2,4-dinitrobenzene (CDNB) and GSH.[11] A similar reaction involves the fluorogenic aromatic electrophile 7-chloro-4-nitrobenzo-2-oxa-1,3-diazole, NBD chloride (Figure 4.10).

A covalent intermediate in the form of a σ-complex (Meisenheimer complex) develops in these conjugations with concomitant loss of the aromaticity of the ring structure. The aromatic character is regained as the leaving group departs from the collapsing σ-complex.

4.4 FEATURES OF THE CATALYTIC FUNCTIONS OF GSTS

4.4.1 Activation of the Thiol Group of Glutathione

Most chemical reactions involving GSH are strongly promoted by ionization of the thiol group to its more reactive thiolate form. A primary function of enzymes catalyzing GSH conjugations is to increase the reactivity of the thiol group by promoting its ionization. This is achieved in part by lowering the pK_a of GSH from 9.2 in aqueous solution to a value <7 in the active site. Exactly how this is accomplished is not known, but proximity of the sulfur to the positive end of an α-helix dipole is probably a contributing factor. The first step in the catalytic cycle is therefore the release of a proton to form the thiolate of GSH (Figure 4.1). In aqueous media the sulfur atom is solvated, and the hydration shell will attenuate the reactivity of GSH. Binding of the thiolate in the active site removes the inhibitory water molecules and lowers the activation energy for the reaction with the second substrate.

A nucleophilic active-site residue (Tyr, Ser, or Cys) is juxtaposed to the sulfur of enzyme-bound GSH, but this residue is not directly involved in the ionization of the thiol group. This is evidenced by the finding that mutation of Tyr 9 into Phe in Alpha-class GSTs has no major effect of the pK_a value of GSH in the active site.[12] Instead, the function of these nucleophilic residues is to promote catalysis by guiding the orbitals of the sulfur for attack on the second substrate. In most GSTs this occurs via hydrogen bonding from a hydroxyl group of Tyr or Ser to a lone electron pair of the sulfur atom, with consequent alignment of a distinct bonding orbital with an unoccupied orbital in the electrophilic substrate. The notion of precise orientation of the reactants forming the activated complex has been recognized as "orbital steering."[13] The accurate steric positioning effected by enzymes is also described as selection of the subpopulation of substrate molecules with the "near attack conformation," thereby promoting approach to the transition state.[14] Thus, the guiding of the reacting sulfur orbital of GSH toward the second substrate is a second common feature of the GSTs. From an energetic point of view, this is essentially an entropic effect.[15]

4.4.2 ACTIVATION OF THE ELECTROPHILIC SUBSTRATE

The actual chemistry of a GST-catalyzed reaction can be regarded as an attack by an incoming electrophile on the reactive sulfur of enzyme-bound GSH. In other words, this aspect of GST catalysis is the productive accommodation of the second substrate with the electrophilic center adjacent to the sulfur of GSH. The electrophilic substrate is generally hydrophobic, at least in a part of the molecule, and its cognate binding pocket has therefore been named the H-site.[16] It is commonly assumed that enzyme catalysis entails transition-state stabilization arising through recognition and high-affinity binding of an activated complex.[17] The dissimilar substrate selectivities displayed by different GSTs rely on differences in the ability of active sites to accommodate particular transition-state structures. Discrimination among alternative substrates is based not only on simple complementarity between a preformed active site and the selected substrate. Structural flexibility of the protein may contribute to the binding of the substrate and invoke structural adjustments that increase the affinity for the activated complex. Such ligand-induced fit is common to many enzymes interacting with their substrates, as well as to receptors binding their cognate agonists. Conformational mobility can also explain how one and the same GST may be active with substrates differing widely in size and other molecular properties. Nevertheless, the H-site is more elaborate than a flexible hydrophobic cavity, because GSTs display distinctive substrate selectivity profiles including regio- and stereospecificity. Clearly, topological and conformational constraints in the binding of substrates govern the selectivity of the catalytic process.

4.5 RATE-DETERMINING PROCESSES

Transition-state stabilization and chemical transformation of the reactants are crucial to the catalytic process, but many GSTs acting on their most active substrates are not rate-limited by chemistry. Instead, conformational transitions and product release restrict the catalytic turnover. Many GST substrates are large hydrophobic molecules and their GSH conjugates can bind with high affinity to both the H-site and the G-site. Like many other enzymes, GSTs depend on flexible regions of their protein structure for efficient catalytic turnover.[18] This aspect of structural agility may not be an obligatory requirement for GST catalysis but appears to be a salient feature for catalytic efficiency of many naturally evolved GSTs.

4.6 NATURAL EVOLUTION OF GSTS

The nature of the "primordial GST" is unknown,[19] but many suggestions have been made about the phylogeny of the GST superfamily. The Theta and Zeta classes are represented in both animals and plants and thus have a monophyletic origin, which is closer to the primordial GST than the Alpha, Mu, and Pi classes, which are not present in plants.[20] A separate evolutionary branch led to the Tau and Phi classes, which have abundant members in many plants but do not exist in animals.[21] Diversification within a given class of GSTs has been accomplished by gene duplications and subsequent mutations and recombinations of DNA segments.

From an evolutionary standpoint it is unlikely that a chain of 200–240 residues (typical of GSTs) could have arisen in a straightforward manner. Instead, fusions and recombinations of fragments have stepwise led to larger structures. Short sequences that rapidly can fold into stable modules composed of a few α-helices or β-strands are considered as key structural building blocks, and tertiary structures can be assembled from such folding units. At the genetic level, duplication of an existing gene provides the DNA that can be modified to encode an altered protein structure. In the general area of GSH-dependent enzymes, the structure of the human glyoxalase I subunit bears evidence of duplications of an α/β-motif.[22] In yeast glyoxalase I this duplication is further duplicated in the same polypeptide chain, thus presenting a "molecular fossil" of these sequential duplications and fusions.[23] In addition, genes may be subject to mutations of individual bases as well as to exchanges, additions, and deletions of larger segments. Exon shuffling, gene convergence, and gene fusions have been suggested as possible mechanisms for extensive DNA redesign.

4.7 THE COMBINATORIAL VIEW OF GST STRUCTURES

4.7.1 The Glutathione-Binding Module

The primary requisite for building a GST molecule is a GSH-binding module. Before any GST structure had been determined, it was suggested that the evolutionary origin of GSTs encompassed the fusion of a GSH-binding module with other structural components that would confer the characteristic enzymatic properties to the different enzymes.[19] The proposed primordial GSH binder seems to have the same origin as the structural α/β-fold that was originally discovered by structural studies of thioredoxin[24] (Figure 4.11).

In the structures of soluble GSTs, this module generally provides the N-terminal domain and is composed of four β-strands sandwiched between two α-helices on the core side and one α-helix on the surface side of the protein molecule. This common fold also contains a characteristic cis-prolyl residue that is highly conserved among GSTs as well as in thioredoxin and thioredoxin-like proteins. In GSTs the conserved cis-proline makes possible a sharp turn of the polypeptide backbone that facilitates the formation of two hydrogen bonds between the main chain of the protein and the cysteinyl moiety of GSH. These structural characteristics contribute to the high specificity for GSH. Strong binding of the GSH tripeptide moiety depends

FIGURE 4.11 (See color insert following page 178.) Thioredoxin (left, PDB: 1xoa), N-terminal domain of GSTM2-2 (center), and the complete subunit of GSTM2-2 (right, C-terminal domain, PDB: 2c4j).

on the presence of the α-carboxyl group of the glutamyl residue,[25,26] which is positioned adjacent to the N-terminal end of the α3-helix.

It is noteworthy that in the protein structures based on the thioredoxin-like fold, the sulfur of GSH (or the redox active sulfur or selenium in related proteins) is juxtaposed to the N-terminus of an α-helix. This conservation of a topologically equivalent α-helix dipole capable of activating the thiol may be an example of mechanism-driven evolution.[27,28] Adopting the thioredoxin-like fold as the GSH-binding module affords both selectivity in binding and activation of the thiol group of the GSH molecule.

The actual binding pocket for the GSH molecule has been called the G-site.[16] There are many variations on a common theme, but the structures representing the major forms of mammalian GSTs (i.e., Alpha, Mu, Pi, and Theta classes) share the features shown in Figure 4.12. Apart from the polar bonds that obviously contribute to the binding of the charged tripeptide, a tyrosine residue with its hydroxyl group within hydrogen-bonding distance from the sulfur of GSH is of importance for catalysis. In many other GSTs, such as members of the Theta, Zeta, Phi, and Tau classes, the hydrogen-bonding hydroxyl group is the side chain of a serine residue. A third variation is a cysteine residue in the bacterial Beta class and in the mammalian Omega class enzymes,[29] which form a mixed disulfide with GSH in the active site. None of these G-site residues appear to be responsible for the actual ionization of the thiol group of GSH, which instead might be due to the proximity of the positive end of a dipole formed by the α1-helix.

The GSH-binding domain has catalytic competence by itself, as evidenced by thioredoxin-like proteins such as thioltransferases, glutaredoxins, etc.[30]

4.7.2 A Module for Binding of the Electrophilic Substrate

Diversification of the catalytic properties and binding of electrophiles of different structures requires a structural complement that would bind the second substrate in a productive manner. This second module, carrying the hydrophobic H-site, would serve not only to deliver the electrophile in proper orientation for reaction with GSH but also to shield the electrophilic center from water and prevent competing hydrolytic reactions. In this manner two separate domains serve to bring the sulfur nucleophile and an electrophile together. The second module could be made in

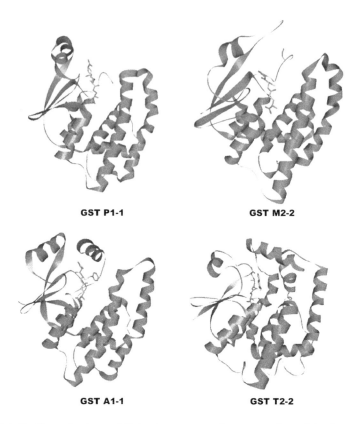

GST P1-1   **GST M2-2**

GST A1-1   **GST T2-2**

FIGURE 4.12 (See color insert.) Subunit structures of representatives of the four classes Pi (PBD: 9gss), Mu (PBD: 1hnc), Alpha (PBD: 1gse), and Theta (PBD: 3ljr) of the mammalian GSTs. The structural differences among the subunits in the vicinity of the bound glutathione structure influence the differential substrate selectivities of the different enzymes.

alternative versions designed to fit substrates of different sizes and to promote varied chemistries. This variability of the second domain is essentially the basis for the division of the enzymes into different classes.[31,32] The modular design of GSTs, in which different domains are joined with the thioredoxin fold, can be looked on as combinatorial protein chemistry at the level of tertiary structure.

4.7.3 BINARY COMBINATIONS OF PROTEIN SUBUNITS

The known soluble GSTs are dimeric structures composed of two equal or similar protein subunits. Under physiologically relevant conditions free monomers have not been observed, and catalytic activity seems to be associated only with the dimer. The dimeric GSTP1-1 has been mutated at the interface, in order to create a stable monomer, but the monomeric subunit does not display any detectable enzyme activity.[33] Our current understanding is that the dimeric GST structure is the norm for enzyme function.

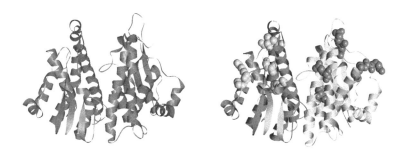

FIGURE 4.13 (See color insert.) GSTs occur as homodimers (M2-2, left, PBD: 2c4j) and heterodimers (M2-3, right, PBD: 3gtu), a manifestation of combinatorics at the level of quaternary structure. Hypervariable residues (space filled) subjected to positive selection in evolution[49] are shown in GSTM2-3.

The functional significance of heterodimer formation is an issue added on top of the question of why the GSTs are dimers in the first place. Early studies indicated that the enzymatic activity and the inhibition pattern of a heterodimer was the result of the properties of the constituent subunits, such that they could be calculated from those of the corresponding homodimers.[34–36] However, incisive investigations with additional substrates revealed that the kinetic properties are not always additive but depend on both the subunits and the substrates used.[37,38] In marked contrast to results obtained in nucleophilic aromatic substitution and steroid isomerization reactions, GSTA1-1 displayed half-of-the-sites reactivity (also known as total negative cooperativity) in the GSH conjugation of nonenal, 4-nitrocinnamaldehyde, and benzyl-isothiocyanate. Similarly, negative subunit interactions in GSTA4-4 result in half-site occupancy in binding of a GSH derivative.[39] GSTP1-1 was reported to exhibit negative cooperativity with respect to glutathione binding at temperatures below 25°C, whereas at temperatures above 35°C, the enzyme demonstrated positive cooperativity.[40]

Most GSTs are homodimers, but some GSTs naturally occur as heterodimers produced by the combination of different polypeptide chains.[34] The human Mu class is represented by five functional genes encoding distinct subunits. Binary combinations of these five subunits could theoretically give rise to a total of 15 different dimers, including the five homodimers (Figure 4.13).

The creation of heterodimeric combinations is an expansion of the multiplicity of GST variants provided by single subunits, even if the number of heterodimers actually produced in living systems does not reach the theoretical maximum. *In vitro*, GST subunits from different organisms can be combined; a heterodimer composed of rat and human GSTM1 subunits has been constructed (unpublished work of Widersten and Mannervik in 1987). Thus, the potential of combinatorial protein chemistry is manifested also at the quaternary structural level.

4.7.4 THE SIGNIFICANCE OF THE QUATERNARY STRUCTURE

The underlying mechanisms for subunit interactions has been further investigated in GSTP1-1.[41] The Pi-, Mu-, and Alpha-class GSTs have a lock-and-key motif that

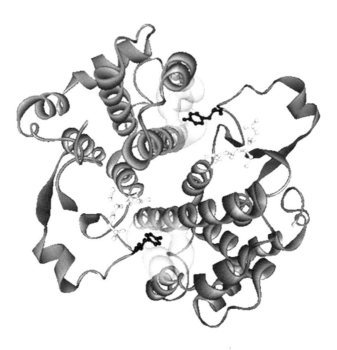

FIGURE 4.14 (See color insert.) Dimer of GSTP1-1 showing the lock-and-key motif inter-locking the two GST subunits. The key residue (Tyr 50, black) is located on a loop of the polypeptide chain that contributes to binding of glutathione in the active site. Fitting the aromatic key residue into the lock promotes binding of glutathione and enhances catalytic efficiency of the enzyme.

is responsible for a highly conserved hydrophobic interaction in the subunit–subunit interface. The aromatic key residue (Tyr50 in human GSTP1-1) in one subunit is wedged into a hydrophobic pocket of the other subunit (Figure 4.14).

Mutation of the key residue Tyr50 into Ala rendered an essentially inactive subunit, which could still form a heterodimer with the fully active wild-type subunit. The k_{cat} value of the heterodimer decreased by about 27-fold rather than the expected 2-fold, in comparison with the wild-type enzyme, indicating that the two active sites of the dimeric enzyme work synergistically. Cooperativity was also suggested by the nonhyperbolic GSH-saturation curves. Crystal structures of wild-type GSTP1-1 demonstrate an internal network of hydrogen-bonded water molecules, connecting the two active sites and the main-chain carbonyl group of Tyr50. This key residue is located on the loop following the α2-helix, which also carries a Lys residue interacting with GSH. The water network thus offers a mechanism for communication between the two active sites. A subunit appears to become catalytically competent by positioning its key residue into the lock of the neighboring subunit, thereby promoting GSH binding. Not all GSTs have the lock-and-key motif, and it remains to be clarified if this structural element is the signature required for subunit cooperativity and tight GSH binding.

4.7.5 COMPLEXES WITH OTHER PROTEINS

Several independent studies indicate that GSTs bind to other cellular proteins. Some of these alternative partners are stress-activated protein kinases involved in signaling of cellular stress and communicating to nuclear activators of transcriptional response.[42,43] A heterodimeric complex between 1-Cys peroxiredoxin and a GSTP1-1 subunit has also been reported as a functional entity.[44] This new dimension of functional quaternary structures bears witness to the versatility of GSTs in biological systems and illustrates how combinatorial protein chemistry can serve as a paradigm for understanding both structure–activity relationships and the design of GSTs with novel properties.

4.8 PROOF OF PRINCIPLES AND APPROACHES TO REDESIGN OF GSTS BY PROTEIN ENGINEERING BASED ON THE COMBINATORIAL DESCRIPTION OF PROTEIN STRUCTURE

The structures of the soluble GSTs illustrate how recombination of building blocks at all structural levels gives rise to novel functional properties.[45] Indeed, genetic engineering of the substrate-contacting segments, involving point mutations, interchange of chain segments, and other structural rearrangements, can give rise to altered substrate selectivities in the GSTs. Even the exchange of a single amino acid residue can cause profound changes.[46–49] Such point mutations can be considered the smallest exchanges of a segment of the primary structure that is possible. Based on the current knowledge of structures and functions of naturally occurring GSTs, it is possible to outline principles for an evolutionary design of novel GSTs.

The molecules of biological systems can be considered as being assembled from a limited number of building blocks. All cellular compounds can essentially be derived from a few dozen simple molecules that are combined and transformed by metabolic reactions.[50] Proteins can be considered as being composed of structural building blocks at all hierarchical levels. The primary structure is a polypeptide chain in which linear segments of amino acid residues are covalently linked together and where the smallest segment is a single amino acid residue. At the level of secondary structure, α-helices, loops, and β-strands are modular units; at the tertiary level, modules in the form of structural domains can often be observed. In oligomeric proteins the subunits are the obvious building blocks of quaternary structure. From the viewpoint of molecular engineering, the design and assembly of proteins can be regarded as combinatorial chemistry in three dimensions.

The natural evolution of novel enzyme properties usually involves stepwise approaches to an optimized function. The starting point is a protein structure, which, by combinations of mutations, additions, or deletions, is redesigned to acquire novel properties. At the protein level, the altered gene may manifest itself as exchange, addition, or deletion of building blocks at all structural levels: sequence of amino acids, secondary structures, domains, and subunits of oligomeric proteins. In protein engineering such combinatorial principles also apply, but the design is not limited to the natural building blocks. Semisynthetic approaches[51] and engineered ribosomal translation systems[52] allow incorporation of novel chemical functionalities and building blocks into recombinant proteins.

4.8.1 DETERMINANTS OF SUBSTRATE SELECTIVITIES

The thiol specificity of GSTs has generally been conserved among the different enzyme variants and is matched by a relatively higher sequence conservation of the N-terminal domain as compared to the C-terminal domain. In contrast, the substrate-activity profiles of GSTs are highly diverse. Binding of the electrophilic substrate involves three or four segments, widely separated in the primary structure, which by proper folding of the polypeptide chain form the H-site (Figure 4.15).

One segment is situated in the loop between the β1-strand and the α1-helix, close to the N-terminus and adjacent to the nucleophilic residue interacting with the sulfur of GSH. A second segment is found in the second half of the α4-helix in the center of the molecule, and a third is in the C-terminal segment of the polypeptide.

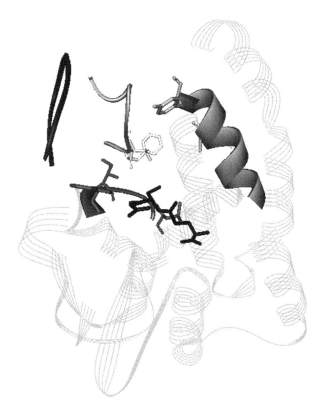

FIGURE 4.15 (See color insert.) Segments of the polypeptide chain contributing to formation of the H-site in GSTM2-2. The different segments are located in the loops between β1 and α1, β2 and α2, the end of the α4-helix, and the C-terminal portion of the structure. The bound glutathione is rendered in black.

The latter two are located in the C-terminal domain. In the Mu class, a fourth segment in the shape of a loop between the β2-strand and the α2-helix contributes to the H-site. It is reasonable to assume that these segments carrying the residues contacting the electrophilic substrate would govern the substrate specificity of a given GST.

4.8.2 RATIONAL REDESIGN

4.8.2.1 Primary Structure Alterations

Structure-based rational redesign of GSTs has given further evidence for the importance of the H-site in the specific recognition of electrophilic substrates. In one case, the H-site of GSTA1-1 was rebuilt to install high activity for alkenal substrates by mimicking the selectivity-determining residues in GSTA4-4.[45] Another example is the transmutation of GSTA2-2 to acquire the steroid double-bond isomerase activity characterizing GSTA3-3.[53] It is also possible to introduce new chemical functionalities in the active site for predicted novel catalytic properties. A His residue in the H-site of GSTA1-1 enables thiolesterase activity,[54] and replacement of the active-site Ser in GSTT2-2 by selenocysteine affords glutathione peroxidase activity.[55] Site-specific replacements of Tyr by fluorinated Tyr derivatives in the active site of GSTA1-1 have also been accomplished.[56] The combinatorial approach to structural and functional variability is consequently applicable to chemical entities beyond the limitations of the genetic code.

4.8.2.2 Exchange of Segments of Primary and Secondary Structure

Recombinations of defined segments of primary structure[47,57,58,59] or elements of secondary structure have also been investigated.[45,60] In some cases the exchanged segments corresponded to discrete exons in the corresponding genes. These experiments have given active enzymes, demonstrating that exon shuffling as an evolutionary mechanism[19,61] could give catalytically competent chimeras, even though the functional properties of the constructs are difficult to predict.

In addition to exon shuffling as a possible evolutionary mechanism, some organisms utilize alternative splicing of exons in real time to generate variant GSTs from a single gene. This is the case for the malaria vector *Anopheles*[62,63] as well as for a human parasite, the filarial nematode *Onchocerca volvolus*.[64] In this manner alternative GSTs with different substrate selectivities can arise from a single gene. It is also possible to imagine trans-splicing of mRNA from different genes as a mechanism for further diversification of the H-site and other regions of the GST structure.

4.8.3 STOCHASTIC REDESIGN

4.8.3.1 Stochastic Recombinations in the Evolution of GSTs with Novel Properties

The most powerful form of recombination of peptide segments is effected by DNA shuffling,[65] in particular family shuffling of fragments from related sequences.[66] By this approach stochastic recombinations are limited to DNA fragments with sequence similarities. An important feature is that a large proportion of the nearest-neighbor residues are conserved, which is expected to favor the stable folding and assembly of the recombined polypeptide chains. In particular, a joint N-cap and hydrophobic staple motif is important for the stability and folding of glutathione transferases[67] and is fully conserved in their evolution.[68,69] DNA shuffling is expected to be robust, because it has high probability of maintaining these conserved and essential structural elements.

Stochastic mutations of GSTs based on DNA shuffling have generated ensembles of variants from which enzymes with novel properties can be selected. Mu-class GSTs were investigated in order to probe structure–activity relationships and to lay a foundation for directed evolution.[70,71] Mosaic sequences were generated by family shuffling[72] of DNA encoding GSTM1-1 and GSTM2-2. The homologous DNA sequences are 89% identical, and at the protein level 34 out of 217 amino acid residues differ in the corresponding primary structures. Of the 34 variant amino acids a limited number are responsible for divergent substrate selectivities of the two enzymes. Approximately 90% of the randomly isolated chimeras displayed enzymatic activity. The presence of segments from both GSTs was demonstrated by sequence analysis, and the amino acid sequences could be divided into ≥11 parental segments that had been recombined. This number translates into approximately 2000 (2^{11}) possible variants of chimeric full-length sequences. GST activities were determined using three alternative electrophilic substrates: 1-chloro-2,4-dinitrobenzene (chloride substitution), 2-cyano-1,3-dimethyl-1-nitrosoguanidine (transnitrosylation), and aminochrome (Michael addition). The activities of the clones analyzed spanned several orders of magnitude with all three substrates, and it should be noted that the three substrates used undergo different types of chemical transformation. CDNB and cyanoDMNG are involved in displacement reactions with different chemistries, whereas aminochrome is conjugated in an addition reaction. Thus, the functional diversity of GSTs can be expanded not only with respect to shape and reactivity of the electrophilic substrate but also with respect to the type of chemical reaction catalyzed.

4.9 RATE-LIMITING CONSTRAINTS

4.9.1 What Limits the Efficiency of GST Catalysis?

Cleland has pointed out that few metabolic enzymes are rate-limited by the chemical transformation per se, but by conformational changes and product release.[73] In the case of GST catalysis, a given enzyme will generally accept a variety of substrates but will not show high activity with all of them. The chemical transformation of slow substrates is clearly rate-limited by chemistry. However, efficient substrates afford k_{cat} values >100 s^{-1} and k_{cat}/K_m values >10^6 M^{-1}s^{-1}, and in these cases the catalytic process is usually, at least to some extent, limited by diffusive properties such as structural transitions and binding or release of reactants. Thus, enhanced catalytic efficiency cannot be expected by optimization of the interactions between the active site and the reactants entering the transition state. The remaining option is modulation of the dynamic properties of the GST molecule to enhance structural flexibility and rate of product release. This is an additional dimension in the design of an efficient catalyst. Furthermore, it should be noted that the catalytic rate is also restricted by thermodynamics via the Haldane relationship, which, in a simple one substrate–one product reaction, equates the equilibrium constant K_{eq} for the overall chemical transformation with the ratio of k_{cat} and K_m for the forward to the ratio for the reverse reaction:[73]

$$K_{eq} = (k_{cat}/K_m)^{forward}/(k_{cat}/K_m)^{reverse}$$

4.9.2 ALTERATION OF THE DYNAMICS IN THE GST STRUCTURE

Several studies of GSTs underscore the significance of mobile regions for the catalytic function.[60,74] Mutant libraries have been created by recombination of Theta-class sequences.[75,76] A mutant with enhanced activity over the parental GSTT2-2 was found to have a single-point mutation (Asn/Asp) far from the active site. The increased catalytic efficiency is likely a result of enhanced mobility due to loss of hydrogen bonding of the Asn side chain to the loop between the α2-helix and β3 in the polypeptide backbone. The catalytic turnover is limited by a conformational change in the protein structure,[77] and the increased flexibility of the loop facilitates product release. The structural agility removed from the active site is functionally distinct from recognition of the substrate.

GSTs with unequal subunits in some cases show catalytic properties that are not additive,[37,39,41] such that they could be deduced from the properties of the corresponding homodimers. This effect can be attributed to long-distance interactions between the active sites and indicates alterations in the dynamics of the enzyme. In this manner combinatorial protein chemistry extends the redesign of functional properties beyond individual GST subunits to the level of quaternary structure.

ACKNOWLEDGMENTS

The Swedish Research Council supported the work in the authors' laboratory. Y.I. was a recipient of stipends from the Sven and Lilly Lawski Foundation.

REFERENCES

1. Fahey, R.C., Novel thiols of prokaryotes. *Annu. Rev. Microbiol.*, 55, 333, 2001.
2. Fahey, R.C., Buschbacher, R.M., and Newton, G.L., The evolution of glutathione metabolism in phototrophic microorganisms. *J. Mol. Evol.*, 25, 818, 1987.
3. Dostal, L.A. et al., Stereoselectivity and regioselectivity of purified human glutathione transferases π, α-ϵ, and μ with alkene and polycyclic arene oxide substrates. *Drug. Metab. Dispos.*, 16, 420, 1988.
4. Ivarsson, Y. and Mannervik, B., Regio- and enantioselectivities in epoxide conjugations are modulated by residue 210 in Mu class glutathione transferases. *Protein Eng. Des. Sel.*, 18, 607, 2005.

5. Carroll, F.A., *Perspectives on Structure and Mechanism in Organic Chemistry.* Brooks-Cole, Pacific Grove, 1998, p. 107.

6. Zhang, Y. et al., Anticarcinogenic activities of sulforaphane and structurally related synthetic norbornyl isothiocyanates. *Proc. Natl. Acad. Sci. USA*, 91, 3147, 1994.

7. Bolton, J.L. et al., Role of quinones in toxicology. *Chem. Res. Toxicol.*, 13, 135, 2000.

8. Segura-Aguilar, J. et al., Human class Mu glutathione transferases, in particular isoenzyme M2-2, catalyze detoxication of the dopamine metabolite aminochrome. *J. Biol. Chem.*, 272, 5727, 1997.

9. Baez, S. et al., Glutathione transferases catalyse the detoxication of oxidized metabolites (o-quinones) of catecholamines and may serve as an antioxidant system preventing degenerative cellular processes. *Biochem. J.*, 324 (Pt. 1), 25, 1997.

10. Dagnino-Subiabre, A. et al., Glutathione transferase M2-2 catalyzes conjugation of dopamine and dopa o-quinones. *Biochem. Biophys. Res. Commun.*, 274, 32, 2000.

11. Graminski, G.F. et al., Formation of the 1-(S-glutathionyl)-2,4,6-trinitrocyclohexadienate anion at the active site of glutathione S-transferase: evidence for enzymic stabilization of sigma-complex intermediates in nucleophilic aromatic substitution reactions. *Biochemistry*, 28, 6252, 1989.

12. Pettersson, P.L. and Mannervik, B., The role of glutathione in the isomerization of Δ^5-androstene-3,17-dione catalyzed by human glutathione transferase A1-1. *J. Biol. Chem.*, 276, 11698, 2001.

13. Mesecar, A.D., Stoddard, B.L., and Koshland, D.E., Jr., Orbital steering in the catalytic power of enzymes: small structural changes with large catalytic consequences. *Science*, 277, 202, 1997.

14. Bruice, T.C. and Lightstone, F.C., Ground state and transition state contributions to the rates of intramolecular and enzymatic reactions. *Acc. Chem. Res.*, 32, 127, 1999.

15. Page, M.I. and Jencks, W.P., Entropic contributions to rate acceleration in enzymic and intramolecular reactions and the chelate effect. *Proc. Natl. Acad. Sci. USA*, 68, 1678, 1971.

16. Mannervik, B. et al. Glutathione conjugation: reaction mechanism of glutathione S-transferase A, in *Conjugation Reactions in Drug Biotransformation,* Elsevier/North Holland, Amsterdam, 1978, pp. 101–110.

17. Pauling, L., Molecular architecture and biological reactions. *Chem. Eng. News*, 24, 1375, 1946.

18. Codreanu, S.G. et al., Local protein dynamics and catalysis: detection of segmental motion associated with rate-limiting product release by a glutathione transferase. *Biochemistry*, 41, 15161, 2002.

19. Mannervik, B., The isoenzymes of glutathione transferase. *Adv. Enzymol. Rel. Areas Mol. Biol.*, 57, 357, 1985.

20. Pemble, S.E. and Taylor, J.B., An evolutionary perspective on glutathione transferases inferred from class-theta glutathione transferase cDNA sequences. *Biochem. J.*, 287, 957, 1992.

21. Soranzo, N. et al., Organisation and structural evolution of the rice glutathione S-transferase gene family. *Mol. Genet. Genomics*, 271, 511, 2004.

22. Cameron, A.D. et al., Crystal structure of human glyoxalase I — evidence for gene duplication and 3D domain swapping. *Embo J.*, 16, 3386, 1997.

23. Ridderström, M. and Mannervik, B., The primary structure of monomeric yeast glyoxalase I indicates a gene duplication resulting in two similar segments homologous with the subunit of dimeric human glyoxalase I. *Biochem. J.*, 316, 1005, 1996.

24. Holmgren, A. et al., Three-dimensional structure of *Escherichia coli* thioredoxin-S_2 to 2.8 Å resolution. *Proc. Natl. Acad. Sci. USA*, 72, 2305, 1975.

25. Adang, A.E. et al., Substrate specificity of rat liver glutathione S-transferase isoenzymes for a series of glutathione analogues, modified at the γ-glutamyl moiety. *Biochem. J.*, 255, 721, 1988.

26. Gustafsson, A. et al., Role of the glutamyl α-carboxylate of the substrate glutathione in the catalytic mechanism of human glutathione transferase A1-1. *Biochemistry*, 40, 15835, 2001.

27. Babbitt, P.C. and Gerlt, J.A., Understanding enzyme superfamilies. Chemistry as the fundamental determinant in the evolution of new catalytic activities. *J. Biol. Chem.*, 272, 30591, 1997.

28. Gerlt, J.A. and Babbitt, P.C., Divergent evolution of enzymatic function: mechanistically diverse superfamilies and functionally distinct suprafamilies. *Annu. Rev. Biochem.*, 70, 209, 2001.

29. Board, P.G. et al., Identification, characterization, and crystal structure of the omega class glutathione transferases. *J. Biol. Chem.*, 275, 24798, 2000.

30. Fernandes, A.P. and Holmgren, A., Glutaredoxins: glutathione-dependent redox enzymes with functions far beyond a simple thioredoxin backup system. *Antioxid. Redox. Signal.*, 6, 63, 2004.

31. Mannervik, B. et al., Identification of three classes of cytosolic glutathione transferase common to several mammalian species: correlation between structural data and enzymatic properties. *Proc. Natl. Acad. Sci. USA*, 82, 7202, 1985.

32. Sinning, I. et al., Structure determination and refinement of human alpha class glutathione transferase A1-1, and a comparison with the mu and pi class enzymes. *J. Mol. Biol.*, 232, 192, 1993.

33. Abdalla, A.M. et al., Design of a monomeric human glutathione transferase GSTP1, a structurally stable but catalytically inactive protein. *Protein Eng.*, 15, 827, 2002.

34. Mannervik, B. and Jensson, H., Binary combinations of four protein subunits with different catalytic specificities explain the relationship between six basic glutathione S-transferases in rat liver cytosol. *J. Biol. Chem.*, 257, 9909, 1982.

35. Danielson, U.H. and Mannervik, B., Kinetic independence of the subunits of cytosolic glutathione transferase from the rat. *Biochem. J.*, 231, 263, 1985.

36. Tahir, M.K. and Mannervik, B., Simple inhibition studies for distinction between homodimeric and heterodimeric isoenzymes of glutathione transferase. *J. Biol. Chem.*, 261, 1048, 1986.

37. Lien, S. et al., Human glutathione transferase A1-1 demonstrates both half-of-the-sites and all-of-the-sites reactivity. *J. Biol. Chem.*, 276, 35599, 2001.

38. Gustafsson, A., Nilsson, L.O., and Mannervik, B., Hybridization of alpha class subunits generating a functional glutathione transferase A1-4 heterodimer. *J. Mol. Biol.*, 316, 395, 2002.

39. Xiao, B. et al., Crystal structure of a murine glutathione S-transferase in complex with a glutathione conjugate of 4-hydroxynon-2-enal in one subunit and glutathione in the other: evidence of signaling across the dimer interface. *Biochemistry*, 38, 11887, 1999.

40. Caccuri, A.M. et al., Temperature adaptation of glutathione S-transferase P1-1. A case for homotropic regulation of substrate binding. *J. Biol. Chem.*, 274, 19276, 1999.

41. Hegazy, U.M., Mannervik, B., and Stenberg, G., Functional role of the lock-and-key motif at the subunit interface of glutathione transferase P1-1. *J. Biol. Chem.*, 279, 9586, 2004.

42. Adler, V. et al., Regulation of JNK signaling by GSTp. *Embo J.*, 18, 1321, 1999.

43. Edalat, M., Persson, M.A.A., and Mannervik, B, Selective recognition of peptide sequences by glutathione transferases: a possible mechanism for modulation of cellular stress–induced signaling pathways. *Biol. Chem.*, 384, 645, 2003.

44. Manevich, Y., Feinstein, S.I., and Fisher, A.B., Activation of the antioxidant enzyme 1-Cys peroxiredoxin requires glutathionylation mediated by heterodimerization with pi GST. *Proc. Natl. Acad. Sci. USA*, 101, 37805, 2004.

45. Nilsson, L.O., Gustafsson, A., and Mannervik, B., Redesign of substrate-selectivity determining modules of glutathione transferase A1-1 installs high catalytic efficiency with toxic alkenal products of lipid peroxidation. *Proc. Natl. Acad. Sci. USA*, 97, 9408, 2000.

46. Widersten, M., Björnestedt, R., and Mannervik, B., Contribution of amino acid residue 208 in the hydrophobic binding site to the catalytic mechanism of human glutathione transferase A1-1. *Biochemistry*, 33, 11717, 1994.

47. Björnestedt, R., Tardioli, S., and Mannervik, B., The high activity of rat glutathione transferase 8-8 with alkene substrates is dependent on a glycine residue in the active site. *J. Biol. Chem.*, 270, 29705, 1995.

48. Sundberg, K. et al., Differences in the catalytic efficiencies of allelic variants of glutathione transferase P1-1 towards carcinogenic diol epoxides of polycyclic aromatic hydrocarbons. *Carcinogenesis*, 19, 433, 1998.

49. Ivarsson, Y. et al., Identification of residues in glutathione transferase capable of driving functional diversification in evolution. A novel approach to protein redesign. *J. Biol. Chem.*, 278, 8733, 2003.

50. Atkinson, D.E., *Cellular Energy Metabolism and Its Regulation*. Academic Press, New York, 1977, pp. 31–35.

51. Nilsson, B.L., Soellner, M.B., and Raines, R.T., Chemical synthesis of proteins. *Annu. Rev. Biophys. Biomol. Struct.*, 34, 91, 2005.

52. Wang, L. and Schultz, P.G., Expanding the genetic code. *Angew. Chem. Int. Ed.*, 44, 34, 2004.

53. Pettersson, P.L., Johansson, A.-S., and Mannervik, B., Transmutation of human glutathione transferase A2-2 with peroxidase activity into an efficient steroid isomerase. *J. Biol. Chem.*, 277, 30019, 2002.

54. Hederos, S. et al., Incorporation of a single His residue by rational design enables thiol-ester hydrolysis by human glutathione transferase A1-1. *Proc. Natl. Acad. Sci. USA*, 101, 13163, 2004.

55. Ren, X. et al., Semisynthetic glutathione peroxidase with high catalytic efficiency. Selenoglutathione transferase. *Chem. Biol.*, 9, 789, 2002.

56. Thorson, J.S. et al., Analysis of the role of the active site tyrosine in human glutathione transferase A1-1 by unnatural amino acid mutagenesis. *J. Am. Chem. Soc.*, 120, 451, 1998.

57. Björnestedt, R. et al., Design of two chimaeric human–rat class alpha glutathione transferases for probing the contribution of C-terminal segments of protein structure to the catalytic properties. *Biochem. J.*, 282 (Pt. 2), 505, 1992.

58. Zhang, P. et al., Modular mutagenesis of exons 1, 2, and 8 of a glutathione S-transferase from the mu class. Mechanistic and structural consequences for chimeras of isoenzyme 3-3. *Biochemistry*, 31, 10185, 1992.

59. Van Ness, K.P., Buetler, T.M., and Eaton, D.L., Enzymatic characteristics of chimeric mY_c/rY_{c1} glutathione S-transferases. *Cancer Res.*, 54, 4573, 1994.

60. Ricci, G. et al., Structural flexibility modulates the activity of human glutathione transferase P1-1. Role of helix 2 flexibility in the catalytic mechanism. *J. Biol. Chem.*, 271, 16187, 1996.

61. Mannervik, B., Glutathione transferase. In *Drug Metabolism — from Molecules to Man*, Benford, D.J., Bridges, J.W., and Gibson, G.G., Editors. Taylor & Francis, London, 1987, p. 30.

62. Ranson, H., Collins, F., and Hemingway, J., The role of alternative mRNA splicing in generating heterogeneity within the *Anopheles gambiae* class I glutathione S-transferase family. *Proc. Natl. Acad. Sci. USA*, 95, 14284, 1998.

63. Jirajaroenrat, K. et al., Heterologous expression and characterization of alternatively spliced glutathione S-transferases from a single *Anopheles* gene. *Insect. Biochem. Mol. Biol.*, 31, 867, 2001.

64. Kampkotter, A. et al., Functional analysis of the glutathione S-transferase 3 from *Onchocerca volvulus* (Ov-GST-3): a parasite GST confers increased resistance to oxidative stress in *Caenorhabditis elegans*. *J. Mol. Biol.*, 325, 25, 2003.

65. Stemmer, W.P.C., Rapid evolution of a protein *in vitro* by DNA shuffling. *Nature*, 370, 389, 1994.

66. Crameri, A. et al., DNA shuffling of a family of genes from diverse species accelerates directed evolution. *Nature*, 391, 288, 1998.

67. Stenberg, G. et al., A conserved "hydrophobic staple motif" plays a crucial role in the refolding of human glutathione transferase P1-1. *J. Biol. Chem.*, 275, 10421, 2000.

68. Cocco, R. et al., The folding and stability of human alpha class glutathione transferase A1-1 depend on distinct roles of a conserved N-capping box and hydrophobic staple motif. *J. Biol. Chem.*, 276, 32177, 2001.

69. Aceto, A. et al., Identification of an N-capping box that affects the α 6-helix propensity in glutathione S-transferase superfamily proteins: a role for an invariant aspartic residue. *Biochem. J.*, 322 (Pt. 1), 229, 1997.

70. Hansson, L.O. et al., Evolution of differential substrate specificities in mu class glutathione transferases probed by DNA shuffling. *J. Mol. Biol.*, 287, 265, 1999.

71. Hansson, L.O. and Mannervik, B., Use of chimeras generated by DNA shuffling: probing structure-function relationships among glutathione transferases. *Methods Enzymol.*, 328, 463, 2000.

72. Ness, J.E. et al., DNA shuffling of subgenomic sequences of subtilisin. *Nat. Biotechnol.*, 17, 893, 1999.

73. Cleland, W.W., What limits the rate of a chemical reaction. *Acc. Chem. Res.*, 8, 145, 1975.

74. Hitchens, T.K., Mannervik, B., and Rule, G.S., Disorder-to-order transition of the active site of human class pi glutathione transferase, GST P1-1. *Biochemistry*, 40, 11660, 2001.

75. Broo, K. et al., An ensemble of theta class glutathione transferases with novel catalytic properties generated by stochastic recombination of fragments of two mammalian enzymes. *J. Mol. Biol.*, 318, 59, 2002.

76. Larsson, A.K. et al., Directed enzyme evolution guided by multidimensional analysis of substrate-activity space. *Protein Eng. Des. Sel.*, 17, 49, 2004.

77. Jemth, P. and Mannervik, B., Fast product formation and slow product release are important features in a hysteretic reaction mechanism of glutathione transferase T2-2. *Biochemistry*, 38, 9982, 1999.

5 Substrates and Reaction Mechanisms of Glutathione Transferases

Piotr Zimniak

CONTENTS

5.1 INTRODUCTION

Four groups of unrelated, or distantly related, proteins are able to catalyze the "signature" reaction of glutathione transferases (GSTs), namely the addition of glutathione (GSH) to an electrophilic center of an acceptor molecule. These are (i) the canonical, soluble GSTs,[1] (ii) the mitochondrial (Kappa-class) GSTs,[2,3] (iii) microsomal GSTs, or MAPEG (membrane-associated proteins in eicosanoid and glutathione metabolism) family,[4] and (iv) the bacterial fosfomycin resistance proteins.[5] The acceptor molecules (electrophilic targets for glutathione addition) differ between the above protein superfamilies, as well as between individual enzymes within these families. Moreover, the details of the catalytic event (insofar as they are known) are distinct for the four enzyme groups. The present chapter focuses on the reaction mechanism of the canonical GSTs, which is characterized in the greatest detail, although the other superfamilies will be briefly mentioned, especially where commonalities in the reaction mechanism are discernible.

GSTs are members of an evolutionarily ancient family derived from an ancestral protein probably most similar to extant thioredoxins and glutaredoxins.[6–8] Some GSTs evolved to be efficient and specialized catalysts of particular reactions. This appears to hold mostly for bacterial GSTs that permit the organism to utilize a bewildering variety of compounds as (often sole) nutrients[9] and for some GSTs from invertebrate multicellular organisms, especially plants.[10] A role in intermediary metabolism is still evident in vertebrates, including humans, for GSTs considered to be evolutionarily "primitive," such as the Sigma and Zeta classes. Unlike microorganisms, animal cells can utilize only a limited set of nutrient molecules. With such metabolism, anything an animal ingests that is not a nutrient becomes a potential toxin. The resulting greater need for detoxification in organisms such as vertebrates is reflected in a shift in GST functions. The Alpha, Mu, and Pi classes of GSTs prevalent in vertebrates catalyze largely detoxification reactions. This metabolic emphasis resulted in structural and functional adaptations of the GST proteins. Cellular detoxification reactions are sometimes compared to an organism's immune system in that they need to deal with a large variety of compounds, including those to which the cell was not previously exposed. At the same time, normal intermediary metabolites of the cell should be spared. The typical solution to these constraints is the evolution of detoxification enzyme families in which the individual enzymes have rather broad and partly over-lapping specificities. This ensures broad coverage and permits selective fine-tuning of particular enzymes if they encroach on normal metabolism. Expansion of protein families is seen not only for GSTs but also for other detoxification enzymes such as cytochromes P450, UDP-glucuronosyltransferases, sulfotransferases, and others.

Mechanistically, broad specificity of an enzyme translates into a permissive structure of the active site. This almost inevitably leads to a penalty in catalytic efficiency: a permissive site cannot be fully optimized for all potential substrates. In some of the detoxification enzyme families this problem is somewhat mitigated by the use of a common cosubstrate (glutathione for GSTs) to which the enzyme is exquisitely adapted. Still, the recognition of the other substrate (electrophile in GSTs) remains suboptimal. The high abundance of GSTs partially compensates for the generally low catalytic efficiency. The evolutionary compromise between versatility and efficiency maximizes the protective benefit to the cell for "general-purpose" GSTs. In contrast, specialized GSTs may reach substantially higher catalytic efficiencies, as discussed later in this chapter on the example of enzymes that metabolize 4-hydroxynon-2-enal (Section 5.3.1).

This discussion of GST substrates will be broad in that it will go beyond the classical enzymologic meaning of the term. In additional to such substrates, non-catalytic but functionally important interactions of GSTs with other molecules will be described.

A number of excellent and exhaustive reviews touching on the topic of this chapter is available including references 1, 11–24, and others. While the basic features of GST function will be recapitulated here, the focus is on recent findings. Therefore, the reader is referred to the earlier reviews for additional details, especially concerning older literature.

5.2 BASIC REACTION MECHANISM OF CANONICAL GSTs

The catalytic event of GSTs encompasses two processes: binding and activation of GSH, and binding of the electrophilic reaction partner. The first process is common to all canonical GSTs, at least in its general characteristics, while the second obviously depends on the chemical nature and structure of the electrophile.

5.2.1 BINDING AND ACTIVATION OF GSH

In canonical GSTs, a GSH-binding site (G-site) is located within the N-terminal domain of each subunit. The characteristic folding motif of the N-terminal domain is evolutionarily ancient. It is derived from bacterial thioredoxin/glutaredoxin and is present, in addition to GSTs, in a variety of other proteins including stress-induced proteins, the eukaryotic translation elongation factor EF1γ,[25,26] the chloride channel CLIC,[27,28] glutathione peroxidases, and others. The active site of thioredoxin contains cysteine residues that form a disulfide as part of the catalytic event (reviewed in references 29 and 30). Glutaredoxins, another class of thiol redox enzymes, have thioredoxin-like activity but can also catalyze a monothiol reaction, which serves to cleave or form mixed disulfides of target proteins with glutathione.[7] The monothiol mechanism involves a transient binding of GSH to an enzyme cysteine as a mixed disulfide (glutathionylation of the enzyme). This mode of GSH binding to the enzyme is retained in GSTs of the Beta and Omega classes.[31–34] In contrast, more recently evolved GSTs utilize GSH primarily (although not exclusively) for conjugative rather than redox reactions. Accordingly, they bind GSH in a different manner (see below). Redox- and other modification-sensitive cysteine residues, such as Cys-47 in the Pi-class hGSTP1-1, may be a vestige of active-site cysteine residues but no longer directly participate in the catalytic event.

Electrophiles react more readily with a thiolate anion than with a sulfhydryl group. GSTs take advantage of this chemical behavior to catalyze a nucleophilic attack of GSH on electrophilic substrates. The pK_a of the sulfhydryl group of GSH in aqueous solution is approximately 9.0. This value is lowered to less than 7 in GSH bound to most GSTs;[35] the thiolate can be directly observed by its characteristic absorption at 239 nm.[36] Therefore, enzyme-bound GSH is largely deprotonated at physiological pH and is thus activated for reaction with an electrophile. The enzyme-bound thiolate of glutathione is stabilized by a hydrogen bond between the proton of a hydroxyl group in the protein and the sulfur atom of the thiolate.[37,38] In most GSTs, the hydrogen-bond donor is the hydroxyl group of a tyrosine close to the N-terminus of the protein, although a serine hydroxyl can carry out the same function in Theta-class GSTs[39] as well as in the insect- and plant-specific GST classes (see

Chapter 2). In Alpha-class GSTs, a secondary interaction of the ε nitrogen atom of Arg-15 with the sulfur of GSH additionally helps to stabilize the thiolate.[40]

The activation of GSH by GSTs can be envisioned as follows. Protonated GSH tightly binds to the enzyme via multiple interactions within the G-site; three distinct but functionally similar binding modes have been described for different classes of GSTs.[41] At least in some GSTs, multiple steps are required for GSH to become catalytically competent. In Pi-class GSTs, GSH initially binds in a pre-catalysis position and is subsequently shifted, in a step that can be rate-limiting for the overall reaction, to a catalysis position.[42,43] The proton of the SH group of GSH is released,[44,45] and the resulting thiolate is stabilized by hydrogen bonding with an active-site tyrosine (or serine) and, in Alpha-class GSTs, by an additional interaction with Arg-15 (Figure 5.1). In Theta-class GSTs, another arginine residue (Arg-107 in hGSTT2-2) may play the same role.[46] Second-sphere effects also contribute to the stabilization of the thiolate.[1] The thiolate is then available for the reaction with

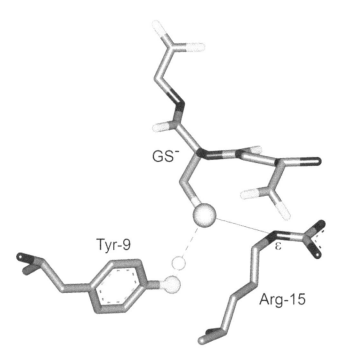

FIGURE 5.1 Activation of glutathione via stabilization of its thiolate form in the active site of a canonical GST. The thiolate is stabilized primarily through a hydrogen bond between the hydroxyl group of a tyrosine (in some GSTs, serine) located close to the N-terminus of the protein and the sulfur atom of glutathione. In Alpha-class GSTs, an additional interaction between the sulfur atom and the ε nitrogen atom of Arg-15 also contributes to the stabilization of the thiolate. The figure is based on the crystal structure of mGSTA1-1 with GSH in its active site (reference 169, pdb accession code 1F3A). Oxygen atoms are shown in light gray, carbon atoms in medium gray, and nitrogen atoms in black; the large sphere denotes the sulfur atom of GSH.

the electrophilic second substrate. Activation of GSH to a thiolate is probably a common feature of enzyme-catalyzed conjugation of electrophiles with GSH because it is also part of the catalytic cycle of the MAPEG enzymes (microsomal GSTs),[47] even though these proteins are structurally unrelated to the canonical GSTs. The mechanism of GSH activation by MAPEG is unknown because the necessary details are not discernible in six Å resolution models obtained by electron diffraction crystallography;[48,49] no high-resolution structure is available as of this writing.

5.2.2 Binding of the Electrophilic Substrate

In contrast to GSH binding and activation, which are not identical but nevertheless are generally conserved among the canonical GSTs, a variety of chemically and structurally dissimilar electrophilic substrates need to bind to the electrophile-binding site (H-site). The H-sites of the various GST classes evolved to accommodate different sets of electrophilic substrates and are therefore highly variable between GST classes and, to a lesser extent, between individual enzymes within a class. The H-site utilizes structural elements from both the N-terminal and the C-terminal domains of a GST subunit. Evolutionarily older enzymes such as those of the Theta class have a rigid structure, while the H-sites of more recently evolved GSTs of the Alpha, Mu, and Pi classes are flexible in ways that optimize substrate binding and catalytic activity.[46,50] The H-site cavities of Mu- and Alpha-class GSTs are generally hydrophobic, while those of Pi-class enzymes are lined with both hydrophobic and hydrophilic surfaces, facilitating the recognition of substrates that have polar and apolar moieties, such as benzo[a]pyrene diol epoxide.[41] Generally, the role of the H-site is to bind the electrophilic substrate and orient it for nucleophilic attack by the GSH thiolate. However, in some cases discussed later in this chapter binding to the H-site activates the electrophile through specific interactions, thus contributing directly to the catalytic event.

5.3 CATALYTIC ACTIVITIES OF GSTs

5.3.1 Glutathione Transferase Activity

The archetypal reaction catalyzed by GSTs is the formation of a thioether between GSH and an electrophilic substrate. The reaction is exemplified by the formation of S-(2,4-dinitrophenyl)glutathione from 1-chloro-2,4-dinitrobenzene (CDNB) (Figure 5.2). CDNB reacts readily with nucleophiles including the thiolate anion of GSH, even in a spontaneous, noncatalyzed reaction. In addition, the CDNB molecule is small compared with the H-site cavity of most GSTs, and it places few steric demands on the enzymes. For these reasons, CDNB is conjugated by most GSTs and is used in the laboratory as an (almost) universal substrate for assaying GSTs.

Conjugation of CDNB is an aromatic nucleophilic substitution that involves an identifiable intermediate of the Meisenheimer-complex type (Figure 5.2). In addition to substitution reactions in which a leaving group (such as chloride) is replaced by glutathione, some GSTs also catalyze nucleophilic addition reactions. An example (Figure 5.3) is the addition of GSH to a double bond in an α,β-unsaturated carbonyl

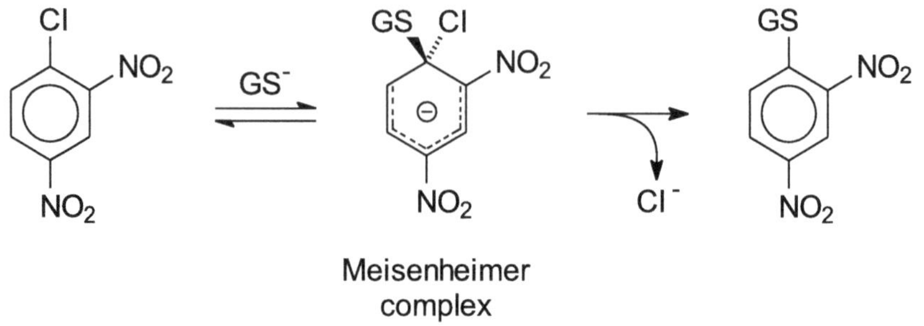

FIGURE 5.2 GST-catalyzed aromatic nucleophilic substitution reaction. The substrate (1-chloro-2,4-dinitrobenzene, CDNB) is converted to S-(2,4-dinitrophenyl)glutathione. The intermediate Meisenheimer complex is shown.

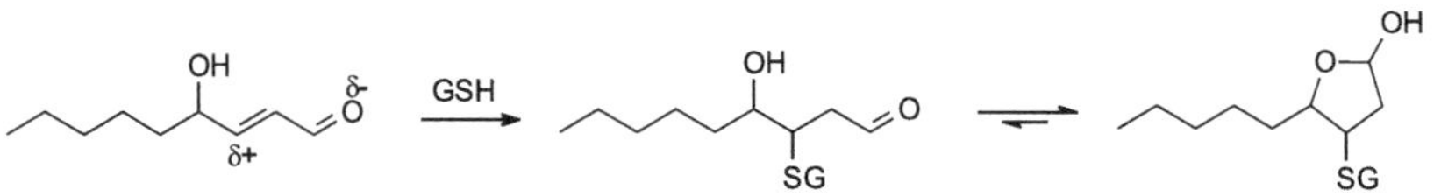

FIGURE 5.3 GST-catalyzed Michael addition of glutathione to an α,β-unsaturated carbonyl compound (exemplified in this figure by 4-hydroxynonenal, or 4-HNE). The aldehyde group of the reaction product, no longer conjugated to the double bond, forms a cyclic hemiacetal with the hydroxyl group in position 4 of 4-HNE.

compound (Michael acceptor). The third major type of glutathione transferase reactions is the opening of a strained oxirane ring (Figure 5.4). In the Michael addition as well as oxirane ring opening, the glutathione thiolate attacks an electrophilic center in the target molecule (Figures 5.3 and 5.4).

An instructive special case is the Michael addition of GSH to the electrophilic carbon atom in position 3 of 4-hydroxynon-2-enal (4-HNE) (Figure 5.3). The reaction is catalyzed by specialized isoforms of GSTs exemplified by the murine mGSTA4-4[51,52] or human hGSTA4-4.[53] These enzymes catalyze the conjugation of 4-HNE with very high catalytic efficiency, exceeding 10^6 $M^{-1}s^{-1}$, which is among the highest catalytic efficiencies of any GST with any substrate. In contrast, catalytic efficiencies of the same enzymes for the conjugation of other substrates, including CDNB, are lower by one to two orders of magnitude. In mGSTA4-4, the high

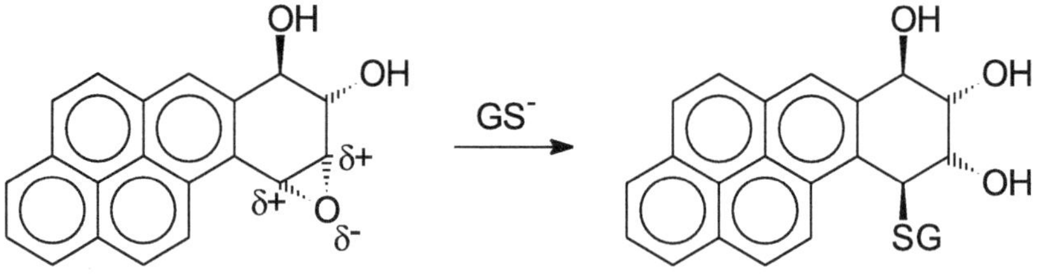

FIGURE 5.4 GST-catalyzed oxirane (epoxide) ring opening and glutathione addition. The substrate shown is (+)-anti-benzo[a]pyrene diol epoxide, the ultimate carcinogen derived from benzo[a]pyrene.

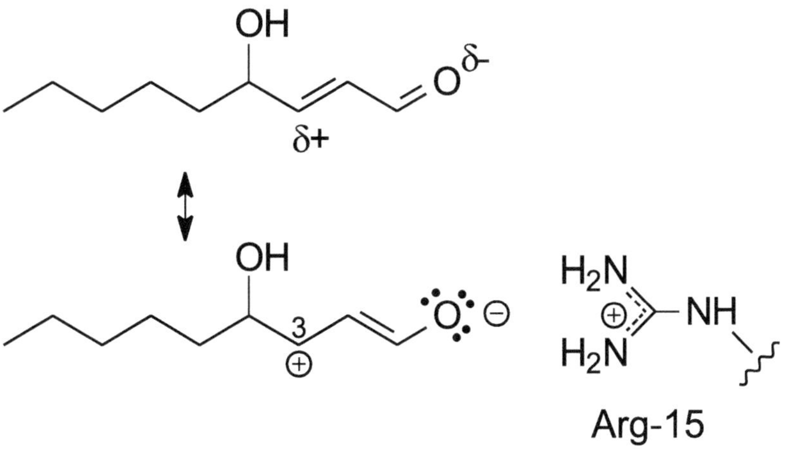

FIGURE 5.5 Postulated mechanism of activation of 4-HNE, the electrophilic reaction partner, by mGSTA4-4 for Michael addition of the thiolate of GSH. The proximity of the positively charged guanidinium group of Arg-15 shifts the π electron cloud of the conjugated double-bond system of 4-HNE toward the carbonyl oxygen, increasing the positive charge and thus the electrophilicity of the carbon atom in position 4 of 4-HNE. Based on reference 55.

catalytic efficiency for 4-HNE is conditional on the accessibility of the guanidinium group of Arg-15.[54,55] In the majority of mammalian Alpha-class GSTs, the guanidinium group of Arg-15 forms an ion pair with the carboxyl group of Glu-104. In contrast, in mGSTA4-4 and its rat ortholog, rGSTA4-4, position 104 is occupied by methionine, and no other negative charge is available to neutralize the positive charge of the guanidinium group of Arg-15. The crystal structure of mGSTA4-4 complexed with the glutathione conjugate of 4-HNE[55] shows that, in one subunit of the enzyme, Arg-15 is positioned in the active site so that its guanidinium group would be in close proximity of the carbonyl group of the 4-HNE substrate. It has been postulated[55] that the positive charge on Arg-15 shifts the π electrons of the conjugated double-bond system of 4-HNE toward the carbonyl group, thus increasing the electrophilicity of carbon-3 (Figure 5.5) and therefore facilitating the Michael addition of the glutathione thiolate to 4-HNE. The activation of both GSH and the electrophilic reaction partner by the GST could explain the unusually high catalytic efficiency of mGSTA4-4 for 4-HNE. The particular mode of interaction between the guanidinium group of Arg-15 and 4-HNE determines the specificity of the effect: the reaction with CDNB, a substrate conjugated by a different mechanism (aromatic nucleophilic substitution), is not augmented in mGSTA4-4.

Known cases of activation of the electrophilic substrate of GSTs are infrequent but have been previously reported. The phenomenon was originally described for a Mu-class GST in which Tyr-115 provides electrophilic assistance both in the Michael addition of GSH to 4-phenylbut-3-en-2-one and in the opening of the oxirane ring and GSH addition to phenanthrene 9,10-oxide.[56]

Interestingly, in the human enzyme hGSTA4-4, which has similar catalytic properties toward 4-HNE as the murine protein mGSTA4-4, the guanidinium group of Arg-15 is engaged in an ion pair with Glu-104,[57] as typically seen in most Alpha-class GSTs and therefore does not participate in the activation of 4-HNE. The high catalytic efficiency of hGSTA4-4 for 4-HNE can be, however, explained by a different mechanism of electrophilic assistance. Unlike mGSTA4-4, in which 4-HNE enters the H-site with its aliphatic chain first ("tail first," see the structure in reference 55), leaving the aldehyde group facing the cleft between subunits and accessible to Arg-15, in hGSTA4-4 the 4-HNE molecule enters the active site "head first."[57] In this orientation, the aldehyde group of 4-HNE is in close proximity to Tyr-212. Tyr-212 is postulated to donate a hydrogen bond to the carbonyl oxygen of 4-HNE and thus to stabilize the partial negative charge on that oxygen. This polarizes the π electron cloud of the conjugated double-bond system of 4-HNE and increases the electrophilicity of carbon-3, much as described above for Arg-15 in mGSTA4-4. Thus, the chemical mechanism of 4-HNE activation is similar in mGSTA4-4 and hGSTA4-4, but its structural basis is quite different. This indicates that mGSTA4-4 and hGSTA4-4 are unlikely to be orthologs. The two enzymes may have acquired 4-HNE-conjugating activity by convergent evolution, as probably did several additional invertebrate GSTs belonging to classes different from Alpha.[58–60]

As mentioned previously, in most mammalian Alpha-class GSTs Arg-15 forms an ion pair with Glu-104. The only known exceptions are the already discussed mGSTA4-4 and rGSTA4-4, as well as mGSTA3-3 and its rat ortholog, rGSTA5-5 (see reference 21 for nomenclature of rodent GSTs). In mGSTA3-3 and rGSTA5-5, residue 104 is hydrophobic (Ile and Leu, respectively). According to the model proposed for mGSTA4-4, the lack of an acidic residue in position 104 should result in a free guanidinium group of Arg-15 and therefore in high catalytic efficiency for 4-HNE. However, the activity of mGSTA3-3 for 4-HNE is low.[3] This seeming inconsistency may be explained by the fact that in both mGSTA3-3 and rGSTA5-5, position 103 is occupied by Glu. While no crystal structure has been reported for either enzyme, theoretical models generated on the basis of structures of related GSTs are available.[61] For example, in the model of mGSTA3-3 the guanidinium and carboxyl groups of Arg-15 and Glu-103 are separated by approximately six to eight Å, but can be brought to juxtaposition at less than three Å, as necessary for ion-pair formation. This requires changing the conformation of the two side chains plus that of Leu-106 but does not violate packing constraints. Thus, mGSTA3-3 may have a low activity for 4-HNE conjugation because the positive charge of the guanidinium group of Arg-15 is neutralized by Glu-103 (instead of Glu-104 typical for other Alpha-class GSTs) or because of other H-site constraints that do not favor 4-HNE binding.

The effect on enzyme activity of seemingly small and conservative changes in the H-site is illustrated by the comparison of two pairs of enzymes: the naturally occurring allelic forms of hGSTP1-1 and of hGSTA1-1/hGSTA2-2. The enzymes catalyze the addition of glutathione to an oxirane structure in (+)-*anti*-benzo[*a*]pyrene diol epoxide (BPDE), the ultimate carcinogen derived from benzo[*a*]pyrene, and therefore play a protective role against the carcinogenicity of benzo[*a*]pyrene and other polycyclic aromatic hydrocarbons.

Allelic hGSTP1-1 variants carrying either isoleucine or valine in position 104[62,63] were characterized and found to differ in their catalytic properties toward CDNB and heat stability[64] as well as in their catalytic parameters for diol epoxides of polycyclic aromatic hydrocarbons.[65,66] The mechanistic basis for this difference in activity was elucidated by solving the crystal structures of hGSTP1-1[Ile-104] and hGSTP1-1[Val-104] complexed with the glutathione conjugate of (+)-*anti*-BPDE.[67] Ile-104 was found to assume a fixed position in the protein that could interfere with the binding of bulky substrates and hinder the alignment of their electrophilic centers with the thiolate of GSH. As a consequence, the bulky (+)-*anti*-BPDE is forced into a suboptimal binding position in the H-site of hGSTP1-1[Ile-104]. On the other hand, Val-104 has considerable conformational freedom and may adjust to accommodate different substrate molecules. The more favorable binding of (+)-*anti*-BPDE to the H-site of hGSTP1-1[Val-104] explains the 3.4-fold higher k_{cat} as compared to hGSTP1-1[Ile-104] with the same substrate.[65]

Human Alpha-class enzymes hGSTA1-1 and hGSTA2-2 are able to conjugate (+)-*anti*-BPDE, albeit with a significantly lower catalytic efficiency than hGSTP1-1.[68] Nevertheless, the two Alpha-class enzymes are biologically relevant since they are abundant in liver, the primary site of detoxification of xenobiotics, while hGSTP1-1 is very low in adult human livers. The catalytic efficiency of (+)-*anti*-BPDE conjugation is approximately 5-fold higher for hGSTA1-1 than hGSTA2-2. However, mutating the active-site residue Ile-11 of hGSTA2-2 to alanine (as present in hGSTA1-1) increases the catalytic efficiency of hGSTA2-2 to the level of hGSTA1-1, and further remodeling of the H-site by mutating an additional three residues increases the activity even further. In fact, the modified hGSTA2-2 has a catalytic efficiency that is 14-fold higher than that of the wild-type enzyme and exceeds the catalytic efficiency of hGSTA1-1 almost 3-fold.[68] Molecular modeling indicates that the activity is determined by contacts between the substrate molecule and H-site residues, which determine the precise positioning of the substrate.

It has been recently demonstrated that two yeast glutaredoxins — Grx1 and Grx2 — have glutathione transferase activity toward CDNB and a related substrate, 1,2-dichloro-4-nitrobenzene, probably via a monothiol mechanism[69] similar to that seen in Omega-class GSTs. Interestingly, the length of the two proteins is only 110 and 109 amino acids, respectively[70] (similar to that of a thioredoxin molecule). Therefore, Grx1 and Grx2 contain the thioredoxin fold and glutathione-binding site but should lack a fully formed H-site that, in GSTs, requires the presence of the C-terminal domain. The H-site may thus become dispensable in the case of small and reactive electrophilic substrates. This is in stark contrast with bulky substrates such as BPDE (see above), which require precise positioning in the active site and accurate alignment with enzyme-bound thiolate for optimal catalytic activity.

The majority of glutathione-conjugation reactions leads to a detoxification of the target electrophile. In rare cases, however, the product can have increased, rather than decreased, toxicity (bioactivation of a toxin). An example of such a reaction is shown in Figure 5.6. Dichloromethane initially undergoes glutathione conjugation, but the product is unstable and gives rise to toxic formaldehyde. Bioactivation to a toxic product is the basis of prodrugs used in cancer chemotherapy.[71–73]

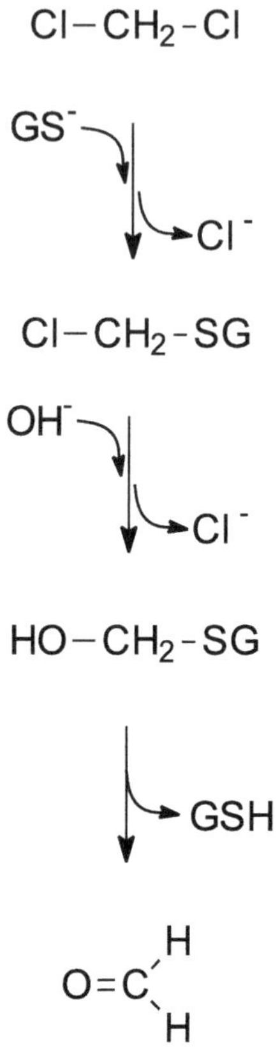

$$Cl-CH_2-Cl$$

$$Cl-CH_2-SG$$

$$HO-CH_2-SG$$

$$O=C\,\substack{H \\ H}$$

FIGURE 5.6 Bioactivation by a GST-catalyzed reaction of dichloromethane to formaldehyde.

5.3.2 GLUTATHIONE PEROXIDASE ACTIVITY

GSTs belonging to the Alpha class have the ability to utilize GSH as a reductant to convert organic hydroperoxides, but not hydrogen peroxide, to the corresponding alcohol.[13] As discussed elsewhere in this volume, the major physiological significance of this reaction is the reduction of phospholipid hydroperoxides without the need for prior hydrolysis of the molecule to liberate the oxidized fatty acid.[74] It has been shown recently that Alpha-class GSTs can account for the majority of cellular glutathione peroxidase activity for phospholipid hydroperoxides, at least in some tissues.[74,75]

5.3.3 ISOMERASE ACTIVITY

The synthesis of steroid hormones such as testosterone and progesterone involves an isomerization step that converts the Δ^5-3-ketosteroid (e.g., androst-5-ene-3,17-dione) to a Δ^4-3-ketosteroid. The isomerization is catalyzed by GSTs,[76] more recently identified as members of the Alpha class, in particular hGSTA3-3.[77] The active-site residues contributing to this reaction have been defined.[78,79] On the

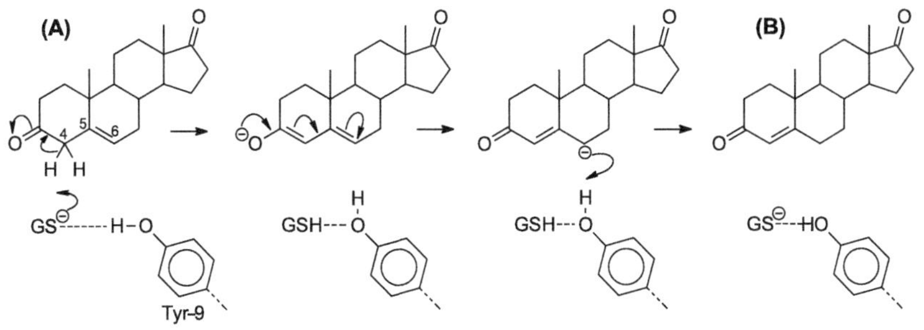

FIGURE 5.7 Mechanism of isomerization of androst-5-ene-3,17-dione (structure A) to androst-4-ene-3,17-dione (structure B) catalyzed by hGSTA3-3. The thiolate anion of glutathione, stabilized by Tyr-9, abstracts a proton from carbon 4 of the steroid molecule. The resulting negative charge is transferred via the conjugated double-bond system to carbon 6, and, in a concerted process, a proton is formally transferred to carbon 6 of the steroid via a proton wire including glutathione and the hydroxyl group of Tyr-9. Based on reference 80.

basis of enzymologic and site-directed mutagenesis studies[78] and a crystal structure of hGSTA3-3 in complex with GSH,[80] a model of the isomerization reaction has been developed (Figure 5.7). According to this model,[80] the thiolate of GSH acts as a base abstracting a proton from position 4 of the steroid. The proton is then formally transferred via a proton wire involving GSH and Tyr-9 to position 6 of the steroid, completing the isomerization. GSH has a catalytic role in this reaction but is not consumed.

Another distinct, physiologically essential isomerization reaction is the conversion of maleylacetoacetate to fumarylacetoacetate catalyzed by GSTZ1-1, a member of the Zeta family of canonical GSTs.[81] The phylogenetically ubiquitous distribution of the enzyme, including plants, fungi, insects, and vertebrates, is consistent with the importance of the reaction, which is part of the catabolism of phenylalanine and tyrosine. The reaction mechanism is proposed to involve transient addition of GSH to the substrate followed by rotation around the now-single bond and glutathione elimination[82] (Figure 5.8). The details of the catalytic event of Zeta-class GSTs are only partially understood. For example, it has been proposed that Ser-17 of the Zeta-class GSTs of *Arabidopsis thaliana* interacts with the sulfur atom of GSH and lowers its pKa, in analogy to the active-site tyrosine in most other GSTs.[82] However, according to the crystal structure,[83] Ser-14 in the human hGSTZ1-1 (which corresponds to Ser-17 of the *A. thaliana* enzyme) points away from the sulfur atom and is thus unlikely to fulfill a catalytic role.[84]

Isomerization of 13-*cis*-retinoic acid to all-*trans*-retinoic acid is catalyzed by hGSTP1-1 and, to a lesser extent, by hGSTM1-1 and hGSTA1-1.[85,86] Strikingly, the isomerization was reported to be GSH-independent. This would require a mechanism distinct from that of steroid or maleylacetoacetate isomerizations, which involve transient addition of GSH to the substrate. However, no rigorous measures beyond dialysis were undertaken to remove GSH bound to the enzymes. Since catalytic rather than stoichiometric amounts of GSH are needed for isomerization reactions, a GSH-mediated mechanism cannot be ruled out with certainty.

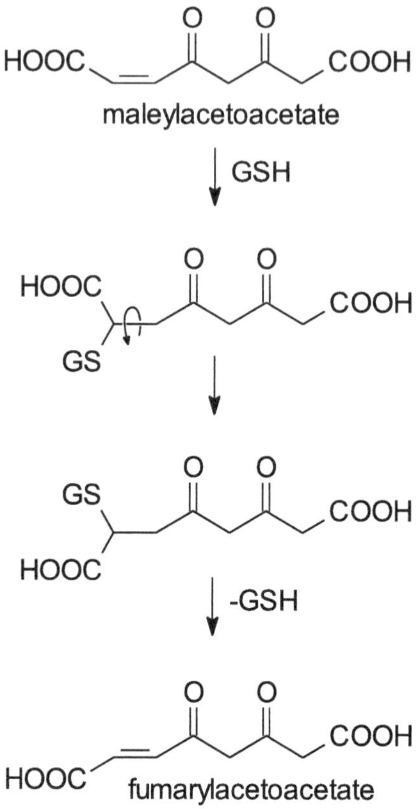

FIGURE 5.8 Isomerization of maleylacetoacetate to fumarylacetoacetate catalyzed by GSTZ1-1.

5.3.4 Role in Metabolism of Eicosanoids

The eicosanoid lipid mediators prostaglandins and leukotrienes are synthesized from arachidonic acid. In prostaglandin synthesis, the common precursor PGH_2 is converted to a number of products, among them PGD_2 (Figure 5.9). The latter reaction can be catalyzed by two prostaglandin D_2 synthases, one of which has been identified as the only mammalian GST belonging to the Sigma class.[87,88] The proposed reaction mechanism involves transient addition of glutathione to the substrate but no net glutathione consumption.[89] An alternative fate of the precursor PGH_2 is the conversion to PGE_2, a reaction that can be catalyzed by the MAPEG enzyme PGE synthase[90] or by the canonical Mu-class GSTM2-2 and GSTM3-3[91] (Figure 5.9). Strikingly, GSH can be replaced in the latter reaction by other thiols. Dithiothreitol, 2-mercaptoethanol, and cysteine were less effective than GSH but nevertheless supported the reaction.[91] The mechanistic implications of this finding could be interesting because compounds other than GSH are unlikely to bind to the rather specific G-site.

PGD_2 and PGE_2 can undergo dehydration, which converts the cyclopentane ring to cyclopentene and results in PGJ_2 and PGA_2, respectively. The resulting

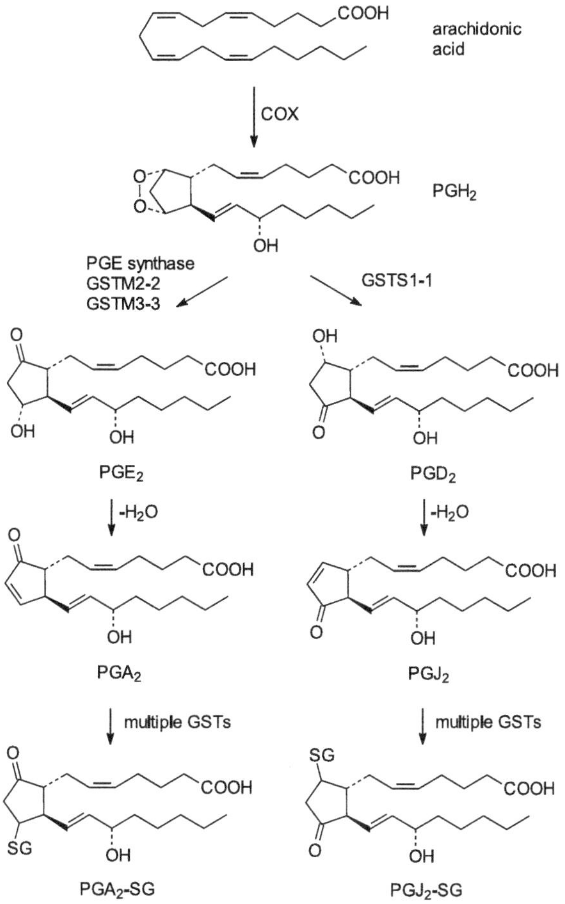

FIGURE 5.9 Role of GSTs in prostaglandin synthesis. Arachidonic acid is converted by cyclooxygenases to the common prostaglandin precursor, PGH_2. Two of the possible fates of PGH_2 are GSTS1-1-catalyzed conversion to PGD_2 and conversion to PGE_2 catalyzed by PGE synthase (a MAPEG enzyme), or by GSTM2-2 or GSTM3-3. Loss of water from PGE_2 and PGD_2 yields PGA_2 and PGJ_2, respectively, which can be conjugated with GSH by several canonical GSTs.

α,β-unsaturated ketone moiety is a Michael acceptor structure to which GSH can be added by a number of GSTs (Figure 5.9), including GSTA1-1, A2-2, M1a-1a, and P1-1, in a highly stereospecific reaction.[92]

A distinct branch of the arachidonate cascade leads to the synthesis of leukotrienes.[93,94] In an initial step, arachidonic acid is converted by 5-lipoxygenase and the 5-lipoxygenase-activating protein FLAP to 5-HPETE (5-hydroperoxyeicosatetraenoic acid) and then to leukotriene A_4 (LTA_4; Figure 5.10). One of the fates of LTA_4 is glutathione conjugation to LTC_4, catalyzed by LTC_4 synthase. Both FLAP and LTC_4 synthase are members of the MAPEG family of GSTs. The mechanism of oxirane ring opening in LTA_4 by LTC_4 synthase is thought to be a nucleophilic

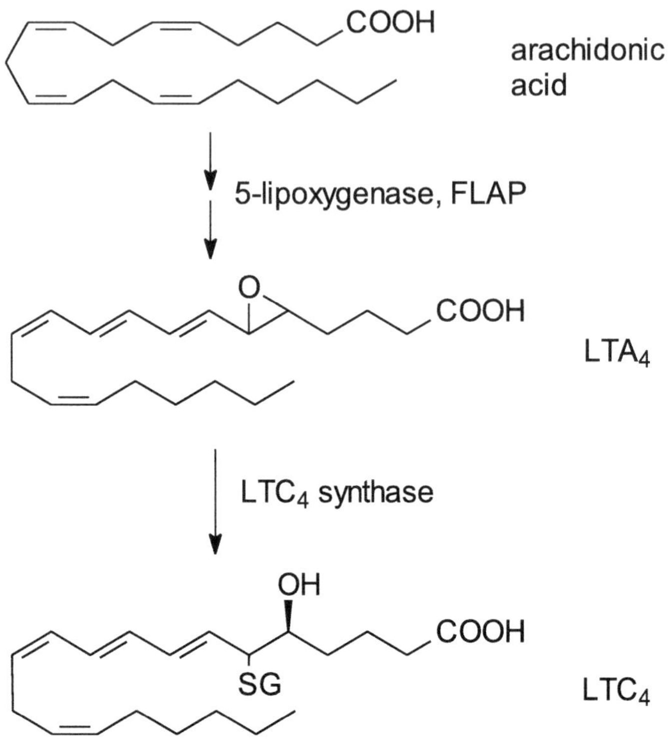

FIGURE 5.10 Role of MAPEG GSTs in the synthesis of cysteinyl leukotrienes. FLAP (5-lipoxygenase-activating protein) and LTC$_4$ synthase belong to the MAPEG family.

attack of the glutathione thiolate on a carbon atom of the oxirane structure whose electrophilicity is enhanced by the proximity of an arginine residue.[95]

5.3.5 OTHER REACTIONS

Some of the more "primitive" GSTs (i.e., those closer to an ancestral protein) have little or no activity for typical GST substrates, including CDNB. Instead, they have other catalytic abilities. For example, dehydroascorbate dehydrogenase and thiol transferase activities are associated with invertebrate as well as mammalian Omega-class GSTs.[34,96] These catalytic properties place the Omega class between the ancestral thioredoxins/glutaredoxins and the "mainstream" vertebrate GSTs.

5.3.6 EFFECT OF GST SUBUNIT DIMERIZATION ON CATALYTIC ACTIVITY

Heterodimers of GST subunits belonging to different classes may form, at least *in vitro*,[97] and it has been proposed that GST monomers with reduced but measurable activity can be generated by mutating interface residues.[98] Nevertheless, canonical GSTs are typically dimers of identical subunits or of subunits belonging to the same class. Even though the active sites in the two subunits of a dimer are distant from each other, blocking the active site of one subunit by affinity labeling affects the other subunit.[99–101] This indicates cross-talk between the subunits. Such cross-talk has been demonstrated by structural and functional means in mGSTA4-4. As mentioned in Section 5.3.1 of this chapter, Arg-15 of mGSTA4-4 activates the electrophilic substrate 4-HNE (but not other substrates such as CDNB) by polarizing the π electron cloud of the conjugated double-bond system of 4-HNE. Analysis of the crystal structure of mGSTA4-4 demonstrated that, at any given time, this electrophilic assistance occurs in only one subunit of the homodimeric enzyme.[55] Positioning of Arg-15 in the active site of one subunit of mGSTA4-4 causes, via a chain of residues spanning the subunit interface, the withdrawal of Arg-15 from the active site of the other subunit, perhaps to facilitate product release. The enzyme thus acts in a half-site reactivity mode, with Arg-15 activating the electrophile in one subunit, while withdrawing from the active site of the other subunit to expel the reaction product. This results in a negative cooperativity of the enzyme with respect to 4-HNE concentration. The chain of residues necessary for intersubunit communication is an "arginine relay." It involves Arg-69 residues from both subunits whose guanidinium groups are stacked at the subunit interface, and, alternatingly, Arg-15 from one or the other subunit. Mutating Arg-69 in both subunits interrupts the cross-talk between the subunits and relieves the negative cooperativity of mGSTA4-4.[55]

The effects of interface contact sites on enzyme stability and on catalytic properties have been further studied by generating heterodimers of a wild-type subunit with the same subunit in which an interface residue was mutated.[102,103] In both hGSTP1-1[102] and rGSTA1-1,[103] homodimers of subunits mutated in an interface residue had a greatly reduced catalytic activity compared with a wild-type enzyme (the residue that was mutated in hGSTP1-1 was Tyr-50, the "key" residue in the key-and-lock interface motif, and in rGSTA1-1 it was Arg-69, which is involved in an electrostatic interaction with the opposite subunit). For catalytically independent subunits, a heterodimer of one wild-type and one mutated subunit would be expected to have an activity equal to the mean of the wild-type and the fully mutated dimer. However, for both hGSTP1-1 and rGSTA1-1, the measured activity of the heterodimer was significantly lower than the calculated mean.[102,103] This indicates that one subunit senses the other's state. However, the simplest explanation for the need to dimerize would be static: to maintain the native conformation, each subunit needs to be in contact with the other (in other words, the two subunits support each other in a mutual fashion). Perturbations of the interface, even if caused by a mutation in only one subunit, could propagate to both subunits and affect their activity. This is different from true catalytic cooperativity, which is dynamic: one subunit senses and adjusts to the substrate/product occupancy of the other subunit at any given time. Catalytic cooperativity, mechanistically mediated by an arginine relay spanning the two subunits, was observed for mGSTA4-4 acting on 4-HNE (see above). Such cooperativity was previously found in hGSTP1-1 where it depends on the temperature and viscosity of the medium and is conditional on the movement of the α2 helix, which is part of the GSH-binding site.[42,102,104] For the latter enzyme, it has been postulated that GSH binding to one subunit, communicated to the other via a strand of ordered water molecules, causes shifts in helix α2, which codetermine whether GSH in the second subunit is in precatalytic or catalytic position.[102]

An unusual activity of GSTP1-1 has been recently reported.[105] The reaction involves adding GSH to a sulfenic acid (R-SOH, formed by oxidation of a sulfhydryl, R-SH, by peroxides) in peroxiredoxin Prx VI. Sulfenic acid is electrophilic and thus an appropriate target of GST-catalyzed glutathionylation. It has been postulated[105] that the reaction proceeds via the formation of a heterodimer of a Prx VI subunit with a GSTP1 subunit. The heterodimerization causes a partial loss of activity of the GST with CDNB as substrate but permits a transfer of GST-bound GSH to the peroxiredoxin. Further work will be needed to elucidate how the sulfenic acid moiety on Prx VI can enter the active site of GSTP1 and whether the interface requirements, such as the key-and-lock structure, are satisfied in the postulated heterodimer.

5.4 NONCATALYTIC ACTIVITIES OF GSTs

5.4.1 LIGANDIN FUNCTION

In addition to binding of reaction substrates, GSTs have the ability to bind mostly apolar and hydrophobic molecules noncatalytically (ligandin function[106,107]). The physiological role of this binding is a sink for compounds that otherwise could interfere with normal cell function. In addition to binding and removal of potentially toxic material, GSTs may provide a buffer, or form of storage, for compounds sequestered for later use. In that sense, GSTs act intracellularly much as albumin does in the circulation. The high abundance of GSTs ensures an appropriate binding capacity.

In special cases, noncatalytic sequestration by GSTs of a bioactive ligand or signaling molecule prevents its action, thus modulating cell response. Several GSTs are able to bind 15-deoxy-$\Delta^{12,14}$-prostaglandin J$_2$ or its glutathione conjugate.[108] This prostaglandin acts as a PPARγ ligand in the nucleus. Its sequestration in the cytoplasm prevents nuclear translocation and inhibits PPARγ activation.[108] Another example is the binding of photosensitizers by GSTs.[109,110] Photosensitizers, used in photodynamic therapy, are typically hydrophobic, planar molecules with extensive conjugated double-bond systems, e.g., porphyrin derivatives. The compounds partition into membranes. When excited with light in the visible wavelength range, in the presence of oxygen they produce singlet oxygen and other reactive oxygen species (ROS) and cause cell injury or apoptosis. Hypericin, a photosensitizer derived from St. John's wort, is a noncatalytic ligand for hGSTA1-1 and hGSTP1-1.[110] The ROS yield of GST-bound hypericin is diminished by a mechanism that is not fully understood but is highly dependent on the GST isoform. Thus, GSTs may provide antioxidant protection via their liganding function, and cells overexpressing GSTs, such as many cancer cells, may be less sensitive to photodynamic therapy than cells with normal GST levels.[110]

Among the noncatalytic ligands of GSTs are steroids, bile acids, heme, bilirubin, and a variety of organic dyes, drugs, and other xenobiotics. Their binding often inhibits the catalytic function of the GSTs, although the mode of inhibition is typically complex, frequently involving noncompetitive effects.

Unlike the catalytic activity of GSTs, which utilizes a well-defined G-site and is less conserved but still a unique H-site, the binding sites for noncatalytic ligands appear to differ in number and location between enzymes and depend on the particular ligand. It is therefore not surprising that less is known about the noncatalytic binding sites. A cavity forms between the subunits of a dimer and has been proposed to serve as a binding site for hydrophobic ligands.[111] A potential binding site, normally occupied by structured water, has been identified in a high-resolution structure of hGSTA1-1.[112] Ligands in the cleft between subunits have been observed in crystal structures of Sigma-class GSTs.[113,114] Since intersubunit contacts are not only highly structured but also relevant to enzyme function, at least in the major Alpha, Mu, and Pi GSTs (Section 5.3.6), binding of a ligand may inhibit GST activity, as is often observed. In addition to, or instead of, utilizing a site in the cleft between subunits, nonsubstrate ligands may bind to sites within or partially over-lapping the H-site of a GST[115] and thus inhibit enzyme activity. Very potent bivalent inhibitors of GSTs have been designed to occupy simultaneously the H-sites of both subunits and the intersubunit cleft.[116] Finally, sites able to bind buffer molecules (MES or HEPES), identified in crystal structures of Pi-class GSTs, may serve as binding sites for noncatalytic ligands,[41,117] although it has been argued that these sites are too small to accommodate many known ligands and are too distant from the active sites to effect enzyme inhibition.[115] Indeed, there was no enzyme inhibition upon binding of bromosulfophthalein to a site in hGSTA1-1, which corresponds to the buffer-binding site of Pi-class GSTs, while concomitant binding of the same ligand to or near the H-site was associated with inhibition.[118]

Given the chemical and structural dissimilarity of GST ligands as well as structural differences between individual GSTs, the above findings are probably not contradictory. There is no single noncatalytic ligand-binding site; the binding is likely to be idiosyncratic with respect to both ligand and GST. The double duty of H-sites (for binding of substrates and noncatalytic ligands) is not surprising and probably inevitable. There is strong selective pressure to evolve permissive substrate-binding sites of broad specificity, to enable metabolism of the constantly changing repertoire of plant, microbial, and — more recently — industrial toxins. Recognition by the H-site is governed mostly by the size of the cavity, a general distribution of hydrophobic and polar surfaces, and, to a much lesser extent, by specific molecular interactions. It is therefore likely that some nonsubstrate molecules will satisfy the above requirements and will bind. In fact, their recognition would be more relaxed than that of substrates because no alignment of an electrophilic center with GSH is involved. Less obvious is the evolutionary origin of ligand-binding sites that do not overlap the H-site. GSTs may be well suited for carrying such sites because they are abundant, ubiquitous, and largely constitutively expressed.

5.4.2 MODULATION OF MAP KINASE SIGNALING VIA PROTEIN–PROTEIN INTERACTIONS

The structure of canonical GSTs may be viewed as a predisposing factor to inter-actions with other proteins. Canonical GSTs are stable dimers, but subunit swapping is readily achieved in the presence of moderate concentrations of organic solvents, which probably shield the hydrophobic intersubunit contact patches when subunits dissociate.[103] Heat shock proteins may replace the organic solvent *in vivo*. GSTs could be thus considered to exist in solution as a mixture of monomers and dimers, with the equilibrium shifted almost completely toward the dimer. Nevertheless, a GST monomer could associate with another protein if the target presents an appropriate contact surface. Differences in the subunit interface between GST classes increase the range of possible binding partners. The high abundance of GSTs in cells makes them attractive candidates for a buffer or sink for regulatory proteins, in analogy to the ligandin function of GSTs for small molecules. Binding of GSTs to unrelated proteins has been observed.

The direct interaction of Pi-class GSTs with c-Jun N-terminal kinase 1 (JNK1)[119,120] was an unexpected finding of considerable physiological interest. A monomeric GSTP1 subunit binds to the C-terminal portion of JNK1 with high affinity ($K_d \approx 200$ nM)[121] and blocks the activity of JNK1. The details of the binding are not fully known, although GSTP1 domains likely to be involved in the process have been identified.[122,123] In response to oxidative stress, the GSTP1 subunits form intermolecular disulfide bridges and oligomerize or aggregate to release active JNK1. Both rodent and human Pi-class GSTs carry out this function.[124] Thus, GSTP1 acts as a redox sensor, activating JNK1-mediated signaling in response to oxidative stress.

Stress-activated MAP kinase signaling is further regulated by a different GST, GSTM1-1, in two distinct ways. First, GSTM1-1 directly binds to the C-terminal portion of MEKK1 and inhibits its function.[125] MEKK1 is one of several MAP kinase kinase kinases in the JNK pathway.[126,127] The inhibitory action of GSTM1-1 on MEKK1 is independent of enzymatic activity of the GSTs because a catalytically inactive mutant is as effective in inhibiting the kinase as the wild-type GST.[125] Second, GSTM1-1 binds and inhibits ASK1,[128] a distinct MAP kinase kinase kinase that can signal through both the JNK and p38 pathways. The ASK1-GSTM1-1 complex dissociates in response to heat shock.[129] ASK1 is also inhibited by thioredoxin. In this case, oxidative stress rather than heat shock causes release of ASK1.[128] The structural details of GSTM1-1 binding to its target kinases have not yet been defined. For example, it is not known whether the binding partner is the monomeric or dimeric form of GSTM1. A scheme of the modulating roles of GSTs in stress-activated MAP kinase signaling is shown in Figure 5.11.

5.4.3 OTHER PROTEIN–PROTEIN INTERACTIONS

The heterodimerization of Pi-class GSTs with peroxiredoxin[105] has already been discussed (Section 5.3.6). A plant GST of the Tau class, named BI-GST, has been found to inhibit the pro-apoptotic action of Bax in yeast cells.[130] BI-GST can form complexes with other proteins but it is not clear whether Bax inhibition is mediated by such interaction or by the catalytic activity of BI-GST toward pro-oxidants.[131] The Fanconi anemia group C protein in hematopoietic cells binds to GSTP1-1 and prevents its oxidation, which in turn prevents apoptosis.[132] *Drosophila melanogaster* DmGSTS1-1,[133] a very abundant protein that constitutes 1–2% of all soluble proteins in the fly,[59] is present in the asynchronous indirect flight muscle where it is attached to the myofibrils.[134] The attachment is via a 46-amino acid N-terminal extension of the GST. The extension is rich in Ala, Pro, Gly, and Glu, and may interact with a homologous C-terminal extension of troponin H.[134] Thus, the mode of binding is distinct from that of GSTP1 to JNK. The function of DmGSTS1-1 in the asynchro-

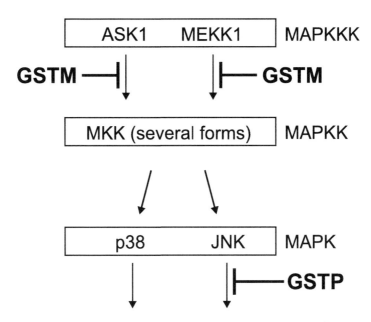

FIGURE 5.11 Inhibitory action of GSTs in stress-activated MAP kinase signaling: MAPK, MAP kinase; MAPKK, MAP kinase kinase; MAPKKK, MAP kinase kinase kinase. Only those MAPKKK that are known to be modulated by GSTs are shown.

nous indirect flight muscle is not known. Its binding to troponin H may place it close to the myosin cross-bridge where it could participate in the stretch-sensing mechanism.[134] Alternatively or in addition, the catalytic activity of DmGSTS1-1 toward 4-HNE[59] could be physiologically important to counter oxidative stress in a highly aerobic tissue. Tissue transglutaminase associates with GSTP1-1 and causes its polymerization.[135] Finally, use of GSTs or GST fragments in unbiased screening techniques such as phage display,[136] yeast two-hybrid system,[131] or direct pull-down[137] have uncovered interactions of GSTs with multiple cellular proteins. Elucidation of the functional significance of such binding will require further study.

5.4.4 GSTs as Nitric Oxide Carriers

Nitric oxide is a relatively short-lived signaling molecule. Its half-life is significantly increased in a complex with glutathione and iron (dinitrosyl-diglutathionyl-iron complex, or DNDGIC), but even then it does not exceed a few minutes. It has been shown that DNDGIC binds with high affinity (K_d between 0.1 and 1 nM) to Alpha-, Mu-, and Pi-class GSTs. The binding stabilizes the complex and increases its half-life to several hours.[138–140] The phenomenon was observed *in vitro* as well as in tissue homogenates that mimic *in vivo* conditions. Binding of DNDGIC induces a negative cooperativity between the subunits of the GST. Depending on the isoform, the second subunit may be functional or inhibited in terms of its ability to catalyze conjugation reactions. On the basis of these results, it has been postulated that GSTs are intracellular nitric oxide carriers,[138–140] in analogy to the function of hemoglobin in the circulation.[141] Storage of complexed nitric oxide resembles the noncatalytic binding of ligands to GSTs (Section 5.4.1). It is discussed separately because, unlike the ligandin function, it appears to involve the G-site as its primary binding site.[138]

5.5   GST INHIBITORS

There is considerable interest in GST inhibitors due to two possible clinical applications. The first is the treatment of parasitic infections including helminths, which affect two billion people worldwide, and malaria, which causes 300 million acute episodes per year. Worm and plasmodial GSTs are distinct from those of the human host and therefore constitute potential selective therapeutic targets. The structures of several GSTs from parasitic organisms have been solved[113,142–146] and are likely to assist in rational design of specific inhibitors. The second possible use of GST inhibitors is in cancer therapy. Many drug-resistant cancers overexpress GSTs, in particular Pi-class proteins. Although increased metabolism of anticancer drugs by GSTs may contribute to drug resistance, often the drug is not a GST substrate. An alternative mechanism by which overexpression of GSTs could protect cancer cells is the modulation of signaling cascades. As described in Section 5.4.2, Pi-class GSTs can associate with JNK and maintain it in an inactive (storage) form. GST inhibitors can disrupt the interaction between GST and JNK, activate the kinase, and thus affect the mitogenic and apoptotic responses.[147,148] Mu-class GSTs similarly inhibit the pro-apoptotic kinase ASK1.[128] Drugs able to dissociate the GST-ASK complex could be useful in cancer therapy.

GST inhibitors fall into three broad groups.[72,116,149,150] The first group is comprised of nonsubstrate ligands that bind to noncatalytic sites in GSTs. As already discussed in Section 5.4.1, such binding frequently inhibits the enzyme. However, these inhibitors typically lack potency and specificity. The other two groups of inhibitors are glutathione analogs that bind to the G-site and hydrophobic compounds that bind to the H-site of the enzyme. Certain compounds, including glutathione conjugates that are GST reaction products, recognize the G- and H-sites simultaneously. This explains why GSTs are often strongly inhibited by their own products, and it emphasizes the need for efficient transport or further metabolism of the conjugates. Inhibitors that target the active site of GSTs achieve relatively low K_i values, typically in the low micromolar range.[149,151] An interesting variant is a bifunctional inhibitor, which binds simultaneously to both active sites of a dimer. This increases binding affinity and GST isoform selectivity of the inhibitor. The most potent in a series of such compounds had a K_i of approximately 40 nM for GSTP1-1 and a selectivity for hGSTP1-1 over hGSTA1-1 of at least three orders of magnitude.[116] The chemistry of GST inhibitors has been recently reviewed.[150]

5.6   ENGINEERED GSTs WITH NEW FUNCTIONS

GSTs are relatively simple enzymes. The role of the protein in catalysis is limited to an activation of GSH and, for the most part, a proper positioning of the electrophilic cosubstrate in a rather permissive H-site. It is therefore not surprising that the catalytic efficiencies of GSTs are several orders of magnitude short of the "perfectly evolved enzymes," which are limited by the diffusion rates of substrates and products rather than by the catalytic event itself.[152,153] What they lack in mechanistic sophistication, GSTs make up for in versatility. While the binding to the G-site is very specific, the requirements of binding to the H-site are quite relaxed. This permits the enzymes to metabolize a wide variety of substrates, including those the cell was not faced with previously (e.g., synthetic drugs or industrial pollutants).

The versatility of the GST scaffold lends itself to modifications that could yield new catalytic activities. To this end, two conceptually distinct approaches have been used: knowledge-based (rational) design and *in vitro* evolution (arational design, as defined in reference 154), although actual experiments typically utilized elements of both of the above extremes. DNA shuffling,[155,156] which mimics homologous recombination, is an example of the arational approach. By this method, enzymes were generated from hGSTM1-1 and hGSTM2-2,[157] hGSTT1-1 and rGSTT2-2,[158] and maize Tau-class ZmGSTU1-1 and ZmGSTU2-2.[159] Screening of these proteins revealed forms with changed or enhanced catalytic properties. Statistical formalisms have been developed to guide and optimize directed evolution based on DNA reshuffling. The method yielded impressive results, such as a 65-fold increase of activity over the parental enzyme.[160] A methodologically different approach to *in vitro* evolution was the generation of a phage display library of active-site mutants of hGSTA1-1, followed by selection on an immobilized transition-state (Meisenheimer-complex) analog.[161] The choice of active-site residues was based on the crystal structure of the enzyme and prior knowledge of the mechanism of the catalytic event. A method has been developed that dispenses with such prior knowledge but requires sequence information on a family of homologous proteins with different substrate specificities.[162] Comparison of the sequences identifies hypervariable "hot spots," which are then targeted for random mutagenesis. A method that comes closest to a fully knowledge-based (rational) design is comparing structures and sequences of two enzymes that differ in specificity and engineering one to carry out the functions of the other. The redesign is based on structural comparisons and involves mutating of amino acids of one protein to those found in corresponding positions in the other, ideally in conjunction with an understanding of which residues and motifs are relevant to the activity of interest. As applied to GSTs, the approach has been used to graft onto hGSTA1-1 an activity (4-HNE conjugation) that is characteristic of hGSTA4-4.[163] In another example, hGSTA2-2, which has high glutathione peroxidase activity, was converted into an enzyme with high steroid isomerase activity by mutations patterned on hGSTA3-3.[79] A single amino acid mutation in the H-site of hGSTA2-2 led to an increase in catalytic efficiency for BPDE to the level of hGSTA1-1 on which the mutation was modeled, and a protein with three additional mutations actually exceeded the activity of hGSTA1-1 (see reference 68 and Section 5.3.1).

In the examples summarized above, GSTs were mutated to acquire the activities of other existing GSTs. GSTs can, however, be engineered to gain entirely new activities. Introduction of a single histidine residue in hGSTA1-1 gives rise to an unprecedented thiol ester hydrolase activity.[164] Chemical conversion of serine residues in rGSTT2-2, presumably including the active-site serine, to selenocysteine converts the enzyme into a glutathione peroxidase able to metabolize H_2O_2, a reaction of which GSTs are not capable, with an efficiency equal to that of natural selenium-dependent glutathione peroxidases.[165] The same shift of activity was achieved by a selective replacement of the active-site Ser-9 by selenocysteine in *Lucilia cuprina* Delta-class GST[166] (note that the *L. cuprina* enzyme, originally designated as a member of the Theta class, has been reclassified as Delta[167]).

The relatively facile laboratory generation, sometimes by a single mutation or a combination of a few mutations, of GSTs with enhanced activities raises an obvious question. Why have such mutations not become fixed in the population of their corresponding species, even though they must have occurred many times on an evolutionary time scale? The answer may lie in the differences in optimization criteria between biological evolution and laboratory experiments. The latter are guided by curiosity regarding mechanisms and by potential technological utility of engineered enzymes. In contrast, biological optimization balances many more factors. Catalytic efficiency for a particular substrate is certainly one of these factors, but others may include activity for other, sometimes dissimilar, substrates, ability to sequester reaction products,[168] minimal interference with intermediary metabolism, stability of the protein, or its ability to interact with and modulate other proteins within the cell. Optimization of a GST with regard to multiple constraints may result in an enzyme that is suboptimal with respect to any single criterion. On a different level, it may well be true that more is not always better. For example, mGSTA4-4 and hGSTA4-4 metabolize 4-HNE, a compound that is both a toxicant and a signaling molecule (as discussed elsewhere in this volume). An optimal activity for GSTA4-4 may be in a range sufficient for detoxification but one low enough not to interfere with signaling. The existence of biological constraints indicates that there is room for "improvement" of GSTs for specific laboratory or industrial applications.

ACKNOWLEDGMENT

The author's work related to GSTs is supported by National Institutes of Health grants AG 18845 and ES 07804 (to P.Z.).

REFERENCES

1. Armstrong, R.N., Structure, catalytic mechanism, and evolution of the glutathione transferases, *Chem. Res. Toxicol.* 10, 2, 1997.
2. Ladner, J.E. et al., Parallel evolutionary pathways for glutathione transferases: structure and mechanism of the mitochondrial class kappa enzyme rGSTK1-1, *Biochemistry* 43, 352, 2004.
3. Jowsey, I.R. et al., Biochemical and genetic characterization of a murine-class Kappa glutathione S-transferase, *Biochem. J.* 373, 559, 2003.
4. Morgenstern, R., Guthenberg, C., and DePierre, J.W., Microsomal glutathione S-transferase. Purification, initial characterization and demonstration that it is not identical to the cytosolic glutathione S-transferases A, B and C, *Eur. J. Biochem.* 128, 243, 1982.
5. Bernat, B.A., Laughlin, L.T., and Armstrong, R.N., Fosfomycin resistance protein (FosA) is a manganese metalloglutathione transferase related to glyoxalase I and the extradiol dioxygenases, *Biochemistry* 36, 3050, 1997.

6. Nakamura, H., Thioredoxin as a key molecule in redox signaling, *Antioxid. Redox Signal.* 6, 15, 2004.

7. Fernandes, A.P. and Holmgren, A., Glutaredoxins: glutathione-dependent redox enzymes with functions far beyond a simple thioredoxin backup system, *Antioxid. Redox Signal.* 6, 63, 2004.

8. Martin, J.L., Thioredoxin — a fold for all reasons, *Structure* 3, 245, 1995.

9. Vuilleumier, S. and Pagni, M., The elusive roles of bacterial glutathione S-transferases: new lessons from genomes, *Appl. Microbiol. Biotechnol.* 58, 138, 2002.

10. Dixon, D.P., Lapthorn, A., and Edwards, R., Plant glutathione transferases, *Genome Biology* 3, reviews 3004.1, 2002.

11. Jakoby, W.B., The glutathione S-transferases: a group of multifunctional detoxification proteins, *Adv. Enzymol. Rel. Areas Mol. Biol.* 46, 383, 1978.

12. Mannervik, B., The isoenzymes of glutathione transferase, *Adv. Enzymol. Rel. Areas Mol. Biol.* 57, 357, 1985.

13. Mannervik, B. and Danielson, U.H., Glutathione transferases — structure and catalytic activity, *CRC Crit. Rev. Biochem.* 23, 283, 1988.

14. Armstrong, R.N., Glutathione S-transferases: reaction mechanism, structure, and function, *Chem. Res. Toxicol.* 4, 131, 1991.

15. Rushmore, T.H. and Pickett, C.B., Glutathione S-transferases, structure, regulation, and therapeutic implications, *J. Biol. Chem.* 268, 11475, 1993.

16. Armstrong, R.N., Glutathione S-transferases: structure and mechanism of an archetypical detoxication enzyme, *Adv. Enzymol. Rel. Areas Mol. Biol.* 69, 1, 1994.

17. Wilce, M.C. and Parker, M.W., Structure and function of glutathione S-transferases, *Biochim. Biophys. Acta* 1205, 1, 1994.

18. Hayes, J.D. and Pulford, D.J., The glutathione S-transferase supergene family: regulation of GST and the contribution of the isoenzymes to cancer chemoprotection and drug resistance, *Crit. Rev. Biochem. Mol. Biol.* 30, 445, 1995.

19. Raha, A. and Tew, K.D., Glutathione S-transferases, *Cancer Treat. Res.* 87, 83, 1996.

20. Armstrong, R.N., Mechanistic imperatives for the evolution of glutathione transferases, *Curr. Opin. Chem. Biol.* 2, 618, 1998.

21. Eaton, D.L. and Bammler, T.K., Concise review of the glutathione S-transferases and their significance to toxicology, *Toxicol. Sci.* 49, 156, 1999.

22. Salinas, A.E. and Wong, M.G., Glutathione S-transferases — a review, *Curr. Med. Chem.* 6, 279, 1999.

23. Sheehan, D. et al., Structure, function and evolution of glutathione transferases: implications for classification of non-mammalian members of an ancient enzyme superfamily, *Biochem. J.* 360 (Part 1), 1, 2001.

24. Hayes, J.D., Flanagan, J.U., and Jowsey, I.R., Glutathione transferases, *Annu. Rev. Pharm. Toxicol.* 45, 51, 2005.

25. Koonin, E.V. et al., Eukaryotic translation elongation factor 1γ contains a glutathione transferase domain — study of a diverse, ancient protein superfamily using motif search and structural modeling, *Protein Sci.* 3, 2045, 1994.

26. Jeppesen, M.G. et al., The crystal structure of the glutathione S-transferase-like domain of elongation factor 1Bγ from *Saccharomyces cerevisiae*, *J. Biol. Chem.* 278, 47190, 2003.

27. Cromer, B.A. et al., From glutathione transferase to pore in a CLIC, *Eur. Biophys. J.* 31 (5), 356, 2002.

28. Tulk, B.M., Kapadia, S., and Edwards, J.C., CLIC1 inserts from the aqueous phase into phospholipid membranes, where it functions as an anion channel, *Am. J. Physiol. Cell Physiol.* 282 (5), C1103, 2002.

29. Watson, W.H. et al., Thioredoxin and its role in toxicology, *Toxicol. Sci.* 78, 3, 2004.
30. Gromer, S., Urig, S., and Becker, K., The thioredoxin system — from science to clinic, *Med. Res. Rev.* 24, 40, 2004.
31. Rossjohn, J. et al., A mixed disulfide bond in bacterial glutathione transferase: functional and evolutionary implications, *Structure* 6, 721, 1998.
32. Caccuri, A.M. et al., GSTB1-1 from *Proteus mirabilis*: a snapshot of an enzyme in the evolutionary pathway from a redox enzyme to a conjugating enzyme, *J. Biol. Chem.* 277, 18777, 2002.
33. Caccuri, A.M. et al., Properties and utility of the peculiar mixed disulfide in the bacterial glutathione transferase B1-1, *Biochemistry* 41, 4686, 2002.
34. Board, P.G. et al., Identification, characterization, and crystal structure of the omega class glutathione transferases, *J. Biol. Chem.* 275 (32), 24798, 2000.
35. Chen, W.J., Graminski, G.F., and Armstrong, R.N., Dissection of the catalytic mechanism of isozyme 4-4 of glutathione S-transferase with alternative substrates, *Biochemistry* 27, 647, 1988.
36. Graminski, G.F., Kubo, Y., and Armstrong, R.N., Spectroscopic and kinetic evidence for the thiolate anion of glutathione at the active site of glutathione S-transferase [published erratum appears in *Biochemistry* 1989 Aug 22; 28 (17):7138], *Biochemistry* 28, 3562, 1989.
37. Parsons, J.F. and Armstrong, R.N., Proton configuration in the ground state and transition state of a glutathione transferase–catalyzed reaction inferred from the properties of tetradeca(3-fluorotyrosyl)glutathione transferase, *J. Am. Chem. Soc.* 118, 2295, 1996.
38. Thorson, J.S. et al., Analysis of the role of the active site tyrosine in human glutathione transferase A1-1 by unnatural amino acid mutagenesis, *J. Am. Chem. Soc.* 120, 451, 1998.
39. Board, P.G. et al., Evidence for an essential serine residue in the active site of the theta class glutathione transferases, *Biochem. J.* 311, 247, 1995.
40. Bjornestedt, R. et al., Functional significance of arginine 15 in the active site of human alpha glutathione transferase A1-1, *J. Mol. Biol.* 247, 765, 1995.
41. Ji, X. et al., Structure and function of the xenobiotic substrate-binding site and location of a potential non-substrate-binding site in a class π glutathione S-transferase, *Biochemistry* 36, 9690, 1997.
42. Ricci, G. et al., Structural flexibility modulates the activity of human glutathione transferase P1-1. Role of helix 2 flexibility in the catalytic mechanism, *J. Biol. Chem.* 271, 16187, 1996.
43. Hitchens, T.K., Mannervik, B., and Rule, G.S., Disorder-to-order transition of the active site of human class Pi glutathione transferase, GSTP1-1, *Biochemistry* 40 (39), 11660, 2001.
44. Caccuri, A.M. et al., Proton release upon glutathione binding to glutathione transferase P1-1 — kinetic analysis of a multistep glutathione binding process, *Biochemistry* 37, 3028, 1998.
45. Caccuri, A.M. et al., Proton release on binding of glutathione to Alpha, Mu, and Delta class glutathione transferases, *Biochem. J.* 344, 419, 1999.
46. Caccuri, A.M. et al., Human glutathione transferase T2-2 discloses some evolutionary strategies for optimization of substrate binding to the active site of glutathione transferases, *J. Biol. Chem.* 276 (8), 5427, 2001.
47. Svensson, R. et al., Kinetic characterization of thiolate anion formation and chemical catalysis of activated microsomal glutathione transferase 1, *Biochemistry* 43, 8869, 2004.

48. Schmidt-Krey, I. et al., The three-dimensional map of microsomal glutathione transferase 1 at 6 angstrom resolution, *Embo. J.* 19, 6311, 2000.

49. Holm, P.J., Morgenstern, R., and Hebert, H., The 3-D structure of microsomal glutathione transferase 1 at 6 angstrom resolution as determined by electron crystallography of p22(1)2(1) crystals, *Biochim. Biophys. Acta* 1594, 276, 2002.

50. Caccuri, A.M. et al., Human glutathione transferase T2-2 discloses some evolutionary strategies for optimization of the catalytic activity of glutathione transferases, *J. Biol. Chem.* 276 (8), 5432, 2001.

51. Zimniak, P. et al., A subgroup of class alpha glutathione S-transferases: cloning of cDNA for mouse lung glutathione S-transferase GST 5.7, *FEBS Lett.* 313, 173, 1992.

52. Zimniak, P. et al., Estimation of genomic complexity, heterologous expression, and enzymatic characterization of mouse glutathione S-transferase mGSTA4-4 (GST 5.7), *J. Biol. Chem.* 269, 992, 1994.

53. Hubatsch, I., Ridderstrom, M., and Mannervik, B., Human glutathione transferase A4-4: an alpha class enzyme with high catalytic efficiency in the conjugation of 4-hydroxynonenal and other genotoxic products of lipid peroxidation, *Biochem. J.* 330, 175, 1998.

54. Nanduri, B. et al., Amino acid residue 104 in an alpha-class glutathione S-transferase is essential for the high selectivity and specificity of the enzyme for 4-hydroxynonenal, *Arch. Biochem. Biophys.* 335, 305, 1996.

55. Xiao, B. et al., Crystal structure of a murine glutathione S-transferase in complex with a glutathione conjugate of 4-hydroxynon-2-enal in one subunit and glutathione in the other: evidence of signaling across the dimer interface, *Biochemistry* 38, 11887, 1999.

56. Ji, X. et al., Structure and function of the xenobiotic substrate binding site of a glutathione S-transferase as revealed by X-ray crystallographic analysis of product complexes with the diastereomers of 9-(S-glutathionyl)-10-hydroxy-9,10,dihydrophenanthrene, *Biochemistry* 33, 1043, 1994.

57. Bruns, C.M. et al., Human glutathione transferase A4-4 crystal structures and mutagenesis reveal the basis of high catalytic efficiency with toxic lipid peroxidation products, *J. Mol. Biol.* 288, 427, 1999.

58. Engle, M.R. et al., Invertebrate glutathione transferases conjugating 4-hydroxynonenal: CeGST 5.4 from *Caenorhabditis elegans*, *Chem. Biol. Interact.* 133, 244, 2001.

59. Singh, S.P. et al., Catalytic function of *Drosophila melanogaster* glutathione S-transferase DmGSTS1-1 (GST-2) in conjugation of lipid peroxidation end products, *Eur. J. Biochem.* 268, 2912, 2001.

60. Sawicki, R. et al., Cloning, expression, and biochemical characterization of one epsilon-class (GST-3) and ten delta-class (GST-1) glutathione S-transferases from *Drosophila melanogaster*, and identification of additional nine members of the Epsilon class, *Biochem. J.* 370, 661, 2003.

61. Kopp, J. and Schwede, T., The Swiss-model repository of annotated three-dimensional protein structure homology models, *Nucl. Acids Res.* 32, D230, 2004.

62. Board, P.G., Webb, G.C., and Coggan, M., Isolation of a cDNA clone and localization of the human glutathione S-transferase 3 genes to chromosome bands 11q13 and 12q13-14, *Ann. Human Genet.* 53, 205, 1989.

63. Ahmad, H. et al., Primary and secondary structural analyses of glutathione S-transferase pi from human placenta, *Arch. Biochem. Biophys.* 278, 398, 1990.

64. Zimniak, P. et al., Naturally occurring human glutathione S-transferase GSTP1-1 isoforms with isoleucine and valine in position 104 differ in enzymic properties, *Eur. J. Biochem.* 224, 893, 1994.

65. Hu, X. et al., Activity of four allelic forms of glutathione S-transferase hGSTP1-1 for diol epoxides of polycyclic aromatic hydrocarbons, *Biochem. Biophys. Res. Commun.* 238, 397, 1997.

66. Sundberg, K. et al., Detoxication of carcinogenic fjord-region diol epoxides of polycyclic aromatic hydrocarbons by glutathione transferase P1-1 variants and glutathione, *FEBS Lett.* 438, 206, 1998.

67. Ji, X.H. et al., Structure and function of residue 104 and water molecules in the xenobiotic substrate-binding site in human glutathione S-transferase P1-1, *Biochemistry* 38, 10231, 1999.

68. Singh, S.V. et al., Structural basis for catalytic differences between α class human glutathione transferases hGSTA1-1 and hGSTA2-2 for glutathione conjugation of environmental carcinogen benzo[*a*]pyrene-7,8-diol-9,10-epoxide, *Biochemistry* 43, 9708, 2004.

69. Collinson, E.J. and Grant, C.M., Role of yeast glutaredoxins as glutathione S-transferases, *J. Biol. Chem.* 278, 22492, 2003.

70. Luikenhuis, S. et al., The yeast *Saccharomyces cerevisiae* contains two glutaredoxin genes that are required for protection against reactive oxygen species, *Mol. Biol. Cell* 9, 1081, 1998.

71. Shami, P.J. et al., JS-K, a glutathione/glutathione S-transferase–activated nitric oxide donor of the diazeniumdiolate class with potent antineoplastic activity, *Mol. Cancer Ther.* 2 (4), 409, 2003.

72. Townsend, D.M. and Tew, K.D., The role of glutathione-S-transferase in anti-cancer drug resistance, *Oncogene* 22 (47), 7369, 2003.

73. Findlay, V.J. et al., Tumor cell responses to a novel glutathione S-transferase–activated nitric oxide-releasing prodrug, *Mol. Pharmacol.* 65, 1070, 2004.

74. Zhao, T. et al., The role of human glutathione S-transferases hGSTA1-1 and hGSTA2-2 in protection against oxidative stress, *Arch. Biochem. Biophys.* 367, 216, 1999.

75. Yang, Y. et al., Role of glutathione S-transferases in protection against lipid peroxidation. Overexpression of hGSTA2-2 in K562 cells protects against hydrogen peroxide–induced apoptosis and inhibits JNK and caspase 3 activation, *J. Biol. Chem.* 276, 19220, 2001.

76. Benson, A.M. et al., Relationship between the soluble glutathione-dependent delta 5-3-ketosteroid isomerase and the glutathione S-transferases of the liver, *Proc. Natl. Acad. Sci. USA* 74, 158, 1977.

77. Johansson, A.S. and Mannervik, B., Human glutathione transferase A3-3, a highly efficient catalyst of double-bond isomerization in the biosynthetic pathway of steroid hormones, *J. Biol. Chem.* 276 (35), 33061, 2001.

78. Johansson, A.S. and Mannervik, B., Active-site residues governing high steroid isomerase activity in human glutathione transferase A3-3, *J. Biol. Chem.* 277 (19), 16648, 2002.

79. Petterson, P.L., Johansson, A.S., and Mannervik, B., Transmutation of human glutathione transferase A2-2 with peroxidase activity into an efficient steroid isomerase, *J. Biol. Chem.* 277 (33), 30019, 2002.

80. Gu, Y. et al., Crystal structure of human glutathione S-transferase A3-3 and mechanistic implications for its high steroid isomerase activity, *Biochemistry* 43, 15673, 2004.

81. Fernandez-Canon, J.M. and Penalva, M.A., Characterization of a fungal maleylacetoacetate isomerase gene and identification of its human homologue, *J. Biol. Chem.* 273, 329, 1998.

82. Thom, R. et al., The structure of a zeta class glutathione S-transferase from *Arabidopsis thaliana*: Characterisation of a GST with novel active-site architecture and a putative role in tyrosine catabolism, *J. Mol. Biol.* 308 (5), 949, 2001.

83. Polekhina, G. et al., Crystal structure of maleylacetoacetate isomerase/glutathione transferase zeta reveals the molecular basis for its remarkable catalytic promiscuity, *Biochemistry* 40 (6), 1567, 2001.

84. Board, P.G. et al., Clarification of the role of key active site residues of glutathione transferase zeta/maleylacetoacetate isomerase by a new spectrophotometric technique, *Biochem. J.* 374, 731, 2003.

85. Chen, H. and Juchau, M.R., Glutathione S-transferases act as isomerases in isomerization of 13-*cis*-retinoic acid to all-*trans*-retinoic acid *in vitro*, *Biochem. J.* 327, 721, 1997.

86. Chen, H. and Juchau, M.R., Recombinant human glutathione S-transferases catalyse enzymic isomerization of 13-cis-retinoic acid to all-trans-retinoic acid *in vitro*, *Biochem. J.* 336, 223, 1998.

87. Meyer, D.J. and Thomas, M., Characterization of rat spleen prostaglandin H D-isomerase as a sigma-class GSH transferase, *Biochem. J.* 311, 739, 1995.

88. Jowsey, I.R. et al., Mammalian class sigma glutathione S-transferases: catalytic properties and tissue-specific expression of human and rat GSH-dependent prostaglandin D2 synthases, *Biochem. J.* 359, 507, 2001.

89. Kanaoka, Y. et al., Cloning and crystal structure of hematopoietic prostaglandin D synthase, *Cell* 90, 1085, 1997.

90. Jakobsson, P.J. et al., Identification of human prostaglandin E synthase: a microsomal, glutathione-dependent, inducible enzyme, constituting a potential novel drug target, *Proc. Natl. Acad. Sci. USA* 96, 7220, 1999.

91. Beuckmann, C.T. et al., Identification of mu-class glutathione transferases M2-2 and M3-3 as cytosolic prostaglandin E synthases in the human brain, *Neurochem. Res.* 25, 733, 2000.

92. Bogaards, J.J., Venekamp, J.C., and van Bladeren, P.J., Stereoselective conjugation of prostaglandin A2 and prostaglandin J2 with glutathione, catalyzed by the human glutathione S-transferases A1-1, A2-2, M1a-1a, and P1-1, *Chem. Res. Toxicol.* 10, 310, 1997.

93. Lam, B.K. and Austen, K.F., Leukotriene C4 synthase: a pivotal enzyme in cellular biosynthesis of the cysteinyl leukotrienes, *Prostaglandins Other Lipid Mediat.* 68–69, 511, 2002.

94. Funk, C.D., Prostaglandins and leukotrienes: advances in eicosanoid biology, *Science* 294, 1871, 2001.

95. Lam, B.K. et al., Site-directed mutagenesis of human leukotriene C4 synthase, *J. Biol. Chem.* 272, 13923, 1997.

96. Girardini, J. et al., Characterization of an omega-class glutathione S-transferase from *Schistosoma mansoni* with glutaredoxin-like dehydroascorbate reductase and thiol transferase activities, *Eur. J. Biochem.* 269, 5512, 2002.

97. Pettigrew, N.E. and Colman, R.F., Heterodimers of glutathione S-transferase can form between isoenzyme classes pi and mu, *Arch. Biochem. Biophys.* 396, 225, 2001.

98. Vargo, M.A., Nguyen, L., and Colman, R.F., Subunit interface residues of glutathione S-transferase A1-1 that are important in the monomer-dimer equilibrium, *Biochemistry* 43, 3327, 2004.

99. Wang, J., Bauman, S., and Colman, R.F., Photoaffinity labeling of rat liver glutathione S-transferase, 4-4, by glutathionyl S-[4-(succinimidyl)-benzophenone], *Biochemistry* 37, 15671, 1998.

100. Wang, J.B., Bauman, S., and Colman, R.F., Probing subunit interactions in alpha class rat liver glutathione S-transferase with the photoaffinity label glutathionyl S-[4-(succinimidyl)benzophenone], *J. Biol. Chem.* 275, 5493, 2000.

101. Vargo, M.A. and Colman, R.F., Affinity labeling of rat glutathione S-transferase isozyme 1-1 by 17β-iodoacetoxy-estradiol-3-sulfate, *J. Biol. Chem.* 276, 2031, 2001.

102. Hegazy, U.M., Mannervik, B., and Stenberg, G., Functional role of the lock and key motif at the subunit interface of glutathione transferase P1-1, *J. Biol. Chem.* 279 (10), 9586, 2004.

103. Vargo, M.A. and Colman, R.F., Heterodimers of wild-type and subunit interface mutant enzymes of glutathione S-transferase A1-1: interactive or independent active sites? *Protein Sci.* 13, 1586, 2004.

104. Ricci, G. et al., Site-directed mutagenesis of human glutathione transferase P1-1. Mutation of Cys-47 induces a positive cooperativity in glutathione transferase P1-1, *J. Biol. Chem.* 270, 1243, 1995.

105. Manevich, Y., Feinstein, S.I., and Fisher, A.B., Activation of the antioxidant enzyme 1-CYS peroxiredoxin requires glutathionylation mediated by heterodimerization with πGST, *Proc. Natl. Acad. Sci. USA* 101, 3780, 2004.

106. Litwack, G., Ketterer, B., and Arias, I.M., Ligandin: a hepatic protein which binds steroids, bilirubin, carcinogens and a number of exogenous organic anions, *Nature* 234, 466, 1971.

107. Habig, W.H. et al., The identity of glutathione S-transferase B with ligandin, a major binding protein of liver, *Proc. Natl. Acad. Sci.* 71, 3879, 1974.

108. Paumi, C.M. et al., Glutathione S-transferases (GSTs) inhibit transcriptional activation by the peroxisomal proliferator-activated receptor γ (PPARγ) ligand, 15-deoxy-Δ12,14prostaglandin J2 (15-d-PGJ2), *Biochemistry* 43, 2345, 2004.

109. Smith, A., Nuiry, I., and Awasthi, Y.C., Interactions with glutathione S-transferases of porphyrins used in photodynamic therapy and naturally occurring porphyrins, *Biochem. J.* 229, 823, 1985.

110. Lu, W.D. and Atkins, W.M., A novel antioxidant role for ligandin behavior of glutathione S-transferases: attenuation of the photodynamic effects of hypericin, *Biochemistry* 43 (40), 12761, 2004.

111. Reinemer, P. et al., The three-dimensional structure of class pi glutathione S-transferase in complex with glutathione sulfonate at 2.3: a resolution, *Embo. J.* 10, 1997, 1991.

112. Le Trong, I. et al., 1.3-Angstrom resolution structure of human glutathione S-transferase with S-hexyl glutathione bound reveals possible extended ligandin binding site, *Proteins* 48 (4), 618, 2002.

113. McTigue, M.A., Williams, D.R., and Tainer, J.A., Crystal structures of a schistosomal drug and vaccine target: glutathione S-transferase from *Schistosoma japonica* and its complex with the leading antischistosomal drug Praziquantel, *J. Mol. Biol.* 246, 21, 1995.

114. Ji, X.H. et al., Location of a potential transport binding site in a sigma class glutathione transferase by x-ray crystallography, *Proc. Natl. Acad. Sci. USA* 93, 8208, 1996.

115. Oakley, A.J. et al., The ligandin (non-substrate) binding site of human pi class glutathione transferase is located in the electrophile binding site (h-site), *J. Mol. Biol.* 291, 913, 1999.

116. Lyon, R.P., Hill, J.J., and Atkins, W.M., Novel class of bivalent glutathione S-transferase inhibitors, *Biochemistry* 42 (35), 10418, 2003.

117. Prade, L. et al., Structures of class pi glutathione S-transferase from human placenta in complex with substrate, transition-state analogue and inhibitor, *Structure* 5, 1287, 1997.
118. Kolobe, D., Sayed, Y., and Dirr, H.W., Characterization of bromosulphophthalein binding to human glutathione S-transferase A1-1: thermodynamics and inhibition kinetics, *Biochem. J.* 382, 703, 2004.
119. Adler, V. et al., Regulation of JNK signaling by GSTp, *Embo. J.* 18, 1321, 1999.
120. Yin, Z.M. et al., Glutathione S-transferase p elicits protection against H_2O_2-induced cell death via coordinated regulation of stress kinases, *Cancer Res.* 60, 4053, 2000.
121. Wang, T. et al., Glutathione S-transferase P1-1 (GSTP1-1) inhibits c-Jun N-terminal kinase (JNK1) signaling through interaction with the C terminus, *J. Biol. Chem.* 276, 20999, 2001.
122. Chie, L. et al., An effector peptide from glutathione-S-transferase-pi strongly and selectively blocks mitotic signaling by oncogenic ras-p21, *Protein J.* 23, 235, 2004.
123. Adler, V. and Pincus, M.R., Effector peptides from glutathione-S-transferase-pi affect the activation of jun by jun-N-terminal kinase, *Ann. Clin. Lab. Sci.* 34, 35, 2004.
124. Ranganathan, P.N., Whalen, R., and Boyer, T.D., Characterization of the molecular forms of glutathione S-transferase P1 in human gastric cancer cells (Kato III) and in normal human erythrocytes, *Biochem. J.* 386, 525, 2005.
125. Ryoo, K. et al., Negative regulation of MEKK1-induced signaling by glutathione S-transferase mu, *J. Biol. Chem.* 279 (42), 43589, 2004.
126. Davis, R.J., Signal transduction by the JNK group of MAP kinases, *Cell* 103, 239, 2000.
127. Pearson, G. et al., Mitogen-activated protein (MAP) kinase pathways: regulation and physiological functions, *Endocr. Rev.* 22 (2), 153, 2001.
128. Cho, S.G. et al., Glutathione S-transferase mu modulates the stress-activated signals by suppressing apoptosis signal-regulating kinase 1, *J. Biol. Chem.* 276, 12749, 2001.
129. Dorion, S., Lambert, H., and Landry, J., Activation of the p38 signaling pathway by heat shock involves the dissociation of glutathione S-transferase mu from Ask1, *J. Biol. Chem.* 277 (34), 30792, 2002.
130. Kampranis, S.C. et al., A novel plant glutathione S-transferase/peroxidase suppresses Bax lethality in yeast, *J. Biol. Chem.* 275, 29207, 2000.
131. Kilili, K.G. et al., Differential roles of tau class glutathione S-transferases in oxidative stress, *J. Biol. Chem.* 279, 24540, 2004.
132. Cumming, R.C. et al., Fanconi anemia group C protein prevents apoptosis in hemato-poietic cells through redox regulation of GSTP1, *Nature Med.* 7, 814, 2001.
133. Beall, C. et al., Isolation of a *Drosophila* gene encoding glutathione S-transferase, *Biochem. Genet.* 30, 515, 1992.
134. Clayton, J.D. et al., Interaction of troponin-H and glutathione S-transferase-2 in the indirect flight muscles of *Drosophila melanogaster*, *J. Muscle Res. Cell Motil.* 19, 117, 1998.
135. Piredda, L. et al., Identification of "tissue" transglutaminase binding proteins in neural cells committed to apoptosis, *FASEB J.* 13, 355, 1999.
136. Edalat, M., Persson, M.A.A., and Mannervik, B., Selective recognition of peptide sequences by glutathione transferases: a possible mechanism for modulation of cellular stress-induced signaling pathways, *Biol. Chem.* 384 (4), 645, 2003.
137. Greetham, D. et al., Evidence of glutathione transferase complexing and signaling in the model nematode *Caenorhabditis elegans* using a pull-down proteomic assay, *Proteomics* 4, 1989, 2004.

138. Lo Bello, M. et al., Human glutathione transferase P1-1 and nitric oxide carriers: new role for an old enzyme, *J. Biol. Chem.* 276 (45), 42138, 2001.

139. Turella, P. et al., Glutathione transferase superfamily behaves like storage proteins for dinitrosyl-diglutathionyl-iron complex in heterogeneous systems, *J. Biol. Chem.* 278 (43), 42294, 2003.

140. De Maria, F. et al., The specific interaction of dinitrosyl-diglutathionyl-iron complex, a natural NO carrier, with the glutathione transferase superfamily — suggestion for an evolutionary pressure in the direction of the storage of nitric oxide, *J. Biol. Chem.* 278 (43), 42283, 2003.

141. Muller, B. et al., Nitric oxide transport and storage in the cardiovascular system, *Ann. N.Y. Acad. Sci.* 962, 131, 2002.

142. Rossjohn, J. et al., Crystallization, structural determination and analysis of a novel parasite vaccine candidate: *Fasciola hepatica* glutathione S-transferase, *J. Mol. Biol.* 273, 857, 1997.

143. Johnson, K.A. et al., Crystal structure of the 28 kDa glutathione S-transferase from *Schistosoma haematobium*, *Biochemistry* 42, 10084, 2003.

144. Fritz-Wolf, K. et al., X-ray structure of glutathione S-transferase from the malarial parasite *Plasmodium falciparum*, *Proc. Natl. Acad. Sci. USA* 100, 13821, 2003.

145. Perbandt, M. et al., Native and inhibited structure of a mu class–related glutathione S-transferase from *Plasmodium falciparum*, *J. Biol. Chem.* 279, 1336, 2004.

146. Perbandt, M. et al., Structure of the major cytosolic glutathione *S*-transferase from the parasitic nematode *Onchocerca volvulus*, *J. Biol. Chem.* 280, 12630, 2005.

147. Ruscoe, J.E. et al., Pharmacologic or genetic manipulation of glutathione S-transferase P1-1 (GST pi) influences cell proliferation pathways, *J. Pharm. Exp. Ther.* 298, 339, 2001.

148. Turella, P. et al., Proapoptotic activity of new glutathione S-transferase inhibitors, *Cancer Res.* 65, 3751, 2005.

149. Burg, D. and Mulder, G.J., Glutathione conjugates and their synthetic derivatives as inhibitors of glutathione-dependent enzymes involved in cancer and drug resistance, *Drug Metab. Rev.* 34 (4), 821, 2002.

150. Mahajan, S. and Atkins, W.M., The chemistry and biology of inhibitors and pro-drugs targeted to glutathione S-transferases, *Cellular Mol. Life Sci.* 62, 1221, 2005.

151. Kunze, T. and Heps, S., Phosphono analogs of glutathione: inhibition of glutathione transferases, metabolic stability, and uptake by cancer cells, *Biochem. Pharmacol.* 59, 973, 2000.

152. Albery, W.J. and Knowles, J.R., Evolution of enzyme function and the development of catalytic efficiency, *Biochemistry* 15, 5631, 1976.

153. Pettersson, G., Effect of evolution on the kinetic properties of enzymes, *Eur. J. Biochem.* 184, 561, 1989.

154. Babbitt, P.C., Reengineering the glutathione *S*-transferase scaffold: a rational design strategy pays off, *Proc. Natl. Acad. Sci. USA* 97, 10298, 2000.

155. Stemmer, W.P., Rapid evolution of a protein *in vitro* by DNA shuffling, *Nature* 370, 389, 1994.

156. Stemmer, W.P., DNA shuffling by random fragmentation and reassembly: *in vitro* recombination for molecular evolution, *Proc. Natl. Acad. Sci. USA* 91, 10747, 1994.

157. Hansson, L.O. et al., Evolution of differential substrate specificities in mu class glutathione transferases probed by DNA shuffling, *J. Mol. Biol.* 287, 265, 1999.

158. Broo, K. et al., An ensemble of theta class glutathione transferases with novel catalytic properties generated by stochastic recombination of fragments of two mammalian enzymes, *J. Mol. Biol.* 318, 59, 2002.

159. Dixon, D.P. et al., Forced evolution of a herbicide detoxifying glutathione transferase, *J. Biol. Chem.* 278, 23930, 2003.

160. Larsson, A.K. et al., Directed enzyme evolution guided by multidimensional analysis of substrate-activity space, *Protein Eng. Design Select.* 17, 49, 2004.

161. Hansson, L.O., Widersten, M., and Mannervik, B., Mechanism-based phage display selection of active-site mutants of human glutathione transferase A1-1 catalyzing SNAr reactions, *Biochemistry* 36, 11252, 1997.

162. Ivarsson, Y. et al., Identification of residues in glutathione transferase capable of driving functional diversification in evolution. A novel approach to protein redesign, *J. Biol. Chem.* 278, 8733, 2003.

163. Nilsson, L.O., Gustafsson, A., and Mannervik, B., Redesign of substrate-selectivity determining modules of glutathione transferase A1-1 installs high catalytic efficiency with toxic alkenal products of lipid peroxidation, *Proc. Natl. Acad. Sci. USA* 97, 9408, 2000.

164. Hederos, S. et al., Incorporation of a single His residue by rational design enables thiol-ester hydrolysis by human glutathione transferase A1-1, *Proc. Natl. Acad. Sci. USA* 101 (36), 13163, 2004.

165. Ren, X. et al., A semisynthetic glutathione peroxidase with high catalytic efficiency: selenoglutathione transferase, *Chem. Biol.* 9, 789, 2002.

166. Yu, H.J. et al., Engineering glutathione transferase to a novel glutathione peroxidase mimic with high catalytic efficiency: incorporation of selenocysteine into a glutathione-binding scaffold using an auxotrophic expression system, *J. Biol. Chem.* 280, 11930–11935, 2005.

167. Chelvanayagam, G., Parker, M.W., and Board, P.G., Fly fishing for GSTs: a unified nomenclature for mammalian and insect glutathione transferases, *Chem. Biol. Interact.* 133, 256, 2001.

168. Meyer, D.J., Significance of an unusally low Km for glutathione in glutathione transferases of the α, μ and π classes, *Xenobiotica* 23, 823, 1993.

169. Gu, Y.J., Singh, S.V., and Ji, X.H., Residue R216 and catalytic efficiency of a murine-class alpha glutathione S-transferase toward benzo[*a*]pyrene 7(*R*),8(*S*)-diol 9(*S*),10(*R*)-epoxide, *Biochemistry* 39, 12552, 2000.

6 Specificity of Glutathione S-Transferases in the Glutathione Conjugation of Carcinogenic Diol Epoxides

Hui Xiao and Shivendra V. Singh

CONTENTS

6.1 INTRODUCTION

Glutathione (GSH) S-transferases (EC 2.5.1.18) play an important role in defense against toxic and carcinogenic effects of a wide variety of electrophilic xenobiotics as well as products of oxidative stress.[1,2] GST activity in most mammalian tissues is expressed by multiple isozymes exhibiting remarkably broad substrate specificity.[1,2] In addition to catalyzing the GSH conjugation of various electrophilic substrates, certain GST isozymes can also function as peroxidases and isomerases[1,2] and have the ability to sequester nonsubstrate drugs and hormones.[3] The electrophilic center in GST substrates may be provided by a carbon, nitrogen, or sulfur. Thus, it is not surprising that the list of GST substrates is quite long and includes carcinogenic (intermediates of) environmental pollutants (e.g., aflatoxin B_1, 2-amino-1-methyl-6-phenylimidazo[4,5-*b*]pyridine, polycyclic aromatic hydrocarbons, 4-nitroquinoline-*N*-oxide, and so forth), pesticides (e.g., alachlor, atrazine, lindane, and methyl parathion), products of oxidative stress (e.g., acrolein, base propenals, cholesterol α-oxide, fatty acid hydroperoxides, and 4-hydroxynonenal), certain anticancer agents (e.g., 1,3-*bis*[2-chloroethyl]-1-nitrosourea, chlorambucil, melpahan, cyclophosphamide, and thiotepa), and a handful of endogenous metabolites (e.g., leukotriene A_4 and prostaglandin H_2).[2] This chapter reviews the experimental evidence supporting contribution of GSTs to the detoxification of activated metabolites of polycyclic aromatic hydrocarbon (PAH) family of environmental carcinogens as well as summarizes the current understanding of the structural basis for catalytic differences between different classes of GSTs in the GSH conjugation of PAH metabolites.

PAHs are widely spread environmental pollutants abundant in cigarette smoke, automobile exhaust, polluted air, and so forth. This class of environmental pollutants is tumorigenic in experimental animals and suspected human carcinogens.[4–7] Tumorigenic and mutagenic effects of many PAHs are attributed to their respective diol epoxides,[6–10] which are generated by sequential epoxidation and hydration reactions catalyzed by cytochrome P450-dependent monooxygenases and epoxide hydrolase, respectively.[11–13] The PAH diol epoxides are generally grouped into two classes: bay-region and fjord-region, based on the location of their epoxide functional group.[11–13] The bay-region diol epoxides (e.g., benzo[*a*]pyrene-7,8-diol-9,10-epoxide; hereafter abbreviated as BPDE) are planar molecules, whereas fjord-region compounds (e.g., benzo[*g*]chrysene-11,12-diol-13,14-epoxide; hereafter abbreviated as BGCDE) are distorted from planarity due to the location of their epoxide functional group in a sterically crowded region (compare Figure 6.1 for structures of representative bay- and fjord-region PAH diol epoxides).[13] The diol precursors metabolically formed in

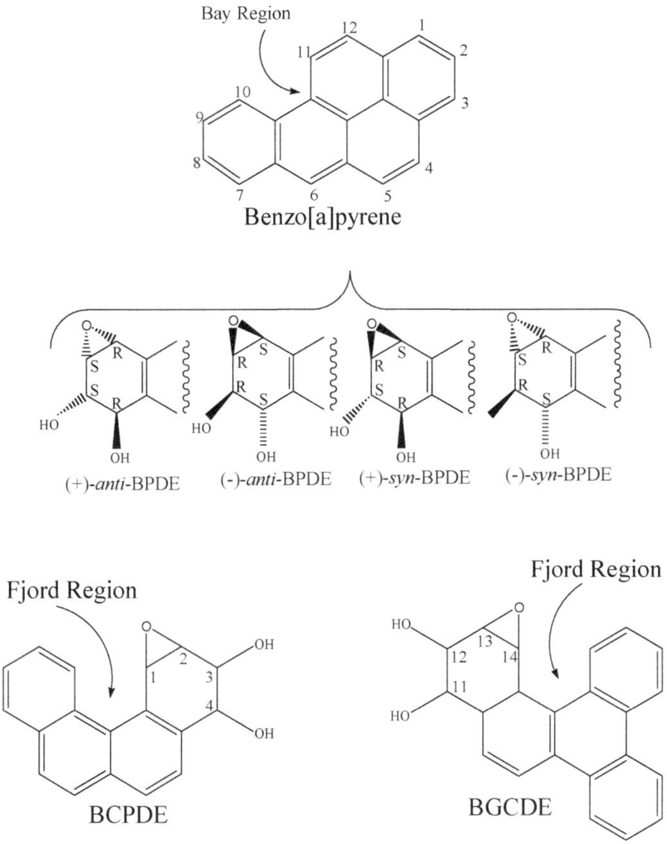

FIGURE 6.1 Structure and absolute configuration of representative bay- and fjord-region PAH diol epoxides.

mammalian systems possess *trans* configuration.[13] Further epoxidation of the diol precursor leads to the formation of (+)- and (−)-enantiomers of *syn-* and *anti-* diastereomers.[12,13] Systematic investigation of optically pure isomers of bay-region diol epoxides has clearly shown that the (+)-*anti*-isomer with (R,S)-diol (S,R)-epoxide absolute configuration (Figure 6.1) is the most potent mutagen in bacterial systems as well as in mammalian cells and is the most potent carcinogen *in vivo*.[6,7,9,10] The tumorigenic profile for fjord-region diol epoxides is strikingly different from that of bay-region compounds. For example, in the case of the fjord-region diol epoxide of benzo[c]phenanthrene (BCPDE), both (−)-*anti*-BCPDE with (R,S)-diol (S,R)-epoxide absolute configuration and the (+)-*syn*-BCPDE isomer possess high skin tumor–initiating activity.[8,14] On the other hand, the (−)-*anti*-BCPDE isomer is a significantly more potent pulmonary carcinogen than other isomers in the newborn mouse model.[14]

Covalent interaction of PAH diol epoxides with nucleophilic sites in DNA is believed to be a critical reaction in the initiation of PAH-induced cancers.[15–18] Studies have shown that *in vitro* chemical modification by *anti*-BPDE leads to activation of c-H-ras proto-oncogene.[19] In addition, within the p53 tumor suppressor gene, *anti*-BPDE preferentially modifies guanine residues at positions (e.g., codon 157) that are mutational hotspots in human cancers.[20,21] Interestingly, one of these mutational hotspots involving codon 157 is specific for human lung cancer.[21] More recently, Harris and coworkers have extended these observations and examined the mutability of p53 hotspot codons following BPDE exposure and shown G:C to T:A transversions at p53 codons 157, 248, and 249.[22] It was also shown that the nonmalignant lung tissues from smokers with lung cancer carried a high load of p53 mutations at the above codons.[22]

Several different mechanisms have been identified that can inactivate PAH diol epoxides and consequently reduce their interaction with DNA. The known mechanisms of diol epoxide inactivation include spontaneous hydrolysis to tetrols and keto-diols, hydration by microsomal epoxide hydrolase, metabolism to triols and triolepoxides, and GST-catalyzed conjugation with GSH.[23–35] Recent studies have suggested that quinone reductase may also provide protection against carcinogenic effects of PAHs.[36] However, direct experimental evidence for tumorigenic activity of quinone metabolites of PAHs is lacking.

6.2 EVIDENCE FOR PROTECTIVE EFFECT OF GSTS AGAINST ACTIVATED PAHS

Initial evidence for contribution of GSTs to defense against DNA binding by activated PAHs was provided by Hesse et al.,[37] who demonstrated that the DNA binding by 7,8-dihydro-7,8,-dihydroxybenzo[a]pyrene (BP-7,8-diol), the precursor of BPDE, is reduced significantly in the presence of rat hepatic cytosol. The protective effect of the cytosolic fraction seemed to be related to its GSH content and could be reproduced in the presence of GSH and a purified rat GST isozyme.[37] Binding of BP-7,8-diol to DNA was inhibited to a small extent by GSH alone but much more effectively in the presence of GSH and cytosol or purified rat GST isoforms.[38] Further work by these investigators confirmed GSH conjugation of BP-7,8-diol or BPDE using isolated rat liver hepatocytes and perfused rat livers.[39–42] GST-mediated GSH conjugation of another bay-region–type PAH diol epoxide (*anti*-chrysene-1,2-diol-3,4-epoxide; hereafter referred as *anti*-CDE) has also been documented.[43] More recent studies including those from our laboratory have provided convincing evidence to indicate that the GSH/GST system is indeed a major contributor to the defense mechanisms against DNA damage caused by PAH diol epoxides. For instance, it has been shown that ectopic expression of GSTs confers significant protection against DNA damage caused by *anti*-BPDE.[44–46] We were the first to provide *in vivo* evidence for GSH-mediated protection against benzo[a]pyrene (BP)-induced carcinogenesis. We showed that D,L-buthionine-*S,R*-sulfoximine–mediated GSH depletion in mouse tissues *in vivo*, which is likely to compromise GST-mediated GSH conjugation of PAH diol epoxides because GSH is the required nucleophilic substrate for GSTs, increases BP-induced tumor multiplicity.[47] Furthermore, Henderson et al.[48] have demonstrated that the GST Pi knockout mice are significantly more sensitive to PAH-induced skin tumorigenesis as compared with wild-type mice.

6.3 SPECIFICITY OF RODENT GSTS IN THE GSH CONJUGATION OF PAH DIOL EPOXIDES

6.3.1 SPECIFICITY OF RAT GSTs

The GST isozymes differ markedly in their ability to catalyze the GSH conjugation of various electrophilic substrates, and PAH diol epoxides are no exception to this substrate selectivity. Jernström et al.[23] were the first to systematically examine specificity of purified rat liver GST isozymes, including GST1-1 (rGSTA1-1 according to the currently accepted nomenclature that was originally proposed for human GSTs but is being used for rat and murine GSTs[49]), GST1-2 (rGSTA1-2), GST2-2 (rGSTA2-2), GST3-3 (rGSTM1-1), GST3-4 (rGSTM1-2), and GST4-4 (rGSTM2-2) for GSH conjugation of racemic *anti*-BPDE. Nonenzymatic conjugation of (±)-*anti*-BPDE with GSH was minimal in the absence of GST protein, but it increased considerably in the presence of purified rat hepatic GST isozymes. For example, it has been estimated that the GSH conjugation of *anti*-BPDE is increased by >60-fold in the presence of 1.6 µM rGSTM2-2 over nonenzymatic reaction.[23] Each rat GST isozyme studied adhered to Michaelis–Menten kinetics when the activity was measured as a function of varying (±)-*anti*-BPDE concentrations at a fixed 1 mM GSH concentration. Based on kinetic data, the homodimers could be ranked in the order of rGSTM2-2 > rGSTA1-1 > rGSTA2-2 > rGSTM1-1 with respect to their V_{max} for (±)-*anti*-BPDE–GSH conjugation.[23] Interestingly, the kinetic data obtained using fixed (±)-*anti*-BPDE concentration and varying GSH concentrations showed a biphasic response for each rat isozyme studied. The nonlinearity is due to an apparent increase in the affinity of the enzyme for GSH with a decrease in the GSH concentration.[23] Subsequent work by these investigators showed that the Pi-class isozyme GST 7-7 (rGSTP1-1), isolated from the rat lung, kidney, and hyperplastic liver nodules, is an even more efficient catalyst of (±)-*anti*-BPDE–GSH conjugation compared with rGSTM2-2 or other rat isozymes.[50] The rat GST isozymes including rGSTP1-1 were shown to be highly selective for the (+)-enantiomer of *anti*-BPDE.[50]

Besides BPDE, rGSTM2-2 is shown to catalyze the GSH conjugation of diastereoisomers of *trans* CDE and *trans* 3,4-dihydroxy-1,2-epoxy-1,2,3,4-tetrahydrobenz[*a*]anthracene (BADE).[24] However, BPDE appears to be a relatively better substrate for rGSTM2-2 when compared with either CDE or BADE.[24] Further studies are needed to determine if similar substrate specificity exists for other classes of rat GSTs. In addition, the relative contribution of other rat GST subunits including rGSTA4, rGSTA5, rGSTM3, rGSTM4, rGSTM5, and Theta-class GSTs (rGSTT1-1or rGSTT2-2) toward GSH conjugation of PAH diol epoxides remains to be determined.

6.3.2 SPECIFICITY OF MURINE GSTs

The specificity of murine GST isozymes, purified from the livers or forestomachs of female A/J mice, for GSH conjugation of optically pure (+)- and (−)-enantiomers of *anti*-BPDE was investigated in our laboratory.[29,51] The murine GST isozymes included mGSTA1-2 (designated as GST9.5 in earlier publications based on its isoelectric point prior to the cDNA cloning of the individual subunits[29,51]), mGSTA3-3, mGSTA4-4, mGSTM1-1, and mGSTP1-1.[29,51] Unlike rat GSTA1-1 or rGSTA2-

2, murine GSTA1-2 is exceptionally efficient in catalyzing the GSH conjugation of (+)-*anti*-BPDE.[29,51] Thus, we have shown that the catalytic efficiency (k_{cat}/K_m) of mGSTA1-2 in the GSH conjugation of (+)-*anti*-BPDE is about 9- to 655-fold higher when compared with other murine GSTs.[29,51] Moreover, the recombinant mGSTA1-1 exhibits about 3.3-fold higher catalytic efficiency toward (+)-*anti*-BPDE compared with its close structural homologue mGSTA2-2.[51] Similar to rat GST isozymes, however, the (+)-*anti*-enantiomer of BPDE is a preferred substrate for murine GSTs compared with the (−)-*anti*-BPDE.[29] Thus, the murine GSTs are between 2- and 20-fold more efficient catalysts of (+)-*anti*-BPDE–GSH conjugation compared with (−)-*anti*-BPDE.[29] Because the expression of the mGSTA1 or mGSTA2 subunit is barely detectable in A/J mouse livers,[52] which is the main site of xenobiotic metabolism, we speculate that the hepatic GSH conjugation of (+)-*anti*-BPDE, at least in A/J mouse strain, may primarily be catalyzed by mGSTP1-1 and mGSTM1-1. We tested this possibility by estimating the relative contribution of different GST isozymes to hepatic and forestomach GSH conjugation of (+)-*anti*-BPDE.[53] Indeed, we found that mGSTP1-1, and to a lesser extent mGSTM1-1, is primarily responsible for the GSH conjugation of (+)-*anti*-BPDE in A/J mouse livers.[53] In the forestomachs of A/J mice, however, the relative contribution of mGSTA1-2 is more or less similar to that of mGSTP1-1[53]. The lung is another known target organ for BPDE-induced tumorigenesis. In an A/J mouse lung, about 77% of the cytosolic GST activity is accounted for by mGSTM1-1 and mGSTP1-1, whereas the mGSTA1 or mGSTA2 subunit is not detectable.[54] Thus, it is reasonable to speculate that the GSH conjugation of (+)-*anti*-BPDE in mouse lungs is primarily catalyzed by the combined activity of mGSTP1-1 and mGSTM1-1. It is important to point out, however, that the specificity of several other murine GST subunits, including mGSTM2, mGSTM3, mGSTM4, mGSTM5, mGSTP2, and mGSTT1, in the GSH conjugation of (+)-*anti*-BPDE or any other PAH diol epoxide has not been determined, and thus the above conclusions regarding relative contribution may not be totally valid. Nonetheless, based on available kinetic data, the murine GST isozymes can be ranked in the order of mGSTA1-1 > mGSTA2-2 > mGSTP1-1 > mGSTM1-1 > mGSTA3-3 > mGSTA4-4 with respect to their catalytic efficiency toward (+)-*anti*-BPDE.[29,51]

We have also determined catalytic efficiencies of purified murine GST isozymes in the GSH conjugation of bay-region class *anti*-CDE and fjord-region class *anti*-benzo[*g*]chrysene-11,12-diol-13,14-epoxide (*anti*-BGCDE).[30,55] Even though the kinetic constants for recombinant mGSTA1-1 or mGSTA2-2 in the GSH conjugation of (±)-*anti*-CDE have not been estimated, tissue-isolated mGSTA1-2 appears to be a much more efficient catalyst of GSH conjugation of this diol epoxide substrate compared with mGSTA3-3, mGSTA4-4, mGSTM1-1, or mGSTP1-1.[30] Similarly, the GSH conjugation of (−)-*anti*- and (+)-*syn*-BGCDE (the two stereoisomers to which the moderate carcinogen benzo[*g*]chrysene is metabolically activated in mouse skin)[56] is catalyzed more efficiently in the presence of recombinant mGSTA1-1 compared with other murine GSTs including recombinant mGSTA2-2.[55] We also found that the murine GST isozymes differ markedly in their enantioselectivity toward both *anti*-CDE and *anti*-BGCDE.[30] In the absence of the GST protein, more or less equal amounts of the GSH conjugates are formed from (+)- and (−)-enantiomers of *anti*-CDE or *anti*-BGCDE, which is expected for racemic substrates.[30]

Interestingly, the Pi-class isoform mGSTP1-1 exhibits a marked preference toward (+)-enantiomer of *anti*-CDE with *R,S*-diol *S,R*-epoxide absolute configuration.[30] Similarly, mGSTP1-1 displays a preference toward *anti*-BGCDE enantiomer with *R,S*-diol *S,R*-epoxide absolute configuration.[30] Relative preference for GSH conjugation of *anti*-CDE and *anti*-BGCDE isomers with *R,S*-diol *S,R*-epoxide absolute configuration is also apparent for mGSTA1-2. On the other hand, mGSTM1-1 shows preference toward (+)-*anti*-BGCDE with *S,R*-diol *R,S*-epoxide absolute configuration with about >70% of the GSH conjugation occurring with this enantiomer.[30]

The enantioselectivity of murine GSTs is also evident for benzo[*c*]phenanthrene-3,4-diol-1,2-epoxide (BCPDE),[57] which is similar to BGCDE and belongs to the fjord-region class. For example, while the Pi-class isozyme mGSTP1-1 is virtually inactive toward BCPDE stereoisomers with 1*S* configuration ([-]-*syn* and [+]-*anti*-BCPDE), mGSTA1-2 is able to catalyze the GSH conjugation of both these stereosiomers.[57] Even though mGSTA1-2, similar to *anti*-CDE and *anti*-BGCDE,[30] is relatively more efficient in catalyzing the GSH conjugation of (+)- and (−)-enantiomers of both *anti*- and *syn*-BCPDE compared with either mGSTP1-1 or mGSTM1-1, it is interesting to note that the catalytic efficiency of mGSTA3-3 toward (−)-*anti*-, (+)-*anti*-, and (+)-*syn*-BCPDE isomers is more or less similar to that of mGSTA1-2.[57] It is important to point out that mGSTA3-3 is a rather poor catalyst of GSH conjugation of other PAH diol epoxides including (±)-*anti*-CDE, (−)-*anti*-BGCDE, or (+)-*syn*-BGCDE.[30,56] Based on these studies, we conclude that the murine GST isozymes exhibit a rather unique enantioselectivity and substrate preference for GSH conjugation of environmentally relevant PAH diol epoxides.

Detailed examination of the specificity of murine GSTs toward PAH diol epoxides and other substrates has also allowed us to draw important conclusions regarding their physiological functions, especially Alpha-class isozymes. For instance, recombinant mGSTA1-1 and mGSTA2-2 as well as tissue-isolated mGSTA1-2 are highly efficient catalysts of GSH conjugation of PAH diol epoxides, whereas mGSTA4-4 is virtually inactive toward this class of substrates.[29,30,51,55,57] On the other hand, mGSTA4-4 is an exceptionally efficient catalyst of GSH conjugation of 4-hydroxynonenal,[58] which is the end-product of lipid peroxidation. Thus, we can conclude that while mGSTA1-1 and mGSTA2-2 may contribute to GSH conjugation of PAH diol epoxides, mGSTA4-4 is the principal isozyme responsible for GSH conjugation of 4-hydroxynonenal. On the other hand, mGSTA3-3 may play a major role in GSH conjugation of aflatoxin B_1-8,9-epoxide as well as reduction of lipid hydroperoxides, which are far better substrates for mGSTA3-3 compared with other murine GSTs.[2,59,60]

6.3.3 STRUCTURAL BASIS FOR CATALYTIC DIFFERENCES BETWEEN mGSTA1-1 AND mGSTA2-2 TOWARD PAH DIOL EPOXIDES

It is interesting to note that despite >95% amino acid sequence similarity, recombinant mGSTA1-1 is approximately 3.3-fold more efficient than recombinant mGSTA2-2 in catalyzing the GSH conjugation of (+)-*anti*-BPDE.[51] The amino acid sequences of mGSTA1-1 and mGSTA2-2 differ at 10 positions including amino acid residues in positions 65 (Ala in mGSTA1, Val in mGSTA2), 95 (Ser in mGSTA1, Thr in mGSTA2), 157 (Ile in mGSTA1, Val in mGSTA2), 162 (Val in mGSTA1, Leu in mGSTA2), 169 (Phe in mGSTA1, Leu in mGSTA2), 207 (Met in mGSTA1, Leu in mGSTA2), 213 (Gln in mGSTA1, Glu in mGSTA2), 218 (Ala in mGSTA1, Val in mGSTA2), 221 (Ile in mGSTA1, Phe in mGSTA2), and 222 (Gln in mGSTA1). Molecular modeling suggested that the amino acid substitutions in positions 207 and 221 may be responsible for catalytic differences between mGSTA1-1 and mGSTA2-2 toward (+)-*anti*-BPDE.[61] We tested this possibility by site-directed mutagenesis. The k_{cat}/K_m of recombinant mGSTA1-1 in the GSH conjugation of (+)-*anti*-BPDE is reduced by about 42% and 61%, respectively, by replacement of Met-207-Leu and Ile-221-Phe, the equivalent residues of mGSTA2-2.[61] The catalytic efficiency of mGSTA1-1 is reduced even further upon combined replacement of both Met-207-Leu and Ile-221-Phe.[61] Thus, the catalytic difference between mGSTA1-1 and mGSTA2-2 in the GSH conjugation of (+)-*anti*-BPDE is dictated by the amino acid residues in positions 207 and 221.

The catalytic efficiency of mGSTA1-1 in the GSH conjugation of (+)-*anti*-BPDE is approximately 77- and 655-fold higher when compared with mGSTA3-3 and mGSTA4-4, respectively.[51] The amino acid sequence of mGSTA1-1 is about 69% and 59% identical to those of mGSTA3-3 and mGSTA4-4, respectively. While the molecular basis for this remarkable difference in catalytic efficiency between mGSTA1-1 and mGSTA3-3 or mGSTA4-4 is not clear, it is interesting to note that only mGSTA1-1 contains an Ile at position 221. This position is occupied by Phe in mGSTA2-2 as well as in mGSTA4-4 (mGSTA3-3 is comprised of 220 amino acid residues). On the other hand, the amino acid residue in position 207 is different among the four Alpha-class murine GSTs. This position is occupied by Met in mGSTA1-1, Leu in mGSTA2-2, Asp in mGSTA3-3, and Pro in mGSTA4-4. It is possible that the variation in position 207 may contribute to catalytic disparity between different Alpha-class murine GSTs in the GSH conjugation of (+)-*anti*-BPDE. Further studies are needed to experimentally verify this speculation. The amino acid residues in positions 207 and 221 seem to be important determinants of differential activity of recombinant mGSTA1-1 and mGSTA2-2 toward (−)-*anti*-BCPDE as well.[62]

6.4   CRYSTAL STRUCTURES OF MGSTA1-1 AND MGSTA2-2 IN COMPLEX WITH GSH CONJUGATE OF (+)-*ANTI*-BPDE

6.4.1   ACTIVE CENTER AND BINDING OF GSH CONJUGATE OF (+)-*ANTI*-BPDE (GSBPD) IN MGSTA1-1

Because mGSTA1-1 is an exceptionally efficient catalyst of (+)-*anti*-BPDE–GSH conjugation, we determined the crystal structure of mGSTA1-1 in complex with GSBpd to gain insights into the structural basis for this substrate selectivity.[63] The crystal structure revealed that the GSH moiety is bound to the G-site via comprehensive electrostatic interactions. Most of the polar atoms in GSH are involved in

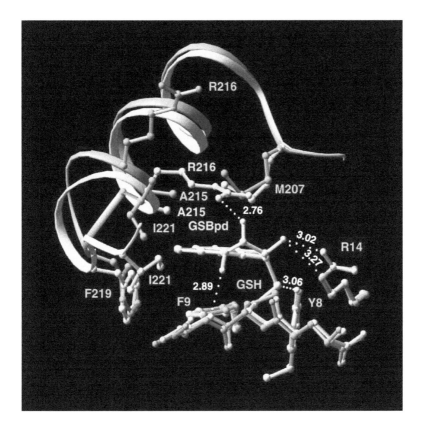

FIGURE 6.2 (See color insert following page 178.) Significant conformational changes of the H-site in mGSTA1-1 upon xenobiotics substrate binding. The side chains are represented as ball-and-stick models, and the electrostatic interactions are shown as dotted white lines. Reprinted with permission from Gu, Y., Singh, S.V., and Ji, X., *Biochemistry*, 39, 12552, 2000. Copyright © 2000 American Chemical Society.

the formation of hydrogen bonds or salt bridges as shown previously for other GSTs,[64] among which the hydrogen bond between atom SG2 of GSH and atom OH of Tyr-8 is shown in Figure 6.2. The H-site in mGSTA1-1 is composed of conserved residues Tyr-8, Phe-F9, and Arg-14 and C-terminal residues Met-207, Ala-215, Phe-219, and Ile-221. These hydrophobic side chains construct a hydrophobic pocket for the binding of xenobiotic substrates such as (+)-*anti*-BPDE. The contact between the BPDE moiety and the H-site residues includes both electrostatic and hydrophobic interactions. There are three hydroxyl groups on the BPDE moiety, among which the C7 hydroxyl forms a hydrogen bond with N3 of the GSH moiety, whereas those of C8 and C9 form hydrogen bonds with Arg-216 and Arg-14, respectively (Figure 6.2). We found that Arg-216 from the C-terminal helix interacts with the product/substrate molecule(s). It appears that Arg-216 has two roles in catalysis: participating in the binding of the substrate/intermediate/product molecule(s) and providing electrostatic assistance in the epoxide ring-opening reaction. The BPDE moiety is sandwiched between the hydrophobic portion of the Arg-216 side chain and the phenyl ring system of F9 (Figure 6.2). The distance from the Phe-9 phenyl ring to the BPDE polycyclic aromatic system is 4.0 Å and that from BPDE to the Arg-216 side chain is 3.8 Å. In addition, the BPDE moiety is surrounded by the hydrophobic side chains of Met-207, Ala-215, Phe-219, and Ile-221 with contact distances of 4.5, 3.9, 3.5, and 3.5 Å, respectively. Among the side chains interacting with the BPDE moiety of GSBpd, those of Arg-216 and Ile-221 exhibit the largest conformational changes upon substrate/product binding[63] (Figure 6.2).

6.4.2 XENOBIOTIC-INDUCED CONFORMATIONAL CHANGES IN mGSTA1-1

The unique C-terminus of Alpha-class GSTs plays an important role in catalysis and therefore has been recognized as the critical component of the active site.[64,65] The active site of mGSTA1-1 exerts significant conformational changes on binding of GSBpd (Figure 6.2). First, the C-terminal α9 helix moves toward the H-site. Second, along with the shift of the C-terminal helix, the Cα atom of Arg-216 moves ~2.3 Å and the side chain shifts from pointing away from the H-site to protruding into the active center. The guanidinium group of Arg-216 moves ~7.7 Å and forms a strong hydrogen bond (2.76 Å) with the C8 hydroxyl group of BPDE (Figure 6.2). This electrostatic interaction may serve to orient and position the substrate (+)-*anti*-BPDE in the H-site and also to facilitate the ring-opening reaction of the epoxide. Third, in mGSTA1-1•GSH, Ile-221 is partially disordered and Gln-222 is completely disordered, whereas in mGSTA1-1•GSBpd, both Ile-221 and Gln-222 are well defined. The φ-φ angles of Lys-220 in mGSTA1-1•GSH are –99° and 114°, whereas in mGSTA1-1•GSBpd, they are 48° and 97°, respectively. Consequently, Ile-221 exhibits significant positional changes upon xenobiotic binding and interacts extensively with the bound substrate[63] (Figure 6.2).

6.4.3 H-SITE COMPARISON OF mGSTA1-1 WITH mGSTA4-4

As discussed above, mGSTA4-4 exhibits high activity in conjugating the lipid peroxidation product 4-hydroxynonenal with GSH.[58] However, (+)-*anti*-BPDE is an extremely poor substrate for this isozyme. mGSTA4-4 and mGSTA1-1 have very similar three-dimensional structures.[63,66] In addition, Arg-14 is available in both mGSTA1-1[63] and mGSTA4-4[66] for catalysis. A closer examination of the two structures reveals that two additional proline residues in mGSTA4-4 — Pro-207 and Pro-210 — may make this isozyme a poor catalyst of (+)-*anti*-BPDE–GSH conjugation. Figure 6.3 depicts H-site comparison of mGSTA1-1 (shown in blue) and mGSTA4-4 (shown in orange). The residue corresponding to Pro-207 in mGSTA1-1 is Met-207, which, as discussed above, plays an important role in the GSH conjugation of (+)-*anti*-BPDE.[61] The change of Met-207 to Pro-207 in mGSTA4-4 results in three consecutive proline residues (Pro-205, Pro-206, and Pro-207), which may alter flexibility of the loop before the C-terminal helix. Residue Lys-210 in mGSTA1-1 corresponds to Pro-210 of mGSTA4-4, which is located at the N-terminus of the

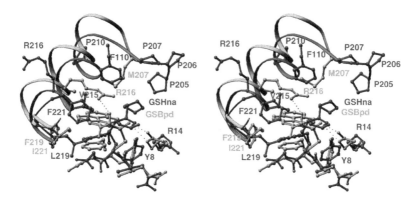

FIGURE 6.3 (See color insert.) H-site comparison of mGSTA1-1 with mGSTA4-4 indicates that Pro-205, Pro-206, Pro-207, and Pro-210 in mGSTA4-4 may be responsible for the rigidity of its C-terminal helix and consequently the narrow H-site. Reprinted with permission from Gu, Y., Singh, S.V., and Ji, X., *Biochemistry*, 39, 12552, 2000. Copyright © 2000 American Chemical Society.

C-terminal helix and thereby alters the direction of the C-terminal helix (Figure 6.3). Consequently, Arg-216, which appears important in the (+)-*anti*-BPDE–GSH conjugation,[67] is blocked by Val-215 (Figure 6.3) and Phe-110 (not shown in Figure 6.3) and therefore it is impossible for Arg-216 to reach the active center of mGSTA4-4.

6.4.4 ROLE OF ARG-216 IN MGSTA1-1-CATALYZED GSH CONJUGATION OF (+)-*ANTI*-BPDE

The crystal structure of the mGSTA1-1·GSBpd complex[63] suggested that Arg-216 may play an important role in mGSTA1-1–catalyzed GSH conjugation of (+)-*anti*-BPDE. We experimentally verified this possibility by determining the effect of Arg-216-Ala mutation on the activity of mGSTA1-1.[67] Indeed, the Arg-216-Ala mutant of mGSTA1-1 is a much less efficient catalyst of (+)-*anti*-BPDE–GSH conjugation compared with the wild-type enzyme. The Arg-216-Ala mutation also caused a marked reduction in specific activity of mGSTA1-1 toward (±)-*anti*-CDE and (±)-*anti*-BCPDE.[67] Thus, these studies confirmed that Arg-216 plays an important role in mGSTA1-1–catalyzed GSH conjugation of PAH diol epoxides.

6.4.5 GSBPD BINDING IN THE ACTIVE CENTER OF MGSTA2-2

We solved the crystal structure of mGSTA2-2 in complex with GSBpd in an attempt to explain catalytic differences between mGSTA2-2 and mGSTA1-1 toward (+)-*anti*-BPDE.[68] Similar to mGSTA1-1,[63] the GSH moiety of GSBpd in mGSTA2-2•GSBpd is bound to the G-site via comprehensive electrostatic interactions. Most of the polar atoms in GSH are involved in the formation of hydrogen bonds or salt bridges. The hydrophobic ring system of GSBpd is located in the H-site constructed by the side chains of Phe-9, Ala-11, Leu-110, Leu-207, Ile-212, Ala-215, Arg-216, and Phe-221. As shown in Figure 6.4a, to the left side of the Bpd ring lie the side

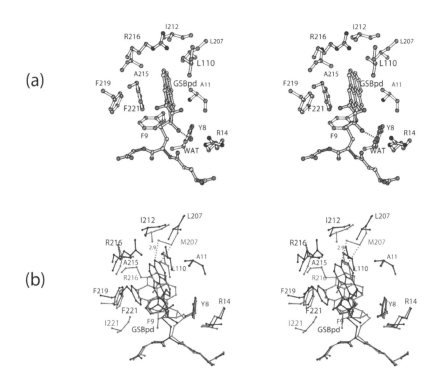

FIGURE 6.4 (See color insert.) (a) Active center architecture of mGSTA2-2·GSBpd. Ball-and-stick model is used with the atomic color scheme: carbon in gray, oxygen in red, nitrogen in blue, and sulfur in yellow. Hydrogen bonds and electrostatic interactions are shown as dashed lines. (b) Active center comparison of mGSTA1-1·GSBpd and mGSTA2-2·GSBpd. Reprinted with permission from Gu et al., *Biochemistry*, 42, 917, 2003. Copyright © 2003 American Chemical Society.

chains of Ala-215 and Phe-221, with the side chain of Phe-221 parallel to the ring system at the average distance of 3.5~4.0 Å; on the right side of the Bpd ring are the hydrophobic side chains of Leu-110 and Leu-207; on the top of the Bpd ring lie the side chains of Arg-216 and Ile-212; and the side chains of Tyr-8, Phe-9, and Ala-11 are located near the bottom of the hydrophobic Bpd moiety. Among all the active-site residues that surround the Bpd moiety, Tyr-8 and Arg-216 have the potential of forming hydrogen bonds with the substrate. However, no hydrogen bonds are formed between Tyr-8 or Arg-216 and GSBpd. The -OH group of Tyr-8 interacts with one of the hydroxyl groups of GSBpd via a water molecule, whereas the side chain of Arg-216 lies perpendicular to the Bpd moiety and the distance between the guanidinium group of Arg-216 and any hydroxyl group of GSBpd is >7.5 Å (Figure 6.4a).

The overall structure of mGSTA2-2•GSBpd is very similar to that of mGSTA1-1•GSBpd.[63,68] However, the conformation of GSBpd is different between mGSTA1-1 and mGSTA2-2 mainly because of the different conformation around the CB2–SG2 bond (Figure 6.4b). In mGSTA2-2•GSBpd, the CB2–SG2 bond points away from the -OH of Tyr-8. In mGSTA1-1•GSBpd structure, however, the CB2–SG2 bond points toward the hydroxyl group of Tyr-8, which forms a hydrogen bond (3.2 Å) with SG2.[68]

Unlike mGSTA1-1, the Bpd ring system is almost perpendicular to the phenyl ring of Phe-9 in mGSTA2-2•GSBpd (Figure 6.4b). Consequently, the side chain of Arg-216 is perpendicular to the Bpd moiety and points away from the active center in mGSTA2-2 (Figure 6.4b). Nevertheless, the stacking interaction of the Bpd ring and a phenyl ring is maintained for mGSTA2-2, of which F221 provides the phenyl ring instead of Phe-9 (Figure 6.4b).[68]

6.5 SPECIFICITY OF HUMAN GSTS TOWARD PAH DIOL EPOXIDES

6.5.1 KINETIC ANALYSIS OF HUMAN GSTs TOWARD DIOL EPOXIDES

Jernström and colleagues were the first to systematically investigate specificity of human GSTs in the GSH conjugation of PAH diol epoxides.[25–28,69] Initial work by this group used a mixture of basic Alpha-class human GSTs (designated as GST α-ε) and a near-neutral μ-type isozyme purified from the liver, and GST π purified from the placenta to examine GSH conjugation of *anti*-BPDE.[25] While GST α-ε and GST μ were able to catalyze the addition of GSH to both (+)- and (−)-enantiomers of *anti*-BPDE, GST exhibited a preference toward (+)-*anti*-BPDE with about 90% of the conjugation occurring with this enantiomer.[25] The μ- and the π-type GSTs were much more efficient catalysts of (+)-*anti*-BPDE–GSH conjugation than GST α-ε.[25] In a follow-up study, the same group of investigators measured kinetic parameters for hGSTA1-1, hGSTM1-1, and hGSTP1-1 by using optically pure isomers of BCPDE. Each human GST isozyme that was studied catalyzed GSH conjugation of all four stereoisomers of BCPDE, albeit less effectively when compared with BPDE.[25,26] Another interesting finding of this study was that the BCPDE isomers with 1*R*,2*S*-epoxide absolute configuration ([+]-*syn* and [−]-*anti*-BCPDE) were generally better substrates for human GSTs than the corresponding 1*S*,2*R*-epoxides.[26] However, hGSTP1-1 was highly selective toward the BCPDE isomer with 4*R*,3*S*-diol 2*S*,1*R*-epoxide absolute configuration ([−]-*anti*-BCPDE).[26] On the other hand, both hGSTA1-1 and hGSTM1-1 exhibited preferences toward the (+)-*syn*-BCPDE isomer with 4*R*,3*S*-diol 2*R*,1*S*-epoxide absolute configuration.[26]

Subsequently, Jernström and colleagues determined kinetic parameters for hGSTA1-1, hGSTM1-1, and hGSTP1-1 toward individual stereoisomers of diol epoxides of chrysene, dibenz[*a,h*]anthracene, benzo[*a*]pyrene, benzo[*c*]phenanthrene, benzo[*c*]chrysene, and benzo[*g*]chrysene.[27,28] While hGSTA1-1 was shown to catalyze the formation of GSH conjugate of each diol epoxide substrate tested, a marked variation in catalytic efficiency was evident.[27] In addition, hGSTA1-1 exhibited a marked preference for conjugation of the *anti*-enantiomer with *R* configuration at the benzylic position of the oxirane ring for both bay- and fjord-region–type compounds.[27] A similar enantioselectivity was observed for hGSTA1-1 toward *syn* isomers of the fjord-region diol epoxides.[27] Moreover, the catalytic efficiency of hGSTA1-1 did not correlate with either chemical reactivity or lipophilicity of the diol epoxides.[27] Similar to hGSTA1-1, hGSTM1-1 and hGSTP1-1 demonstrate considerable GSH conjugation activity toward PAH diol epoxides belonging to both bay- and fjord-region classes.[28] The hGSTM1-1 exhibits an enantioselectivity ranging from no preference (50%) to high preference for GSH conjugation of the enantiomers with *R* configuration at the benzylic position of the oxirane ring.[28] In contrast to hGSTA1-1 and hGSTM1-1, hGSTP1-1 displays an exclusive preference toward the enantiomers with the *R* configuration at the benzylic oxirane.[28] The chemically most reactive isomer — (+)-*syn*-BPDE — is the best substrate for both hGSTM1-1 and hGSTP1-1.[28]

The specificity of all four well-characterized Alpha-class human GSTs, including hGSTA1-1, hGSTA2-2, hGSTA3-3, and hGSTA4-4, toward carcinogenic (–)-*anti*- and (+)-*syn*-diol epoxides derived from the nonplanar dibenzo[*a,l*]pyrene (DBPDE) was investigated by Dreij et al.[69] While hGSTA4-4 was inactive, both isomers of DBPDE were substrates for other Alpha-class human GSTs. Interestingly, the catalytic efficiency of hGSTA1-1 for (–)-*anti*-DBPDE–GSH conjugation is approximately 4- and 20-fold higher when compared with hGSTA3-3 and hGSTA2-2, respectively.[69] Similarly, hGSTA1-1 is between 2.5- and 15-fold more efficient a catalyst of GSH conjugation of (+)-*syn*-DBPDE when compared with other Alpha-class human GSTs.[69] It is interesting to note that (+)-*syn*-DBPDE is a much better substrate for hGSTA1-1 than the (–)-*anti* isomer of this diol epoxide. A comparison of the catalytic efficiency values for hGSTA1-1 toward different fjord-region–type diol epoxides also suggests that an increase in complexity of the aromatic ring system and thus an increase in lipophilicity of the substrate favors diol epoxide–GSH conjugation. For example, an extension of the aromatic ring system from four in BCPDE to five in DBPDE causes >100-fold increase in catalytic efficiency of hGSTA1-1.[69] Dreij et al.[69] were also able to speculate on the structural basis for the difference in activity of hGSTA1-1 toward DBPDE stereoisomers through molecular modeling, which revealed more favorable interactions between the substrate and the enzyme–GSH complex for (+)-*syn*-DBPDE than for the (–)-*anti*-isomer.[69]

6.5.2 Structural Basis for Catalytic Differences between hGSTA1-1 and hGSTA2-2 toward (+)-*anti*-BPDE

The hGSTA1-1 and hGSTA2-2 are structurally closely related but the estimated k_{cat}/K_m of hGSTA1-1 toward (+)-*anti*-BPDE (1.1 mM^{-1} s^{-1}) is approximately 5-fold higher when compared with hGSTA2-2.[70] Based on the three-dimensional structure of hGSTA1-1,[64] the GSH-binding site (G-site) residues in hGSTA1-1 include Tyr-8, Arg-14, Arg-44, Gln-53, Val-54, Gln-66, Thr-67, Asp-100, Arg-130, and Phe-219 (numbering excludes initiator methionine). The amino acid residues assigned to the hydrophobic substrate-binding H-site of hGSTA1-1 are Phe-9, Ala-11, Gly-13, Glu-103, Leu-106, Leu-107, Pro-109, Val-110, Met-207, Leu-212, Ala-215, and Phe-221. The amino acid sequence alignment of hGSTA1-1 and hGSTA2-2 reveals 10 substitutions, but only four of these changes are located in the H-site, including residues in positions 9 (Phe in hGSTA1-1, Ser in hGSTA2-2), 11 (Ala in hGSTA1-1, Ile in hGSTA2-2), 110 (Val in hGSTA1-1, Phe in hGSTA2-2), and 215 (Ala in hGSTA1-1, Ser in hGSTA2-2). On the other hand, the G-site–defining residues are conserved between hGSTA1-1 and hGSTA2-2. We hypothesized that the H-site amino acid substitutions might contribute to catalytic differences between hGSTA1-1 and hGSTA2-2 toward (+)-$anti$-BPDE. We tested the above hypothesis by determining (+)-$anti$-BPDE–GSH-conjugating activities of the mutants of hGSTA2-2 in which amino acid residues in positions 9 (Ser to Phe), 11 (Ile to Ala), 110 (Phe to Val), and 215 (Ser to Ala) were mutated to the corresponding residues of hGSTA1-1 (hereafter designated as hGSTA2/S9F, hGSTA2/I11A, hGSTA2/F110V, and hGSTA2/S215A mutants). The catalytic efficiency of hGSTA2-2 for GSH conjugation of (+)-$anti$-BPDE was not affected by S9F mutation despite an approximate 50% reduction in the V_{max} and K_m values of the hGSTA2/S9F mutant. On the other hand, the I11A mutant was a significantly more efficient catalyst of (+)-$anti$-BPDE–GSH conjugation than hGSTA2-2 and exhibited a catalytic efficiency value similar to that of wild-type hGSTA1-1. The V_{max} and the K_m values for the F110V mutant of hGSTA2-2 were lower by about 35% and 63%, respectively, in comparison with the wild-type hGSTA2-2, leading to an approximate 1.5-fold increase in catalytic efficiency of the mutant protein. On the other hand, the S215A mutation caused a marked decrease in catalytic efficiency of hGSTA2-2 due to an approximate 61% reduction in the V_{max} and about a 1.5-fold increase in the K_m of the mutant protein.[70] We also generated a quadruple mutant of hGSTA2-2 to mimic the H-site of hGSTA1-1 by mutating S9F, I11A, F110V, and S215A together (hGSTA2/SIFS mutant). Interestingly, the hGSTA2/SIFS mutant was a more efficient catalyst of (+)-$anti$-BPDE–GSH conjugation than either the wild-type hGSTA1-1 or wild-type hGSTA2-2.[70] In addition, similar to the wild-type hGSTA1-1, the hGSTA2-2/SIFS mutant protein exhibited a preference toward GSH conjugation of the (+)-$anti$-BPDE enantiomer compared with (−)-$anti$-BPDE. These results clearly indicate that the amino acid residue in position 11 is responsible for catalytic differences between hGSTA1-1 and hGSTA2-2 in the GSH conjugation of (+)-$anti$-BPDE.

We attempted to rationalize the catalytic differences between hGSTA1-1 and hGSTA2-2 toward (+)-$anti$-BPDE through molecular modeling by docking the GSH conjugate of BPDE (GSBpd) into the active site of hGSTA1-1 in complex with the GSH conjugate of ethacrynic acid (GSEa).[65] A comparison of the crystal structures of murine GSTA1-1 and GSTA2-2 in complex with GSBpd,[63,68] hGSTA1-1·GSEa, and allelic variants of hGSTP1-1 in complex with GSBpd[71] reveals two distinct binding modes of GSBpd: one in mGSTA2-2 and hGSTP1-1 (Ile-104, Ala-113) (binding mode 1) and the other in mGSTA1-1 and hGSTP1-1 (Val-104, Ala-113) (binding mode 2).[63,65,68,71] Binding mode 1 is associated with lower catalytic activity, most likely because the GSH sulfur is beyond the hydrogen-bond distance of the hydroxyl group of the GST's catalytic tyrosine residue.[68] As discussed above, the H-site residues of hGSTA1-1 and hGSTA2-2 differ in four positions (residues 9, 11, 110, and 215), among which residue 9 in hGSTA2-2 is smaller than that in hGSTA1-1 but the rest become bulkier. The Phe-9, Ala-11, Val-110, and Ala-215 of hGSTA1-1 provide favorable hydrophobic interactions with the BPDE moiety. Therefore, the bulkier side chains of Ile-11, Phe-110, and Ser-215 in hGSTA2-2 impose unfavorable steric interactions with the BPDE moiety; although Ser-9 of hGSTA2-2 is smaller than Phe-9 of hGSTA1-1, Ile-11 and Ser-215 prevent the

BPDE moiety from moving towards Ser-9. Consequently, hGSTA2-2 has an even lower catalytic activity toward (+)-*anti*-BPDE, which is consistent with the kinetic data of the hGSTA2-2 mutants. Among residues 11, 110, and 215, residue 11 appears to have the biggest impact on the catalytic activity of the enzyme toward (+)-*anti*-BPDE, which is readily explainable when the mobilities of these three residues in the structure are taken into account. The mobility of a residue, suggested by its B factor, is correlated with the rigidity of the structural motif in which it is located. In the crystal structure of the hGSTA1-1·GSEa complex,[65] the B factors of the Cα atoms of residues 11, 110, and 215 are 19.6, 48.3, and 47.6 Å, respectively. Therefore, residue 11 is located in a more rigid structural motif and the Ala-11-Ile mutation should have much more profound steric impact on the BPDE moiety than the Val-110-Phe and Ala-215-Ser mutations. The kinetic data using H-site mutants of hGSTA2-2 support this prediction.

The kinetic data discussed above have made it possible to speculate on the role of different classes of human GSTs in the GSH conjugation and detoxification of PAH diol epoxides. It is reasonable to postulate that the majority of the inactivation (GSH conjugation) of activated PAHs in extrahepatic tissues including some of the target organs for PAH-induced cancers (e.g., lung) is most likely handled by hGSTP1-1, especially in individuals lacking hGSTM1-1 due to gene deletion, because of the higher abundance and reasonably good activity of hGSTP1-1. On the other hand, hGSTA1-1 may play an important role in GSH conjugation of diol epoxides in the liver, which is the main site of xenobiotic transformation, because hGSTP1-1 is very low in adult human livers and about 50% of Caucasians lack hGSTM1-1.

The substrate specificity profiling of Alpha-class human GSTs has also allowed us to draw important conclusions regarding their physiological roles. For example, both hGSTA1-1 and hGSTA2-2 exhibit high selenium-independent GSH peroxidase activity toward fatty acid hydroperoxides and phospholipid hydroperoxides,[72] whereas hGSTA3-3 is a highly efficient catalyst of double-bond isomerization of intermediates in steroid hormone biosynthesis (e.g., Δ[5]-androstene-3,17-dione).[73,74] The GSH conjugation of the lipid peroxidation product 4-hydroxynonenal is preferentially catalyzed by hGSTA4-4.[75,76] Distribution of Alpha-class GSTs is also highly variable in human tissues. For instance, hGSTA1-1 and hGSTA2-2 isoforms are the predominant Alpha-class GSTs in human livers and lungs,[77,78] whereas the message for hGSTA3 is not detectable in human livers.[73] Thus, it is reasonable to postulate that while hGSTA1-1 or hGSTA2-2 may contribute to inactivation of PAH diol epoxides as well as lipid hydroperoxides, hGSTA3-3 and hGSTA4-4 may respectively handle the double-bond isomerization of steroid biosynthesis intermediates and GSH conjugation of 4-hydroxynonenal.

6.5.3 Human GSTP1-1 Polymorphism and GSH Conjugation of PAH Diol Epoxides

The Pi-class human isozyme is polymorphic in human populations.[79–83] The hGSTP1-1 polymorphism involves base substitutions A313G in exon 5 (codon 104) and C341T in exon 6 (codon 113) that result in amino acid substitutions of Ile to Val and Ala to Val, respectively.[79–83] All four combinations of these substitutions (i.e., Ile-104/Ala-113, designated as *hGSTP1*A* in some publications and hereafter abbreviated as IA variant; Val-104/Ala-113, *hGSTP1*B* or VA variant; Val-104/Val-113, *hGSTP1*C* or VV variant; and Ile-104/Val-113, *hGSTP1*D* or IV variant) have now been identified in human populations.[78–83] The allele frequency varies between ethnic groups with IA being most frequent, but IV seems to be a rare allele.[82,83] Initial work in the laboratories of Zimniak and Awasthi indicated that the amino acid substitution in position 104 markedly changed the activity of hGSTP1-1 toward the model GST substrate 1-chloro-2,4-dinitrobenzene (CDNB), with IA being relatively more active than VA.[84] We extended these observations and determined the effects of amino acid substitutions in positions 104 or 113 on the efficacy of hGSTP1-1 for diol epoxide–GSH conjugation using planar *anti*-BPDE and *anti*-CDE.[85–87] In some publications, the variant sites are identified as 105 and 114 due to inclusion of initiator methionine.[88] We were the first to report that, unlike CDNB, the VA and VV variants were relatively more efficient than either IA or IV in catalyzing the GSH conjugation of *anti*-BPDE and *anti*-CDE.[85–87] For example, the estimated k_{cat}/K_m for VV variant (48 mM^{-1} s^{-1}) toward (+)-*anti*-BPDE–GSH conjugation was >3.5-fold higher when compared with IA and IV.[86] Similarly, the VA variant displayed >6-fold higher catalytic efficiency toward *anti*-CDE when compared with the IA isoforms.[87] Our results were subsequently confirmed by other laboratories.[88,89] For example, Sundberg et al.[88] observed between 1.4- and 2.7-fold higher k_{cat}/K_m for the VA variant when compared with IA using (+)-*anti*- and (+)-*syn*-isomers of bay-region CDE, BPDE, and DBADE.

In a separate study, Coles et al.[89] examined hGSTP1-1 variants purified from normal human lung samples for their activity toward (+)-*anti*-BPDE. Interestingly, the differences between the native hGSTP1-1 variants were much greater than those observed with recombinant proteins.[89] However, the order of differential catalytic efficiency of the native proteins was the same as for the recombinant proteins.[89]

In a follow-up study, we compared the kinetic parameters for hGSTP1-1 variants using 1*R*,2*S*-dihydroxy-3*S*,4*R*-epoxy-1,2,3,4-tetrahydro-5-methylchrysene ([+]-*anti*-5-MeCDE), a diol epoxide derived from 5-methylchrysene.[90] We were surprised to observe a reversal in catalytic efficiency ranking of hGSTP1-1 variants toward (+)-*anti*-5-MeCDE.[90] Thus, unlike *anti*-BPDE or *anti*-CDE, the estimated k_{cat}/K_m of IA for (+)-*anti*-5-MeCDE-GSH conjugation was ~1.7-fold higher compared with VA.[90] These results suggested that the specificity of the hGSTP1-1 variants may depend on the molecular shape of the diol epoxide substrate. Subsequently, we demonstrated that the reversal of activity ranking for hGSTP1-1 variants observed with (+)-*anti*-5-MeCDE was not restricted to this diol epoxide substrate because the IA variant was a relatively more efficient catalyst than VA or VV toward fjord-region *anti*-BCPDE and *anti*-BGCDE.[91] The IA and VA variants were found to have comparable activity toward (–)-*anti*-BGCDE or (–)-*anti*-BCPDE in a study conducted by Sundberg et al.[92] The maximum difference in catalytic efficiency toward *anti*-BCPDE in our study was observed between IA and VV;[91] the later was not included in the analysis by Sundberg et al.[92] Even though the reasons for the discrepancy between the results from our laboratory[91] and those by Sundberg et al.[92] are not yet clear, we speculated that the molecular shape of the diol epoxide substrate may be an important determinant of catalytic specificity of hGSTP1-1.

However, this conclusion was based on activity measurements with diol epoxides having different ring structures.

More recently, we addressed the question of whether differential catalytic efficiency ranking of hGSTP1-1 variants is due to differences in the ring structures of the diol epoxide substrates by determining kinetic constants for hGSTP1-1 variants toward a pair of *anti*-diol epoxide isomers of benzo[*c*]chrysene.[93] Benzo[*c*]chrysene is interesting because, unlike many other PAHs, it can be activated to yield a bay-region–type diol epoxide (*anti*-benzo[*c*]chrysene-1,2-diol-3,4-epoxide; abbreviated as *anti*-BCCDE-1) as well as a fjord-region–type isomer (*anti*-benzo[*c*]chrysene-9,10-diol-11,12-epoxide; abbreviated as *anti*-BCCDE-2).[94] The two *anti*-BCCDE isomers have identical ring structures but differ with respect to the locations of the epoxide group. The epoxide group in *anti*-BCCDE-1 is in a bay region, whereas *anti*-BCCDE-2 has fjord-region characteristics due to the location of its epoxide functional group in a sterically hindered region.[94] The valine-104 isoforms were ~1.7- to 2.3-fold more efficient than IA in catalyzing the GSH conjugation of the bay-region *anti*-BCCDE-1 isomer.[93] In contrast, the catalytic efficiency toward the fjord-region *anti*-BCCDE-2 isomer was significantly higher for the IA isoform than for either VA or VV.[93] These results clearly indicated that the reversal of activity ranking for hGSTP1-1 variants is independent of the difference in ring structure of the diol epoxide.

6.5.4 CRYSTAL STRUCTURES OF HGSTP1/IA·GSBPD AND HGSTP1/VA·GSBPD COMPLEXES

We were able to solve crystal structures of IA and VA variants of hGSTP1-1 in complex with the GSH conjugate of (+)-*anti*-BPDE at 2.1- and 2.0-Å resolution, respectively.[71] The structures revealed that the H-site of hGSTP1-1 is half-hydrophobic and half-hydrophilic. A hydrogen-bond network involving Arg-13, Asn-204, and five water molecules defines the hydrophilic portion of the H-site. This was the first such description that was sufficient to distinguish the H-site in hGSTP1-1 from that in other isozymes whose H-sites are hydrophobic cavities. A comparison of the crystal structures revealed that residue 104 in the H-site dictates the binding mode of the product molecule with three consequences. First, the distance between the hydroxyl group of Tyr-7 and the sulfur atom of GSBpd is 5.9 Å in the hGSTP1/IA·GSBpd complex versus 3.2 Å in hGSTP1/VA·GSBpd. Second, one of the hydroxyl groups of GSBpd forms a direct hydrogen bond with Arg-13 in the hGSTP1/VA·GSBpd complex, which is not observed in the hGSTP1/IA·GSBpd complex.[71] Third, in the hydrophilic portion of the H-site of the hGSTP1/IA·GSBpd complex, five water molecules are observed, whereas two of the five water molecules are displaced by the Bpd moiety of GSBpd in the structure of the hGSTP1/VA·GSBpd complex.[71] In conclusion, these structures revealed that the binding mode of the product of GSH conjugation of (+)-*anti*-BPDE is significantly different between hGSTP1/VA·GSBpd and hGSTP1/IA·GSBpd due to the lack of conformational freedom of Ile-104. Thus, Val-104 is able to assume two distinct conformations depending on the size and shape of the xenobiotic substrate, consequently having a broader substrate specificity than IA.

TABLE 6.1

Formation of DNA Adducts of (+)-*anti*-BPDE in HepG2 Cells Transfected with Variants of hGSTP1-1

	DNA Adducts of (+)-*anti*-BPDE (pmol/mg DNA)		
Cell Line	0.1 µM (+)-*anti*-BPDE	1.0 µM (+)-*anti*-BPDE	5.0 µM (+)-*anti*-BPDE
HepG2-vector	3.2 ± 0.6[a]	37.2 ± 2.3	106 ± 6
HepG2(IA)	2.9 ± 0.5	21.4 ± 5.0[b]	93 ± 14
HepG2(VA)	1.6 ± 0.3[b,c]	25.3 ± 4.8[b]	73 ± 7[b]
HepG2(VV)	1.0 ± 0.3[b,c]	16.9 ± 3.9[b]	62 ± 6[b,c]

[a] Data are mean ± SD of three determinations.

[b] Significantly different compared with vector-transfected HepG2 cells; $P < 0.05$.

[c] Significantly different compared with HepG2(IA) cells; $P < 0.05$.

Reprinted with permission from Hu et al., *Cancer Research*, 59, 2358, 1999. Copyright © 1999 American Association for Cancer Research.

6.5.5 EFFICACY OF HGSTP1-1 VARIANTS AGAINST (+)-*ANTI*-BPDE–INDUCED DNA DAMAGE IN CELLS

To determine the cellular significance of kinetic data with purified hGSTP1-1 variants, we stably transfected HepG2 cells with allelic variants of hGSTP1-1.[45] A matched set of cell lines with comparable hGSTP1-1 variant protein levels (1275 ± 100, 1259 ± 144, and 1232 ± 148 ng of the corresponding hGSTP1-1 allelic protein/mg of total cellular protein in clonal cell lines HepG2/hGSTP1[IA]-20, HepG2/hGSTP1[VA]-4, and HepG2/hGSTP1[VV]-21, respectively) were examined for (+)-*anti*-BPDE–DNA adduction. The hGSTP1-1 expression was not detectable in vector-transfected control cells (HepG2-vect cells).[45]

Data on BPDE–DNA adduct formation in hGSTP1-1 variant-transfected cells are summarized in Table 6.1. Exposure of vector-transfected control cells and hGSTP1-1 variant-transfected cells to (+)-*anti*-BPDE resulted in the formation of DNA adducts of BPDE in a concentration-dependent manner.[45] The levels of BPDE–DNA adducts were significantly lower in HepG2(VA) and HepG2(VV) cells compared with vector-transfected control cells. In agreement with kinetic data using purified proteins,[85,86] the IA variant was least effective in protecting against (+)-*anti*-BPDE–induced DNA damage, whereas maximum protection was afforded by the VV isoform.[45] The differences were more pronounced at lower concentrations of (+)-*anti*-BPDE than at higher concentrations. These results clearly indicate that the VA and VV variants were more effective than IA in protecting against BPDE-induced DNA damage in an experimental system that is closer to a physiological situation than *in vitro* studies but is still well defined and controlled. However, the *in vivo* efficacy of hGSTP1-1 variants for protection against DNA-damaging effects of fjord-region diol epoxides remains to be determined. Sundberg et al.[95] have also shown that overexpression of hGSTP1-1(VA) confers significant protection against DNA adduct formation by (+)-*anti*-BPDE.

6.5.6 ASSOCIATION OF hGSTP1 POLYMORPHISM WITH CANCER SUSCEPTIBILITY

The association between hGSTP1 polymorphism and cancer susceptibility has been studied extensively.[82,83,96–100] However, the results of these studies are inconclusive. For example, Ryberg et al.[82] examined the association between polymorphism at codon 104 and lung cancer incidence. These investigators showed that the lung cancer patients had a significantly higher frequency of VA homozygotes (15.9%) and a lower frequency of IA homozygotes (38.4%) than the controls (9.1% and 51.5%, respectively).[82] Such an association was not observed in other studies.[97,98] Results are also inconclusive for other types of cancers. Thus, the results to date on the association between hGSTP1 polymorphism and cancer susceptibility are variable.

There are two likely reasons for the poor penetrance of the hGSTP1 genotype in terms of cancer incidence. One is the presence of uncontrolled variability inherent to the human population, including both genetic background and environmental variables such as diet and drug regimens that may induce GSTs including hGSTP1-1 and other phase I and phase II drug-metabolizing enzymes. The second is the fact that human exposure to PAH-derived carcinogens differs in duration, level, and, above all, composition of PAHs. Our results show that bay-region PAH diol epoxides are more readily conjugated by the Val-104–containing hGSTP1-1 variants than by Ile-104 enzymes, but this activity ranking is reversed for the fjord-region PAH diol epoxides.[85–87,90,91,93] Thus, a given set of hGSTP1 gene products present in an individual can either afford protection or lead to increased vulnerability, depending on the source of PAH exposure, which is largely unknown and uncontrollable.

ACKNOWLEDGMENTS

The work in our laboratory was supported by USPHS grants CA076348 awarded by the National Cancer Institute and ES09140 awarded by the National Institute of Environmental Health Sciences. The authors thank past and present members of the Singh laboratory for their contributions to the work summarized in this chapter and to Kamayani Singh for preparation of the figures.

REFERENCES

1. Mannervik, B. and Danielson, U.H., Glutathione transferases — structure and catalytic activity, *Crit. Rev. Biochem. Mol. Biol.*, 23, 283, 1988.
2. Hayes, J.D. and Pulford, D.J., The glutathione S-transferase supergene family: regulation of GST* and the contribution of the isoenzymes to cancer chemoprotection and drug resistance, *Crit. Rev. Biochem. Mol. Biol.*, 30, 445, 1995.
3. Listowsky, I. et al., Intracellular binding and transport of hormones and xenobiotics by glutathione S-transferases, *Drug Metab. Rev.*, 19, 305, 1988.

4. International Agency for Research on Cancer, *IARC Monographs on the Evaluation of Carcinogenic Risk of Chemicals to Man, Certain Polycyclic Aromatic Hydrocarbons and Heterocyclic Compounds*, IARC, Lyon, France, Vol. 3, 1973.

5. International Agency for Research on Cancer, *IARC Monographs on the Evaluation of Carcinogenic Risk of Chemicals to Man, Polynuclear Aromatic Compounds, Part I, Chemical, Environmental, and Experimental Data*, IARC, Lyon, France, Vol. 32, 1983.

6. Buening, M.K. et al., Tumorigenicity of the optical enantiomers of the diastereomeric benzo[*a*]pyrene-7,8-diol-9,10-epoxides in newborn mice: exceptional activity of (+)-7β,8α-dihydroxy-9α,10α-epoxy-7,8,9,10-tetrahydrobenzo[a]pyrene, *Proc. Natl. Acad. Sci., USA*, 75, 5358, 1978.

7. Slaga, T.J. et al., Marked differences in the skin tumor–initiating activities of the optical enantiomers of the diastereomeric benzo[*a*]pyrene-7,8-diol-9,10-epoxides, *Cancer Res.*, 39, 67, 1979.

8. Amin, S. et al., Tumorigenicity of fjord-region diol epoxides of polycyclic aromatic hydrocarbons, *Polycycl. Arom. Hydrocarb.*, 11, 365, 1996.

9. Wood, A.W. et al., Differences in mutagenicity of the optical enantiomers of the diastereomeric benzo[*a*]pyrene-7,8-diol-9,10-epoxides, *Biochem. Biophys. Res. Commun.*, 77, 1389, 1977.

10. Glatt, H. et al., Fjord- and bay-region diol-epoxides investigated for stability, SOS induction in *Escherichia coli*, and mutagenicity in *Salmonella typhimurium* and mammalian cells, *Cancer Res.*, 51, 1659, 1991.

11. Sims, P. and Grover, P.L., Epoxides of polycyclic aromatic hydrocarbons. Metabolism and carcinogenesis, *Adv. Cancer Res.*, 20, 165, 1974.

12. Thakker, D.R. et al., Metabolism of benzo[*a*]pyrene: stereoselective metabolism of benzo[*a*]pyrene and benzo[*a*]pyrene 7,8-dihydrodiol to diol epoxides, *Chem. Biol. Interact.*, 16, 281, 1977.

13. Thakker, D.R., Yagi, H., Levin, W., Wood, A.W., Conney, A.H., and Jerina, D.M., Polycyclic aromatic hydrocarbons: metabolic activation to ultimate carcinogens, in *Bioactivation of Foreign Compounds*, Anders, M.W., Ed., Academic Press, New York, 1985, 177.

14. Levin, W. et al., Tumorigenicity of optical isomers of the diastereomeric 3,4-diol-1,2-epoxides of benzo(*c*)phenanthrene in murine tumor models, *Cancer Res.*, 46, 2257, 1986.

15. Weinstein, I.B. et al., Benzo[*a*]pyrene diol epoxides as intermediates in nucleic acid binding *in vitro* and *in vivo*, *Science*, 193, 592, 1976.

16. Cheng, S.C. et al., DNA adducts from carcinogenic and non-carcinogenic enantiomers of benzo[*a*]pyrene dihydrodiol epoxide, *Chem. Res. Toxicol.*, 2, 334, 1989.

17. Geacintov, N.E. et al., NMR solution structures of stereoisomeric covalent polycyclic aromatic carcinogen-DNA adducts: principles, patterns, and diversity, *Chem. Res. Toxicol.*, 10, 111, 1997.

18. Szeliga, J. and Dipple, A., DNA adduct formation by polycyclic aromatic hydrocarbon dihydrodiol epoxides, *Chem. Res. Toxicol.*, 11, 1, 1998.

19. Marshall, C.J., Vousden, K.H., and Phillips, D.H., Activation of c-H-ras-1 proto-oncogene by *in vitro* modification with a chemical carcinogen benzo[*a*]pyrene diol-epoxide, *Nature*, 310, 586, 1984.

20. Denissenko, M.F. et al., Preferential formation of benzo[*a*]pyrene adducts at lung cancer mutational hotspots in P53, *Science*, 274, 430, 1996.

21. Hollstein, M. et al., p53 mutations in human cancers, *Science*, 253, 49, 1991.

22. Hussain, S.P. et al., Mutability of p53 hotspot codons to benzo(*a*)pyrene diol epoxide (BPDE) and the frequency of p53 mutations in nontumorous human lung, *Cancer Res.*, 61, 6350, 2001.

23. Jernström, B. et al., Glutathione conjugation of the carcinogenic and mutagenic electrophile (±)-7β,8α-dihydroxy-9α,10α-oxy-7,8,9,10-tetrahydrobenzo[*a*]pyrene catalyzed by purified rat liver glutathione transferases, *Carcinogenesis*, 6, 85, 1985.

24. Robertson, I.G.C. and Jernström, B., The enzymatic conjugation of glutathione with bay-region diol epoxides of benzo[*a*]pyrene, benzo[*a*]anthracene and chrysene, *Carcinogenesis*, 7, 1633, 1986.

25. Robertson, I.G.C. et al., Differences in stereoselectivity and catalytic efficiency of three human glutathione transferases in the conjugation of glutathione with 7β,8α-dihydroxy-9α,10α-oxy-7,8,9,10-tetrahydrobenzo[*a*]pyrene, *Cancer Res.*, 46, 2220, 1986.

26. Jernström, B. et al., Glutathione conjugation of trans-3,4-dihydroxy 1,2-epoxy 1,2,3,4-tetrahydrobenzo[*c*]phenanthrene isomers by human glutathione transferases, *Carcinogenesis*, 13, 1549, 1992.

27. Jernström, B. et al., Glutathione S-transferase A1-1-catalyzed conjugation of bay- and fjord-region diol epoxides of polycyclic aromatic hydrocarbons with glutathione, *Carcinogenesis*, 17, 1491, 1996.

28. Sundberg, K. et al., Glutathione conjugation of bay- and fjord-region diol epoxides of polycyclic aromatic hydrocarbons by glutathione transferases M1-1 and P1-1, *Chem. Res. Toxicol.*, 10, 1221, 1997.

29. Hu, X. et al., An alpha class mouse glutathione S-transferase with exceptional catalytic efficiency in the conjugation of glutathione with 7β,8α-dihydroxy-9α,10α-oxy-7,8,9,10-tetrahydrobenzo(*a*)pyrene, *J. Biol. Chem.*, 271, 32684, 1996.

30. Hu, X. and Singh, S.V., Differential catalytic efficiency and enantioselectivity of murine glutathione S-transferase isoenzymes in the glutathione conjugation of carcinogenic anti-diol epoxides of chrysene and benzo(*g*)chrysene, *Arch. Biochem. Biophys.*, 345, 318, 1997.

31. Yang, S.K. and Gelboin, H.V., Nonenzymatic reduction of benzo[*a*]pyrene diol-epoxides to trihydroxypentahydrobenzo[*a*]pyrenes by reduced nicotinamide adenine dinucleotide phosphate, *Cancer Res.*, 36, 4185, 1976.

32. Dock, L. et al., Studies on the further activation of benzo[*a*]pyrene diol epoxides by rat liver microsomes and nuclei, *Chem. Biol. Interact.*, 58, 301, 1986.

33. Lu, A.Y.H., Jerina, D.M., and Levin, W., Liver microsomal epoxide hydrolase, *J. Biol. Chem.*, 252, 3715, 1977.

34. Thakker, D.R. et al., Stereospecificity of microsomal and purified epoxide hydrolase from rat liver, *J. Biol. Chem.*, 252, 6328, 1977.

35. Penning, T.M., Dihydrodiol dehydrogenase and its role in polycyclic aromatic hydrocarbon metabolism, *Chem. Biol. Interact.*, 89, 1, 1993.

36. Long, D.J. et al., NAD(P)H:quinone oxidoreductase 1 deficiency increases susceptibility to benzo(*a*)pyrene-induced mouse skin carcinogenesis, *Cancer Res.*, 60, 5913, 2000.

37. Hesse, S. et al., Inhibition of binding of benzo(*a*)pyrene metabolites to nuclear DNA by glutathione and glutathione S-transferase B, *Biochem. Biophys. Res. Commun.*, 94, 612, 1980.

38. Hesse, S. et al., Inactivation of DNA-binding metabolites of benzo[*a*]pyrene and benzo[*a*]pyrene-7,8-dihydrodiol by glutathione and glutathione S-transferases, *Carcinogenesis*, 3, 757, 1982.

39. Jernström, B. et al., Glutathione conjugation and DNA-binding of (+/–)-trans-7,8-dihydroxy-7,8-dihydrobenzo[a]pyrene and (+/–)-7 beta, 8 alpha-dihydroxy-9 alpha, 10 alpha-epoxy-7,8,9,10-tetrahydrobenzo[a]pyrene in isolated rat hepatocytes, *Carcinogenesis*, 3, 861, 1982.

40. Jernström, B., Brigelius, R., and Sies, H., Glutathione conjugation of benzo[a]pyrene-7,8-dihydrodiol and benzo[a]pyrene-7,8-dihydrodiol-9,10-oxide in the perfused rat liver, *Chem. Biol. Interact.*, 44, 185, 1983.

41. Morrison, H., Hammarskiold, V., and Jernström, B., Status of reduced glutathione in primary cultures of rat hepatocytes and the effect on conjugation of benzo[a]pyrene-7,8-dihydrodiol-9,10-oxide, *Chem. Biol. Interact.*, 45, 235, 1983.

42. Jernström, B. et al., Metabolism of benzo[a]pyrene-7,8-dihydrodiol and benzo[a]pyrene-7,8-dihydrodiol-9,10-epoxide to protein-binding products and glutathione conjugates in isolated rat hepatocytes, *Carcinogenesis*, 5, 1079, 1984.

43. Hodgson, R.M. et al., Metabolism of the bay-region diol-epoxide of chrysene to a triol-epoxide and the enzyme catalysed conjugation of these epoxides with glutathione, *Carcinogenesis*, 7, 2095, 1986.

44. Fields, W.R. et al., Overexpression of stably transfected human glutathione S-transferase P1-1 protects against DNA damage by benzo[a]pyrene diol-epoxide in human T47D cells, *Mol. Pharmacol.*, 54, 298, 1998.

45. Hu, X. et al., Differential protection against benzo[a]pyrene-7,8-dihydrodiol-9,10-epoxide–induced DNA damage in HepG2 cells stably transfected with allelic variants of π class human glutathione S-transferase, *Cancer Res.*, 59, 2358, 1999.

46. Guo, J., Pan, S.S., and Singh, S.V., Exceptional activity of murine glutathione transferase A1-1 against (7R,8S)-dihydroxy-(9S,10R)-epoxy-7,8,9,10-tetrahydrobenzo[a]pyrene-induced DNA damage in stably transfected cells, *Mol. Carcinogenesis*, 36, 67, 2003.

47. Srivastava, S.K. et al., Potentiation of benzo[a]pyrene-induced pulmonary and forestomach tumorigenesis in mice by D,L-buthionine-S,R-sulfoximine-mediated tissue glutathione depletion, *Cancer Lett.*, 153, 35, 2000.

48. Henderson, C.J. et al., Increased skin tumorigenesis in mice lacking pi class glutathione S-transferase, *Proc. Natl. Acad. Sci. USA*, 95, 5275, 1998.

49. Mannervik, B. et al., Nomenclature for human glutathione transferases, *Biochem. J.*, 282, 305, 1992.

50. Robertson, I.G.C. et al., Glutathione transferases in rat lung: the presence of transferase 7-7, highly efficient in the conjugation of glutathione with the carcinogenic (+)-7β,8α-dihydroxy-9α,10α-oxy-7,8,9,10-tetrahydrobenzo[a]pyrene, *Carcinogenesis*, 7, 295, 1986.

51. Xia, H. et al., Cloning, expression, and biochemical characterization of a functionally novel alpha class glutathione S-transferase with exceptional activity in the glutathione conjugation of (+)-anti-7,8-dihydroxy-9,10-oxy-7,8,9,10-tetrahydrobenzo(a)pyrene, *Arch. Biochem. Biophys.*, 353, 337, 1998.

52. Hu, X. et al., Glutathione S-transferases of female A/J mouse liver and forestomach and their differential induction by anti-carcinogenic organosulfides from garlic, *Arch. Biochem. Biophys.*, 336, 199, 1996.

53. Hu, X. et al., Induction of glutathione S-transferase π as a bioassay for the evaluation of potency of inhibitors of benzo(a)pyrene-induced cancer in a murine model, *Int. J. Cancer*, 73, 897, 1997.

54. Hu, X. and Singh, S.V., Glutathione S-transferases of female A/J mouse lung and their induction by anticarcinogenic organosulfides from garlic, *Arch. Biochem. Biophys.*, 340, 279, 1997.

55. Pal, A. et al., Specificity of murine glutathione S-transferase isozymes in the glutathione conjugation of (–)-*anti*- and (+)-*syn*-stereoisomers of benzo[*g*]chrysene 11,12-diol 13,14-epoxide, *Carcinogenesis*, 20, 1997, 1999.

56. Giles, A.S., Seidel, A., and Phillips, D.H., Covalent DNA adducts formed in mouse epidermis by benzo[*g*]chrysene, *Carcinogenesis*, 17, 1331, 1996.

57. Hu, X. et al., Differential enantioselectivity of murine glutathione S-transferase isoenzymes in the glutathione conjugation of trans-3,4-dihydroxy-1,2-oxy-1,2,3,4-tetrahydrobenzo[*c*]phenanthrene stereoisomers, *Arch. Biochem. Biophys.*, 358, 40, 1998.

58. Zimniak, P. et al., Estimation of genomic complexity, heterologous expression, and enzymatic characterization of mouse glutathione S-transferase mGSTA4-4 (GST 5.7), *J. Biol. Chem.*, 269, 992, 1994.

59. McLellan, L.I. et al., Regulation of mouse glutathione S-transferases by chemoprotectors: molecular evidence for the existence of three distinct alpha-class glutathione S-transferase subunits, Ya_1, Ya_2 and Ya_3, in mouse liver, *Biochem. J.*, 276, 461, 1991.

60. Hayes, J.D. et al., Molecular cloning and heterologous expression of a cDNA encoding a mouse glutathione S-transferase Yc subunit possessing high catalytic activity for aflatoxin B_1-8,9-epoxide, *Biochem. J.*, 285, 173, 1992.

61. Xia, H. et al., Amino acid substitutions at positions 207 and 221 contribute to catalytic differences between murine glutathione S-transferase A1-1 and A2-2 toward (+)-*anti*-7,8-dihydroxy-9,10-epoxy-7,8,9,10-tetrahydrobenzo[*a*]pyrene, *Biochemistry*, 38, 9824, 1999.

62. Pal, A. et al., C-terminal region amino acid substitutions contribute to catalytic differences between murine class alpha glutathione transferases mGSTA1-1 and mGSTA2-2 toward anti-diol epoxide isomers of benzo[*c*]phenanthrene, *Biochemistry*, 40, 7047, 2001.

63. Gu, Y., Singh, S.V., and Ji, X., Residue R216 and catalytic efficiency of a murine class alpha glutathione S-transferase toward benzo[*a*]pyrene 7(*R*),8(*S*)-diol 9(*S*),10(*R*)-epoxide, *Biochemistry*, 39, 12552, 2000.

64. Sinning, I. et al., Structure determination and refinement of human alpha class glutathione transferase A1-1, and a comparison with the mu and pi class enzymes, *J. Mol. Biol.*, 232, 192, 1993.

65. Cameron, A.D. et al., Structural analysis of human alpha-class glutathione transferase A1-1 in the apo-form and in complexes with ethacrynic acid and its glutathione conjugate, *Structure*, 3, 717, 1995.

66. Xiao, B. et al., Crystal structure of a murine glutathione S-transferase in complex with a glutathione conjugate of 4-hydroxynon-2-enal in one subunit and glutathione in the other: evidence of signaling across the dimer interface, *Biochemistry*, 38, 11887, 1999.

67. Pal, A. et al., Role of arginine-216 in catalytic activity of murine alpha class glutathione transferases mGSTA1-1 and mGSTA2-2 toward carcinogenic diol epoxides of polycyclic aromatic hydrocarbons, *Carcinogenesis*, 22, 1301, 2001.

68. Gu, Y. et al., Residues 207, 216, and 221 and the catalytic activity of mGSTA1-1 and mGSTA2-2 toward benzo[*a*]pyrene-(7*R*,8*S*)-diol-(9*S*,10*R*)-epoxide, *Biochemistry*, 42, 917, 2003.

69. Dreij, K. et al., Catalytic activities of human alpha class glutathione transferases toward carcinogenic dibenzo[*a,l*]pyrene diol epoxides, *Chem. Res. Toxicol.*, 15, 825, 2002.

70. Singh, S.V. et al., Structural basis for catalytic differences between alpha class human glutathione transferases hGSTA1-1 and hGSTA2-2 for glutathione conjugation of environmental carcinogen benzo[*a*]pyrene-7,8-diol-9,10-epoxide, *Biochemistry*, 43, 9708, 2004.

71. Ji, X. et al., Structure and function of residue 104 and water molecules in the xenobiotic substrate-binding site in human glutathione S-transferase P1-1, *Biochemistry*, 38, 10231, 1999.

72. Zhao, T.J. et al., The role of human glutathione S-transferases hGSTA1-1 and hGSTA2-2 in protection against oxidative stress, *Arch. Biochem. Biophys.*, 367, 216, 1999.

73. Johansson, A. and Mannervik, B., Human glutathione transferase A3-3, a highly efficient catalyst of double-bond isomerization in the biosynthetic pathway of steroid hormones, *J. Biol. Chem.*, 276, 33061, 2001.

74. Gu, Y. et al., Crystal structure of human glutathione S-transferase A3-3 and mechanistic implications for its high steroid isomerase activity, *Biochemistry*, 43, 15673, 2004.

75. Board, P.G., Identification of cDNAs encoding two human alpha class glutathione transferases (GSTA3 and GSTA4) and the heterologous expression of GSTA4-4, *Biochem. J.*, 330, 827, 1998.

76. Hubatsch, I., Ridderstrom, M., and Mannervik, B., Human glutathione transferase A4-4: an alpha class enzyme with high catalytic efficiency in the conjugation of 4-hydroxynonenal and other genotoxic products of lipid peroxidation, *Biochem. J.*, 330, 175, 1998.

77. Rowe, J.D., Nieves, E., and Listowsky, I., Subunit diversity and tissue distribution of human glutathione S-transferase: interpretations based on electrospray ionization-MS and peptide sequence-specific antisera, *Biochem. J.*, 325, 481, 1997.

78. Singhal, S.S. et al., Glutathione S-transferase of human lung: characterization and evaluation of the protective role of the alpha-class isozymes against lipid peroxidation, *Arch. Biochem. Biophys.*, 299, 232, 1992.

79. Board, P.G., Webb, G.C., and Coggan, M., Isolation of a cDNA clone and localization of the human glutathione S-transferase 3 genes to chromosome bands 11q13 and 12q13–14, *Ann. Hum. Genet.*, 53, 205, 1989.

80. Ahmad, H. et al., Primary and secondary structural analyses of glutathione S-transferase from human placenta, *Arch. Biochem. Biophys.*, 278, 398, 1990.

81. Ali-Osman, F. et al., Molecular cloning, characterization, and expression in *Escherichia coli* of full-length cDNAs of three human glutathione S-transferase pi gene variants: evidence for differential catalytic activity of the encoded proteins, *J. Biol. Chem.*, 272, 10004, 1997.

82. Ryberg, D. et al., Genotypes of glutathione transferase M1 and P1 and their significance for lung DNA adduct levels and cancer risk, *Carcinogenesis*, 18, 1285, 1997.

83. Watson, M.A. et al., Human glutathione S-transferase P1 polymorphisms: relationship to lung tissue enzyme activity and population frequency distribution, *Carcinogenesis*, 19, 275, 1998.

84. Zimniak, P. et al., Naturally occurring human glutathione S-transferase GSTP1-1 isoforms with isoleucine and valine in position 104 differ in enzymic properties, *Eur. J. Biochem.*, 224, 893, 1994.

85. Hu, X. et al., Active site architecture of polymorphic forms of human glutathione S-transferase P1-1 accounts for their enantioselectivity and disparate activity in the glutathione conjugation of 7β,8α-dihydroxy-9α,10α-oxy-7,8,9,10-tetrahydrobenzo(a)pyrene, *Biochem. Biophys. Res. Commun.*, 235, 424, 1997.

86. Hu, X. et al., Activity of four allelic forms of glutathione S-transferase hGSTP1-1 for diol epoxides of polycyclic aromatic hydrocarbons, *Biochem. Biophys. Res. Commun.*, 238, 397, 1997.

87. Hu, X. et al., Mechanism of differential catalytic efficiency of two polymorphic forms of human glutathione S-transferase P1-1 in the glutathione conjugation of carcinogenic diol epoxide of chrysene, *Arch. Biochem. Biophys.*, 345, 32, 1997.

88. Sundberg, K. et al., Differences in the catalytic efficiencies of allelic variants of glutathione transferase P1-1 towards carcinogenic diol epoxides of polycyclic aromatic hydrocarbons, *Carcinogenesis*, 19, 433, 1998.

89. Coles, B. et al., Expression of *hGSTP1* alleles in human lung and catalytic activity of the native protein variants towards 1-chloro-2,4-dinitrobenzene, 4-vinylpyridine and (+)-anti benzo[*a*]pyrene-7,8-diol-9,10-epoxide, *Cancer Lett.*, 156, 167, 2000.

90. Hu, X. et al., Specificities of human glutathione S-transferase isozymes toward *anti*-diol epoxides of methylchrysenes, *Carcinogenesis*, 19, 1685, 1998.

91. Hu, X. et al., Catalytic efficiencies of allelic variants of human glutathione S-transferase P1-1 toward carcinogenic *anti*-diol epoxides of benzo[*c*]phenanthrene and benzo[*g*]chrysene, *Cancer Res.*, 58, 5340, 1998.

92. Sundberg, K. et al., Detoxication of carcinogenic fjord-region diol epoxides of polycyclic aromatic hydrocarbons by glutathione transferase P1-1 variants and glutathione, *FEBS Lett.*, 438, 206, 1998.

93. Pal, A. et al., Location of the epoxide function determines specificity of the allelic variants of human glutathione transferase pi toward benzo[*c*]chrysene diol epoxide isomers, *FEBS Lett.*, 486, 163, 2000.

94. Desai, D.H. et al., Syntheses and identification of benzo[*c*]chrysene metabolites, *J. Poly. Aromatic Compd.*, 16, 255, 1999.

95. Sundberg, K. et al., Glutathione conjugation and DNA adduct formation of dibenzo[*a,l*]pyrene and benzo[*a*]pyrene diol epoxides in V79 cells stably expressing different human glutathione transferases, *Chem. Res. Toxicol.*, 15, 170, 2002.

96. Harries, L.W. et al., Identification of genetic polymorphisms at the glutathione S-transferase Pi locus and association with susceptibility to bladder, testicular, and prostate cancer, *Carcinogenesis*, 18, 641, 1997.

97. Harris, M.J. et al., Polymorphism of the pi class glutathione S-transferase in normal populations and cancer patients, *Pharmacogenetics*, 8, 27, 1998.

98. Saarikoski, S.T. et al., Combined effect of polymorphic GST genes on individual susceptibility to lung cancer, *Int. J. Cancer*, 77, 516, 1998.

99. Shepard, T.F. et al., No association between the I105V polymorphism of the glutathione S-transferase P1 gene (GSTP1) and prostate cancer risk: a prospective study, *Cancer Epidemiol. Biomar. Prevent.*, 9, 1267, 2000.

100. Van Lieshout, E.M et al., Polymorphic expression of the glutathione S-transferase P1 gene and its susceptibility to Barrett's esophagus and esophageal carcinoma, *Cancer Res.*, 59, 586, 1999.

7 GST Polymorphism: Where to Now? Clinical Applications and Functional Analysis

S.L. Holley, A.A. Fryer, W. Carroll,
P.R. Hoban, and R.C. Strange

CONTENTS

7.1   INTRODUCTION

Glutathione S-transferases (GSTs; EC 2.5.1.18) are a supergene family of detoxifying dimeric enzymes found in cells from virtually all life forms.[1–3] Traditionally, GSTs are considered to be phase II detoxifying enzymes that catalyze the conjugation of reduced glutathione with a wide variety of electrophilic substrates.[4] It is now clear that the functions of GST are much broader than just the classical concept of xenobiotic detoxification. For example, they have peroxidase and isomerase activities, they can inhibit the c-Jun N-terminal kinase (JNK) (as a monomer), and they can bind noncatalytically with a wide range of endogenous and exogenous ligands. Consequently, they could plausibly mediate cellular responses to exogenous or endogenous electrophiles, inflammation and the consequent oxidative stress, response to treatment with a variety of therapeutic agents, and control of the cell cycle. Thus, GSTs demonstrate potentially multiple functions. The true biological function of the GST remains unclear though it could be any of these activities.

The pleiotropic roles of the GST have prompted much interest, not least in the possibilities that allelic variation in these genes will have important biological and clinical consequences.[5] This chapter will first briefly review organization of the GST gene family and GST regulation and subcellular localization and will then describe the putative functions of GST, discuss the known allelic sites, and assess the clinical implications of GST polymorphism. Special attention will be given to GSTP1, as it is probably, with respect to polymorphism, the gene most convincingly associated with a clinical phenotype and dual functionalities.

7.2   STRUCTURE OF THE GST SUPERGENE FAMILY

Two evolutionarily distinct supergene families of GST isoenzymes exist; one is composed of cytosolic soluble proteins and another is composed of microsomal proteins.[6–8] The soluble proteins exist as monomeric subunits ranging from 22 kDa to 27 kDa, with catalytic activities as homo- or heterodimers.[9] At least 18 genes have been identified in the human cytosolic GST supergene family. These have been subdivided into eight classes designated Alpha, Mu, Pi, Theta, Zeta, Kappa, Sigma, and Omega.[7] Based on sequence homology and protein activity, a GST-like protein Chi, or χ, related to class Omega has been identified in mice as a stress-responsive protein.[10] Furthermore, the Kappa-class GST is expressed in the mitochondria.[11] The members of these families and their chromosomal localizations and tissue distributions are shown in Table 7.1. The second group is the evolutionarily distinct membrane-associated proteins in eicosanoid and glutathione metabolism (MAPEG) family.[12] Members of the MAPEG family have been described elsewhere[12–14] and are not covered in this chapter.

TABLE 7.1
Glutathione S-Transferase Polymorphisms, Chromosomal Localization, Tissue Expression and Functional Consequence

Class (Chromosome)	Genes	Alleles	Nucleotide Change	Amino Acid Change	Tissue Expression	Functional Consequence
Alpha (6p12)	A1–A5	A1*A	−69 C	Promoter	Liver, testis, kidney, adrenal, pancreas, lung	Reference[17]
		A1*B	−69 T	Promoter	Liver, testis, kidney	↓ Expression[17]
		A2*A	332 C, 335 C, 589 G, 629 A	Pro110, Ser112, Lys196, Glu210		Reference[18]
		A2*B	332 C, 335 C, 589 G, 629 C	Pro110, Ser112, Lys196, Ala210	Adrenal, pancreas, lung, brain	No change[18]
		A2*C	332 C, 335 G, 589 G, 629 A	Pro110, Thr112, Lys196, Glu210		No change[18]
		A2*E	332 T, 335 C, 589 G, 629 A	Ser110, Ser112, Lys196, Glu210		↓ Activity[18]
		A3			Placenta, testis, ovary, adrenal	
		A4			Small intestine, spleen, liver, kidney, brain, colon, heart, skin	
		A5			Liver	
Mu (1p13.3)	M1–M5	M1*A	519 G	Lys173	Liver, testis, kidney	Reference[19]
		M1*B	519 C	Asn173	Adrenal, lung, brain	No change[19]
		M1*0	Deletion	No protein		No activity[20]
		M1*Ax2	Duplication	Overexpression		↑ Activity[21]
		M2			Brain, skeletal muscle, testis, heart, kidney	
		M3*A	AAG	Intron 6	Brain, testis, skeletal muscle	Reference[22,23]

TABLE 7.1 (*Continued*)
Glutathione S-Transferase Polymorphisms, Chromosomal Localization, Tissue Expression and Functional Consequence

Class (Chromosome)	Genes	Alleles	Nucleotide Change	Amino Acid Change	Tissue Expression	Functional Consequence
		M3*B	AAG deletion	Intron 6		Altered YY1 motif[24]
		M4*A	2517 T	Intron 6	Brain, skeletal muscle, heart	Reference[25]
		M4*B	2517 C	Intron 6	Brain, skeletal muscle, heart, lung	No change[25]
		M5				
Pi (11q13)	P1	P1*A	313 A, 341 C, 555 C	Ile105, Ala114, Ser185	Lung, brain, liver, kidney, testis	Reference[26,27]
		P1*B	313 G, 341 C, 555 T	Val105, Ala114, Ser185	Pancreas, skeletal muscle, heart	Substrate dependent[26,27]
		P1*C	313 G, 341 T, 555 T	Val105, Val114, Ser185		Substrate dependent[26,27]
		P1*D	313 A, 341 T	Ile105, Val114		No change[26,27]
Theta (22q11)	T1–T2	T1*A	310 A	Thr104	Liver, kidney, brain, prostate	Reference[28]
		T1*B	310 C	Pro104	Prostate, small intestine	↓ Activity[28]
		T1*0	Gene deletion	No protein		No activity[28
		T2*A	481 G	Met139	Liver	Reference[28]
		T2*B	481 A	Ile139		Unknown[28]

Class	Gene	Allele	Nucleotide	Amino acid	Tissue	Status
Sigma (4q21-22)	S1				Fetal liver, bone marrow	
Zeta (14q24.3)	Z1	Z1*A	94 A, 124 A, 245 C	Lys32, Arg42, Thr82	Fetal liver, skeletal muscle	Reference[29,30]
		Z1*B	94 A, 124 G, 245 C	Lys32, Gly 42, Thr82		↓ Activity[30,31]
		Z1*C	94G, 124 G, 245 C	Glu32, Gly42, Thr82		↓ Activity[30,31]
		Z1*D	94 G, 124 G, 245 T	Glu32,Gly42, Met82		↓ Activity[29]
Omega (10q23-25)	O1–O2	O1*A	419 C, 464+1 AAG	Ala140, Glu155	Liver, breast, macrophages, brain	Reference[32]
		O1*B	419 C, 464+1 del	Ala140, del115		↑ Activity[32]
		O1*C	419 A, 464+1 AAG	Asp140, Glu155		↓ Activity[32]
		O1*D	419 A, 464+1 del	Asp140, del115		Unknown[32]
		O1*E	650 C	Thr217		Reference[32]
		O1*F	650 A	Asn 217		↓ Activity[32]
		O2*A	424 A	Asn142	Testis, liver, heart, prostate	Reference[32]
		O2*B	424 G	Asp 142	Skeletal muscle, kidney	Unknown[32]
		O3p (pseudogene)				
Kappa (unknown)	K1				Liver (mitochondria)	

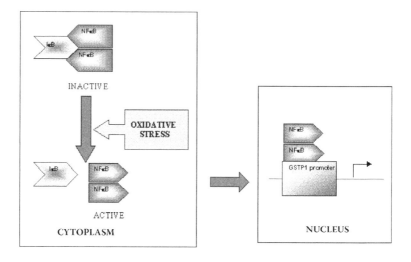

FIGURE 7.1 Regulation of GSTP1 expression by the NF-κB (nuclear factor-kappa B) transcription factor. Dimeric NF-κB is inactive when bound to IκB. Oxidative stress causes dissociation of IκB from NF-κB. This reactivates NF-κB, allowing it to induce expression of NF-κB responsive genes including GSTP1.[36]

7.3 GST REGULATION

Regulation of GST expression varies widely between classes, with response elements to a range of transcription factors identified in the promoters of both human and rodent GST genes. These include the antioxidant response element (ARE); xenobiotic, glucocorticoid, and Barbie box response elements; and GSTP enhancer 1 and motifs for AP-1, Jun, and NK-κB. These control elements suggest an evolutionary adaptive response to their functional roles. Antioxidant compounds transcriptionally activate GST genes through the ARE.[15,16] Exactly how the cell detects different chemicals and the information is transduced to the nucleus is unclear.

In the human GSTP1 gene, many transcription-factor recognition sites have been identified: NF-κB, a TATA box, a consensus AP-1 binding site, and two SP1 sites (G/C boxes).[33–36] In most cell types, NF-κB, a heterodimeric protein, is located in the cytoplasm where it is bound to IκB, an inhibitory subunit. Dissociation of IκB from the NF-κB dimer activates NF-κB, leaving it free to translocate to the nucleus where it expresses downstream genes containing NF-κB sites, including GSTP1[36] (Figure 7.1).

7.4 GST SUBCELLULAR LOCALIZATION

GSTs such as Alpha, Pi, and particularly Theta, are able to translocate from cytosol to the nucleus in response to drug treatment or other cellular stresses, possibly as transporters of prostaglandins and as protein kinase modulators.[37–39] Indeed, translocation of GSTs, possibly occurring in direct response to oxidative stress, has been observed with GSTT1 in response to the chemopreventive agent oltipraz[37] and with GSTP1 with acute lead exposure.[40] It is unclear whether a protein chaperone is needed to move GSTP1 into the nucleus or if diffusion occurs through nuclear membrane pores. This action suggests GSTs are useful as it is otherwise difficult to explain why such a response to an external threat would occur. This also raises the question of whether translocation of GST proteins occurs during development when profound time-dependent changes are needed in the level of expression of different GST genes.

7.5 GST FUNCTION

As we will see, both association studies between GST polymorphism and disease risk and functional analysis of specific polymorphism relate back to the plurality of function of GSTs detailed below.

7.5.1 GSTs and Xenobiotic Metabolism

Classically, GSTs are involved in phase II enzymatic detoxification, rendering the resultant compound more water soluble and thus more readily excreted.[41] The phase II reaction occurs through a lock-and-key active site involving, in GSTP1 for example, Tyr at position 50.[42] GSTs catalyze the nucleophilic attack of the sulphur atom of the tripeptide glutathione (GSH) on electrophilic groups of a range of hydrophobic substrates, both endogenous (e.g., by-products of reactive oxygen species [ROS] activity) and exogenous (e.g., polycyclic aromatic hydrocarbons [PAH]).[6,43] GSTP1 is associated with detoxification of the products of oxidative damage to nucleic acids,[44] and GSTA1 and GSTM1 with that of lipid hydroperoxides.[44] Most substrates are inactivated by conjugation with GSH but some are converted to more reactive compounds. For example, Theta-class GSTs are capable of activating dichloromethane to form the reactive S-chloromethylglutathione.[45]

7.5.2 GSTs and Cell Cycle Regulation

GSTs can also modulate protein kinase activity in the nucleus. Indeed, GSTP1 functions as an endogenous inhibitor of c-Jun N-terminal kinase (JNK) signaling in both rodent and human cells.[39,46] Under nonstressed conditions, the GSTP1 monomer functions as a JNK inhibitor by forming a complex with JNK (Figure 7.2a). Exposure to oxidative stress causes GSTP1 dissociation from JNK, and GSTP1 then forms covalently linked (enzymatically inactive) dimers or oligomers (Figure 7.2b). Liberation of JNK restores full kinase activity, as it can then be phosphorylated by a dual specific mitogen-activated protein (MAP) kinase, JKK. JNK can then bind and phosphorylate c-Jun, leading to c-Jun activation and increased transcription of AP-1 responsive genes. This activates signaling pathways for stress response and apoptosis.[47] Recent evidence has substantiated these original findings.[48,49]

Further studies suggest that the interaction between GSTP1 and JNK influences cell proliferation. Suppression of GSTP1 leads to elevated JNK activity, increased proliferation, and reduced apoptosis.[50]

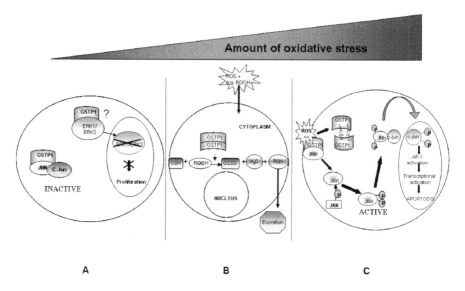

FIGURE 7.2 The role of GST at different levels of oxidative stress. (a) At low levels of oxidative stress or unstressed conditions GSTP1 is found catalytically active in complex with JNK thus inactivating JNK activity. GSTP1 is also thought to play a role in inhibition of cell proliferation involving ERK kinases — is this through direct binding? (b) As the level of oxidative stress increases, GSTP1 performs its classical role as a phase II detoxification enzyme (catalytically active dimer) involving conjugation with glutathione (GSH). (c) Higher levels of oxidative stress, induced by UV irradiation or H_2O_2, results in dissociation of the GSTP1:JNK complex, reactivation of the kinase by JKK, and formation of disulphide bonds between subunits of GSTP1 to form a GSTP1 catalytically inactive covalent dimer. Activated JNK catalyzes phosphorylation of c-Jun, leading to c-Jun activation and increased transcription of AP-1 responsive genes and ultimately apoptosis.[39]

7.5.3 GSTs, Oxidative Stress, and Apoptosis

GSTs demonstrate activity toward a number of by-products of oxidative stress including lipid and DNA hydroperoxides and 4-hydroxynonenal. The detoxification of such compounds is critical because it prevents their involvement in free radical propagation reactions, which, if unchecked, can damage cellular components. At physiological concentrations, GSTs regulate ROS levels,[6] thereby influencing cellular proliferation, preventing ROS accumulation, and protecting cells against ROS-mediated death.[51]

The glutathione peroxidase activity of Alpha-class GSTs makes them candidates for diseases in which oxidative stress is a component. Furthermore, Theta-class substrates include lipid hydroperoxides, thus polymorphism in GSTT1 and GSTT2 may be implicated in susceptibility to inflammatory pathologies.

Omega-class GSTs are thought to have a novel ion channel modulation role. Dulhunty et al.[52] hypothesized that the product of the GSTO1-1 gene could protect cells containing ryanodine receptors from apoptosis induced by calcium mobilization from intracellular stores.

GSTM genes could be potentially important in oxidative stress-related diseases due to their activity toward DNA hydroperoxides. Further, GSTM1 plays a regulatory role in the heat shock–sensing pathway by binding to and inhibiting ASK1 (a MAP kinase) activity.[53,54] ASK1 activity is low in nonstressed cells due to its sequestration via protein–protein interactions with GSTM1 to form a GSTM1–ASK1 complex. Oxidative stress and heat shock lead to the dissociation of the complex resulting in liberation and activation of ASK1.[55,56]

As described above, GSTP1 influences signaling pathways for apoptosis by modulating JNK kinase activity.[39] Much of this work has now been substantiated. For example, Bernardini et al.[57] observed dimeric GSTP1 upon increased apoptosis of Jurkat cells following etoposide or H_2O_2 treatment and reduction of 1-chloro-2, 4-dinitrobenzene (CDNB) conjugation by GSTs.

7.5.4 GSTs as Mediators of Other Pathways

Eicosanoids such as prostaglandins (PG) and leukotrienes (LT) play key roles in inflammatory diseases such as asthma. They are involved in tissue injury and bronchial constriction and can indirectly stimulate cells to release ROS and proteases.[58] This leads to the induction of antioxidant enzyme expression and apoptotic/necrotic cell death.

GSTs are involved in eicosanoid synthesis.[6] GSTA1 and GSTA2 catalyze the formation of prostaglandin $F_{2\alpha}$ ($PGF_{2\alpha}$) and PGD_2, respectively, from PGH_2.[59,60] Sigma-class GSTs catalyze isomerization of PGH_2 to either PGD_2 or PGE_2[7] while PGJ_2 and PGA_2, formed from PGD_2 and PGE_2, respectively, can be inactivated by Alpha-, Mu-, and Pi-class GSTs via their conjugation with GSH.[61] Thus, increased conjugation of prostaglandins due to GST overexpression could increase cell proliferation because of the inhibition of the antiproliferative mode of action.

7.5.5 GSTs and Drug Resistance

As classical phase II enzymes, GSTs protect cells from exogenous cytotoxic agents. However, this protective mechanism is also activated on exposure to cytotoxic chemotherapeutic agents and expression of several GSTs is induced in response to such exposure, contributing to anticancer drug resistance.[62] For example, increased immunohistochemical expression of GSTP1 is related to clinical drug resistance in nonsmall-cell lung cancer.[63,64] GSTs can serve two distinct roles in the development of drug resistance either via direct detoxification or in cell cycle control as an inhibitor of the MAPK pathway.[8] As an inhibitor of JNK activity, it is possible that GSTP1 overexpression results in cellular resistance to apoptosis during anticancer therapy.[39,65,66]

A variety of GST inhibitors derived from a group of rationally designed glutathione analogs has been shown to modulate drug resistance by sensitizing tumor cells to anticancer drugs.[67–70] Selective targeting of susceptible tumor phenotypes is a strategy that should result in the release of more active drugs in malignant cells compared with normal tissue, thereby achieving an improved therapeutic index. Such drugs include TLK199, which is a selective inhibitor of the protein product from

GSTP1*A (Ile[105]–Ala[114]).[71] The GSTP1 Ile[105] allele has been associated with reduced disease-free survival in patients treated with platinum; the resistance to therapy in these patients may be due to the formation of platinum–glutathione conjugates.[72,73] Patients homozygous for GSTP1*B (reduced catalytic activity) could have a treatment advantage as they will have a diminished capacity to detoxify platinum-based anticancer agents.[73]

Alpha-class GSTs have also been evaluated in the context of anticancer drug resistance,[8] though the effect of polymorphism has not so far been examined.

7.5.6 GSTP1 IN DEVELOPMENT

GSTs are thought to have a multitude of functions as described above, but are GST enzymes important? Supporting evidence comes from studies of GST expression during development. Indeed, the impact of GSTs on cell cycle control and the regulatory role of low levels of ROS suggest a possible role in tissue growth and development. Further, GSTs exhibit marked time- and tissue-dependent levels of expression. For example, in human lungs, Pi-class GSTs decrease markedly after 15 weeks of gestation, and at birth the level of activity of the isoenzyme is only about 10–20% to those obtained during the first trimester. Alpha- and Mu-class GSTs are weakly expressed in lungs during development and each class is responsible for less than 10% of cytosolic activity.[74] Thus, in lungs, GSTP1 is the predominant GST with the highest levels of expression in the epithelium.[74–76] Similar gestational changes in GSTP1 expression are observed in hepatocytes.

GSTP1 may have a particularly important role in lung development and repair mechanisms. Cells from GSTP1 knockout mice (GSTP1[-/-]) have significantly faster doubling times than those from wild-type mice.[50] Furthermore, null mice demonstrate significantly larger lungs than wild-type mice, implying a role for this gene in lung growth. A role for GSTP1 in human lung biology is supported by GSTP1 polymorphism studies in asthmatic patients. Lung function can be used as an indirect measure of structure in studies of humans. The maximum volume of breath exhaled during a forced expiration reflects the anatomy of the airways.[77–79] Two easily measured components of a forced full expiration are particularly valuable in determining lung function. The forced expiratory volume in one second (FEV_1) is the maximum amount of air that can be exhaled following a full inspiration in one second. This value reflects the elasticity and caliber of the individual airways. Forced vital capacity (FVC) is a measure of the maximum volume that can be exhaled from full inspiration to full expiration. This measurement is more reflective of the total lung volume and is correlated with total alveolar number. Individual measurements and longitudinal changes offer insight into how the GSTs might determine lung development and repair in humans. Estimates of heritability suggest an important but variable genetic contribution to adult lung function values.[77,78,80] The relationship of these lung function values to each other is also important. Lower-than-expected FEV_1/FVC ratios are seen in lungs with narrowed conducting airways, a pattern that is characteristic of asthma.

Human GSTP1 is located on chromosome 11q13, a hotspot for asthma susceptibility. Longitudinal studies suggest that the GSTP1 Val[105]Val[105] genotype is asso-

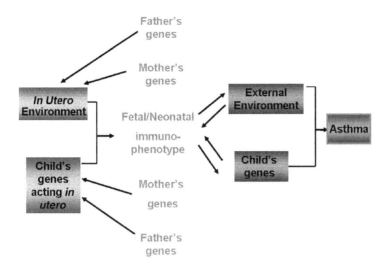

FIGURE 7.3 Factors affecting the development of asthma.

ciated with decreased lung function growth in children.[81] We and others have also shown that GSTP1 Val[105]Val[105] is associated with lung function in children[82] and with bronchial hyperresponsiveness (BHR) in both children and adults with atopic and occupational asthma.[83–85] The associations between GSTP1 polymorphism and lung function do not appear to be limited to either individuals' lifetimes or even their own genotypes. There is mounting evidence in humans that GSTP1 has an important role before birth *in utero*. We have shown that the maternal (but not paternal) GSTP1 genotype is associated with both BHR and lung function in their children.[86] From our data, this association appears independent of transmission of specific maternal alleles. This observation suggests that maternal GSTP1 can exert a lasting influence on lung function and that these changes are likely to happen before birth.

GST gene dosage, haploinsufficiency, and deletion could directly affect development. Further, GST gene variants via cell proliferation and apoptotic pathways could influence lung development during pre- and postnatal lung growth and may reduce the lung's ability to effectively repair itself during later life. Maternal GSTP1 may influence fetal lung growth (Figure 7.3).

7.6 POLYMORPHISM IN GSTS

The number of allelic variants in genes has increased rapidly with the Human Genome project and the availability of Web-based single-nucleotide polymorphism (SNP) and expressed sequence tag (EST) databases. But it is estimated that up to 85% of such theoretical SNPs cannot be confirmed in population studies. For example, EST searches have identified 26 potential SNPs in GSTO1, yet only one of these has been subsequently confirmed in population studies.[87] We will now describe the known and characterized polymorphisms within GSTs and then discuss the clinical implications of GSTs.

7.6.1 How Important Are Polymorphisms Within GSTs — Does It All Matter?

Polymorphisms are the low-penetrance mutations that make us all so subtlely different. Are GST polymorphisms important in mediating clinical phenotypes? One problem in answering this question is that we do not yet know the effect on gene function of each of the polymorphisms. That being said, as described above, the GSTs are very widely expressed and their expressions are elegantly controlled during human development. In addition GST involvement in any of the previously mentioned functions is likely to be important.

We can answer the question of the importance of polymorphism by looking at the potentially most severe polymorphism, that which causes loss of the gene; for example, in those individuals homozygous for the GSTM1 null allele. The importance of deletion or loss of GST expression and possible repercussions is evident through the effects observed in knockout mice and reports of methylation of the GSTP1 gene.

7.6.2 GST Knockout Mice

GSTP1 knockout mice (GSTP1$^{-/-}$) were originally described by Henderson et al.[88] Using a promoter trap technique they "knocked out" both GSTP1 and GSTP2 in the mice. Initial findings suggested GSTP1 was not essential for survival. The phenotypes of GSTP1$^{-/-}$ mice were essentially the same as that of nontransgenic littermates in terms of litter size, reproductive capabilities, and general health. Apart from larger lungs, the major organs revealed no significant changes. However, male GSTP1$^{-/-}$ mice did have a significantly greater body weight than their wild-type counterparts. Perhaps unsurprisingly, GST activity toward CDNB was markedly reduced in GSTP1$^{-/-}$ lung tissue, but significant pulmonary CDNB activity remained, indicating other GSTs are important in the lungs. Henderson et al. also studied the development of skin tumors in these mice as GSTP1 is the only GST found in mouse skin.[89] Skin tumorigenesis induced by the PAH, 7,12 dimethylbenz anthracene and the tumor-promoting agent, 12-0- tetradecanoylphorbol-13-acetate, resulted in a highly significant increase in skin papillomas in GSTP1$^{-/-}$ mice. GSTP1 expression also had a profound effect on the sensitivity of mice to acetaminophen (an analgesic) hepatotoxicity, which can often be fatal.[90] GSTP1$^{-/-}$ mice were, paradoxically, resistant to the effects of this compound.

Null-mouse models have also been generated for GSTZ1 and GSTA4. While none of these systems appear fatal to the organism, there is evidence of significant long-term effects, particularly on exposure to specific stresses. A GSTA4 knockout mouse exhibited overexpression of Alpha-, Mu-, and Pi-class GSTs, suggesting an adaptive mechanism responding to chemical exposure.[91] Deletion of the Zeta-class GST in a knockout-mouse model also demonstrated overexpression of these GSTs.[92] Phenotypically, GSTZ1 knockout mice had enlarged livers, as well as splenic atrophy, multifocal hepatitis, and ultrastructural changes in the kidneys,[92] suggesting a significant role for this gene perhaps in mesodermal development.

7.6.3 Methylation of GSTs

A further important epigenetic change to human GSTP1 expression is methylation. Little data exist on methylation in other GSTs apart from limited studies in the rat.[93] Methylation could account for much of the variation of GSTP1 expression: 82 CpG islands are located in the 5′ flanking region of GSTP1. Hypermethylation or "silencing" of GSTP1 is associated with the development of various cancers,[94–97] particularly prostate cancer, where associations with GSTP1 alleles have been shown to double the risk (GSTP1 Val[105]Val[105], OR 1.80) of developing the disease at an early age.[98] Methylation has also been shown to modulate GSTP1 expression in breast cancer cells.[98] The GSTP1 CpG island has been shown to be more densely methylated in estrogen receptor (ER) positive cells[99] while GSTP1 mRNA was more stable in ER negative cells.[100] Applicable to GSTP1 methylation-dependent silencing is that methyl-CpG binding proteins may mediate repression of methylated genes indirectly.[101–103]

The two genetic changes described above indicate the importance of tightly controlled GST expression. This also suggests polymorphisms are likely to be important, but which ones? Polymorphisms have been identified in the majority of GST classes.

7.7 POLYMORPHISMS IDENTIFIED WITHIN GSTS

7.7.1 Alpha Class

The GST Alpha-class genes show significant homology, making SNP identification and genotyping assay design problematic. The GSTA1 promoter is composed of four SNPs in complete linkage disequilibrium.[104] The GSTA1*B allele (T at position −69, A at position −52) alters a Sp1 binding site resulting in reduced expression. There is no evidence that two nonsynonymous substitutions (Thr112Ser and Glu210Ala) in GSTA2 alter GST activity (Table 7.1). No polymorphisms have been confirmed in GSTA3–A5.

7.7.2 Mu Class

Complete deletion of GSTM1 (GSTM1*0) is common (allele frequency 0.70 in Caucasians) and individuals not expressing GSTM1 thus have complete loss of GSTM1 activity.[105,106] Gene duplication can also occur (GSTM1*1x2 allele); individuals carrying such a duplication display ultrarapid GSTM1 activity.[107]

A nonsynonymous substitution in exon 7 of GSTM1 (Lys173Asp) was identified, although the differential associations of the GSTM1*A and GSTM1*B alleles with disease risk are likely to result from linkage disequilibrium with a neighboring SNP.[7] Indeed, GSTM1*A is in linkage disequilibrium with GSTM3*B, a 3-bp deletion that may generate a YY1 recognition motif.[7] The functional effect of an SNP in GSTM4 is unknown. No significant polymorphisms have been characterized in GSTM2 or GSTM5.

7.7.3 THETA CLASS

GSTT1 is commonly deleted (GSTT1*0) with an allele frequency in Caucasians of 0.40. An SNP (Thr104Pro) in GSTT1 has been described but little studied. Although three GSTT2 SNPs have been described (Table 7.1), assay design is hampered by homology with a pseudogene, GSTT2P. Consequently, there is little data on the clinical effect of these alleles. Both GSTT1*0 and the GSTT1 Pro[104] variant demonstrate reduced activity toward haloalkanes commonly used in paint stripper and in the synthesis of plastics.

7.7.4 PI CLASS

PupaSNP (http://pupasnp.bioinfo.cnio.es/; Conde et al.[108]) and Ensemble (http://www.ensemble.org) list 42 SNPs to exist in the GSTP1 gene; 1 in the 5′UTR, 24 intronic, 14 coding, and 3 in the 3′UTR. Of these potential SNPs, four alleles have been extensively studied: the wild-type GSTP1*A (Ile[105]–Ala[114]) and three variant alleles GSTP1*B (Val[105]–Ala[114]), GSTP1*C (Val[105]–Val[114]), and GSTP1*D (Ile[105]–Val[114]).[109–111] Combinations of these four alleles give rise to ten different genotypes: AA, AB, AC, AD, BB, BC, BD, CC, CD, and DD (Table 7.1). A further polymorphism in exon 7 (C/T) appears to be in tight linkage disequilibrium with the GSTP1 Ile[105]Val substitution.[86]

Many studies have investigated the effect of allelism in GSTP1 on detoxification as two characterized amino acid changes are in the electrophile-binding active site of the GSTP1 peptide.[109] While both the GSTP1 Val[105] variant and GSTP1 Ile[105] variant have an identical affinity for glutathione,[112] the GSTP1 Val[105] variant, compared to GSTP1 Ile[105], appears to confer a 7-fold higher catalytic efficiency for PAH diol epoxides but a 3-fold lower efficiency for CDNB.[109,113,114] The crystal structure of GSTP1 Val[105] reveals that it is likely to fit less bulky substrates but has broader substrate specificity than the GSTP1 Ile[105] variant of the enzyme.[115] In addition, heat stability of the proteins is different with half-lives of 19 minutes (GSTP1 Ile[105]) and 51 minutes (GSTP1 Val[105]). Amino acid 105 lies in close proximity to the active center and it is therefore not surprising that this residue influences catalytic activity.[115] Amino acid 114 is not located close to the catalytic site and the effect of the GSTP1 Ala[114]–Val[114] substitution is therefore unclear, though there is some evidence that it augments the efficacy of the GSTP1 Ile[105]–Val[105] substitution.[113]

7.7.5 ZETA CLASS

Expressed sequence tag (EST) database analysis has identified three functional substitutions in GSTZ1 (Table 7.1).[116] The GSTZ*A allele (Lys32, Arg42, Thr82) shows the highest activity toward dichloroacetic acid (a treatment for lactic acidosis) and the lowest toward fluoroacetate.

7.7.6 OMEGA CLASS

Polymorphisms in GSTO1 and GSTO2 have been described.[87] No significant difference was observed in GSTO1 allele activity toward CDNB. But thioltransferase activity has been shown to be decreased by 75% in GSTO1 Asp[140] (GSTO1*C) and by 40%

in GSTO1 Asn[217] variants compared to the wild-type GSTO1, but it was increased when the GSTO1*B allele (del 115) was examined. One polymorphism in GSTO1 alters a splice junction and causes the deletion of E155. This appears to contribute toward both a loss of heat stability and increased enzymatic activity.[87] An A>G substitution in GSTO2 at position 424 gives rise to an amino acid change from asparagine to aspartic acid in exon 4, but the functional consequence of this is unknown.

7.8 GST POLYMORPHISMS: CLINICAL IMPLICATIONS

When assessing the potential clinical potential of GST polymorphisms, genetic variation in CYP2D6 provides a good example of how polymorphisms can influence patient treatment. CYP2D6 genotypes give rise to individuals who are "extensive" (e.g., CYP2D6*2xn) and "poor" (e.g., CYP2D6*5) metabolizers of therapeutic agents including antidepressants, antipsychotics, and cancer drugs such as tamoxifen.[117] A recent article in *Pharmaceutical Discovery*[118] further highlights how allelic variants can shed light on tailor-made therapeutics. The heart failure drug BiDil (NitroMed, Lexington, MA, USA) is a combination drug that enhances nitric oxide levels in the body. BiDil was shown to reduce mortality rates from heart disease among African Americans, who suffer from a greater deficiency of nitric oxide than non-African Americans. This may be partly due to differences in the polymorphic endothelial nitric oxide synthase gene.[119,120] Cote et al.[121] have highlighted associations between GSTP1 and susceptibility to lung cancer only in African Americans compared to Caucasians.

Association studies have identified links between GST polymorphisms and susceptibility and outcome of cancers. For example, GSTP1 Ile[105] is associated with increased risk in head and neck cancer[122] and breast cancer,[123] yet other studies show protective effects with this allele. The reason for these differences is unclear though the effect of polymorphism appears substrate- and tissue-specific and concentration dependent.

In response to the question of which polymorphic sites are important, we can turn to the vast amount of data from GST molecular epidemiology studies. If we were to be critical, many early studies lack statistical power and have not been independently replicated. There has been little examination of GST linkage disequilibrium, tagged SNPs, or construction of haplotype blocks. Haplotype blocks are sites of closely located SNPs that are inherited in blocks. This gives rise to a few common haplotypes, which can be far more informative than the singular polymorphisms examined in some past association studies. While most significant associations between GST genotypes and clinical phenotypes have not been replicated, a minority have now been independently confirmed. Indeed, GSTP1 genotype association with asthma is a perfect example. GSTP1 Val[105]Val[105] has been associated with the reduced risk of airway hyperresponsiveness (OR 0.23–0.38) in three studies in asthmatic adults and children.[83–85]

7.8.1 Tobacco-Related Diseases

GSTM1 and GSTT1 have been studied as risk candidates for tobacco-related cancers.[124] While some studies show a significantly increased risk associated with

GSTM1*0 and GSTT1*0 homozygotes, others have not replicated these findings, prompting the use of meta-analysis. Analysis using 43 studies indicates that GSTM1*0 was not associated with increased lung cancer risk and that there was no evidence for an interactive effect between the genotype and tobacco consumption.[124] In addition, analysis of 31 studies of the influence of GSTM1, GSTT1, and GSTP1 polymorphism on head and neck cancer risk found only modest associations between risk and GSTM1*0 (odds ratio about 1.30) and GSTT1 with greater risk (odds ratio 2.06) associated with combinations of the genotypes.[125] Analysis of six studies have also failed to identify significant associations between GSTM1*0 and colorectal cancer risk.[126] Furthermore, no GST polymorphisms were found to interact with smoking to modify risk of breast cancer.[127]

Overall, the concept that GST polymorphisms, by reflecting a detoxication-deficient phenotype, identify individuals at increased risk of xenobiotic-related cancers is not generally supported by available data. This could also reflect case heterogeneity or the choice of polymorphic sites. The impact of polymorphisms may be more readily observed using alternative end points such as the formation of DNA adducts. Indeed, Perera et al.[128] found that DNA adducts in white blood cells from smokers were significant predictors of lung cancer risk and that adduct levels were higher in individuals with combinations of GSTM1 and GSTP1 genotypes.

Interestingly, cigarette smoking is thought to reduce the risk of Parkinson's disease. GSTP1 polymorphism may interact with cigarette smoking to influence the risk of this disease.[129] GST variants could also mediate risk of nonmalignant diseases in which tobacco is a causative factor. However, the interactive effect of GSTM1*0 and cigarette smoking was not found to be significant in the etiology of severe coronary artery disease in 868 patients with angiographically characterized disease.[130] Smoking is also associated with disease severity in women with rheumatoid arthritis. We have previously found that smoking was associated with the most severe disease in patients with GSTM1*0 polymorphism.[131]

7.8.2 OXIDATIVE STRESS-RELATED DISORDERS

Genetic variation in detoxifying enzymes such as GSTs has attracted interest as a possible mechanism for observed differences in disease susceptibility. Much research has focused on cancer, but other studies have included diseases such as asthma[83] and other pro-inflammatory conditions such as chronic obstructive pulmonary disease (COPD),[132] rheumatoid arthritis (RA),[131] and multiple sclerosis (MS).[133] A common underlying feature for diseases such as RA, MS, COPD, and asthma is the implication of ROS and therefore the ability of GSTs to detoxify cytotoxic products generated by ROS.[83,131–133] Variation in an individual's ability to detoxify products of ROS could accordingly be important, justifying the study of the relationship of these highly polymorphic enzymes to these diseases.

Oxidative stress is implicated in the inflammatory demyelination that characterizes MS, suggesting that GST polymorphisms may be associated with disability. In 177 patients with disease duration over 10 years, GSTM3 AA (OR=2.4) and homozygosity for both GSTM1*0 and the GSTP1*Ile[105]-encoding allele (OR=5.0) were linked with severe disability, suggesting that long-term prognosis in MS is influenced by the GST-mediated ability to remove toxic products of oxidative stress.[7] Exposure to ultraviolet radiation also results in local oxidative stress in skin. Response to such exposure, examined as minimal erythema dose, has been shown to be mediated by GSTM1 and GSTT1 genotypes in a gene dosage–dependent manner.[134] Furthermore, nonmelanoma skin cancer has been linked to these polymorphisms.[7]

A recent report from our group[135] demonstrated that the effect of GST variants on nonmelanoma skin cancer development in transplant recipients depends on immunosuppressant dose. This study highlights that the relative importance of the dual roles of GST may vary depending on external stimuli and the dose of exposure (e.g., the degree of ultraviolet radiation–induced oxidative stress or exposure to glucorticoids). Furthermore, there is increasing evidence of an immune-related role of some GSTs, which may reflect (a) a direct effect on T-cell function,[136] (b) an effect on detoxification of ROS thereby indirectly influencing immune function, or (c) linkage disequilibrium with a nearby immune-related gene.

7.8.3 ASSOCIATION OF GST GENOTYPES AND DISEASE SEVERITY

In a study performed in our laboratory, an association between polymorphism in GSTP1 and asthma and airway hyperresponsiveness (AHR) was identified.[83] We found that the GSTP1 Val[105]Val[105] genotype conferred a 6-fold decrease in asthma risk compared with GSTP1 Ile[105]Ile[105]. Trend analysis showed that GSTP1 Val[105]Val[105] frequency correlated with decreasing severity of AHR. Individuals with this genotype showed a 4- and 10-fold reduced risk of atopy, defined by skin test positivity and IgE levels. The association with AHR was subsequently confirmed in adults and children in several independent populations from the United Kingdom,[84] Italy,[85] and Turkey.[137] A recent report by Carroll et al.[86] presents interesting data indicating that the maternal GSTP1 genotype is associated with lung function in children, suggesting an *in utero* effect of GSTP1 on lung growth. Polymorphisms in GSTM1 have also been shown to influence lung growth in children.[81,82,86]

Polymorphisms within GSTM3 have also been shown to contribute to clinical severity of cystic fibrosis and have not only prognostic significance but also could lead to earlier targeted therapy in younger patients.[138]

7.8.4 RESPONSE TO TREATMENT AND GST POLYMORPHISMS

As mentioned previously, studies have shown that GSTs can operate in synergy with efflux transporters and multidrug resistance proteins to confer resistance to several carcinogens, mutagens, and anticancer drugs. In addition, interindividual variations in GSTs lead to differences in drug responses and toxicities. Thus, pharmacogenetic screening prior to anticancer drug administration may lead to identification of specific populations predisposed to drug toxicity or poor drug responses.

While there is abundant data on GST expression and drug response, there is little confirmed data on the effect of polymorphisms on response to treatment. In 148 women with epithelial ovarian cancer, those who were both GSTM1 and GSTT1 null exhibited decreased survival and reduced response to primary chemotherapy compared to those with active GSTM1 or GSTT1.[139,140] Depeille et al.[141] have shown that GSTM1 was involved in melanoma resistance to chlorambucil while Stoehlma-cher et al.[73] demonstrated that the GSTP1 Ile[105]Val polymorphism was associated in a dose-dependent fashion with increased survival of patients with advanced colorec-tal cancer receiving 5-fluorouracil/oxaliplatin chemotherapy. Yang et al.[142] described how the GSTP1 Ile[105] polymorphism has a potential role in reducing mortality risk and thus predicting the clinical outcomes of patients with breast carcinoma who are treated with the chemotherapy drug tamoxifen. Sweeney et al.[143] also described an apparent difference in survival after treatment for breast cancer according to GSTA1 genotype. Further details on cancer drugs and genetic variation within the GSTs are described in an elegant review by Townsend and Tew.[8]

7.8.5 UNCHARACTERIZED POLYMORPHISMS IN GSTS

We have to concede that we may not have investigated the most important GST polymorphic sites to date. Perhaps we should consider the functional effects of polymorphisms before use in molecular epidemiological studies. In addition to the changes in xenobiotic metabolism observed for some polymorphic sites in GSTs such as GSTP1,[105] there are more potential polymorphisms that could have functional effects remaining to be investigated. Many Web-based tools are available to allow identification of polymorphisms with potential function. Other GSTP1 polymor-phisms may affect catalytic function through altered protein folding. For example, the polymorphism creating the GSTP1 D147Y amino acid change theoretically causes loss of a hydrogen bond (http://www.snps3d.org).

Regulation of GSTs may also be affected. One SNP exists in the 5′ promoter region of GSTP1 (rs8191439), which consists of an A>G change but does not look likely to alter any transcription factor–binding sites (TFSearch: http://www.cbrc.jp/research/db/TFSEARCH.htm). However, it is positioned three bases downstream of a CAP signal for transcription initiation. Eight polymorphisms in GSTP1 lie in exonic splicing–enhancer regions: in five of these instances this could lead to loss of a particular splicing motif. This could potentially affect the splicing of GSTP1 and function of the gene. Analysis of the GSTP1 sequence also reveals putative nuclear localization signals (NLS) (RPKLK) and nuclear export signals (LX3LX3LXL). A polymorphism exists four bases from a putative NLS sequence, which may affect binding of other proteins such as "chaperones" needed for protein transportation, thus affecting the polymorphism's activity in the nucleus, though this is nonsynon-ymous. Three SNPs in GSTP1 are speculated to exist in the 3′UTR region according to public databases. The 3′UTR is an untapped resource for SNP function including mRNA stability, translation efficiency, and polyadenylation. None of the three spec-ulated SNPs, however, appear to be located in one of these AU-rich regions.

In addition, how polymorphisms in GSTM1 may influence the efficacy of the GSTM1–ASK1 complex described in a previous section has also not been investigated.

7.9 CONCLUSION

The wide range of *in vitro* GST substrates has resulted in their being considered almost universal candidates for susceptibility and outcome in disease.[7] However, critical analysis of some or even most of the early GST association studies lack statistical power or often a clear a priori hypothesis. Thus, most GST polymorphism work is based on individual variants, particularly GSTM1*0 and GSTT1*0: even in our most well-studied GST (GSTP1) few SNPs have been studied. Further, the possible association of GSTM1 with cancer may need more examination as GSTM1*0 is the only putative risk allele described. Little work has also been performed around tagged polymorphisms allowing construction of distinct GST haplotypes that describe allelic variation in the individual gene and thus give rise to haplotype blocks. A good example where this has been performed is the VDR gene where three distinct haplotype blocks have been established.[144]

But, it is important to continue examining GST polymorphism both functionally and through association analysis, as the potential of GST polymorphisms for clinical medicine is great. The impact is 2-fold in response to drug treatment through the ability to target specific GST alleles shown through the differential effects of drugs, including TER199 on GSTP1 alleles. Further, study of polymorphisms in a more subtle way can help to understand how low-penetrance genetic variation affects tissue growth and cell cycle control. Indeed, if genetic polymorphisms reflect *in vivo* functional differences at the cellular level such as described for GSTP1, it is likely that events occurring early in life may induce long-lasting clinical effects in suscep- tible individuals.

However, we should seriously consider whether to study a polymorphism if it is not potentially functional or is linked with one that is. The caveat is that it is difficult to choose the disease to study with a functional polymorphism when then the biological function of even the GST genes as a whole regardless of polymorphism is unclear. In conclusion, to truly use genetic variation to its full potential, we need to investigate both avenues of association and function to identify individuals at risk and then find a way to improve their treatment.

REFERENCES

1. Sheehan, D. et al., Structure, function, and evolution of glutathione transferases: implications for classification of non-mammalian members of an ancient enzyme superfamily, *Biochem. J.*, 360, 1–16, 2001.
2. Dixon, D.P., Lapthorn, A., and Edwards, R., Plant glutathione transferases, *Genome Biol.*, 3, REVIEWS 3004, 2002.
3. Hayes, J.D., Flanagan, J.U., and Jowley, I.R., Glutathione transferases, *Annu. Rev. Pharmacol. Toxicol.*, 45, 51–88, 2005.
4. Hayes, J.D. and Pulford, D.J., The glutathione S-transferase supergene family: reg- ulation of GST and the contribution of the isoenzymes to cancer chemoprotection and drug resistance, *Crit. Rev. Biochem. Mol. Biol.*, 30, 445–600, 1995.
5. Strange, R.C. and Fryer, A.A., Glutathione S-transferase genotype, In *Encyclopedia of Medical Genomics and Proteomics*, Marcel Dekker, New York, 2004, 536–542.
6. Hayes, J.D. and McLellan, L.I., Glutathione and glutathione-dependent enzymes represent a co-ordinately regulated defence against oxidative stress, *Free Radic. Res.*, 31, 273–300, 1999.
7. Hayes, J.D. and Strange, R.C., Glutathione S-transferase polymorphisms and their biological consequences, *Pharmacology*, 61, 154–166, 2000.

8. Townsend, D. and Tew, K., Cancer drugs, genetic variation, and the glutathione-S-transferase gene family, *Am. J. Pharmacogenomics*, 3, 157–172, 2003.

9. Guengerich, F.P., Purification and characterization of xenobiotic-metabolizing enzymes from lung tissue. *Pharmacol. Ther.*, 45, 299–307, 1990.

10. Kodym, R., Calkins, P., and Story, M., The cloning and characterization of a new stress response protein. A mammalian member of a family of theta class glutathione s-transferase-like proteins, *J. Biol. Chem.*, 274, 5131–5137, 1999.

11. Pemble, S.E., Wardle, A.F., and Taylor, J.B., Glutathione S-transferase class kappa: characterization by the cloning of rat mitochondrial GST and identification of a human homologue, *Biochem. J.*, 319, 749–754, 1996.

12. Jakobsson, P.J. et al., Membrane-associated proteins in eicosanoid and glutathione metabolism (MAPEG). A widespread protein superfamily, *Am. J. Respir. Crit. Care Med.*, 161, S20–S24, 2000.

13. Iida, A. et al., Catalog of 46 single-nucleotide polymorphisms (SNPs) in the microsomal glutathione S-transferase 1 (MGST1) gene, *J. Hum. Genet.*, 46, 590–594, 2001.

14. Hayes, J.D., Flanagan, J.U., and Jowsey, I.R., Glutathione transferases, *Annu. Rev. Pharmacol. Toxicol.*, 45, 51–88, 2005

15. Rushmore, T.H., Morton, M.R., and Pickett, C.B., The antioxidant responsive element. Activation by oxidative stress and identification of the DNA consensus sequence required for functional activity, *J. Biol. Chem.*, 266, 11632–11639, 1991.

16. Nguyen, T., Rushmore, T.H., and Pickett, C.B., Transcriptional regulation of a rat liver glutathione S-transferase Ya subunit gene. Analysis of the antioxidant response element and its activation by the phorbol ester 12-O-tetradecanoylphorbol-13-acetate, *J. Biol. Chem.*, 269, 13656–13662, 1994.

17. Morel, F. et al., The human glutathione transferase alpha locus: genomic organization of the gene cluster and functional characterization of the genetic polymorphism in the hGSTA1 promoter, *Pharmacogenetics*, 12, 277–286, 2002.

18. Ning, B. et al., Human glutathione S-transferase A2 polymorphisms: variant expression, distribution in prostate cancer cases/controls and a novel form, *Pharmacogenetics*, 14, 35–44, 2004.

19. Widersten, M., Holmstrom, E., and Mannervik, B., Cysteine residues are not essential for the catalytic activity of human class mu glutathione transferase M1a-1a, *FEBS Lett.*, 293, 156–159, 1991.

20. Smith, G. et al., Metabolic polymorphisms and cancer susceptibility, *Cancer Surv.*, 25, 27–65, 1995.

21. McLellan, R.A. et al., Characterisation of a human glutathione S-transferase mu cluster containing a duplicated GSTM1 gene that causes ultrarapid enzyme activity, *Mol. Pharmacol.*, 52, 958–965, 1997.

22. Van Cong, N. et al., Glutathione S-transferases: tissue distribution, number of loci, polymorphism, chromosome localisation, *Cytogenet. Cell Genet.*, 37, 554, 1984.

23. Campbell, E. et al., A distinct human testis and brain mu-class glutathione S-transferase: molecular cloning and characterisation of a form present even in individuals lacking hepatic type mu isoenzymes, *J. Biol. Chem.*, 265, 9188–9193, 1990.

24. Inskip, A. et al., Identification of polymorphism at the glutathione S-transferase, GSTM3 locus: evidence for linkage with GSTM1*A, *Biochem J.*, 312, 713–716, 1995.

25. Liloglou, T.W.M. et al., A T2517C polymorphism in the GSTM4 gene is associated with risk of developing lung cancer, *Lung Cancer*, 37, 143–146, 2002.

26. Strange, R.C. et al., Glutathione S-transferase family of enzymes. *Mutat. Res.*, 482, 21–26, 2001.

27. Strange, R.C. and Fryer, A.A., The glutathione S-transferases: influence of polymorphism on cancer susceptibility, *IARC Sci. Publ.*, 148, 231–249, 1999.

28. Coggan, M. et al., Structure and organisation of the human theta-class glutathione S-transferase and D-dopachrome tautomerase gene complex, *Biochem. J.*, 334, 617–623, 1998.

29. Blackburn, A.C. et al., GSTZ1d: a new allele of glutathione transferase zeta and maleylacetoacetate isomerase, *Pharmacogenetics*, 11, 671–678, 2001.

30. Blackburn, A.C. et al., Discovery of a functional polymorphism in human glutathione transferase zeta by expressed sequence tag database analysis, *Pharmacogenetics*, 10, 49–57, 2000.

31. Board, P. et al., Identification of novel glutathione transferases and polymorphic variants by expressed sequence tag database analysis, *Drug Metab. Dispos.*, 29, 544–547, 2001.

32. Whitbread, A.K. et al., Characterization of the human omega class glutathione transferase genes and associated polymorphisms, *Pharmacogenetics*, 13, 131–144, 2003.

33. Cowell, I.G. et al., The structure of the human glutathione S-transferase pi gene, *Biochem. J.*, 255, 79-83, 1988.

34. Morrow, C.S., Cowan, K.H., and Goldsmith, M.E., Structure of the human genomic glutathione S-transferase-pi gene, *Gene*, 75, 3–11, 1989.

35. Moffat, G.J., McLaren, A.W., and Wolf, C.R., Involvement of Jun and Fos proteins in regulating transcriptional activation of the human pi class glutathione S-transferase gene in multidrug-resistant MCF7 breast cancer cells, *Biol. Chem. J.*, 269, 16397–16402, 1994.

36. Xia, C. et al., The organization of the human GSTP1-1 gene promoter and its response to retinoic acid and cellular redox status, *Biochem. J.*, 313, 155–161, 1996.

37. Sherratt, P.J. et al., Increased bioactivation of dihaloalkanes in rat liver due to induction of class theta glutathione S-transferase T1-1, *Biochem. J.*, 335, 619–630, 1998.

38. van Iersel, M.L. et al., Interactions of prostaglandin A2 with the glutathione-mediated biotransformation system, *Biochem. Pharmacol.*, 57, 1383–1390, 1999.

39. Adler, V. et al., Regulation of JNK signalling by GSTp, *Embo. J.*, 18, 1321–1334, 1999.

40. Daggett, D.A. et al., Effects of lead on rat kidney and liver: GST expression and oxidative stress, *Toxicology*, 128, 191–206, 1998.

41. Nebert, D.W., McKinnon, R.A., and Puga, A., Human drug-metabolizing enzyme polymorphisms: effects on risk of toxicity and cancer, *DNA Cell Biol.*, 15, 273–280, 1996.

42. Hegazy, U.M., Mannervik, B., and Stenberg, G., Functional role of the lock and key motif at the subunit interface of glutathione transferase p1-1, *J. Biol. Chem.*, 279, 9586–9596, 2004.

43. Armstrong, R.N., Structure, catalytic mechanism, and evolution of the glutathione transferases, *Chem. Res. Toxicol.*, 10, 2–18, 1997.

44. Berhane, K., Detoxication of base propenals and other alpha, beta-unsaturated aldehyde products of radical reactions and lipid peroxidation by human glutathione transferases, *Proc. Natl. Acad. Sci. USA*, 91, 1480–1484, 1994.

45. Sherratt, P.J. et al., Evidence that human class theta glutathione S-transferase T1-1 can catalyse the activation of dichloromethane, a liver and lung carcinogen in the mouse. Comparison of the tissue distribution of GST T1-1 with that of classes alpha, mu, and pi GST in human, *Biochem. J.*, 326, 837–846, 1997.

46. Ranganathan, P.N., Whalen, R., and Boyer, T.D., Characterization of the molecular forms of glutathione S-transferase P1 in human gastric cancer cells (Kato III) and in normal human erythrocytes, *Biochem. J.*, 386, 525–533, 2005.

47. Tew, K.D. and Ronai, Z., GST function in drug and stress response, *Drug Resist. Updat.*, 2, 143–147, 1999.

48. Elsby, R. et al., Increased constitutive c-Jun N-terminal kinase signalling in mice lacking glutathione S-transferase pi, *J. Biol. Chem.*, 278, 22243–22249, 2003.

49. Ishii, T., Teramoto, S., and Matsuse, T., GSTP1 affects chemoresistance against camptothecin in human lung adenocarcinoma cells, *Cancer Lett.*, 216, 89–102, 2004.

50. Ruscoe, J.E. et al., Pharmacologic or genetic manipulation of glutathione S-transferase P1-1 (GSTpi) influences cell proliferation pathways, *J. Pharmacol. Exp. Ther.*, 298, 339–345, 2001.

51. Yin, Z. et al., Glutathione S-transferase p elicits protection against H2O2-induced cell death via coordinated regulation of stress kinases, *Cancer Res.*, 60, 4053–4057, 2000.

52. Dulhunty, A. et al., The glutathione transferase structural family includes a nuclear chloride channel and a ryanodine receptor calcium release channel modulator, *Biol. Chem.*, 276, 3319–3323, 2001.

53. Cho, S.G. et al., Glutathione S-transferase mu modulates the stress-activated signals by suppressing apoptosis signal-regulating kinase 1, *J. Biol. Chem.*, 276, 12749–12755, 2001.

54. Ichijo, H. et al., Induction of apoptosis by ASK1, a mammalian MAPKKK that activates SAPK/JNK and p38 signalling pathways, *Science*, 275, 90–94, 1997.

55. Saitoh, M. et al., Mammalian thioredoxin is a direct inhibitor of apoptosis signal-regulating kinase (ASK) 1, *Embo. J.*, 17, 2596–2606, 1998.

56. Dorion, S., Lambert, H., and Landry, J., Activation of the p38 signalling pathway by heat shock involves the dissociation of glutathione S-transferase Mu from Ask1, *J. Biol. Chem.*, 277, 30792–30797, 2002.

57. Bernardini, S. et al., Modulation of GST P1-1 activity by polymerization during apoptosis, *J. Cell Biochem.*, 77, 645–653, 2000.

58. Halliwell, B. and Gutteridge, J.M.C., *Free Radicals in Biology and Medicine*, 2nd edition, Clarendon Press, Oxford, 1989.

59. Burgess, J.R. et al., Amino acid substitutions in the human glutathione S-transferases confer different specificities in the prostaglandin endoperoxide conversion pathway, *Biochem. Biophys. Res. Commun.*, 158, 497–502, 1989.

60. Urade, Y. et al., Structural and functional significance of cysteine residues of glutathione-independent prostaglandin D synthase. Identification of Cys65 as an essential thiol, *J. Biol. Chem.*, 270, 1422–1428, 1995.

61. Bogaards, J.J., Venekamp, J.C., and van Bladeren, P.J., Stereoselective conjugation of prostaglandin A2 and prostaglandin J2 with glutathione, catalyzed by the human glutathione S-transferases A1-1, A2-2, M1a-1a, and P1-1, *Chem. Res. Toxicol.*, 10, 310–317, 1997.

62. Zhang, K., Mack, P., and Wong, K.P., Glutathione-related mechanisms in cellular resistance to anticancer drugs, *Int. J. Oncol.*, 12, 871–882, 1998.

63. Arai, T. et al., Immunohistochemical expression of glutathione transferase-pi in untreated primary non-small cell lung cancer, *Cancer Detect. Prev.*, 24, 252–257, 2000.

64. Nakanishi, Y. et al., Expression of p53 and glutathione S-transferase-pi relates to clinical drug resistance in non-small cell lung cancer, *Oncology*, 57, 318–323, 1999.

65. Lo, H.W. and Ali-Osman, F., Cyclic AMP mediated GSTP1 gene activation in tumour cells involves the interaction of activated CREB-1 with the GSTP1 CRE: a novel mechanism of cellular GSTP1 gene regulation, *J. Cell Biochem.*, 87, 103–116, 2002.

66. Hara, T. et al., Glutathione S-transferase P1 has protective effects on cell viability against camptothecin, *Cancer Lett.*, 203, 199–207, 2004.

67. Tew, K.D., Bomber, A.M., and Hoffman, S.J., Ethacrynic acid and piriprost as enhancers of cytotoxicity in drug-resistant and sensitive cell lines, *Cancer Res.*, 48, 3622–3625, 1988.
68. Hall, A. et al., Possible role of inhibition of glutathione S-transferase in the partial reversal of chlorambucil resistance by indomethacin in a Chinese hamster ovary cell line, *Cancer Res.*, 49, 6265–6268, 1989.
69. Ford, J.M. et al., Modulation of resistance to alkylating agents in cancer cell by gossypol enantiomers, *Cancer Lett.*, 56, 85–94, 1991.
70. Kauvar, L.M., Peptide mimetic drugs: a comment on progress and prospects, *Nat. Biotechnol.*, 14, 709, 1996.
71. Lyttle, M.H. et al., Glutathione-S-transferase activates novel alkylating agents, *J. Med. Chem.*, 37, 1501–1507, 1994.
72. Goto, S. et al., Overexpression of glutathione S-transferase pi enhances the adduct formation of cisplatin with glutathione in human cancer cells, *Free Radic. Res.*, 31, 549–558, 1999.
73. Stoehlmacher, J. et al., Association between glutathione S-transferase P1, T1, and M1 genetic polymorphism and survival of patients with metastatic colorectal cancer, *J. Natl. Cancer Inst.*, 94, 936–942, 2002.
74. Fryer, A.A., Hume, R., and Strange, R.C., The development of glutathione S-transferase and glutathione peroxidase activities in human lung, *Biochim. Biophys. Acta*, 883, 448–453, 1986.
75. Terrier, P. et al., An immunohistochemical study of pi class glutathione S-transferase expression in normal human tissue, *Am. J. Pathol.*, 137, 845–853, 1990.
76. Whalen, R. and Boyer, T.D., Human glutathione S-transferases, *Semin. Liver Dis.*, 18, 345–358, 1998.
77. Hubert, H.B. et al., Genetic and environmental influences on pulmonary function in adult twins, *Am. Rev. Respir. Dis.*, 125, 409–415, 1982.
78. Chen, Y., Genetics and pulmonary medicine 10, Genetic epidemiology of pulmonary function, *Thorax*, 54, 818–824, 1999.
79. Zach, M.S., The physiology of forced expiration, *Paed. Respir. Rev.*, 1, 36–39, 2000.
80. Sears, M.R. et al., A longitudinal, population-based cohort study of childhood asthma followed into adulthood, *N. Engl. J. Med.*, 349, 1414–1422, 2003.
81. Gilliland, F.D. et al., Effects of glutathione S-transferase M1, T1, and P1 on childhood lung function growth, *Am. J. Respir. Crit. Care Med.*, 166, 710–716, 2002.
82. Carroll, W.D. et al., Effects of glutathione S-transferase P1, M1, and T1 and lung function in asthmatic families, *Clin. Expl. Allergy*, 35, 1155–1161, 2005.
83. Fryer, A.A. et al., Polymorphism at the glutathione S-transferase GSTP1 locus. A new marker for bronchial hyperresponsiveness and asthma, *Am. J. Respir. Crit. Care Med.*, 161, 1437–1442, 2000.
84. Child, F. et al., The association of maternal but not paternal genetic variation in GSTP1 with asthma phenotypes in children, *Respir. Med.*, 97, 1247–1256, 2003.
85. Mapp, C.E. et al., Glutathione S-transferase GSTP1 is a susceptibility gene for occupational asthma induced by isocyanates, *J. Allergy Clin. Immunol.*, 109, 867–872, 2002.
86. Carroll, W.D. et al., Maternal glutathione S-transferase GSTP1 genotype is a specific predictor of phenotype in children with asthma, *Pediatr. Allergy Immunol.*, 16, 32–39, 2005.
87. Whitbread, A.K. et al., Characterization of the human omega class glutathione transferase genes and associated polymorphisms, *Pharmacogenetics*, 13, 131–144, 2003.
88. Henderson, C.J. et al., Increased skin tumorigenesis in mice lacking pi class glutathione S-transferases, *Proc. Natl. Acad. Sci. USA*, 95, 5275–80, 1998.
89. Raza, H. et al., Glutathione S-transferases in human and rodent skin: multiple forms and species-specific expression, *J. Invest. Dermatol.*, 96, 463–467, 1991.

90. Henderson, C.J. et al., Increased resistance to acetaminophen hepatotoxicity in mice lacking glutathione S-transferase pi, *Proc. Natl. Acad. Sci. USA*, 97, 12741–12745, 2000.

91. Hayes, J.D., Flanagan, J.U., and Jowsey, I.R., Glutathione transferases, *Annu. Rev. Pharmacol. Toxicol.*, 45, 51–88, 2005.

92. Fernandez-Canon, J.M. et al., Maleylactoacetate isomerase (MAAI/GSTZ)-deficient mice reveal a glutathione-dependent nonenzymatic bypass in tyrosine catabolism, *Mol. Cell Biol.*, 22, 4943–4951, 2002.

93. Johnson, J.A. et al., Characterisation of methylation of rat liver cytosolic glutathione S-transferase by using reverse-phase H.P.L.C. and chromatofocusing, *Biochem. J.*, 270, 483–489, 1990.

94. Esteller, M. et al., Inactivation of glutathione S-transferase P1 gene by promoter hypermethylation in human neoplasia, *Cancer Res.*, 58, 4515–4518, 1998.

95. Harden, S.V. et al., Quantitative GSTP1 methylation and the detection of prostate adenocarcinoma in sextant biopsies, *J. Natl. Cancer Inst.*, 95, 1634–1637, 2003.

96. Harden, S.V. et al., Quantitative GSTP1 methylation clearly distinguishes benign prostatic tissue and limited prostate adenocarcinoma, *J. Urol.*, 169, 1138–1142, 2003.

97. Lee, S. et al., Aberrant CpG island hypermethylation along multistep hepatocarcinogenesis, *Am. J. Pathol.*, 163, 1371–1378, 2003.

98. Kote-Jarai, Z. et al., Relationship between glutathione S-transferase M1, P1, and T1 polymorphisms and early onset prostate cancer, *Pharmacogenetics*, 11, 325–330, 2001.

99. Jhaveri, M.S. and Morrow, C.S., Methylation-mediated regulation of the glutathione S-transferase P1 gene in human breast cancer cells, *Gene*, 210, 1–7, 1998.

100. Morrow, C.S., Chiu, J., and Cowan, K.H., Posttranscriptional control of glutathione S-transferase pi gene expression in human breast cancer cells, *Biol. Chem.*, 267, 10544–10550, 1992.

101. Antequera, F., Macleod, D., and Bird, A.P., Specific protection of methylated CpGs in mammalian nuclei, *Cell*, 58, 509–517, 1989.

102. Boyes, J. and Bird, A., Repression of genes by DNA methylation depends on CpG density and promoter strength: evidence for involvement of a methyl-CpG binding protein, *Embo. J.*, 11, 327–333, 1992.

103. Meehan, R.R. et al., Identification of a mammalian protein that binds specifically to DNA containing methylated CpGs, *Cell*, 58, 499–507, 1989.

104. Coles, B.F. et al., Effect of polymorphism in the human glutathione S-transferase A1 promoter on hepatic GSTA1 and GSTA2 expression, *Pharmacogenetics*, 11, 663–669, 2001.

105. Seidegard, J. et al., Hereditary differences in the expression of the human glutathione transferase active on trans-stilbene oxide are due to a gene deletion, *Proc. Natl. Acad. Sci. USA*, 85, 7293–7297, 1988.

106. Xu, S. et al., Characterization of the human class mu glutathione S-transferase gene cluster and the GSTM1 deletion, *Biol. Chem.*, 273, 3517–3527, 1998.

107. McLellan, R.A. et al., Characterization of a human glutathione S-transferase mu cluster containing a duplicated GSTM1 gene that causes ultrarapid enzyme activity, *Mol. Pharmacol.*, 52, 958–965, 1997.

108. Conde, L. et al., PupaSNP finder: a web tool for finding SNPs with putative effect at transcriptional level, *Nucleic Acids Res.*, 32, W242–W248, 2004.

109. Ali-Osman, F. et al., Molecular cloning, characterization, and expression in *Escherichia coli* of full-length cDNAs of three human glutathione S-transferase pi gene variants. Evidence for differential catalytic activity of the encoded proteins, *J. Biol. Chem.*, 272, 10004–10012, 1997.

110. Harries, L.W. et al., Identification of genetic polymorphisms at the glutathione S-transferase pi locus and association with susceptibility to bladder, testicular, and prostate cancer, *Carcinogenesis*, 18, 641–644, 1997.

111. Watson, M.A. et al., Human glutathione S-transferase P1 polymorphisms: relationship to lung tissue enzyme activity and population frequency distribution, *Carcinogenesis*, 19, 275–280, 1998.

112. Zimniak, P. et al., Naturally occurring human glutathione S-transferase GSTP1-1 isoforms with isoleucine and valine in position 104 differ in enzymic properties, *Eur. J. Biochem.*, 224, 893–899, 1994.

113. Hu, X. et al., Activity of four allelic forms of glutathione S-transferase hGSTP1-1 for diol epoxides of polycyclic aromatic hydrocarbons, *Biochem. Biophys. Res. Commun.*, 238, 397–402, 1997.

114. Sundberg, K. et al., Differences in the catalytic efficiencies of allelic variants of glutathione transferase P1-1 towards carcinogenic diol epoxides of polycyclic aromatic hydrocarbons, *Carcinogenesis,* 19, 433–436, 1998.

115. Ji, X. et al., Structure and function of residue 104 and water molecules in the xenobiotic substrate-binding site in human glutathione S-transferase P1-1, *Biochemistry,* 38, 10231–10238, 1999.

116. Blackburn, A.C. et al., Discovery of a functional polymorphism in human glutathione transferase zeta by expressed sequence tag database analysis, *Pharmacogenetics,* 10, 49–57, 2000.

117. Ingelman-Sundberg, M., Genetic polymorphisms of cytochrome P450 2D6 (CYP2D6): clinical consequences, evolutionary aspects, and functional diversity, *Pharmacogenomics*, 5, 6–13, 2005.

118. Gonzalez, J., News hits — the research and business of drug discovery — genetic variants shed new light on tailor-made therapeutics, *Pharmaceutical Discovery*, 1, 12–14, 2005.

119. Kalinowski, L., Dobrucki, I.T., and Malinski, T., Race-specific differences in endothelial function: predisposition of African Americans to vascular diseases, *Circulation*, 109, 2511–2517, 2004.

120. Chen, W. et al., Nitric oxide synthase gene polymorphism (G894T) influences arterial stiffness in adults: The Bogalusa Heart Study, *Am. J. Hypertens.*, 17, 553–559, 2004.

121. Cote, M.L. et al., Combinations of glutathione S-transferase genotypes and risk of early-onset lung cancer in Caucasians and African Americans: a population-based study, *Carcinogenesis*, 26, 811–819, 2005.

122. Matthias, C. et al., The glutathione S-transferase GSTP1 polymorphism: effects on susceptibility to oral/pharyngeal and laryngeal carcinomas, *Pharmacogenetics*, 8, 1–6, 1998.

123. Mitrunen, K. et al., Glutathione S-transferase M1, M3, P1, and T1 genetic polymorphisms and susceptibility to breast cancer, *Cancer Epidemiol. Biomarkers Prev.*, 10, 229–236, 2001.

124. Benhamou, S. et al., Meta- and pooled analyses of the effects of glutathione S-transferase M1 polymorphisms and smoking on lung cancer risk, *Carcinogenesis,* 23, 1343–1350, 2002.

125. Hashibe, M. et al., Meta- and pooled analyses of GSTM1, GSTT1, GSTP1, and CYP1A1 genotypes and risk of head and neck cancer, *Cancer Epidemiol. Biomarkers Prev.,* 12, 1509–1517, 2003.

126. Smits, K.M. et al., Interaction between smoking, GSTM1 deletion, and colorectal cancer: results from the GSEC study, *Biomarkers*, 8, 299–310, 2003.

127. Vogl, F.D. et al., Glutathione S-transferases M1, T1, and P1 and breast cancer: a pooled analysis, *Cancer Epidemiol. Biomarkers Prev.*, 13, 1473–1479, 2004.

128. Perera, F.P. et al., Associations between carcinogen-DNA damage, glutathione S-transferase genotypes, and risk of lung cancer in the prospective Physicians' Health Cohort Study, *Carcinogenesis*, 23, 1641–1646, 2002.

129. Deng, Y. et al., Case-only study of interactions between genetic polymorphisms of GSTM1, P1, T1, and Z1 and smoking in Parkinson's disease, *Neurosci. Lett.*, 366, 326–331, 2004.

130. Wang, X.L. et al., Glutathione S-transferase mu1 deficiency, cigarette smoking, and coronary artery disease, *J. Cardiovasc. Risk*, 9, 25–31, 2002.

131. Mattey, D.L. et al., Association of polymorphism in glutathione S-transferase loci with susceptibility and outcome in rheumatoid arthritis: comparison with the shared epitope, *Ann. Rheum. Dis.*, 58, 164–168, 1999.

132. Ishii, T. et al., Glutathione S-transferase P1 (GSTP1) polymorphism in patients with chronic obstructive pulmonary disease, *Thorax*, 54, 693–696, 1999.

133. Mann, C.L. et al., Glutathione S-transferase polymorphisms in MS: their relationship to disability, *Neurology*, 54, 552–557, 2000.

134. Kerb, R. et al., Influence of GSTT1 and GSTM1 genotypes on sunburn sensitivity, *Am. J. Pharmacogenomics*, 2, 147–154, 2002.

135. Fryer, A.A. et al., Polymorphisms in glutathione S-transferases and non-melanoma skin cancer risk in Australian renal transplant recipients, *Carcinogenesis*, 26, 185–191, 2005.

136. Li, D.W. et al., Refolding and characterization of recombinant human GST-PD-1 fusion protein expressed in *Escherichia coli*, *Acta Biochim. Biophys. Sin. (Shanghai)*, 36, 141–146, 2004.

137. Tamer, L. et al., Glutathione-S-transferase gene polymorphisms (GSTT1, GSTM1, GSTP1) as increased risk factors for asthma, *Respirology*, 9, 493–498, 2004.

138. Flamant, C. et al., Glutathione-S-transferase M1, M3, P1, and T1 polymorphisms and severity of lung disease in children with cystic fibrosis, *Pharmacogenetics*, 14, 295–301, 2004.

139. Howells, R.E. et al., Association of glutathione S-transferase GSTM1 and GSTT1 null genotypes with clinical outcome in epithelial ovarian cancer, *Clin. Cancer Res.*, 4, 2439–2445, 1998.

140. Howells, R.E. et al., Glutathione S-transferase GSTM1 and GSTT1 genotypes in ovarian cancer: association with p53 expression and survival, *Int. J. Gynecol. Cancer.*, 11, 107–112, 2001.

141. Depeille, P. et al., Glutathione S-transferase M1 and multidrug resistance protein 1 act in synergy to protect melanoma cells from vincristine effects, *Mol. Pharmacol.*, 65, 897–905, 2004.

142. Yang, G. et al., Genetic polymorphisms in glutathione-S-transferase genes (GSTM1, GSTT1, GSTP1) and survival after chemotherapy for invasive breast carcinoma, *Cancer*, 103, 52–58, 2005.

143. Sweeney, C. et al., Association between a glutathione S-transferase A1 promoter polymorphism and survival after breast cancer treatment, *Int. J. Cancer*, 103, 810–814, 2003.

144. Nejentsev, S. et al., Comparative high-resolution analysis of linkage disequilibrium and tag single nucleotide polymorphisms between populations in the vitamin D receptor gene, *Hum. Mol. Genet.*, 13, 1633–1639, 2004.

8 Glutathione and Glutathione S-Transferases as Targets for Anticancer Drug Development

Victoria J. Findlay, Danyelle M. Townsend, and Kenneth D. Tew

CONTENTS

8.1 INTRODUCTION

There has been consistent interest in developing pharmacological approaches to manipulate thiol and redox pathways, particularly with respect to possible cancer treatments. For many years N-acetyl cysteine (NAC) has been used as a means of reversing the hepatotoxicity of acetaminophen and as an inhalant mucolytic agent (Mucomyst) in the management of a number of lung diseases. This molecule is essentially designed to provide a biologically available and nontoxic form of the amino acid cysteine. As might be expected, the thiol acts as a nucleophile in scavenging electrophilic metabolites of acetaminophen and assists in minimizing disulfide bond formation in proteins in the mucus membranes of the lung. Manipulation of blood glutathione (GSH) levels by using NAC has also been shown to

155

effect immune responses and has been applied to the management of HIV-infected patients.[1,2] Direct therapeutic use of GSH has also been considered, and GSH is frequently adopted by the dietary supplement sector with claims of decreasing aging effects, advancing immune response, and protecting against free radical damage. As will be discussed later in this chapter, oxidized glutathione (GSSG) also has some interesting therapeutic activity. These approaches demonstrate in a most straightforward way the validity of thiol manipulation as a therapeutic strategy.

Direct modulation of GSH levels and of glutathione S-transferase (GST) activity were also attempted as ways to improve response to cancer drugs. Uses of buthionine sulfoximine (BSO) as an inhibitor of γ-glutamyl cysteine synthetase (γ-GCS) and of ethacrynic acid as a small molecule inhibitor of GSTs, while effective in their experimental effects on each pathway, were not successful enough in the clinic to merit continued development.[3,4] One consequence of these early approaches was the conceptual design of a peptidomimetic inhibitor of GSTπ, TLK199 (γ-glutamyl-S-[benzyl]cysteinyl-R[–] phenyl glycine diethyl ester). Preclinical and mechanism of action studies with this agent revealed an unexpected effect in animals, namely that the drug possessed myeloproliferative activity.[5,6] Subsequent studies identified a protein–protein interaction between GSTπ and a critical stress-activated kinase, c-Jun NH$_2$-terminal kinase (JNK).[7,8] From such data, a model for how TLK199 produced proliferative effects in the marrow compartment was formulated.[6] Also of relevance to thiol modulation, a modified form of GSSG, NOV-002, has also been shown to affect bone marrow, with increases in circulating lymphocyte, monocyte, T cell, and NK cell counts. It would be significant to elucidate a plausible general mechanism linking glutathione with immune regulation and correlate such results with the development of drugs that affect these pathways.

There are now a significant number of studies focused on GST-targeted agents. The rationale for such efforts lies with accumulated observations about GST expression in tumors and normal tissues. In particular, the association between high levels of expression of GST isozymes and malignancy or drug resistance provided an ideal rationale for the design of a GSTπ-activated prodrug. In many instances the GSTπ isozyme can accumulate in tumor cells to levels that make it one of the more prevalent cytosolic proteins. In addition, in drug-resistant cells, even when the selecting drug is not a substrate for GSTπ, its expression can be most readily enhanced. Such data complicated interpretation of the connection between GSTπ and drug resistance in cell culture[9] and in clinical trials.[10] Largely because of the more recent connection between GSTπ, JNK, and apoptosis pathways,[11] there is now a clearer understanding of why increased GSTπ may be associated with so many divergent acquired drug-resistant situations.

This observation broadens the functional importance of GSTπ into other arenas, emphasizing how redox-active proteins have evolved functions in excess of reactive oxygen species removal. Some appear to be central to the signaling processes required in the cell's response to stress. Changes in redox conditions can trigger cellular responses through a number of different pathways. The nature and extent of the ROS insult may determine the threshold of the cellular response manifesting as proliferation, stress response, and damage repair or apoptosis. With further understanding, the link between thiol active proteins, GSTs, and stress-activated protein kinases may become an expansive series of interconnected pathways. In an unstressed cell, JNK is kept suppressed by the presence of one or more repressors. Under conditions of oxidative stress, GSTπ dissociates from JNK and forms dimers or multimeric complexes.[11] Meanwhile, the liberated JNK regains its functional capacity to become phosphorylated and to phosphorylate c-Jun. This process activates the stress cascade and many of the sequential downstream transcription factors. In GSTπ overexpressing cells, constitutively active MEKK1 effectively phosphorylated both MKK4 (immediately upstream of JNK) and JNK but did not result in c-Jun phosphorylation, confirming the specificity of the GSTπ–JNK association.[12] Mouse embryo fibroblast (MEF) cell lines from mice[13] engineered to be null for GSTπ expression have high basal levels of JNK activity that can be reduced if cells have transgenic GSTπ. Treatment of GST wild-type cells (but not null cells) with the specific GSTπ inhibitor TLK199 activates JNK activity. Also, human HL60 cells chronically exposed to this inhibitor develop tolerance to the drug and also overexpress JNK, presumably as a means of compensating for the constancy of GST inhibition and the perceived chronic stress.[5] These findings provide evidence that GSTπ has a nonenzymatic, regulatory role in controlling cellular response to external stimuli.

8.2 S-GLUTATHIONYLATION

In the burgeoning era of proteomics, it becomes clear that the central dogma of genetic determinism can be influenced by a number of processes that include polymorphic variants, gene splicing events, exon shuffling, protein domain rearrangements, and the large number of post-translational modifications that contribute to alterations in tertiary and quaternary protein structure. Among these, phosphorylation, glycosylation, methylation, and acetylation account for a large proportion of modifications. More recently, however, adding GSH to available cysteine residues has been shown to be of consequence. The importance of modifying cysteine residues is not necessarily restricted to redox regulation but now seems to lead to changes in signaling processes, particularly as a response to a divergent number of stress responses. By adding the GSH tripeptide to a target protein, an additional negative charge is introduced (as a consequence of the glu residue) and a change in protein conformation is made likely. The implication from this analysis is that cells actively participate in the stochastic production of multiple protein-building blocks with the intent of realizing functional nonredundancy.

Sulfur has the essential chemical properties to exist in a biologically reduced sulfhydryl state where the pKa of the thiol group is ~9.65, accounting for the nucleophilicity of reduced glutathione (GSH). GSH homeostasis is maintained in cells by a complex series of balanced pathways, some of which are illustrated in Figure 8.1. *De novo* synthesis can occur through the γ-glutamyl cycle, where the three constituent amino acids (glu–cys–gly) are combined with rate-limiting catalysis through γ-GCS. Salvage of GSH can occur through the cleavage activity of the membrane-associated γ-glutamyl transpeptidase (GGT) that can recycle constituents of the molecule. While intracellular concentrations of GSH may vary considerably, 0.1 to 10μM are not uncommonly found in mammalian cells (10 to 30μM in plasma). Glutathione can occur in reduced (GSH), oxidized (GSSG), or in mixed disulfide

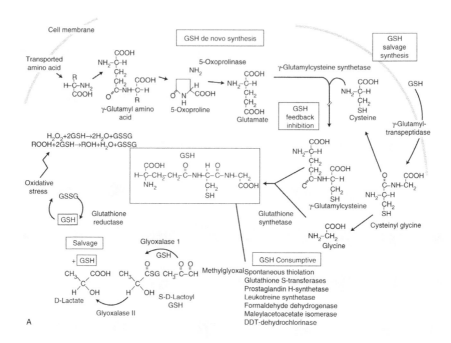

FIGURE 8.1 Pathways of GSH homeostasis. Adapted from *Basic Science of Cancer*, Kruh, G.D. and Tew, K.D., Eds., Current Medicine, Inc., 2000. With permission.

forms, and its ubiquitous abundance is testament to its biological importance. More recently, glutathionylation of proteins has been recognized as an important post-translational modification, triggered by cell exposure to oxidative or nitrosative stress. There remain significant questions about the biological consequences of this modification but specific examples suggest that glutathionylation can influence conformation of structural proteins such as actin[14] or enzyme function such as protein kinase C_α.[15] A scheme for how oxidative stress may initiate attachment of GS- to acceptor cysteine residues is shown in Figure 8.2, where electrophilic attack on the thiol group of cysteine residues can produce a number of intermediates that might act as proximal donors to glutathionylation reactions. Oxidative stress may induce glutathionylation of protein thiols by many different routes. Those highlighted include the direct oxidation of protein cysteines to generate a reactive cysteinyl radical or sulfenic acid that can further react with GSH to form a mixed disulfide. Further oxidation of the sulfenic acid produces sulfinic and sulfonic species. The latter is not readily reversed, except by protein degradation, making it an unlikely glutathionylation donor. Alternatively, a mixed disulfide is formed through reaction with oxidized forms of GSH (i.e., GS-OH [glutathione sulfenate] or GS[O]SG [glutathione disulfide S-oxide]). These latter species may also be the proximal donors for the reaction. Based on these pathways, a mechanism-oriented approach to understand how drugs such as PABA/NO and NOV-002 influence glutathionylation may be informative (see discussion below).

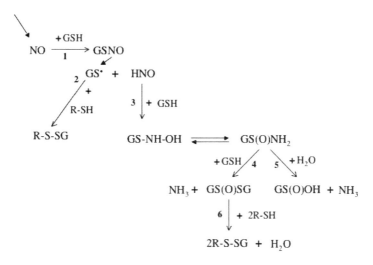

FIGURE 8.2 Glutathionylation pathway intermediates. PABA/NO releases NO that can react with GSH in the cell (1) to form a GSNO and subsequently a glutathionyl radical GS$^\bullet$ and nitroxyl (HNO), which in turn reacts with GSH (3) to give N-hydroxysulfenamide (GS-NH-OH), which can rearrange to generate a sulfinamide (GS[O]NH$_2$). Reaction of GS[O]NH$_2$ with GSH (4) forms glutathione disulfide-S-oxide (GS[O]SG) and NH$_3$ and with H$_2$O (5) to form sulfinic acid (GS[O]OH) and NH$_3$. The key intermediates leading to the synthesis of GS(O)SG are the sulfinamides (GS[O]NH$_2$ and GS[O]-NH-SG). The reaction of GS(O)SG (6) or GS$^\bullet$ (2) with a reduced protein thiol (R-SH) leads to the formation of mixed disulfide.

8.3 GLUTAREDOXIN AND SULFIREDOXIN

The reversibility of glutathionylation provides flexibility that allows the reaction to act as a molecular switch. Figure 8.2 shows how nitrosative stress can create intermediates similar to those for oxidative stress. In both instances, proteins such as glutaredoxin or sulfiredoxin can "catalytically" reverse the thiol modification. For example, thioredoxin is involved in the reduction of protein disulfides and protein sulfenic acid intermediates. Glutaredoxin (Grx) enzymatically deglutathionylates a number of specific protein substrates[16] but can also catalyze the glutathionylation of several proteins (e.g., GAPDH, actin, PTP1B) in the presence of a GS-radical generating system.[17] Hence, Grx is capable of catalyzing both glutathionylation and deglutathionylation of proteins via different mechanisms. Grx contains a conserved motif (CXXC) containing two cysteine residues, both of which are required for its reductive deglutathionylation function via a dithiol mechanism of disulfide exchange. Studies with mutant Grx containing only the N-terminal cysteine residue within the conserved motif (CXXS) showed that Grx can function as an oxidase through a monothiol mechanism.[18,19]

Sulfiredoxin is a small redox protein belonging to a conserved family of antioxidant proteins present in all eukaryotes,[20] and is specifically involved in the reversal of glutathionylation.[21] Human Srx1 contains only one cysteine residue within its entire sequence and is involved specifically in the reductive deglutathionylation of proteins. The cysteine 99 residue within Srx1 (Cys 99) is essential for the deglutathionylation reaction. However, unlike Grx whose conserved cysteine residue is an acceptor for the GS⁻ during the disulfide exchange reaction, Srx1 is not glutathionylated during this reaction. As well as catalyzing deglutathionylation, Srx1 also appears to be able to bind to and inhibit glutathionylation of target proteins. Mechanistically, this appears to be through steric hindrance since preincubation of proteins with wild-type and mutant Srx1 prevents the modification, whereas incubation of wild-type, but not mutant, Srx1 can deglutathionylate premodified proteins *in vitro*. This observation lends further weight to the idea that Srx1 is involved specifically in the reversal of glutathionylation.[21]

Adding a further layer of complexity is the understanding that proteins do not act in isolation in a cellular milieu. Rather, essential protein–protein interactions govern how cellular events unfold.[22] This process has proved to be significant for the regulation of JNK signaling by GSTπ.[7,8] This same paradigm seems to hold for thioredoxin (Trx), Grx, and GSTμ, which together modulate apoptosis signal-regulating kinase, ASK1,[23] implying the possible existence of a general regulatory mechanism for kinases that may involve GSH and associated pathways.[11,24] In addition, emergent literature suggests that direct glutathionylation of critical signaling molecules may trigger cellular events that are influenced by oxidative stress.[25,26] It would seem that GSH and GSTs have roles that extend much further than simple detoxification reactions. Indeed, it does not seem unreasonable to predict that glutathionylation may provide regulatory control complementary to other well-studied and established post-translational modifications.

Other recent studies have suggested that small redox active protein families such as peroxiredoxin have the potential to heterodimerize with GSTπ and to be activated by GST-mediated catalytic addition of GSH to a critical cysteine residue in a sterically protected region of peroxiredoxin.[27] This observation broadens the functional importance of GSTπ into yet another arena.

In subsequent sections, a discussion of some of the drugs that impact glutathione pathways and influence redox homeostasis and are currently under preclinical and clinical investigation will be provided.

8.4 GLUTOXIM (NOV-002)

The structure of NOV-002 is shown in Figure 8.3. It is oxidized glutathione complexed with cis-platinum at an approximate 1,000:1 ratio. It is believed that the platinum may serve to stabilize the GSSG. Novelos, Inc. (Newton, MA), has shown that standard animal and patient dosing results in a cumulative total of cis-platinum that is equivalent to <2% of a typical standard of care in an oncology single dose. As such, it seems unlikely that the platinum component contributes substantially to the pharmacology of the compound.

In terms of mechanism of action, NOV-002 presents a therapeutic profile that on the surface is a conundrum to the belief that oxidized glutathione is detrimental to cell survival. While there is no definitive published report of preclinical data, a

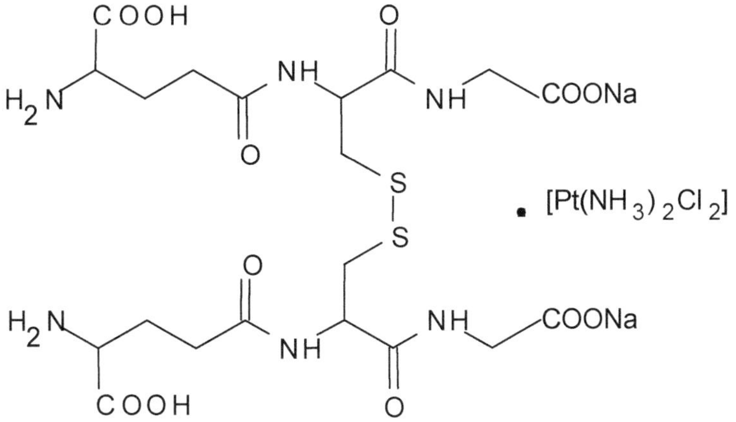

FIGURE 8.3 Structure of Glutoxim (NOV-002).

number of plausible hypotheses are testable. Foremost, NOV-002 has a growth-enhancing effect on bone marrow progenitors, a therapeutic property shared with TLK199. It is also known that GSSG can induce glutathionylation and NOV-002 shares this capacity. Figure 8.2 shows how oxidized glutathione may participate in the pathways that lead to glutathionylation of target proteins. Functional changes in proteins post-translationally modified in this manner could be a significant link to the therapeutic activity of NOV-002. This will no doubt be an area of research of some importance to the eventual registration of this drug with the FDA.

Pharmacological modulation of the glutathione pathways can have multiple (and parallel) effects, with the overall functional consequences dependent on the target cells involved and their physiological states. Because of this complexity, the specific molecular mechanisms and targets for this drug are currently under investigation. It is known that administration of the drug *in vivo* delivers a stabilized form of GSSG reflected by a sustained elevation of serum and tissue levels of GSSG to an extent that cannot be achieved by a commercial preparation of GSSG. The net effect of this elevation is an alteration of the GSSG–GSH ratio and hence of the redox state of the cell. From both *in vitro* and *in vivo* preclinical data, findings with NOV-002 are consistent with a variety of known effects of modulating the glutathione pathway, including cell protection, modulation of cytokine production, including those known to control production of blood cells, apoptosis, and immune system modulation.

One of the advantages of this drug platform is the existing clinical data generated by Novelos, the holding company, in a number of trials carried out in Russia. Evidence of efficacy was obtained in both lung and ovarian cancer patients, with an overall increased one-year survival rate for lung cancer patients of 63% versus 17% in the control arm. Improved quality of life (Karnofsky score) and increased tolerance to chemotherapy (permitting an increased number of chemotherapy cycles) were also registered in these patients. Consistent with some of the other agents discussed here, improved hematological parameters were also recorded following NOV-002 treatment, perhaps again reflecting the importance of thiol pathways in this compartment. While it should be emphasized that none of these clinical data have been

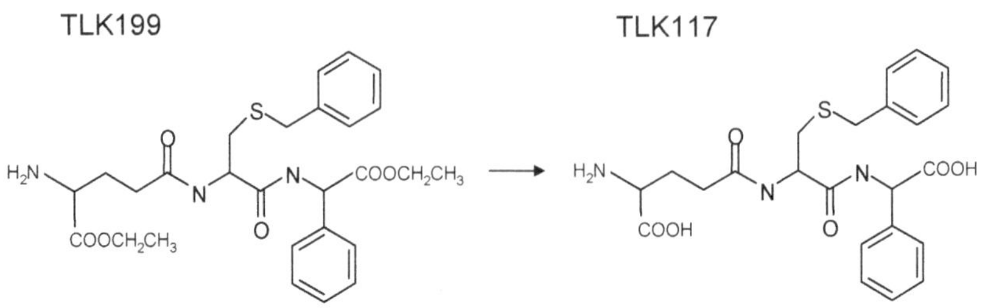

FIGURE 8.4 Structure of TLK199 and the active GSTπ inhibitor, TLK117.

audited (they are available through company documentation), there is an encouraging pattern that serves as a platform for further development in this country.

8.5 TELINTRA (TLK199)

One consequence of earlier attempts to modulate GSH or GSTs was the conceptual design of a peptidomimetic inhibitor of GSTπ named TLK199 (γ-glutamyl-S-[ben-zyl]cysteinyl-R[–]phenyl glycine diethyl ester; see Figure 8.4). This molecule was the end point of a systematic SAR analysis of a diverse series of peptide analogues that could bind to the G-sites of GSTs. Because the original goal was to synthesize inhibitors and not substrates, sulfur functionalization was used as a mechanism to decrease the inherent nucleophilic reactivity of sulfur while maintaining site-directed specificity. In the early 1990s Telik used solution-phase peptide synthesis to produce a series of such molecules that combined numerous aryl and alkyl functionalities, together with various C-terminal amino acid modifications. Synthesis was achieved with a series of N_α-Fmoc-S-benzylcysteine derivatives as intermediates.[28] After sep-aration and purification, TLK117 was the major product. The addition of a phenyl-glycine group to the C-terminus provided selectivity for GSTπ. S-functionalization through replacement of a hexyl group with benzene had a lesser but measureable effect on π inhibition specificity. TLK117 possesses a Ki for GSTπ of 0.45μM, a value approximately one log better than ethacrynic acid (~4μM). To enhance cell membrane permeability and make the molecule into a prodrug, diethyl ester groups were added to produce the final product, TLK199, now called Telintra.

Preclinical and mechanism of action studies with this agent revealed an unex-pected effect in animals, namely that the drug possessed myeloproliferative activ-ity.[5,6] In unstressed cells under normal growth conditions, GSTπ acts as an endog-enous regulatory switch for JNK, at least partly as a consequence of binding to the C-terminal region of the kinase.[8] When exposed to mild levels of oxidative stress, there is an interference with the protein–protein interactions between GSTπ and JNK, resulting in multimeric GSTπ and an activation of the JNK cascade[7] (Figure 8.5). Given the link between GSTπ and the kinase pathways, other model systems have helped to elucidate how TLK199 can produce proliferative effects in the marrow compartment. For example, mouse embryo fibroblasts (MEF) generated from 11-, 13-, or 15-day embryos from GSTπ knockout mice had a significantly faster doubling time than those from wild type (26 versus 34 hours, respectively). Transfection of

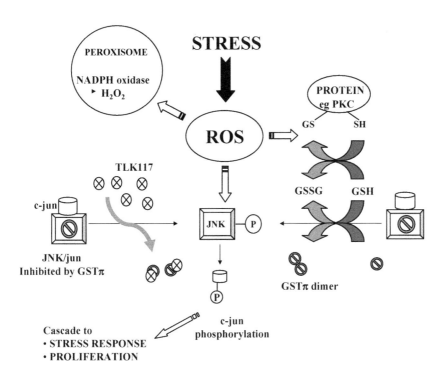

FIGURE 8.5 Model scheme for proposed action of TLK117 on GSTπ–JNK complex.

GSTπ into the knockout MEFs restored the doubling time to a range similar to wild type.[5] *In vivo* studies showed that total white blood cell (WBC) counts in GSTπ knockout mice were approximately 2-fold higher than in wild-type animals. Intra-peritoneal administration of 75 mg/kg TLK199 produced a 2-fold increase in WBC counts in wild-type mice three days after treatment, while having no effect on knockout animals (Table 8.1). Similarly TLK199 caused a dose-dependent increase in the number of bone marrow granulocyte/macrophage progenitor colonies only in wild-type animals;[5] TLK199 enhanced myeloproliferation induced by interleukin (IL)-3, granulocyte (G)-colony stimulating factor (CSF), and granulocyte-macroph-age (GM)-CSF in wild-type mice, but not in knockout animals.[6] TLK199 increased

TABLE 8.1
In Vivo Effect of TLK199 (72 Hours) and JNK Inhibitor
SP600125 on Blood Cell Count in GST Wild-Type Mice

	Vehicle	TLK199 (75 mg/kg)	SP600125 (15 mg/kg)	TLK199 + SP600125
Erythrocytes (× 10^9/ml)	5.4 ± 0.9	7.6 ± 1.7	4.4 ± 0.8	4.2 ± 0.6
Leucocytes (× 10^6/ml)	7.1 ± 1.4	11.6 ± 1.1	5.8 ± 1.6	7.0 ± 1.9

mobilization of GM progenitors from the bone marrow to spleen and peripheral blood.[29] Treatment with inhibitors of the ERK pathway reduced cytokine-induced myeloproliferation but had no effect on enhanced myeloproliferation caused by TLK199. On the other hand, the JNK inhibitor SP-600125 interfered with the myeloproliferation induced by either cytokines or TLK199.[6]

Overall, data discussed above support the model that the myeloproliferative effects of TLK199 depend on both GSTπ levels and JNK activity, with TLK199-induced inhibition of binding of GSTπ to JNK responsible for bone marrow progenitor cell myeloproliferation. It seems likely that these pathways are distinct from those affected by standard cytokines.[6] This effect highlights the potential differences in the tissue tropic influences of kinase activation. For example, while activation of ERKs has generally been linked with proliferation and JNKs with apoptosis, there is evidence that in the bone marrow compartment JNK may have a more direct link with myeloproliferation. This is consistent with recent reports showing a possible role for JNK in proliferation. For example, inhibition of JNK1 or JNK2 expression in human prostate cancer cells was associated with a decrease in cell proliferation.[30] In context, JNK phosphorylates and induces the transactivation of the transcription factors c-Jun, JunB, and JunD.[31,32] Increased expression or activation of c-Jun and JunD has been linked with both cell proliferation and transformation.[33,34] While JNK activation has also been associated with induction of apoptosis,[35] more recent data showed that, following UV exposure, JNK1 was more likely pro-apoptotic while JNK2 was pro-survival.[36] The discrimination between these two functions of JNK may correlate with the level and duration of the enzyme activation. A strong and sustained activation is associated with apoptosis, while a weaker and transient phosphorylation is correlated with proliferation.[24] For example, in mouse hematopoietic BaF3 cells, JNK activity was three times lower when cells were exposed to mitogenic concentrations of IL-3 than when exposed to cytotoxic concentrations of anisomycin.[37] Thus, the enhanced JNK activation observed in GSTπ[-/-] bone marrow cells would be consistent with the increased proliferation of these cells. In addition, JNK expression was lower in GSTπ-deficient bone marrow–derived mast cells (BMMC) than in wild type but the level of activation of this kinase (pJNK versus total JNK) was higher in the knockout cells.[6] Because of the pro-apoptotic properties of JNK, lower levels of the kinase may serve to maintain a higher level of viability in BMMC from GSTπ[-/-] animals. However, lower expression of JNK was observed only in mast cells, and its exact significance remains as yet unclear.

Commensurate with the increased proliferation of GSTπ[-/-] bone marrow and mast cells was an association with a sustained activation of STAT proteins, molecules that have been linked with the proliferation and differentiation of hematopoietic cells.[38] For example STAT5-deficient animals present a decreased proliferation of hematopoietic cells, which is correlated with a defect in response to various cytokines including IL-3, GM-CSF, and G-CSF.[39] Constitutive activation of STAT5 has been observed in various leukemias.[40] In addition, forced expression of a constitutively active form of STAT5 allows IL-3-dependent BaF3 cells to grow in medium without cytokines, while mock transfected cells died by apoptosis.[41] Because of the crucial role played by the JAK–STAT pathway in hematopoiesis, the increased STAT activation in the GST-deficient cells could be a critical component of their accelerated rate of proliferation. This observation could also explain the higher number of circulating white blood cells in the GSTπ[-/-] animals.

In bone marrow–derived mast cells, elevated STAT protein activation was associated with a downregulation of mRNA for various negative regulators of JAK–STAT pathways. For example, CIS-1 and SOCS-1, -2, and -3 have been shown to inhibit the phosphorylation of JAK and STAT proteins, and their expression is usually induced by cytokine treatment. Mechanistically, SOCS-1 and SOCS-3 inhibit JAK catalytic activity while CIS-1 blocks STAT binding to the receptor.[42] In the GSTπ-deficient mast cells, the basal expression of the SOCS family members was similar or slightly decreased (especially SOCS-3[6]). However, IL-3 treatment stimulated their expression more in the knockout than in wild-type cells, perhaps as a consequence of the stimulation of SOCS expression by STAT proteins. Because activation of the latter is more sustained in GSTπ-deficient BMMC, their overall transcriptional activity is also increased. SHP-1 and SHP-2 are phosphatases that can inhibit IL-3-induced phosphorylation of STAT5.[43] SHP-2 can bind STAT5 and dephosphorylate this protein.[44] Negative regulation of these phosphatases has been observed in various lymphomas and is associated with increased activation of the JAK–STAT pathway.[45] Similarly, transmembrane protein tyrosine phosphatase CD45 inhibits cytokine signaling by dephosphorylating JAK.[46] Taken together, these results suggest that increased activation of STAT proteins in GSTπ[-/-] cells might be a consequence of the decreased expression of their endogenous inhibitors of phosphorylation. Also pertinent to these observations, IL-3-induced proliferation of wild-type and GSTπ-deficient mast cells was potentiated by the tyrosine phosphatase inhibitor orthovanadate. This was associated with a more sustained activation of JAK–STAT pathways in both cell lines, suggesting that negative regulation of phosphatase expression in GSTπ-deficient cells was at least partly responsible for the more sustained activation of the JAK–STAT pathway and consequently the increased proliferation of the GSTπ-deficient cells.[6]

Telintra is now being evaluated in clinical trials. A Phase I/IIa clinical trial has been completed in patients with myelodysplastic syndromes (MDS). The drug was administered intravenously over a 60-minute interval daily for five days every two weeks. From this trial, it was concluded that TLK199 demonstrated improvement in all three hematological cell lineages in MDS patients, as well as activating hematopoietic progenitor cells. Dose-limiting toxicities were mild. Two Phase IIa trials using a liposomal preparation of TLK199 have also been reported. Overall, TLK199 was well tolerated at doses of 50 to 400 mg/m² with 50% and 45% (trials 1 and 2, respectively) of patients showing hematological responses, primarily where bone marrow examination showed improvements in maturation, differentiation, M/E ratios, and overall decreased dysplastic morphology.[47,48] These clinical studies are early in development and ongoing. Significantly, existing treatments for MDS are not extremely effective and thus Telintra may provide a meaningful opportunity for therapeutic intervention in this disease.

Recently, selected 7-nitro-2,1,3-benzoxadiazole derivatives have been found to be very efficient inhibitors of GSTπ.[49] Unlike TLK117, these inhibitors are not glutathione-peptidomimetics and their log P values are in a range suitable for traversing membranes. Cytotoxicity of these compounds in cell lines was reported, with equivalent GSTπ inhibition and cytotoxicity IC_{50} values. The lead compound, 6-(7-nitro-2,1,3-benzoxadiazol-4-ylthio) hexanol, in a similar fashion to TLK199, causes the dissociation of the GSTπ–JNK complex. This caused an ROS-independent activation of the JNK-mediated pathway that resulted in apoptosis. In addition, the drug induced an ROS-mediated apoptosis that involved the p38[MAPK] signal transduction pathway. The effective micromolar activity of this lead compound together with the reported limited toxicity in rodents suggest that this type of agent may be worthy of continued evaluation as an anticancer drug. There is presently no indication of whether these agents have any myeloproliferative activity.

8.6   TELCYTA (TLK286)

Telcyta was initially designed as a latent prodrug. TER286 became TLK286 when Terrapin Technologies, Inc. (Palo Alto, CA), changed its name to Telik, Inc. The chemical structure (γ-glutamyl-α-amino-β[2-ethyl-N,N,N′,N′-tetrakis (2-chloroethyl)phosphorodiamidate]-sulfonyl-propionyl-(R)-(−)phenylglycine emerged as the lead candidate from a group of rationally constructed glutathione analogues designed to exploit high GSTπ levels associated with malignancy, poor prognosis, and the development of drug resistance.[50,51] By doing this, selective targeting of susceptible tumor phenotypes defined a strategy to release a more active drug in malignant cells compared to normal tissue, seeking to achieve a more optimal therapeutic index.

In TLK286, the sulfhydryl of glutathione was oxidized to a sulfone. The tyrosine-7 in GSTπ promotes a β-elimination reaction that cleaves the compound (see Figure 8.6), releasing a glutathione analogue and a phosphorodiamidate, which spontaneously forms aziridinium species, the presumed alkylating moieties. The cytotoxic component has tetrafunctionality with properties similar in concept to bifunctional nitrogen mustards. With a short half-life, each can react with cellular nucleophiles.[52,53] The other component of the molecule contains the glutathione backbone and an electrophilic vinyl sulfone is also released. The contribution of the vinyl sulfone to either drug efficacy or toxicity has not been entirely determined. Glutathione conjugates of the sulfone are possible and these could be substrates for ABC transporters such as MRP1. The vinyl sulfone could also be a factor in chain reactions leading to lipid peroxidation and eventual production of hydrogen peroxide or other reactive oxygen species.[54] This could explain the enhanced expression of catalase in an HL60 cell line selected for resistance to TLK286.[55]

TLK286 exhibited cytotoxic activity against a number of different tumors and tumor cell lines. *In vitro* studies have shown that elevated GSTπ in transfected cells correlates with enhanced sensitivity to TLK286 and that drug-resistant cells that overexpress GSTπ are more sensitive to the drug.[56] In a clonogenic assay against human solid tumors, TLK286 showed activity against 15 of 21 lung tumors and 11 of 20 breast tumors. In addition, antitumor activity was shown *in vivo* against human xenografts in nude mice, with only mild bone marrow toxicity.[56]

To study the pharmacology of TLK286, three *in vitro* model systems were employed. The ultimate goal was to gain proof of principle with respect to mechanism of action. The first entailed establishment of a TLK286-resistant cell line, a

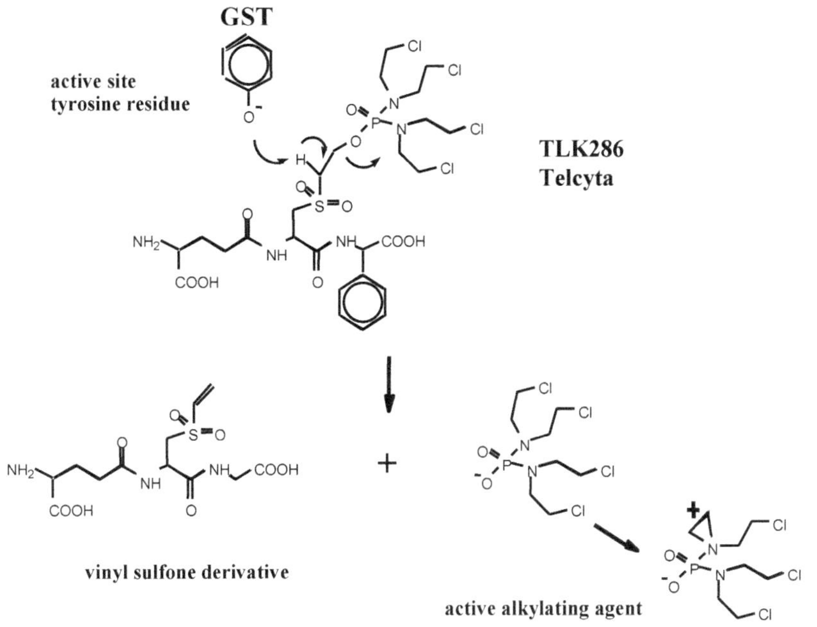

FIGURE 8.6 Activation of TLK286 (Telcyta) by GSTπ.

task that proved unexpectedly difficult to achieve.[55] While cells usually survived an initial low concentration of TLK286, recovery to full viability did not readily occur. Considering the large number of HL60 cell lines resistant to a variety of anticancer drugs, this result is unusual. A plausible explanation is that resistance to TLK286 is governed by multiple factors or that survival response pathways are not rapidly induced following exposure to the drug. A characteristic of GSTπ as a GSH-conjugating enzyme is its low catalytic efficiency and broad substrate specificity. Because GSTπ is directly involved in the regulation of JNK-mediated stress response[7] emphasizing the ligand binding (noncatalytic function of the protein), this may provide a partial explanation for the high GSTπ levels seen in many tumors, where kinase cascade pathways involving JNK may be imbalanced. Even though the β-elimination reaction for TLK286 catalyzed by GSTπ does not inactivate the protein, it may serve to reduce its JNK–ligand binding function. In turn, this may alter the stoichiometry that controls kinase-mediated proliferative/apoptotic pathways and may be a factor in the difficulty experienced in establishing a TLK286-resistant cell line. With eventual establishment of 5-fold resistance, a resultant *decrease* in expression of GSTπ was consistent with a mechanism of action based on the rational design of the drug.[55] Two other approaches corroborated these results. For example, increased resistance to TLK286 in the MEF cell lines derived from GSTπ[-/-] mice[13] was consistent with reduced activation of the drug. Finally, increased sensitivity in an NIH3T3 cell line transfected to overexpress GSTπ also confirmed a direct involvement of the isozyme in determining cytotoxicity.[57] The resistant cell line also had increased glutathione levels, not uncommon with alkylating agent resistance.[50]

Enhanced GSH levels were not a consequence of overexpression of the enzymes responsible for *de novo* γ-(GCS) or salvage (γ-glutamyl transpeptidase; γ-GT) synthesis of the tripeptide. Similarly, MRP1 expression was unaltered in resistant cells. A coordinated, increased expression of γ-GCS, GSTπ, and MRP1 has been shown in cells made resistant to ethacrynic acid, a drug with Michael addition properties.[58] While TLK286 produces aziridinium moieties characteristic of other nitrogen mustards, these are distinct in electrophilic properties from Michael acceptors. This may account for the absence of induced expression of the cadre of GSH-related detoxification gene products in TLK286-resistant cells.[59] The electrophilic characteristics of the vinyl sulfone would predict reactivity with cellular nucleophiles, and because GS conjugates are primary substrates for MRP1,[60] a GS-vinyl sulfone could prove to be an effective substrate for this transporter. This could have pharmacological significance if it serves to interfere with other efflux functions for MRP1.

Many of these preclinical studies were instrumental in clinical trial design. For Phase I analysis,[61,62] patients with advanced malignancies were treated with weekly IV infusions of Telcyta in escalating doses from 60–960 mg/m^2. One treatment cycle was three weekly administrations, and patients were assessed for disease status on the forty-third day, with those receiving clinical benefit continuing until disease progression or dose-limiting toxicity. Thirty-seven patients received 111 cycles of Telcyta at eight dose levels with the outcome that a weekly dose of 960 mg/m^2 was well tolerated without dose-limiting toxicities. Toxicities attributed to Telcyta included grade 1–2 nausea and vomiting, fatigue, and anemia. At doses >960 mg/m^2 other minor toxicities included pancreatitis and bladder symptoms consisting of hematuria, dysuria, and urinary frequency. All patients with dose-limiting toxicities continued one dose below MTD (960 mg/m^2) without recurrence.[61] Toxicities were noncumulative, transient, and resolved without sequelae. Nine of 31 evaluable patients continued therapy for >43 days (median of five cycles), experiencing durable stable disease or tumor regression. Conclusions were that Telcyta was well tolerated when administered weekly at doses up to 960 mg/m^2. Overall, from the two published Phase I trials[61,62] the frequency and severity of adverse events were equivalent. In both studies, Telcyta was well tolerated with no deaths, grade 4 toxicities, clinically significant myelosuppression, or thrombocytopenia due to drug treatments. The type, frequency, and severity of adverse events were comparable, with the indication that Telcyta toxicity was not schedule dependent. Safety and efficacy analysis supported the subsequent Phase II disease-specific investigations.

In a series of American Society of Clinical Oncology meeting abstracts, a 46% objective response rate was reported for a Phase II trial of the combination of Telcyta and liposomal Doxorubicin in platinum refractory ovarian cancer patients; a 56% objective response rate for Telcyta and Carboplatin in platinum-resistant ovarian cancer. Telcyta is also under active testing in a clinical Phase III setting for ovarian cancer. At the time of this writing, enrollment was complete for the ASSIST-1 (Assessment of Survival in Solid Tumors-1) Phase III randomized study of Telcyta versus liposomal Doxorubicin or Topotecan as third-line therapy in cis-platinum-refractory or resistant ovarian cancer. A further Phase III randomized multicenter study (ASSIST-3) compares Telcyta plus Carboplatin with liposomal

Doxorubicin as second-line therapy in cis-platinum-refractory or resistant ovarian cancer and is ongoing.

In nonsmall-cell lung cancer (NSCLC), a 27% objective response rate for a Phase II study of combinations of Telcyta and Docetaxel in platinum-resistant disease was reported. ASSIST-2 is an open enrollment, randomized Phase III study of Telcyta versus Gefitinib as third-line therapy in locally advanced or metastatic nonsmall-cell lung cancer. Whether the overall analysis of these results will be complicated by the recent decision to remove Gefitinib from the approved drugs list for NSCLC remains to be seen.

The Special Protocol Assessment process administered by the FDA reviewed the protocols for ASSIST-1 and ASSIST-2, and both indications received an FDA fast-track designation. Because the ASSIST Phase III ovarian trial is event driven, patients have to die to allow evaluation. Therefore, final analysis will be delayed by the apparent success of the therapy. The long-term success of the NDA and eventual drug registration will depend heavily on the outcome of these trials, a process that is anticipated to reach completion in 2006.

8.7 PABA/NO

As with TLK286, high GST levels in human cancers and drug-resistant disease, together with the knowledge that nitric oxide (NO) has therapeutic potential, provided the rationale for the design of the NO-releasing GST-activated prodrug, O^2-(2,4-dinitro-5-[N-methyl-N-4-carboxyphenylamino]phenyl 1-N,N-dimethylamino)-diazen-1-ium-1,2-diolate (PABA/NO).[63] GSTπ-catalyzed conjugation of GSH to PABA/NO releases a diazeniumdiolate ion, with subsequent release of NO (Figure 8.7). Other NO donors of the diazeniumdiolate class are known to release NO in an enzyme-catalyzed manner, with activation by cytochrome P450[64] or esterases.[63]

In order to confirm the GSTπ activation requirements of PABA/NO, a number of model systems have been used. For example, MEFs from GSTπ$^{-/-}$ mice showed a decreased sensitivity to PABA/NO. The effects on cytotoxicity of eliminating GSTπ can be influenced by slow spontaneous activation of PABA/NO through noncatalytic GSH conjugation and by the expression of other GST isoforms that can also activate the drug at rates different to GSTπ. Components involved in metabolism and detoxification pathways were analyzed in a cell system stably transfected with GSTπ, γGCS, and MRP1, confirming a role for GST and GSH in the activation of PABA/NO.[65] Forced expression of γGCS and MRP1 also provided insight into cellular resistance mechanisms toward PABA/NO. Increased resistance to PABA/NO was conferred by the overexpression of MRP1, supporting the concept that this efflux pump with affinity for GS conjugates is involved in the removal of PABA/NO or active metabolites. Inhibition of transport of a known substrate for MRP1, LTC_4 was observed in the presence of both PABA/NO and GSH but not with either alone. GSTπ stimulated PABA/NO–induced inhibition of LTC_4 transport, suggesting that metabolites of PABA/NO may be substrates for and effluxed by this transporter.[65]

The *in vivo* efficacy of PABA/NO has been shown in tumor-bearing animals.[65,66] Doses of PABA/NO with limited toxicity produced a significant growth delay in a

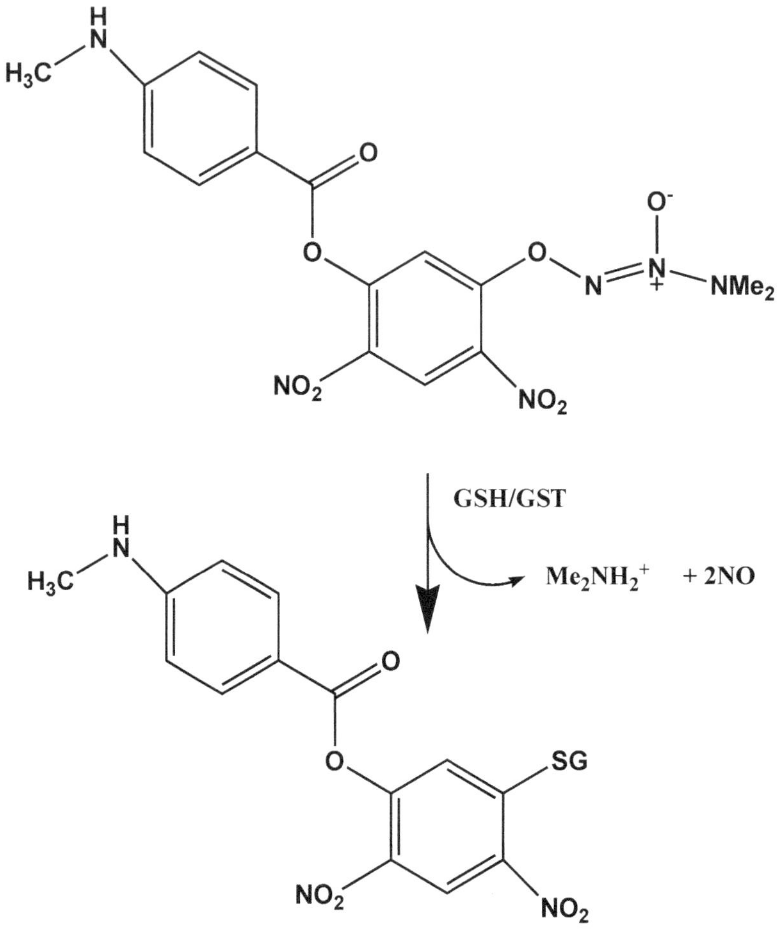

FIGURE 8.7 Activation of PABA/NO and release of nitric oxide by GST and GSH.

human ovarian cancer model in SCID mice (Figure 8.8), with results comparable to those seen with cisplatin (the standard of care for management of ovarian cancer). Of interest, GSTπ overexpression has been associated (although not necessarily causally linked) with resistance to platinum drugs.[67] PABA/NO exerts an apoptotic effect by induction of p38 and JNK. As noted earlier, this observation carries greater significance because GSTπ functions as an endogenous negative regulatory switch for these same regulatory kinase pathways.[7] NO has diverse roles in a variety of physiological processes and augmentation of endogenous with exogenous NO provides the foundation for a broad range of therapeutic applications.[68,69] Bifurcating pathways of inhibition or induction of apoptosis are tissue specific and depend on the amount, duration, and site of NO production.[70] In support of both cellular antioxidant and pro-oxidant actions of NO *in vivo*, low doses of NO can protect cells against peroxide-induced death, where higher concentrations result in increased killing.[71] PABA/NO has a dose- and time-dependent effect for activation of kinases and MRP1 abrogates or delays the

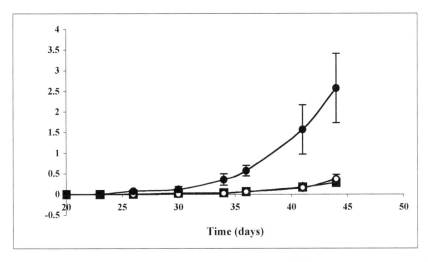

FIGURE 8.8 Treatment with PABA/NO delays growth of A2780 human ovarian tumors in SCID mice. The average tumor volumes ± SE in control (●), PABA/NO- (○), and cisplatin- (■) treated mice are reported. Mice were treated with 0.8% DMSO (control), 3.36 mg/kg PABA/NO, and 2.5 mg/kg cisplatin, respectively. From Findlay, V.J. et al., *Mol. Pharmacol.,* 2004. With permission.

effect, primarily as a consequence of reducing the effective intracellular concentration of the drug.[65] Whether kinase activation occurs as a result of direct NO interaction (e.g., nitrosylation/nitration of residues in JNK) or by the impact of PABA/NO (or metabolites including GSNO) effects on the GSTπ–JNK complex is presently not known. In the United States, PABA/NO remains in early preclinical testing but has the potential disadvantage of poor solubility and stability in aqueous solution.[72] However, reasonable *in vivo* antitumor data suggest that it is a good lead compound for further structure activity and drug discovery efforts.[68]

8.8   AMIFOSTINE (ETHYOL)

S-2-(3-aminopropylamino) ethyl phosphorothioic acid (Amifostine) gained approval by the FDA for indications surrounding protection of normal tissues from radiation and chemotherapy damage. This aminothiol was first investigated as a radioprotective agent following early studies at the Walter Reed Army Institute. Aminothiols prove to have significant impact on cellular redox balance and mechanism of action studies with the active form of the drug, WR1065 (Figure 8.9), helped to consolidate a mechanism of action and to enable clinical development of the drug.

The radio and chemoprotective effects of WR1065 have been ascribed to the nucleophilicity of the sulfhydryl group in the scavenging of reactive oxygen species and in proton donation during DNA repair reactions.[73] A therapeutic advantage is achieved by selective protection of normal tissues when combinations of standard anticancer drugs are used with Amifostine.[74] Specific mechanisms for drug effects have been implied from the drug's capacity to activate the DNA-binding activities

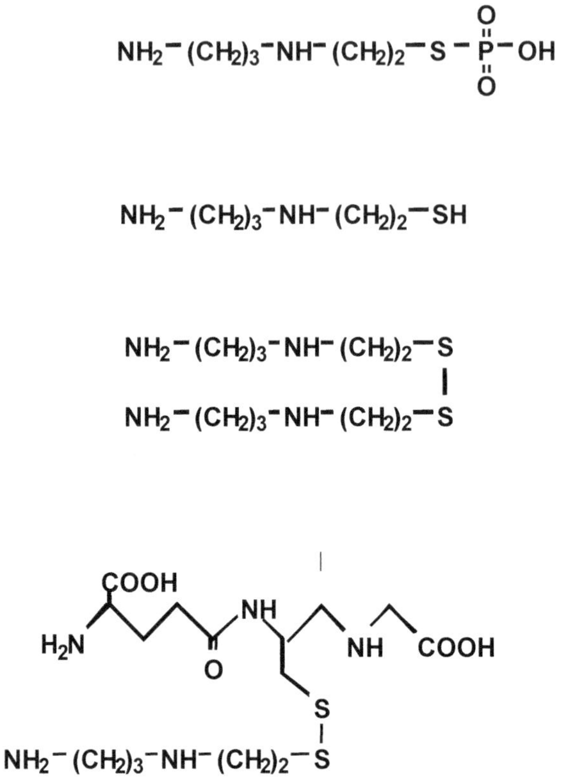

WR2721

WR1065

Disulfide

WR1065-SG conjugate

FIGURE 8.9 Structure of Amifostine (WR2721) and its active metabolites.

of NF-κB, AP-1, and p53. These transcription factors are regulated by intracellular redox status.[75–77] *In vitro* analyses showed that WR1065 enhanced the DNA-binding activity of NF-κB and AP-1 both as a single agent and in combination with paclitaxel. While free glutathione (GSH) also showed an enhancing effect on the DNA-binding activity of NF-κB and AP-1, a variety of agents that lack the free thiol moiety had no effect, supporting the importance of this moiety in the activation/binding of WR1065 to these proteins.

The mechanism for this activation relates to the direct binding to these proteins, all of which contain critical cysteine residues that are sensitive to redox. The Cys 62 residue in the p50 subunit of NF-κB is involved in disulfide bond and dimer formation in an oxidative environment. That this cysteine residue is critical to NF-κB function was demonstrated when its replacement with serine was shown to result in a loss of DNA-binding activity.[78] Similarly, a conserved motif (Lys–Cys–Arg) in the c-Jun and c-Fos subunits of AP-1 is also critical for the DNA-binding activity of these proteins.[79] WR1065 may provide reducing equivalents, thus maintaining these critical cysteine residues in a reactive state. WR1065 may form a mixed disulfide bond with these proteins in such a way that bulkier proteins that may function as endogenous suppressors are displaced. This would enhance DNA binding and transcriptional activation.

NF-κB and p53 can cooperate to induce apoptosis in tumor cells. Specifically, p53 can induce activation of NF-κB, whereas loss of NF-κB activity was shown to abrogate the ability of p53 to induce cell death, while its ability to induce growth arrest was unaffected.[80] A coordinated activation of p53, NF-κB, and AP-1 by WR1065 provides evidence that WR1065 can protect normal cells but not tumor cells from the cytotoxic effects of chemotherapeutic agents. Perhaps this is a consequence of a differential response of these transcription factors and these signaling pathways in tumors and normal tissues.

p53 activation by WR1065 leads to increased levels of functional p53 protein.[81] Redox sensitivity of p53 may involve regulation of the coordination of zinc between critical cysteine residues.[82,83] Of the twelve cysteine residues in p53, replacement of Cys 173, Cys 235, or Cys 239 reduced DNA binding by this protein and completely blocked transcriptional activation by p53.[84] Mixed disulfide binding of WR1065 to a cysteine residue in p53 may contribute to protein stability or increased DNA-binding activity. WR1065 treatment can also produce increased levels of the p53-response genes bax, killer/DR5, fas, p21/waf1, and HDM2, confirming that this drug activated the p53 pathway. WR1065-treated nontransformed fibroblasts induced growth arrest in a p53-dependent manner. In human tumor cell lines, p53 status had no impact on the ability of WR1065 to function as a cytoprotective agent.[85] Radiation protection by WR1065 in human glioma cell lines was also independent of their p53 status.[86] This conclusion is consistent with the principle that downstream p53 pathways may not be operational in these glioma cells. These findings support the premise that WR1065 functions as a cytoprotective agent in normal cells through the combined activation of a set of redox-sensitive transcription factors, but activation of these pathways in tumor cells does not lead to growth arrest. The consequence is a preferential killing of tumor cells by cytotoxic agents.

8.9 SUMMARY

Glutathione (GSH) provides a major source of thiol homeostasis generally believed critical to the maintenance of a reduced cellular environment conducive to cell survival. Glutathione S-transferases (GST) are prevalent in eukaryotes and have been ascribed catalytic functions that involve detoxification of exogenous and endogenous electrophiles through thioether bond formation with the cysteine thiol of GSH. The neutralizing impact of these reactions on products of reactive oxygen has contributed to the significant evolutionary conservation and adaptive functional redundancy of the GSH system. Nevertheless, emergent studies indicate that GSH pools can also provide the substrate for a post-translational modification of low pK cysteine residues in target proteins through S-glutathionylation. In addition, GST isozymes can act as important ligand-binding proteins to regulate signaling kinases through direct protein–protein interactions. The functional importance of these reactions in governing how cells respond to alterations in redox balance exemplifies the broad importance of the GSH/GST in diseases such as cancer and has provided a platform for therapeutic drug development.

REFERENCES

1. Peterson, J.D. et al. Glutathione levels in antigen-presenting cells modulate Th1 versus Th2 response patterns. *Proc. Natl. Acad. Sci. USA*, 95:3071–3076. 1998.
2. Herzenberg, L.A., De Rosa, S.C., Dubs, J.G., Roederer, M., Anderson, M.T., Ela, S.W., and Deresinski, S.C. Glutathione deficiency is associated with impaired survival in HIV disease. *Proc. Natl. Acad. Sci. USA*, 94:1967–1972. 1997.
3. O'Dwyer, P.J. et al. Phase I study of thiotepa in combination with the glutathione transferase inhibitor ethacrynic acid. *Cancer Res.*, 51:6059–6065. 1991.
4. Bailey, H.H. et al. Phase I study of continuous-infusion L-S,R-buthionine sulfoximine with intravenous melphalan. *J. Natl. Cancer Inst.*, 89:1789–1796. 1997.
5. Ruscoe, J.E. et al. Pharmacologic or genetic manipulation of glutathione S-transferase P1-1 (GSTpi) influences cell proliferation pathways. *J. Pharmacol. Exp. Ther.*, 298:339–345. 2001.
6. Gate, L. et al. Increased myeloproliferation in glutathione S-transferase pi-deficient mice is associated with a deregulation of JNK and Janus kinase/STAT pathways. *J. Biol. Chem.*, 279:8608–8616. 2004.
7. Adler, V. et al. Regulation of JNK signaling by GSTp. *Embo. J.*, 18:1321–1334. 1999.
8. Wang, T. et al. Glutathione S-transferase P1-1 (GSTP1-1) inhibits c-Jun N-terminal kinase (JNK1) signaling through interaction with the C terminus. *J. Biol. Chem.*, 276:20999-21003. 2001.
9. Tew, K.D. et al. Glutathione-associated enzymes in the human cell lines of the National Cancer Institute Drug Screening Program. *Mol. Pharmacol.*, 50:149–159. 1996.
10. Schisselbauer, J.C. et al. Characterization of glutathione S-transferase expression in lymphocytes from chronic lymphocytic leukemia patients. *Cancer Res.*, 50:3562–3568. 1990.
11. Adler,V. et al. Role of redox potential and reactive oxygen species in stress signaling. *Oncogene*, 18:6104–6111. 1999.
12. Yin, Z. et al. Glutathione S-transferase p elicits protection against H2O2-induced cell death via coordinated regulation of stress kinases. *Cancer Res.*, 60:4053–4057. 2000.
13. Henderson, C.J. et al. Increased skin tumorigenesis in mice lacking pi class glutathione S-transferases. *Proc. Natl. Acad. Sci. USA*, 95:5275–5280. 1998.
14. Wang, J. et al. Reversible glutathionylation regulates actin polymerization in A431 cells. *J. Biol. Chem.*, 276:47763–47766. 2001.
15. Ward, N.E., Chu, F., and O'Brian, C.A. Regulation of protein kinase C isozyme activity by S-glutathiolation. *Methods Enzymol.*, 353:89–100. 2002.
16. Gravina, S.A. and Mieyal, J.J. Thioltransferase is a specific glutathionyl mixed disulfide oxidoreductase. *Biochemistry*, 32:3368–3376. 1993.
17. Starke, D.W., Chock, P.B., and Mieyal, J.J. Glutathione-thiyl radical scavenging and transferase properties of human glutaredoxin (thioltransferase). Potential role in redox signal transduction. *J. Biol. Chem.*, 278:14607–14613. 2003.
18. Xiao, R. et al. Catalysis of thiol/disulfide exchange. Glutaredoxin 1 and protein-disulfide isomerase use different mechanisms to enhance oxidase and reductase activities. *J. Biol. Chem.*, 280:21099–21106. 2005.

19. Yang, Y. et al. Reactivity of the human thioltransferase (glutaredoxin) C7S, C25S, C78S, C82S mutant and NMR solution structure of its glutathionyl mixed disulfide intermediate reflect catalytic specificity. *Biochemistry*, 37:17145–17156. 1998.

20. Biteau, B., Labarre, J., and Toledano, M.B. ATP-dependent reduction of cysteine-sulphinic acid by S. cerevisiae sulphiredoxin. *Nature*, 425:980–984. 1998.

21. Findlay, V.J. et al. A novel role for human sulfiredoxin in the reversal of glutathio-nylation. Submitted for publication. 2005.

22. Golemis, E.A., Tew, K.D., and Dadke, D. Protein interaction-targeted drug discovery: evaluating critical issues. *Biotechniques*, 32:636–638, 640, 642 passim. 2002.

23. Saitoh, M. et al. Mammalian thioredoxin is a direct inhibitor of apoptosis signal-regulating kinase (ASK) 1. *Embo. J.*, 17:2596–2606. 1998.

24. Davis, W. Jr., Ronai, Z., and Tew, K.D. Cellular thiols and reactive oxygen species in drug-induced apoptosis. *J. Pharmacol. Exp. Ther.*, 296:1–6. 2001.

25. Adachi, T. et al. S-glutathiolation of Ras mediates redox-sensitive signaling by angio-tensin II in vascular smooth muscle cells. *J. Biol. Chem.*, 279(28):29857–29862, 2004.

26. Cross, J.V. and Templeton, D.J. Oxidative stress inhibits MEKK1 by site-specific glutathionylation in the ATP binding domain. *Biochem. J.*, 381(Pt3):675–683, 2004.

27. Manevich, Y., Feinstein, S.I., and Fisher, A.B. Activation of the antioxidant enzyme 1-CYS peroxiredoxin requires glutathionylation mediated by heterodimerization with pi GST. *Proc. Natl. Acad. Sci. USA*, 101:3780–3785. 2004.

28. Lyttle, M.H. et al. Isozyme-specific glutathione-S-transferase inhibitors: design and synthesis. *J. Med. Chem.*, 37:189–194. 1994.

29. Kauvar, L.M., Sanderson, P.E., and Henner, W.D. Glutathione-based approaches to improving cancer treatment. *Chem. Biol. Int.*, 111:225–238. 1998.

30. Yang, Y.M. et al. C-Jun NH(2)-terminal kinase mediates proliferation and tumor growth of human prostate carcinoma. *Clin. Cancer Res.*, 9:391–401. 2003.

31. Binetruy, B., Smeal, T., and Karin, M. Ha-Ras augments c-Jun activity and stimulates phosphorylation of its activation domain. *Nature*, 351:122–127. 1991.

32. Kallunki, T. et al. c-Jun can recruit JNK to phosphorylate dimerization partners via specific docking interactions. *Cell*, 87:929–939. 1996.

33. Lamb, J.A. et al. JunD mediates survival signaling by the JNK signal transduction pathway. *Mol. Cell*, 11:1479–1489. 2003.

34. Shaulian, E. and Karin, M. AP-1 in cell proliferation and survival. *Oncogene*, 20:2390–2400. 2001.

35. Tournier, C. et al. Requirement of JNK for stress-induced activation of the cytochrome c-mediated death pathway. *Science*, 288:870–874. 2000.

36. Hochedlinger, K., Wagner, E.F., and Sabapathy, K. Differential effects of JNK1 and JNK2 on signal specific induction of apoptosis. *Oncogene*, 21:2441–2445. 2002.

37. Terada, K., Kaziro, Y., and Satoh, T. Ras-dependent activation of c-Jun N-terminal kinase/stress-activated protein kinase in response to interleukin-3 stimulation in hematopoietic BaF3 cells. *J. Biol. Chem.*, 272:4544–4548. 1997.

38. Ward, A.C., Touw, I., and Yoshimura, A. The Jak–STAT pathway in normal and perturbed hematopoiesis. *Blood*, 95:19–29. 2000.

39. Teglund, S. et al. Stat5a and Stat5b proteins have essential and nonessential, or redundant, roles in cytokine responses. *Cell*, 93:841–850. 1998.

40. Weber-Nordt, R.M. et al. Stat3 recruitment by two distinct ligand-induced, tyrosine-phosphorylated docking sites in the interleukin-10 receptor intracellular domain. *J. Biol. Chem.*, 271:27954–27961. 1996.

41. Nosaka, T. et al. STAT5 as a molecular regulator of proliferation, differentiation, and apoptosis in hematopoietic cells. *Embo. J.*, 18:4754–4765. 1999.

42. Larsen, L. and Ropke, C. Suppressors of cytokine signalling: SOCS. *Apmis.*, 110:833–844. 2002.

43. Paling, N.R. and Welham, M.J. Role of the protein tyrosine phosphatase SHP-1 (Src homology phosphatase-1) in the regulation of interleukin-3–induced survival, proliferation, and signalling. *Biochem. J.*, 368:885–894. 2002.

44. Yu, C.L., Jin, Y.J., and Burakoff, S.J. Cytosolic tyrosine dephosphorylation of STAT5. Potential role of SHP-2 in STAT5 regulation. *J. Biol. Chem.*, 275:599–604. 2000.

45. Wu, C. et al. The function of the protein tyrosine phosphatase SHP-1 in cancer. *Gene*, 306:1–12. 2003.

46. Irie-Sasaki, J. et al. CD45 is a JAK phosphatase and negatively regulates cytokine receptor signalling. *Nature*, 409:349–354. 2001.

47. Emanuel, P.D. et al. TLK199 (Telintra), a novel glutathione analog inhibitor of GST P1-1, causes proliferation and maturation of bone marrow precursor cells and correlates with clinical improvement in myelodysplastic syndrome (MDS) patients in a Phase 2a study. *Blood*, 104:2372. 2004.

48. Callander, N. et al. Hematologic improvement following treatment with TLK199 (Telintra), a novel glutathione analog inhibitor of GST P1-1, in myelodysplastic syndrome (MDS): interim results of a dose-ranging Phase 2a study. *Blood*, 104:1428. 2004.

49. Turella, P. et al. Proapoptotic activity of new glutathione S-transferase inhibitors. *Cancer Res.*, 65:3751–3761. 2005.

50. Tew, K.D. Glutathione-associated enzymes in anticancer drug resistance. *Cancer Res.*, 54:4313–4320. 1994.

51. Hayes, J.D. and Pulford, D.J. The glutathione S-transferase supergene family: regulation of GST and the contribution of the isoenzymes to cancer chemoprotection and drug resistance. *Crit. Rev. Biochem. Mol. Biol.*, 30:445–600. 1995.

52. Lyttle, M.H. et al. Glutathione-S-transferase activates novel alkylating agents. *J. Med. Chem.*, 37:1501–1507. 1994.

53. Satyam, A. et al. Design, synthesis, and evaluation of latent alkylating agents activated by glutathione S-transferase. *J. Med. Chem.*, 39:1736–1747. 1996.

54. Comporti, M. Three models of free radical–induced cell injury. *Chem. Biol. Interact.*, 72:1–56. 1989.

55. Rosario, L.A. et al. Cellular response to a glutathione S-transferase P1-1 activated prodrug. *Mol. Pharmacol.*, 58:167–174. 2000.

56. Morgan, A.S. et al. Tumor efficacy and bone marrow–sparing properties of TER286, a cytotoxin activated by glutathione S-transferase. *Cancer Res.*, 58:2568–2575. 1998.

57. O'Brien, M., Kruh, G.D., and Tew, K.D. The influence of coordinate overexpression of glutathione phase II detoxification gene products on drug resistance. *J. Pharmacol. Exp. Ther.*, 294:480–487. 2000.

58. Ciaccio, P.J. et al. Effects of chronic ethacrynic acid exposure on glutathione conjugation and MRP expression in human colon tumor cells. *Biochem. Biophys. Res. Commun.*, 222:111–115. 1996.

59. Chen, Z.J. et al. Sensitivity and fidelity of DNA microarray improved with integration of Amplified Differential Gene Expression (ADGE). *BMC Genomics*, 4:28. 2003.

60. Keppler, D., Leier, I., and Jedlitschky, G. Transport of glutathione conjugates and glucuronides by the multidrug resistance proteins MRP1 and MRP2. *Biol. Chem.*, 378:787–791. 1997.

61. Rosen, L.S. et al. Phase I study of TLK286 (glutathione S-transferase P1-1 activated glutathione analogue) in advanced refractory solid malignancies. *Clin. Cancer Res.*, 9:1628–1638. 2003.

62. Rosen, L.S. et al. Phase 1 study of TLK286 (Telcyta) administered weekly in advanced malignancies. *Clin. Cancer Res.*, 10:3689–3698. 2004.

63. Saavedra, J.E. et al. Esterase-sensitive nitric oxide donors of the diazeniumdiolate family: *in vitro* antileukemic activity. *J. Med. Chem.*, 43:261–269. 2000.

64. Saavedra, J.E. et al. Targeting nitric oxide (NO) delivery *in vivo*. Design of a liver-selective NO donor prodrug that blocks tumor necrosis factor-alpha-induced apoptosis and toxicity in the liver. *J. Med. Chem.*, 40:1947–1954. 1997.

65. Findlay, V.J. et al. Tumor cell responses to a novel glutathione S-transferase–activated nitric oxide–releasing prodrug. *Mol. Pharmacol.*, 65:1070–1079. 2004.

66. Shami, P.J. et al. JS-K, a glutathione/glutathione S-transferase–activated nitric oxide donor of the diazeniumdiolate class with potent antineoplastic activity. *Mol. Cancer Ther.*, 2:409–417. 2003.

67. Townsend, D. and Tew, K. Cancer drugs, genetic variation and the glutathione-S-transferase gene family. *Am. J. Pharmacogenomics*, 3:157–172. 2003.

68. Keefer, L.K. Progress toward clinical application of the nitric oxide–releasing diazeniumdiolates. *Annu. Rev. Pharmacol. Toxicol.*, 43:585–607. 2003.

69. Napoli, C. and Ignarro, L.J. Nitric oxide–releasing drugs. *Annu. Rev. Pharmacol. Toxicol.*, 43:97–123. 2003.

70. Umansky, V. and Schirrmacher, V. Nitric oxide–induced apoptosis in tumor cells. *Adv. Cancer Res.*, 82:107–131. 2001.

71. Joshi, M.S., Ponthier, J.L., and Lancaster, J.R., Jr. Cellular antioxidant and pro-oxidant actions of nitric oxide. *Free Radic. Biol. Med.*, 27:1357–1366. 1999.

72. Srinivasan, A. et al. PABA/NO as an anticancer lead: analogue synthesis, structure revision, solution chemistry and reactivity towards glutathione. *J. Med. Chem.*, 2005. In press.

73. van der Vijgh, W.J. and Peters, G.J. Protection of normal tissues from the cytotoxic effects of chemotherapy and radiation by amifostine (Ethyol): preclinical aspects. *Semin. Oncol.*, 21:2–7. 1994.

74. Valeriote, F. and Tolen, S. Protection and potentiation of nitrogen mustard cytotoxicity by WR-2721. *Cancer Res.*, 42:4330–4331. 1982.

75. Gius, D. et al. Intracellular oxidation/reduction status in the regulation of transcription factors NF-kappaB and AP-1. *Toxicol. Lett.*, 106:93–106. 1999.

76. Kamata, H. and Hirata, H. Redox regulation of cellular signalling. *Cell Signal.*, 11:1–14. 1999.

77. Sen, C.K. and Packer, L. Antioxidant and redox regulation of gene transcription. *Faseb. J.*, 10:709–720. 1996.

78. Matthews, J.R. et al. Thioredoxin regulates the DNA binding activity of NF-kappa B by reduction of a disulphide bond involving cysteine 62. *Nucleic Acids Res.*, 20:3821–3830. 1992.

79. Abate, C., Patel, L., and Rauscher, F.J. III, and Curran T. Redox regulation of fos and jun DNA-binding activity *in vitro*. *Science*, 249:1157–1161. 1990.

80. Ryan, K.M. et al. Role of NF-kappaB in p53-mediated programmed cell death. *Nature*, 404:892–897. 2000.

81. North, S. et al. The cytoprotective aminothiol WR1065 activates p21waf-1 and down regulates cell cycle progression through a p53-dependent pathway. *Oncogene*, 19:1206–1214. 2000.

82. Jayaraman, L. et al. Identification of redox/repair protein Ref-1 as a potent activator of p53. *Genes Dev.*, 11:558–570. 1997.

83. Hainaut, P. and Milner, J. Redox modulation of p53 conformation and sequence-specific DNA binding *in vitro*. *Cancer Res.*, 53:4469–4473. 1993.

84. Rainwater, R. et al. Role of cysteine residues in regulation of p53 function. *Mol. Cell. Biol.*, 15:3892–3903. 1995.
85. Shen, H. et al. Binding of the aminothiol WR-1065 to transcription factors influences cellular response to anticancer drugs. *J. Pharmacol. Exp. Ther.*, 297:1067–1073. 2001.
86. Kataoka, Y. et al., Cytoprotection by WR-1065, the active form of amifostine, is independent of p53 status in human malignant glioma cell lines. *Int. J. Radiat. Biol.*, 76: 633–639, 2000.

FIGURE 4.11 Thioredoxin (left, PDB: 1xoa), N-terminal domain of GSTM2-2 (center, red), and the complete subunit of GSTM2-2 (right, C-terminal domain blue, PDB: 2c4j).

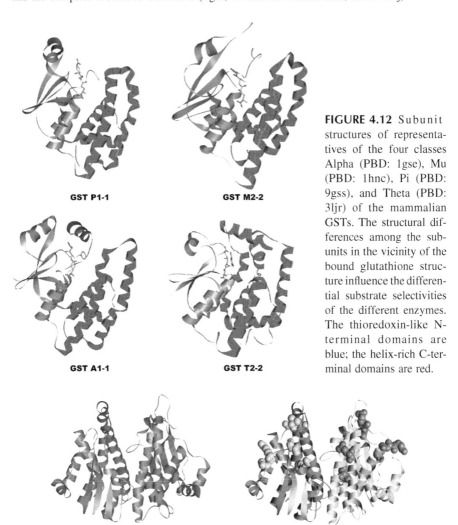

GST P1-1

GST M2-2

GST A1-1

GST T2-2

FIGURE 4.12 Subunit structures of representatives of the four classes Alpha (PBD: 1gse), Mu (PBD: 1hnc), Pi (PBD: 9gss), and Theta (PBD: 3ljr) of the mammalian GSTs. The structural differences among the subunits in the vicinity of the bound glutathione structure influence the differential substrate selectivities of the different enzymes. The thioredoxin-like N-terminal domains are blue; the helix-rich C-terminal domains are red.

FIGURE 4.13 GSTs occur as homodimers (M2-2, left, PBD: 2c4j) and heterodimers (M2-3, right, PBD: 3gtu), a manifestation of combinatorics at the level of quaternary structure. The M2 and the M3 subunits are rendered blue and yellow, respectively. Hypervariable residues (space filled) subjected to positive selection in evolution 49 are shown in GSTM2-3.

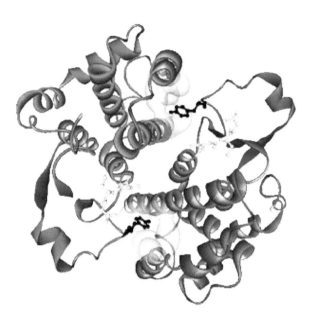

FIGURE 4.14 Dimer of GSTP1-1 showing the lock-and-key motif interlocking the two GST subunits (red and blue). The key residue (Tyr 50, black) is located on a loop of the polypeptide chain that contributes to binding of glutathione (yellow ball and stick) in the active site. Fitting the aromatic key residue into the lock (yellow surface) promotes binding of glutathione and enhances catalytic efficiency of the enzyme.

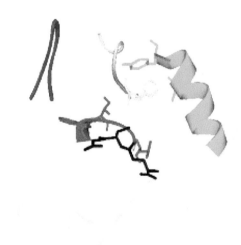

FIGURE 4.15 Segments of the polypeptide chain contributing to formation of the H-site in GSTM2-2. The different segments are located in the loops between β1 and α1 (red), β2 and β3 (blue), the end of the α4-helix (green), and the C-terminal portion (yellow) of the structure. The bound glutathione is rendered in black.

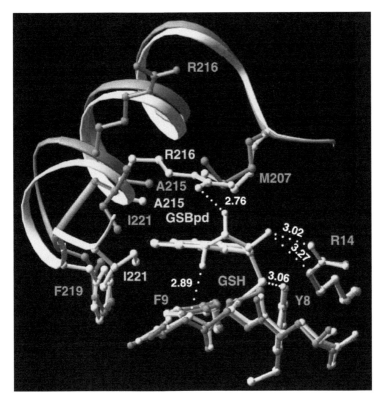

FIGURE 6.2 Significant conformational changes of the H-site in mGSTA1-1 upon xenobiotics substrate binding. The H-sites of mGSTA1-1·GSH and mGSTA1-1·GSBpd complexes are shown in orange and blue colors, respectively. The side chains are represented as ball-and-stick models, and the electrostatic interactions are shown as dotted white lines. Reprinted with permission from Gu, Y., Singh, S.V., and Ji, X., *Biochemistry*, 39, 12552, 2000. Copyright © 2000 American Chemical Society.

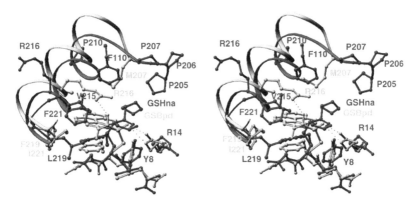

FIGURE 6.3 H-site comparison of mGSTA1-1 (blue) with mGSTA4-4 (orange-brown) indicates that Pro-205, Pro-206, Pro-207, and Pro-210 in mGSTA4-4 may be responsible for the rigidity of its C-terminal helix and consequently the narrow H-site. Reprinted with permission from Gu, Y., Singh, S.V., and Ji, X., *Biochemistry*, 39, 12552, 2000. Copyright © 2000 American Chemical Society.

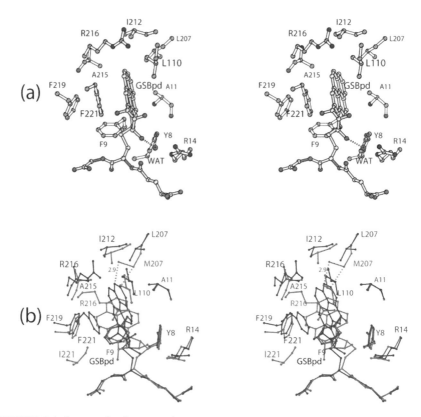

FIGURE 6.4 See text for figure caption.

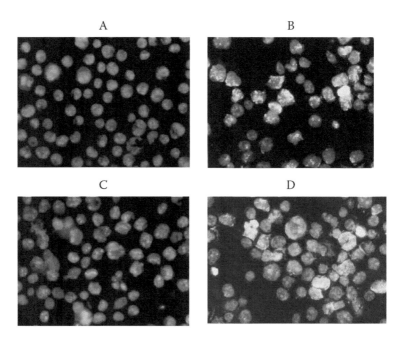

FIGURE 10.6 See text for figure caption.

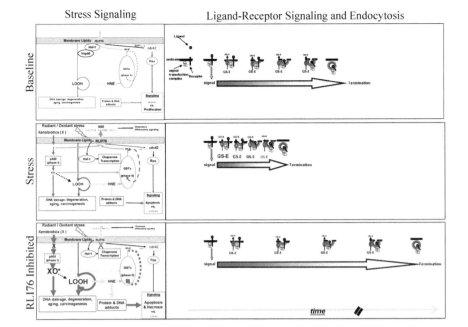

FIGURE 11.1 The role of RLIP76 in stress signaling and receptor-ligand signaling and endocytosis under baseline, stressed, and RLIP76-inhibited conditions. Please see text for complete figure caption.

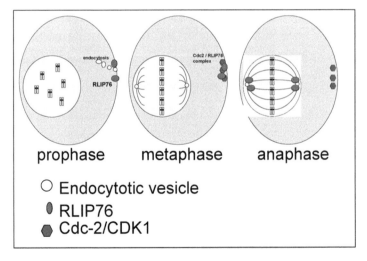

FIGURE 11.5 Role of RLIP76 in the mitotic spindle. Before anaphase, cdc2 binding to RLIP76 terminates endocytosis and RLIP76 is translocated from a membrane location to the mitotic spindle in the centrosomes. Blockade of this event results in a mitotic cell death.

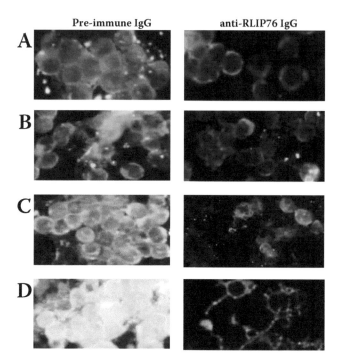

FIGURE 11.7 See text for figure caption.

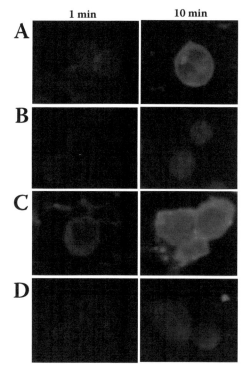

FIGURE 11.8 See text for figure caption.

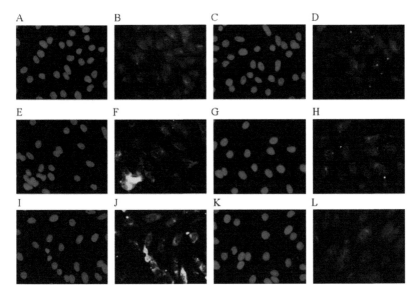

FIGURE 12.2 See text for figure caption.

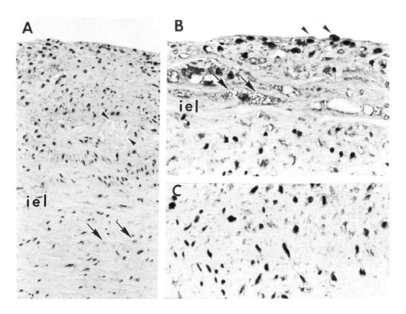

FIGURE 12.4 See text for figure caption.

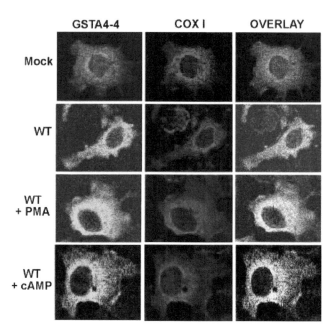

FIGURE 13.2 Confocal immunofluorescence microscopic studies on mitochondrial accumulation of GSTA4-4 in COS cells. COS cells were mock transfected with empty vector or with cDNA for GSTA4-4 cloned in pCMV. Cells were treated with cAMP or phorbol myristyl acetate (PMA) and mitochondrial translocated GSTA4-4 was visualized by using an antibody against GSTA4-4 (green). A bonafide mitochondrial protein, cytochrome c oxidase (COX) was also visualized (red) under similar conditions. The overlapping nature of the expression of these proteins suggests increased mitochondrial accumulation of GSTA4-4 by kinases.

9 Physiological Substrates of Glutathione S-Transferases

Rajendra Sharma, G.A. Shakeel Ansari, and Yogesh C. Awasthi

CONTENTS

9.1 INTRODUCTION

Glutathione S-transferases (GSTs) are primarily viewed as the enzymes involved in detoxification of xenobiotics, but in recent years compelling evidence has emerged that strongly suggests that these enzymes are also responsible for important physiologic functions. GSTs are multifunctional and besides catalyzing the conjugation of electrophiles to glutathione (GSH), these enzymes also display isomerase and glutathione peroxidase (GPx) activities. Isomerase activity of GSTs catalyzing the conversion of androst-5-ene-3,17-dione to androst-4-ene-3,17-dione[1] and of maley-lacetoacetate to fumarylacetoacetate[2,3] has been reported earlier, but only recently the physiological significance of these reactions is being recognized. Likewise, the GPx activity of the cytosolic[4] and microsomal GSTs[5] has been known for a long time, but their roles as antioxidant enzymes have been recognized only recently. GST-mediated detoxification of xenobiotics through conjugation to GSH, chemical removal of xenobiotics by ligandin-type activity,[2] and bioactivation of certain xeno-biotics leading to toxic intermediates have been extensively covered in this volume. This and the succeeding chapter deal with our current understanding of the known and predicted physiological functions of GSTs, highlighting their putative roles as antioxidant enzymes and as modulators of the pathways for biosynthesis of pros-taglandins, leukotrienes, steroid hormones, and stress-mediated signaling.

GSTs are inducible enzymes but, unlike CYP450s, their constitutive expression in most mammalian tissues is high. In major metabolic organs such as the liver and kidney, GSTs comprise a substantial proportion of the cellular protein. In the brain, where exposure to xenobiotics is limited because of the blood–brain barrier, GST isozymes of different classes are still prevalent. Likewise, in reproductive organs such as testes and ovaries high constitutive levels of GSTs are present. Other organs also contain constitutive GSTs in relatively high abundance as compared to CYP450s, suggesting that GSTs may perhaps be involved in physiologic functions in addition to their well-established pharmacological roles. It stands to reason that unless a persistent abundance of GSTs is needed for some important physiologic functions, it would be wasteful for cells to spend their resources to have consistently high levels of these enzymes simply to cope with occasional exposures to pharma-cological or toxicological agents.

9.2 PHYSIOLOGICAL SUBSTRATES OF GSTS

Over the years a number of physiological substrates of GSTs have been identified. A glance at the substrates of various GST isozymes presented in Table 9.1 reveals that these enzymes are involved in the metabolism of the compounds generated during the enzymatic or nonenzymatic oxidative degradation of arachidonic acid and pathways of steroid hormone synthesis, indicating their relevance to physiolog-ical processes. For example, these enzymes are involved in the lipoxygenase (LOX) pathway and could perhaps modulate the levels of leukotrienes that are involved in the signaling for inflammation response. Likewise, these enzymes may modulate the cyclooxygenase (COX) pathway and influence the levels of various prostaglandins, prostacyclins, and thromboxanes. GSTs can reduce lipid hydroperoxides and con-jugate 4-hydroxynonenal (4-HNE) and its homologues to GSH, and thereby attenuate the uncontrolled nonenzymatic lipid peroxidation and regulate 4-HNE levels. Thus GSTs can influence cell-cycle signaling by limiting lipid peroxidation and regulating the intracellular concentrations of 4-HNE.[6,7] Potential physiological significance of the steroid isomerase and prostaglandin isomerase activity of GSTs is also indicated by recent studies.[8,9] In addition, recent studies suggest that GSTs may influence the activity of kinases involved in signal transduction cascades by protein–protein inter-actions. For example, it has been shown that GSTP1-1 through its interaction with

TABLE 9.1
Some of the Prominent Physiological Substrates of GSTs

Substrate	Product	GST Isozymes Involved	References
Androst-5-ene–3,17-dione	Androst-4-ene-3,17-dione	Alpha-class GSTs, particularly GSTA3-3	1, 6, 7
Pregn-5-ene-3,20-dione	Pregn-4-ene-3,20-dione	GSTA3-3	7
Maleyl acetoacetate	Fumarylacetoacetate	GST-Zeta	2, 12
Prostaglandin H_2	Prostaglandin E_2	hGSTM2-2, hGSTM3-3	14, 87
PGH_2	PGF_2	rGSTA1-1, rGSTA3-3	13–15
PGF_2	PGD_2	hGST sigma	13–15
PGH_2	$PGH_2\alpha$	Alpha-class GSTs	13–15
PGJ_2, PGA_2	GSH-conjugates	Alpha, Mu, Pi	16, 17
5-HPETE	5-HETE	Microsomal GSTs, hGSTA1-1, hGSTA2-2	26
Leukotriene A_4	Leukotriene C_4	Alpha-, Mu-, and Pi-class GSTs	27, 88
9-Hydroperoxylinoleic acid	9-Hydroxylinoleic acid	Alpha-class GSTs GSTA1-1 and GSTA2-2	26
13-Hydroperoxylinoleic acid	13-Hydroxylinoleic acid	Alpha-class GSTs GSTA1-1 and GSTA2-2	26
PC-OOH	Reduced (PC-OH)	Alpha-class GSTs GSTA1-1 and GSTA2-2	26, 38–41
PE-OOH	Reduced (PE-OH)	Alpha-class GSTs	26
4-HNE	GSH conjugate	rGST8-8, mGSTA4-4, hGSTA4-4, hGST5.8	43–46, 96, 11, 58
Cholesterol alpha epoxide	GSH conjugate	Alpha-class GSTs	48
Malonaldialdehyde	GSH conjugate	Alpha-class GSTs	48
Acrolein	GSH conjugate	Pi-class GSTs	48
Cytosine propenal	GSH conjugate	Pi-class GSTs	48
Thymine propenal	GSH conjugate	Pi-class GSTs	48
Uracil propenal	GSH conjugate	Pi-class GSTs	48
Phenyl propenal	GSH conjugate	GSTA1-1, M1-1, and P1-1	48
Cholesterol alpha oxide	GSH conjugate	Rat livers	89

c-Jun N-terminal kinase (JNK) can influence stress-mediated signaling for apopto-sis.[10] The mechanisms and physiological significance of GST-mediated catalysis of physiological substrates are discussed in the following sections.

9.2.1 ISOMERASE ACTIVITY OF GSTS AND STEROID HORMONE BIOSYNTHESIS

The isomerase activity of GSTs catalyzing the conversion of androst-5-ene-3,17-dione to androst-4-ene-3,17-dione was first demonstrated about 30 years ago.[2] Even though the isomerization reaction occurs in the absence of the co-factor GSH, the activity is much higher in the presence of GSH.[1,2] Because GSH is not ultimately consumed in the isomerase reaction, this reaction is considered to be different from

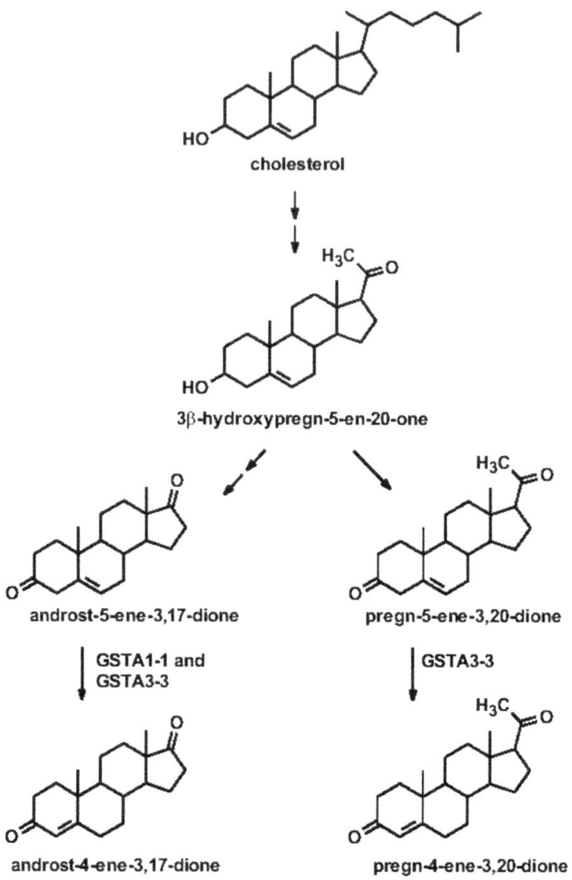

FIGURE 9.1 GST-catalyzed isomerization of steroid hormones.

the conjugation reactions. The α-class GSTs are the most efficient at catalyzing this reaction.[8,9] In particular, hGSTA3-3 has very high isomerase activity and its catalytic efficiency is reported[9] to be approximately 230-fold higher for the isomerization of the C-5 double bond to C-4 as compared to the isomerase activity associated with 3β-hydroxysteroid dehydrogenase, which is believed to be the main enzyme to catalyze the formation of androst-4-ene-3,17-dione (Figure 9.1). The physiological significance of this activity of hGSTA3-3 is suggested by its selective expression in tissues involved in steroid hormone biosynthesis. Relatively higher expression of hGSTA3-3 has been demonstrated in testes, ovaries, and adrenal glands, but in tissues such as livers, skeletal muscles, fetal brains, and the thymus, where steroid hormone synthesis does not occur, no detectable expression of this enzyme is found.[9]

GSTs are known for their remarkably low catalytic efficiency toward xenobiotic substrates,[2] but the catalytic efficiency of GSTA3-3 for androst-5-ene-3,17-dione[9] is higher than the activity of most of the known GST isozymes for their preferred substrates and is comparable to that of human GST isozymes with substrate pref-

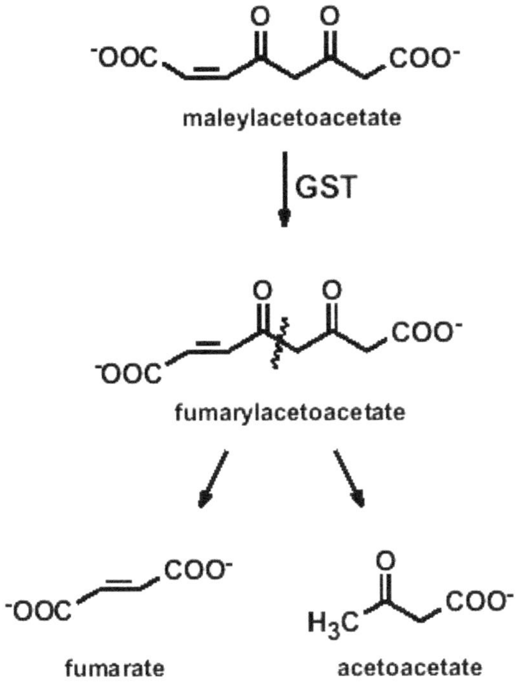

FIGURE 9.2 GST-catalyzed isomerization of maleylacetoacetate to fumarylacetoacetate.

erence for 4-HNE,[11] further suggesting physiological significance of this activity. Other α-class GSTs also show activity toward androst-5-ene-3,17-dione, but their catalytic efficiency is relatively low. For example, the activities of hGSTA1-1 and hGSTA2-2 are approximately 5- and 1,000-fold less than that of hGSTA3-3, respectively, despite high homologies among these enzymes. Higher activity of hGSTA3-3 has been attributed to altered residues in its H-site, and it has been proposed that this enzyme evolved to catalyze the isomerization reactions in the biosynthesis of steroid hormones and to complement the isomerase activity associated with 3β-hydroxysteroid dehydrogenase in steroid hormone synthesis.[8] GSTs also catalyze the isomerization of maleylacetoacetate to fumarylacetoacetate (Figure 9.2), a reaction leading to the formation of fumarate and acetoacetate,[2] and are involved in the catabolism of tyrosine in the liver. The ζ-class GSTs seem to have substrate preference for maleylacetoacetate.[12]

9.3 GSTS AND OXIDATIVE METABOLISM OF ARACHIDONIC ACID

9.3.1 CYCLOOXYGENASE PATHWAY

GSTs seem to be involved in the biosynthesis as well as metabolism of prostanoids. GSH-dependent isomerase activity of GSTs can catalyze isomerization of H-series

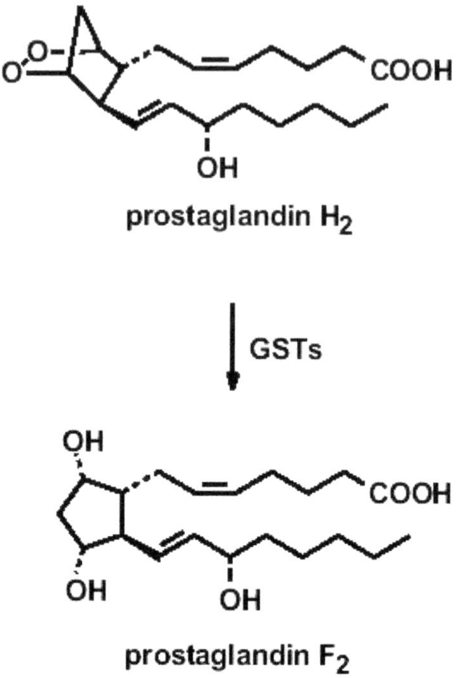

FIGURE 9.3 Reduction of prostaglandin H_2 to prostaglandin F_2 by GSTs.

prostaglandin (PG) to D- and E-series and GPx activity can catalyze the reduction of H-series to F-series (Figure 9.3; Table 9.1). Even though GSH is required, it is not consumed in the isomerase reaction. However, in reduction of H- to F-series, stoichiometric consumption of GSH has been demonstrated. Thus the exact mechanism for this reaction is not clear. All of the major GST isozymes purified from different rat tissues catalyze both these reactions.[13] GST-catalyzed isomerization of PGH_2 to PGD_2 and reduction of PGE_2 to $PGF_{2\alpha}$ has been demonstrated,[13–15] suggesting that GSTs could influence the biosynthesis of these important signaling molecules. The role of GSTs in the biosynthesis of prostaglandins D_2, E_2, and $F_{2\alpha}$ has been extensively investigated by Ujihara et al.[13] These authors systematically examined prostaglandin H_2-converting activity by evaluating prostaglandin H_2E isomerase activity, prostaglandin H_2D isomerase activity, and prostaglandin H_2F reductase activity of the various GST homo- and heterodimers expressed in different rat tissues. More important, these studies also showed that a significant amount of the isomerase/reductase activity exhibited by various rat tissues could be immunoprecipitated by antibodies specific to various GSTs, strongly suggesting an *in vivo* role of GSTs in the regulation of biosynthesis of these prostanoids. The homo- or heterodimeric isozymes of the α-class subunits had higher prostaglandin H_2E isomerase activity as compared to the μ- or π-class GST isozymes of rat tissues. Likewise, prostaglandin H_2D isomerase activity was high in the α-class GSTs while the μ-class was comparatively less active.[13] Prostaglandin H_2F reductase activity was predominant in the α-class GSTs, which appears to be consistent with their high GPx activity toward lipid hydroperoxides. Because GSH consumption in these reactions was stoichiometric, this activity could be attributed to the Se-independent GP_x activity of GSTs. However, a positive correlation between GPx activity and PG reductase activity was not found in these studies. Reasons for this apparent anomaly are not clear but it is possible that tissue-purified enzymes used in these studies may not represent homogeneous preparations. The role of the isomerase and GPx activites of individual GST isozymes in regulation of biosynthetic pathways for prostanoids needs to be investigated further using enzyme preparations of unquestionable purity, preferably recombinant enzymes. The fact that both isomerase and reductase activities are present in livers, kidneys, and spleens, and could be differentially immunoprecipitated by antibodies specific to various GST subunits[13] suggests that the GST isozyme composition of tissues could be a determinant of these activities and the tissue-specific expression of GSTs may be relevant to the regulation of biosynthetic pathways of prostanoids in tissues related to their specific functions. Physiological significance of these activities of GSTs is further suggested by the fact that K_m values for GSH and PGH_2[13] for these reactions are within the suggested physiological range of their concentrations. The μ-class isozymes M2-2 and M3-3 have also been suggested to be involved in prostaglandin biosynthesis, and GSTM2-2 has been characterized in brain cortex as prostaglandin E-synthase. However, not all μ-class GSTs catalyze this reaction because no detectable prostaglandin E-synthase activity was observed in GSTM4-4.[14]

PGD_2, PGE_2, and PGF_2 are known to bind to G protein-coupled receptors involved in the regulation of hormones and neuroreceptors and PGF_2 is known to activate mitogen-activated protein kinase (MAP kinase). It is possible that GSTs may influence these signaling pathways by influencing levels of prostanoids. Cyclopentenone prostanoids PGA_2 and PGJ_2 can nonenzymatically conjugate to GSH. GSTs can catalyze this reaction with some stereoselectivity, which may also be relevant to the regulation of biological activity of these prostanoids[16] (Figure 9.4). 15d-PGJ_2, which is a downstream metabolite of PGD_2, acts as a ligand for the peroxisome proliferator-activated receptor γ (PARPγ). GSTs can modulate 15-d-PGJ_2 and PARPγ-mediated signaling by inactivating 15d-PGJ_2 through its conjugation to GSH as indicated by studies showing that GST overexpression can inhibit 15d-PGJ_2 and PARPγ-mediated activation of genes by conjugating 15d-PGJ_2 to GSH.[17] 15d-PGJ_2 has multiple biological activities,[18–22] including its ability to affect Nrf2 (nuclear factor-erythroid 2p45 related factor 2)[20] and NF-κB (nuclear factor κB)[22]–dependent gene expression. These processes may also be regulated by GSTs through inactivation of 15d-PGJ_2 by GST-catalyzed conjugation to GSH.[23]

Because of their electrophilic nature, cyclopentenone prostanoids of PGA_2 and PGJ_2 series can react with GSH nonenzymatically to form conjugates resulting in modulation of their biological activities. Therefore it is difficult to assess the relative roles of the nonenzymatic and GST-catalyzed conjugation of these prostanoids in regulation of their activities. In this respect, observed stereoselectivity of GSTs in conjugation of PGA_2 and PGJ_2 may have physiologic significance.[16] The nonenzymatic reaction between PGJ_2 results in the formation of racemic mixture of R- and S-conjugates, but GST-catalyzed conjugation seems to have stereoselectivity. It has

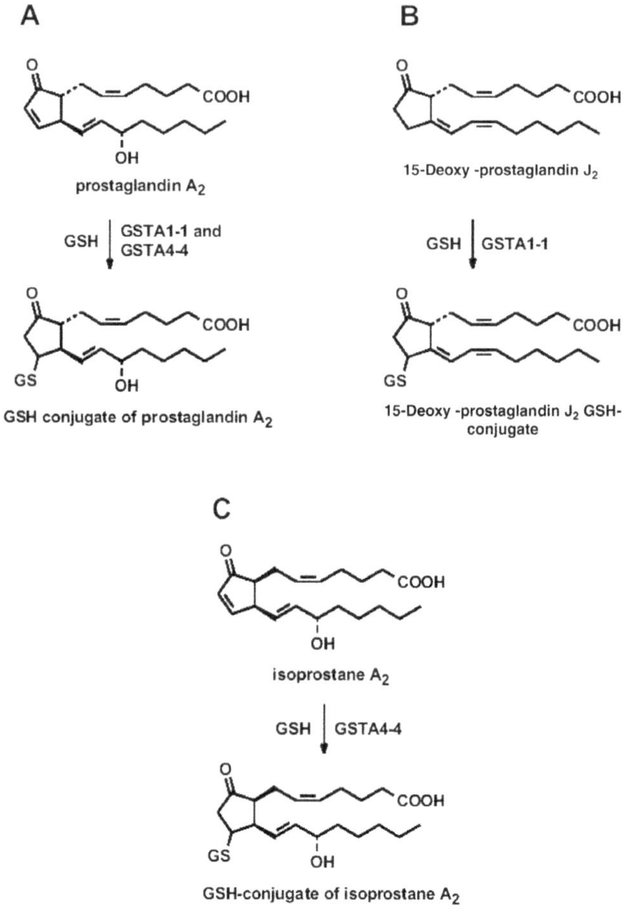

FIGURE 9.4 GST-mediated conjugation of GSH with (a) Prostaglandin A_2, (b) Prostaglandin J_2, and (c) Isoprostane A_2.

been shown[16] that GST isozymes belonging to three major classes of human GSTs (α, μ, and π) catalyze the conjugation of PGJ_2 and PGA_2 to GSH. GSTA1-1–catalyzed conjugation of PGA_2 and PGJ_2 predominantly results in the formation of the R-GSH conjugates. GSTP1-1 on the other hand seems to preferentially catalyze the formation of the S-GSH conjugate of PGA_2 but not of PGJ_2. The μ-class GST, GSTM 1_a-1_a, shows no stereoselectivity and the α-class isozyme GSTA2-2, unlike GSTA1-1, does not favor the formation of α R-GSH conjugate of PGJ_2. These differences in stereoselectivity of different GSTs may have a bearing to the mechanisms regulating the biological activity of these prostaglandins by specific GST isozymes. Induction of GSTs by prostaglandins may also be relevant to the regulation of their biological effects. Compounds having a Michael acceptor group are known to induce GSTs and other Phase II enzymes.[24] Induction of GSTs by 15d-PGJ_2 has been demonstrated,[25] and other prostanoids having the α,β-unsaturated carbonyls are potential inducers of GSTs. It appears that GSTs can influence prostaglandin-mediated signaling through regulation of their levels as well as through their biological activity. Further studies on the role of GSTs in the regulation of prostaglandin-mediated signaling need to be pursued.

9.3.2 LIPOOXYGENASE (LOX) PATHWAY

GSTs can also influence the lipooxygenase (LOX) pathway by reducing the important intermediate, 5-hydroperoxyeicosatetetraenoic acid (5-HPETE) to the corresponding alcohol 5-HETE through its GPx activity[26] (Table 9.1). 5-HPETE is the key intermediate in the biosynthesis of leukotriene B_4 (LTB$_4$) and formation of LTC$_4$, LTD$_4$, and LTE$_4$. 5-HPETE enzymatically formed from arachidonic acid through a reaction catalyzed by 5-LOX yields leukotriene A_4, which can be converted to LTB$_4$ through a reaction catalyzed by a hydrolase enzyme. Alternatively, LTA$_4$ can conjugate with glutathione through a reaction catalyzed by GSTs to form LTC$_4$ (Figure 9.5), which can be further metabolized by sequential actions of mercapturic acid

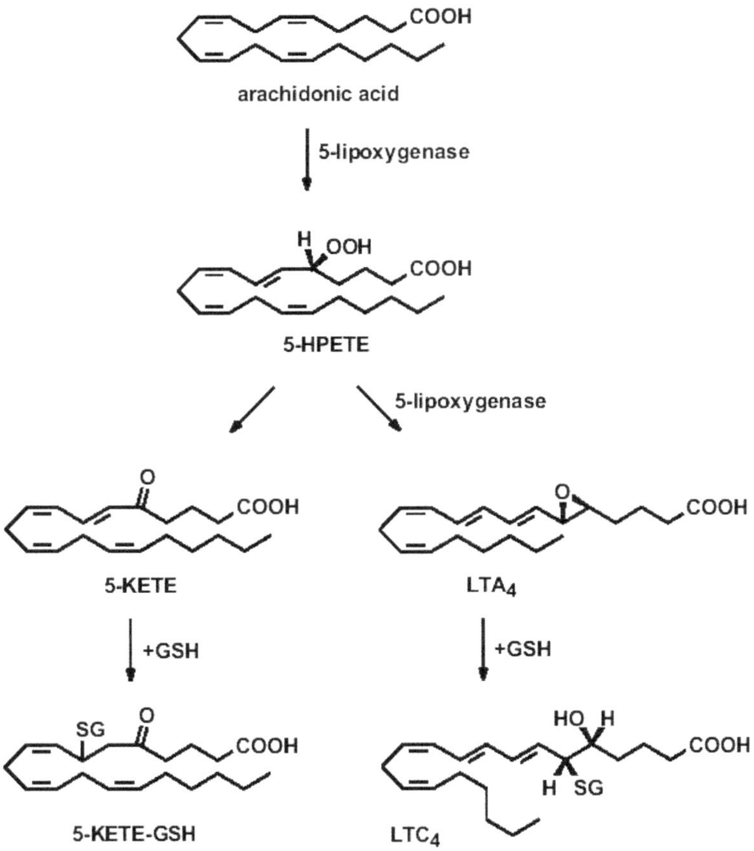

FIGURE 9.5 GSTs in biosynthesis of leukotrienes.

pathway enzymes to LTD$_4$ and LTE$_4$.[27] These chemotactic agents consisting of an LTC$_4$ and LTD$_4$ mixture constitute the classical slow-reacting substance of anaphylaxis (SRSA). The α-class GSTs A1-1 and A2-2 can efficiently reduce the hydroperoxides 5-HPETE and 12-HPETE to corresponding alcohols. The α-class GSTs are relatively more abundant in some organs, including livers, kidneys, lungs, and testes, where the biosynthesis of leukotrienes may be modulated by GSTs through channeling of 5-HPETE to the leukotriene pathway.

Microsomal GSTs have also been shown to catalyze this reaction but their activity seems to be much lower when compared to GSTs A1-1 and A2-2.[28] GST isozymes belonging to the μ- and π-classes have minimal GPx activity toward hydroperoxides and do not utilize 5-HPETE or 5-HETE and therefore may not be involved in the LOX pathway.

9.4 GSTS AS ANTIOXIDANT ENZYMES

The GPx activity of the α-class GSTs that constitutes the bulk of GST protein in several tissues, including livers, kidneys, and testes, appears to serve important physiological roles. In human livers, GSTs account for up to 3% of total soluble protein.[2,29] Also, in other tissues such as kidneys and lungs a significant portion of GST protein is comprised of the α-class isozymes. Surprisingly, in humans, GSTπ, which has a relatively high catalytic efficiency for various xenobiotics, is absent or minimally expressed in the liver.[30] In GSTM 1–null variants, the α-class GSTs may constitute up to 90% of the total GST protein of liver.[31] Because approximately 50% of Caucasians who lack the μ-class gene GSTM 1 are phenotypically normal, the putative physiologic functions of GSTs in the liver, if any, may be primarily attributed to the α-class GSTs. The liver is the primary site of xenobiotic metabolism and a high abundance of GSTs in this organ is consistent with the pharmacological role of GSTs in protection from electrophilic xenobiotics. Due to its high metabolic activity, the liver is continually exposed to relatively high levels of reactive oxygen species (ROS) and the GPx activity of the α-class GSTs could serve an important role in protecting this organ from ROS-induced lipid peroxidation by reducing lipid hydroperoxides. Likewise, in the testes, kidneys, and lungs, where the α-class GSTs are present in substantial amounts, the GPx activity of these enzymes toward lipid hydroperoxides may serve as an important defense mechanism against oxidative stress and ensuing lipid peroxidation.

9.4.1 GSH-DEPENDENT REDUCTION OF LIPID HYDROPEROXIDES

Reduction of lipid hydroperoxides generated during lipid peroxidation is one of the major defense mechanisms against lipid peroxidation caused by oxidative, chemical, and photochemical stress or due to ROS generated in metabolic processes. Lipid peroxidation, initiated by ROS and autocatalytically propagated by lipid hydroperoxides, leads to amplification of oxidative stress. Se-dependent GPx (Se-GPxs) are generally considered as the major defense against ROS-induced lipid peroxidation because these enzymes can catalyze not only the GSH-dependent reduction of H$_2$O$_2$ but also of various organic hydroperoxides including lipid hydroperoxides. GSH-

TABLE 9.2
Se-Dependent Glutathione Peroxidases (GPx) of Mammalian Tissues

Name; Subunit Structure	Other Name	Location	Substrates	References
GPx1 Tetramer	Classical GPx	Cytosolic	H_2O_2, FA-OOH	35, 90, 91
GPx2 Tetramer	G1GPx	Cytosolic	H_2O_2, FA-OOH	92
GPx3 Tetramer	Plasma GPx	Extracellular	H_2O_2, FA-OOH	93
GPx4 Monomer	Phospholipid hydroperoxide GPx	Membrane-bound	PL-OOH, Cholesterol hydroperoxide	94, 95

dependent reduction of fatty acid hydroperoxides by the classical Se-GPx (GPx$_1$) was demonstrated in 1968 and its role in defense against lipid peroxidation was proposed.[32] At least four Se-dependent GPx isozymes (GPx$_1$–GPx$_4$) are known and all these enzymes can utilize fatty acid hydroperoxides (FA-OOH) as substrates.[33] Table 9.2 shows the substrate preference and tissue localization of various Se-GPx isozymes of mammalian tissues. Unlike tetrameric GPx$_1$, GPx$_2$, and GPx$_3$, GPx$_4$ is a monomer and does not utilize H_2O_2 as a substrate. While other isozymes require cleavage of the peroxidized fatty acid moiety from phospholipids before they can be reduced, only GPx$_4$ can reduce intact phospholipidhydroperoxides (PL-OOH) present in membranes.[34] It has been suggested that GPx$_4$ provides protection against oxidative stress by reducing membrane PL-OOH *in situ* and thereby limiting lipid peroxidation in membranes.[34] GPx$_1$, also referred to as the classical GPx first described by Mills,[35] is the most abundant GPx isozyme and is considered to be one of the major antioxidant enzymes in organisms. In addition to the reduction of H_2O_2, this enzyme can also protect against lipid peroxidation by reducing FA-OOH liberated from PL-OOH.

9.4.2 GPx ACTIVITY OF GSTs

A Se-independent GPx activity designated as GPxII was first described by Lawrence and Burk in the rat liver.[36] GPxII differed from the Se-dependent GPx$_1$ in its substrate profile as it could not use H_2O_2 as a substrate and catalyzed GSH-dependent reduction of only organic hydroperoxides. GPxII was later shown to be identical with GSTs, which could catalyze GSH-dependent reduction of nonphysiologic hydroperoxides such as cumene hydroperoxide and t-butyl hydroperoxides as well as FA-OOH.[4] Subsequently, it was shown that in human livers, GPxII activity was displayed only by the cationic GST isozymes.[31] Consistent with this observation it has now been established that the GPx activity of human GSTs is mainly confined to the α-class isozymes. Based on the ability of GSTs to catalyze the reduction of FA-OOH (Figure 9.6a) it was proposed that GSTs provide protection against lipid peroxidation. Earlier studies suggested that the peroxidized fatty acid moiety at sn-2 position in membrane phospholipids had to be released by phospholipase A$_2$ and only the free FA-OOH released in cytosol could be reduced by GSTs.[37] This contention was found to be not necessarily true because later studies showed that human α-class

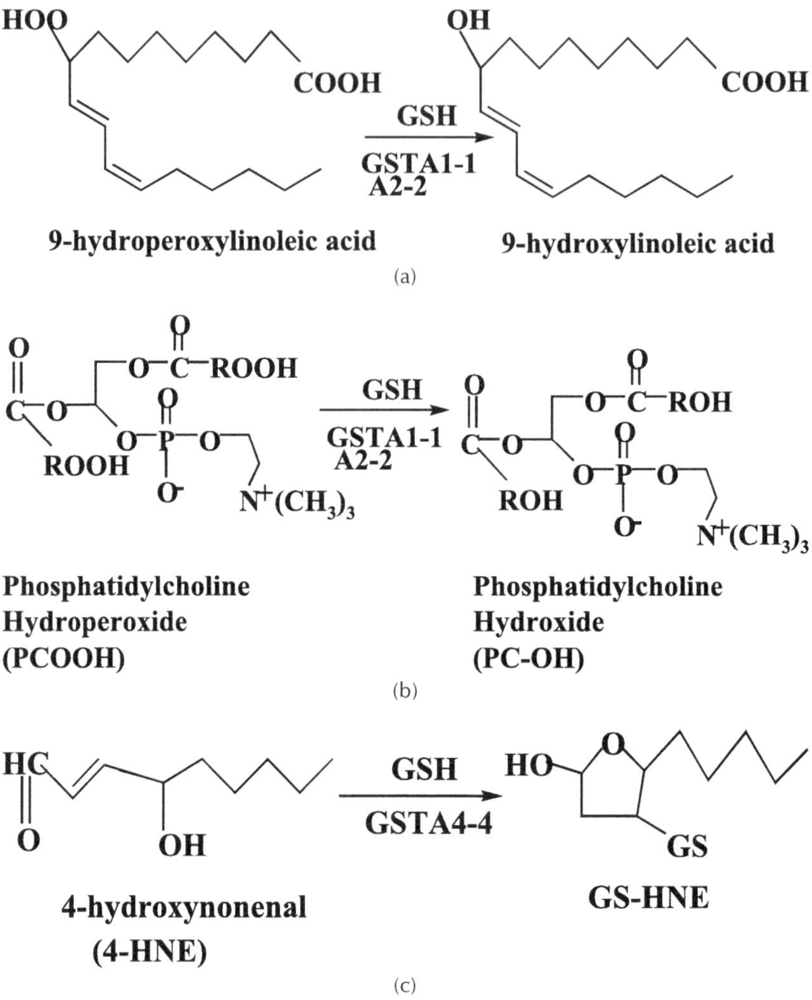

9-hydroperoxylinoleic acid   9-hydroxylinoleic acid

(a)

Phosphatidylcholine   **Phosphatidylcholine**
Hydroperoxide   **Hydroxide**
(PCOOH)   **(PC-OH)**

(b)

4-hydroxynonenal   **GS-HNE**
(4-HNE)

(c)

FIGURE 9.6 GSTs in detoxification of lipid peroxidation products. (a) reduction of FA-OOH; (b) reduction of PL-OOH; (c) conjugation of 4-HNE.

GSTs purified from the lung could catalyze GSH-dependent reduction of intact phosphatidylcholine hydroperoxide (PC-OOH) and phosphatidylethanolamine hydroperoxide (PE-OOH)[38–40] (Figure 9.6b). These findings have since been confirmed by studies showing that the α-class human GSTs obtained by heterologous expression in *E. coli* can utilize PL-OOH as well as FA-OOH as substrates,[26,41] and PL-OOH generated in membranes during oxidative stress can be reduced *in situ* by these enzymes.[26,38,42]

9.4.3 Conjugation Activity of GSTs

The α-class GSTs also protect against the toxic end-products of lipid peroxidation. In this regard their activity for conjugating the α,β-unsaturated alkenals such as 4-

HNE to GSH is particularly important. Even though 4-HNE and its homologues are sufficiently electrophilic to react nonenzymatically with GSH, this reaction is catalyzed by GSTs[43] and the resulting GSH–HNE conjugate mostly occurs in a stable hemiacetal form (Figure 9.6c). A subgroup of the α-class GSTs has a very high catalytic efficiency for catalyzing this reaction.[44–47] These GST isozymes have high activity toward 4-HNE and its homologues, malonaldialdehyde, acrolein, and propenal bases such as cytosinepropenal, thyminepropenal, and uracylpropenal formed during oxidative DNA degradation.[43–50] The role of GSTs in negating the deleterious effects of oxidative stress is therefore important because the α,β-unsaturated carbonyls being capable of Michael addition can interact with cellular nucleophiles to cause toxicity. For example, MDA, acrolein, and 4-HNE are known to form protein adducts that are believed to be involved in the etiology of degenerative disorders.[51] The physiological significance of the activity of these enzymes toward base-propenal is not understood but their role in detoxification of the toxic lipid aldehydes, which are the stable end-products of lipid peroxidation, has been extensively studied in recent years.[52–61] Thus, even though GSTs do not decompose the ROS (e.g., H_2O_2) per se, these enzymes constitute an all-important second line of defense and provide protection against oxidative stress by attenuating lipid peroxidation and by detoxifying the toxic end-products of lipid peroxidation.

Activities and kinetic constants for the α-class GSTs toward various hydroperoxides are presented in Table 9.3. The activity and K_m of these enzymes for FA-OOH and PL-OOH are in the same range, suggesting that these enzymes can utilize both these substrates with similar catalytic efficiencies. Also the K_m of these enzymes for lipid hydroperoxides seem to be in the estimated range of lipid hydroperoxide concentration in cells during oxidative stress. There appears to be a large difference between the activity of these enzymes toward PL-OOH when determined by HPLC[41] and spectrophotometrically.[38] The reason for this discrepancy is not clear.

9.4.4 REDUCTION OF MEMBRANE PL-OOH BY α-CLASS GSTs

GSTs are cytosolic enzymes and their ability to reduce membrane PL-OOH has been questioned. Phospholipase A_2-mediated release of FA-OOH from peroxidized PL-OOH and subsequent reduction of FA-OOH by GSTs was suggested due to localization of GSTs in the cytosol.[37] Studies showing that GSTs can reduce PL-OOH suspended in an aqueous assay medium do not necessarily demonstrate that these enzymes can reduce PL-OOH present in biological membranes. To address this question, peroxidized membrane preparations have been used as substrates of GSTs. Results of these studies[38,42] show that PL-OOH present in peroxidized membranes prepared from human erythrocytes and other cells can be utilized as substrates by hGSTA1-1 and hGSTA2-2.[38,40] Reduction of hydroperoxides present in these membranes has been demonstrated spectrophotometrically by quantitating the residual hydroperoxides in the reaction mixture and by thin-layer chromatography.[42] Using similar approaches, reduction of membrane hydroperoxides by the membrane-bound seleno-enzyme GPx_4 has also been demonstrated.[34] Thus both GSTs and Se-GPx can mitigate lipid peroxidation by reducing PL-OOH formed in membrane as well as the FA-OOH released in cytosol by the action of phospholipase A_2. This protective role

TABLE 9.3
Kinetic Constants of hGSTA1-1 and hGSTA2-2 against Lipid Peroxidation Products

Substrates	hGSTA1-1				hGSTA2-2			
	Specific Activity (μmol/min/mg)	K_m (mM)	k_{cat}[a] (s^{-1})	k_{cat}/k_m (s$^{-1} \cdot$ mM^{-1})	Specific Activity (μmol/min/mg)	K_m (mM)	k_{cat}[a] (s^{-1})	k_{cat}/k_m (s$^{-1} \cdot$ mM^{-1})
4-hydroxynonenal[b]	2.5	0.05	2.94	58.8	1.76	0.08	2.1	26.3
Cumene hydroperoxide[c]	13.5	0.06	14.8	224.2	14.95	0.06	17.7	295
Dilinoleoyl phosphatidylcholine hydroperoxide[c]	12.50	0.08	14.5	181.3	14.58	0.05	16.6	353
Dilinoleoyl phosphatidylethanolamine hydroperoxide[c]	11.6	0.057	11.4	200	15.23	0.04	12.7	318
5-hydroperoxyeicosatetraenoic acid[c]	6.2	0.005	5.92	1183	7.52	0.0066	9.1	1379

[a] Compiled from reference 26.

[b] Transferase activity.

[c] Peroxidase activity.

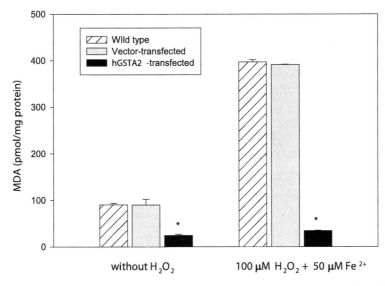

FIGURE 9.7 Effect of hGSTA2-2 overexpression on LPO in K562 cells. 1×10^7 K562 cells were incubated with RPMI complete medium or RPMI complete medium containing 100 µM H_2O_2 and 50 µM $FeSO_4$ for 30 minutes. The cells were pelleted by centrifugation, washed with PBS, and homogenized in 10 mM potassium phosphate buffer, pH 7.0, containing 0.4 mM butylated hydroxytoluene. The whole homogenate was immediately taken for thiobarbituric acid reactive substances (TBARS) assay as described in the text. The values (means ± S.D., n=3) are presented in the bar graph. Asterisk indicates significant difference from the controls (P < 0.01). (From Yang, Y. et al. Role of glutathione S-transferases in protection against lipid peroxidation. *Am. Soc. Biochem. Mol. Biol.* 270:22, 19220–19230, 2001. With permission.

of α-class GSTs in attenuation of oxidative stress-induced lipid peroxidation is reaffirmed by the studies[38,40] showing that Fe^{++}/H_2O_2–induced lipid peroxidation in cells can be prevented by transfecting the cells with hGSTA1 or hGSTA2 (Figure 9.7).

9.4.5 MEMBRANE ASSOCIATION OF THE α-CLASS GSTS

Even though GSTs are cytosolic enzymes, recent studies suggest that the concentration of α-class GSTs may be higher in the proximity of cell membranes. Immunohistochemical studies suggest that these enzymes may be associated with membranes in K562 cells. It has been shown that K562 cells do not express any detectable amounts of hGSTA1-1 or hGSTA2-2, but immunohistological studies with K562 cells after transfection with hGSTA1-1 reveal that the enzyme is concentrated in close proximity to the plasma membrane (Figure 9.8). Murine α-class enzyme mGSTA4-4, which uses 4-HNE as the preferred substrate, has also been shown to be weakly associated with membranes.[47] This association has been attributed to the positively charged lysine residues located on the surface of the protein on loops that flank the active site of mGSTA4-4 to the membrane because substitution of the key lysine 115 residue abolishes its membrane association. Localization of GSTA4-4 in the mitochondrial membrane has also been demonstrated[62] where its role may be important due to the ever-present ROS generated during electron transport.

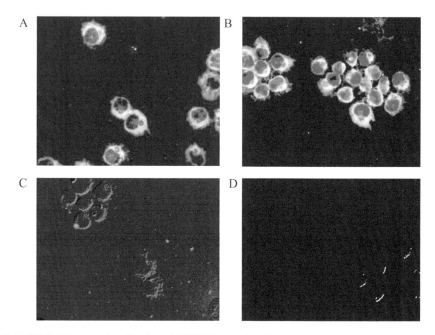

FIGURE 9.8 Immunolocalization of hGSTA1-1 and hGSTA2-2 in K562 cells. hGSTA1 and A2 overexpressing cells were incubated with primary anti-Alpha-class GST antibodies or preimmune serum (diluted 1:100 with 1% BSA in PBS) overnight at 4°C followed with secondary fluorescein isothiocyanate (FITC)-conjugated antibodies, which were diluted 1:80 with 1% BSA in PBS at room temperature for two hours. The slides were examined under fluorescence microscope (Nikon, Eclipse E600, Japan). Panel A: hGSTA1-1; Panel B: hGSTA2-2 overexpressing K562 cells.

The mechanisms as to how GSTs are able to reduce lipid hydroperoxides in the membrane core are not completely understood. Even if GSTs catalyzing the reduction of membrane PL-OOH are attached to the cytoplasmic periphery of the membrane, how does the hydrophilic and charged cosubstrate GSH participate in this reaction, and how are lipid peroxides that are hydrophobic pulled out of the membrane to interact with the active site of GST? The answers to these puzzling questions are not clear. Nonetheless, the available evidence strongly suggests that GSTs do indeed reduce membrane lipid hydroperoxides and provide protection against lipid peroxidation. The three-dimensional structures of the α-class GSTs are now known and should help in understanding the mechanisms through which GSTs express their GPx activity toward membrane PL-OOH.

9.4.6 RELATIVE CONTRIBUTIONS OF GSTs AND SE-GPxs IN REDUCTION OF PL-OOH

Since the Se-enzymes, particularly GPx_4, can also use PL-OOH as a substrate, the relative contribution of GSTs in GSH-dependent reduction of lipid hydroperoxides in cells has been studied. These studies strongly suggest that in human and rat livers, where α-class GSTs are abundant, the majority of GPx activity toward membrane PL-OOH is contributed by these enzymes. Immunotitration studies to quantitate the contribution of the α-class GSTs in the reduction of membrane PL-OOH indicate that more than half of the GPx activity of human[38] and rat livers[42] toward PL-OOH is contributed by the α-class GSTs. These studies show that in human and rat livers up to 60% of the GPx activity toward PL-OOH can be immunoprecipitated by antibodies against GSTA1-1 and GSTA2-2. While it is not certain, the remaining activity is perhaps contributed by the membrane-bound Se-GPx$_4$, which can also utilize intact PL-OOH as a substrate.[34] Nevertheless, these studies suggest that in mammalian livers and in the tissues such as kidneys, testes, and lungs, where the α-class GSTs are present in substantial amounts, these enzymes play a major role in protecting membrane phospholipids from oxidative stress. Thus, as long as cellular GSH is not a limiting factor, GSTs may be one of the main deterrents to oxidative stress-induced lipid peroxidation. Depleted pools of GSH during GST-mediated detoxification of hydroperoxides and 4-HNE generated during the lipid peroxidation are replenished by more effective recycling of GSSG to GSH or by GSH biosynthesis. This is suggested by studies showing induction of GSH biosynthesis by 4-HNE through upregulation of glutamate cysteine ligase.[63,64] Likewise, the hexose monophosphate shunt pathway is activated during oxidative stress to provide a sufficient pool of NADPH required for the reduction of GSSG to GSH.[65,66]

Exposure to xenobiotics usually results in the induction of GSTs. In particular, the xenobiotics, which either contain a Michael acceptor group or can be metabolized to such a compound by the CYP450 system, cause induction of GSTs. Both conjugating and GPx activity of GSTs depend on GSH. Therefore, induction of glutamate cysteine ligase usually occurs simultaneously and GSTs are able to provide sustained protection against xenobiotic electrophiles as well as against ROS generated during metabolism of xenobiotics in liver. High levels of constitutive expression of α-class GSTs in livers seem to be important for the protection of this vital organ from oxidant toxicity. In tissues such as erythrocytes, where α-class GSTs are virtually absent, Se-GPxs may be the major defense against lipid peroxidation. Specific roles of various Se-GPxs and GSTs in attenuating LPO in extrahepatic tissues are not clear, and further studies are needed to delineate the physiologic roles of these enzymes.

9.4.7 Physiological Significance of the GPx Activity of GSTs

While the physiological roles of GSTs toward the intermediates of the COX, LOX, and steroid hormone synthesis pathways are as yet not completely understood, their role as antioxidant enzymes is firmly established. Numerous studies suggest that GSTs negate the deleterious effects of oxidative stress by detoxifying hydroperoxides and 4-hydroxynonenal generated during lipid peroxidation.[37–40,52–61] Cells overexpressing GSTA1-1 or GSTA2-2, which reduce hydroperoxides, are protected against oxidative stress caused by exposure to H_2O_2 and other oxidants that promote lipid peroxidation.[38,39] Likewise, cells overexpressing GSTA4-4, which preferentially detoxifies 4-HNE and also has some GPx activity toward hydroperoxides, are more resistant to oxidant toxicity.[51,54–56] Lipid peroxidation caused by ROS is believed to be involved in the etiology of age-related degenerative disorders such as atherosclerosis,[51,67–70] Alzheimer's disease,[71] cataract,[53,54] Parkinson's disease,[72] and cancer.[73,74] Recent studies suggest that overexpression of GSTs can prevent the ocular lens from oxidative stress-induced cataractogenesis.[53,60,61] In atherosclerotic plaques, deposition of 4-HNE-protein adducts has been shown. The role of GSTs in protection mechanisms against plaque formation is suggested in an animal model of atherosclerosis where an induction of 4-HNE-metabolizing GSTA4-4 is observed.[70] *In vitro* studies show that endothelial cells transfected with hGSTA4 are protected against the oxidative stress-induced apoptosis.[69] This role of GSTs in protective mechanisms against cardiovascular diseases is covered in detail elsewhere in this volume. Studies showing that the toxicity of H_2O_2 and other oxidants *in vitro* and *in vivo* can be attenuated by the overexpression of the α-class GSTs underscore the importance of these enzymes in defense against oxidative stress.[38–40,53–61] Together these studies suggest that GSTs, particularly those belonging to the α-class, are crucial to the overall defense mechanisms against oxidative stress and may be relevant to the etiology of age-related degenerative diseases associated with chronic oxidative stress. Micronutrients that cause induction of GSTs may be considered as potential agents for the prevention or retardation of these diseases.

Anticarcinogenic activity of antioxidant micronutrients such as butylated hydroxyanisole (BHA), butylated hydroxytoluene (BHT), and ethoxyquin has been attributed at least in part to induction of GSTs by these compounds.[75] It is likely that GST activities, the conjugating activity, and GPx activity contribute to their chemoprotective effect. Metabolism of xenobiotics, including carcinogens, by the CYP450 system invariably results in a generation of ROS. Thus GSTs can provide protection against carcinogens not only by detoxifying the ultimate carcinogenic metabolites but also by attenuating oxidative insult due to the ROS generated during xenobiotic metabolism.

As antioxidant enzymes, GSTs complement the role of primary defense enzymes in protecting organisms from the deleterious effects of ROS. While superoxide anion and H_2O_2 are effectively disposed of by the cells through highly efficient enzymes (superoxide dismutase, catalase, and GPx), the ROS escaping this line of defense can initiate the autocatalytic chain of lipid peroxidation through the generation of free radicals capable of abstracting a single hydrogen atom from unsaturated fatty acids. GSTs prevent propagation of lipid peroxidation by reducing lipid hydroperoxides, and GSTs also detoxify the toxic electrophilic end-products of lipid peroxidation. Thus constitute the all-important second and third lines of defense against oxidant toxicity. This role of GSTs as antioxidant enzymes outlined in Figure 9.8 is consistent with the results of *in vitro* studies showing that cells transfected with α-class GSTs are protected from toxicity of oxidants.[38,39] *In vitro* studies show that overexpression of GSTA1-1 or GSTA2-2 suppresses lipid peroxidation in cultured cells and that even during conditions of oxidative stress the extent of lipid peroxidation in these cells is maintained below the basal level.[38,42] The protective role of GSTs *in vivo* is indicated by studies showing that oxidative stress-induced cataractogenesis and atherosclerotic plaque formation can be attenuated by the induction of GSTs.[60,61,69,70]

9.5 NONENZYMATIC PHYSIOLOGICAL ROLES OF GSTS

Besides their enzymatic activity, GSTs can protect cells from toxicants through chemical removal of the agents by noncatalytic binding. Ligandin, a protein binding to physiologic and exogenous ligands, was characterized earlier[76] and its identity with GSTs was established.[2] Studies leading to the identification of ligandin as GSTs have been covered in detail in Chapter 1. Based on the relatively high abundance of GSTs (ligandin), particularly in livers, its role as the intracellular carrier of certain organic molecules had been suggested.[2,77] In recent years there has been a renewed interest in studies on ligandin including structural studies on its binding site and energetics of inorganic anion binding.[78–80] GSTs and ligandin bind toxic xenobiotics and their metabolites and it is believed that this noncatalytic activity of GSTs provides protection to the cellular environment by acting as a shield to protect DNA, proteins, and lipids from the deleterious effects of xenobiotics. It has been shown that toxic xenobiotics such as 4-dimethylaminoazobenze bind to GSTs.[81] Inactivation of GSTs by its substrates in the absence of GSH occurs due to irreversible binding, and this role of GSTs in noncatalytic chemical removal or immobilization of toxicants has been suggested as a "suicidal" protective mechanism.[2] Systematic studies on the binding of xenobiotics to individual GST isozymes are lacking but binding of carcinogenic metabolites of benzo(a)pyrene (B[a]P) to several isozymes of rodent lungs has been demonstrated.[82,83] The lung is one of the major target tissues in B(a)P-induced chemical carcinogenesis in mice. Therefore, binding of the B(a)P or its activated metabolites to lung GSTs and protection against B(a)P-induced carcinogenesis by micronutrients that induce GSTs are consistent with the nonenzymatic protective role of GSTs.[75] It is, however, difficult to predict the relative extent of this nonenzymatic protective role of GSTs and to delineate it from their enzymatic role in converting the activated carcinogens into inactive GSH conjugates. Nonetheless, the role of GSTs in providing a chemical shield to a cellular constituent appears to be important and is consistent with their relatively high constitutive expression in mammalian tissues. The physiological significance of the irreversible binding of steroids, hematin, and bilirubin[84] to GSTs is not completely understood. The cationic isozymes of human livers, all of which belong to the α-class, bind bilirubin and hematin, and the isozymes' role in storage and transport of these compounds has been suggested.

GSTs can modulate functionality of some proteins through binding. For example, it has been shown that GSTP1-1 binds to c-Jun N-terminal kinase (JNK) and prevents its activation.[10] This binding occurs between monomeric enzymes, and during stress conditions such as UV exposure the enzyme oligomerizes and dissociates from JNK, leading to its activation and signaling for apoptosis.[10,85] This role of GSTs in regulation of stress-mediated signaling and the interactions between various kinases and GSTs needs to be studied further, particularly in view of studies showing that signaling for apoptosis can be modulated by overexpression of GSTs[56–60] and that certain adherent cell lines in culture can be transformed by transfection with hGSTA4-4.[86] Whereas it is likely that the enzymatic activity of GSTs may be primarily responsible for the ability of GSTs to modulate stress-mediated signaling

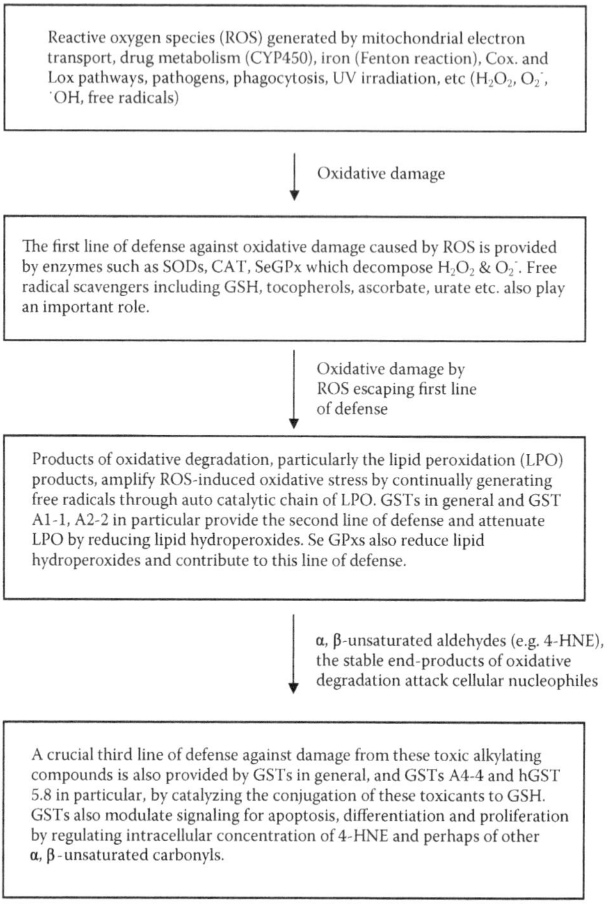

Reactive oxygen species (ROS) generated by mitochondrial electron transport, drug metabolism (CYP450), iron (Fenton reaction), Cox. and Lox pathways, pathogens, phagocytosis, UV irradiation, etc (H_2O_2, O_2^-, $^{\cdot}OH$, free radicals)

↓ Oxidative damage

The first line of defense against oxidative damage caused by ROS is provided by enzymes such as SODs, CAT, SeGPx which decompose H_2O_2 & O_2^-. Free radical scavengers including GSH, tocopherols, ascorbate, urate etc. also play an important role.

↓ Oxidative damage by ROS escaping first line of defense

Products of oxidative degradation, particularly the lipid peroxidation (LPO) products, amplify ROS-induced oxidative stress by continually generating free radicals through auto catalytic chain of LPO. GSTs in general and GST A1-1, A2-2 in particular provide the second line of defense and attenuate LPO by reducing lipid hydroperoxides. Se GPxs also reduce lipid hydroperoxides and contribute to this line of defense.

↓ α, β-unsaturated aldehydes (e.g. 4-HNE), the stable end-products of oxidative degradation attack cellular nucleophiles

A crucial third line of defense against damage from these toxic alkylating compounds is also provided by GSTs in general, and GSTs A4-4 and hGST 5.8 in particular, by catalyzing the conjugation of these toxicants to GSH. GSTs also modulate signaling for apoptosis, differentiation and proliferation by regulating intracellular concentration of 4-HNE and perhaps of other α, β-unsaturated carbonyls.

FIGURE 9.9 GSTs as antioxidant enzymes.

processes, the possibility of nonenzymatic binding with proteins involved in these cascades may not be ruled out.

ACKNOWLEDGMENT

Research for this chapter was supported in part by NIH grants EY04396 and ES012171.

REFERENCES

1. Benson, A.M. et al., Relationship between the soluble glutathione-dependent delta 5-3-ketosteroid isomerase and the glutathione S-transferases of the liver, *Proc. Natl. Acad. Sci. USA*, 74, 158, 1977.

2. Jakoby, W.B., The glutathione S-transferases: a group of multifunctional detoxification proteins, *Adv. Enzymol. Relat. Areas Mol. Biol.,* 46, 383, 1978.
3. Keen, J.H. and Jakoby, W.B., Glutathione transferases. Catalysis of nucleophilic reactions of glutathione, *J. Biol. Chem.,* 253, 5654, 1978.
4. Prohaska, J.R. and Ganther, H.E., Glutathione peroxidase activity of glutathione-S-transferases purified from rat liver, *Biochem. Biophys. Res. Commun.,* 76, 437, 1976.
5. Mosialou, E. et al., Evidence that rat liver microsomal glutathione transferase is responsible for the glutathione-dependent protein against lipid peroxidation, *Biochem. Pharmacol.,* 45, 1645, 1993.
6. Awasthi, Y.C. et al., Regulation of 4-hydroxynonenal mediated signaling by S-glutathione S-transfeases, *Free Radical Biol. Med.,* 37, 607, 2004.
7. Awasthi, Y.C. et al., Role of 4-hydroxynonenal in stress mediates signaling, *Mol. Aspects Med.,* 24, 219, 2003.
8. Pettersson, P.L. and Mannervik, B., The role of glutathione in the isomerization of delta 5-androstene-3, 17-dione catalyzed by human glutathione transferase A1-1, *J. Biol. Chem.,* 276, 11698, 2001.
9. Johansson, A.S. and Mannervik, B., Human glutathione transferase A3-3, a highly efficient catalyst of double-bond isomerization in the biosynthetic pathway of steroid hormones, *J. Biol. Chem.,* 276, 33061, 2001.
10. Alder, V. et al., Regulation of JNK signaling by GSTp, *Embo. J.,* 18, 1321, 1999.
11. Singhal, S.S. et al., Several closely related glutathione S-transferase isozymes catalyzing conjugation of 4-hydroxynonenal are differentially expressed in human tissues, *Arch. Biochem. Biophys.,* 311, 242, 1994.
12. Fernandez-Caron, J.M. and Penalava, M.A., Characterization of a fungal maleylacetoacetate isonerase gene and identification of its human homologue, *J. Biol. Chem.,* 273, 329, 1998.
13. Ujihara, M. et al., Biochemical and immunological demonstration of prostaglandins D2, E2, and F2 alpha formation from prostaglandins H2 by various rat glutathione S-transferease isozymes, *Arch. Biochem. Biophys.,* 264, 428, 1988.
14. Beuckmann, C.T. et al., Identification of mu-class glutathione transferases M2-2 and M3-3 as cytosolic prostaglandin E synthases in the human brain, *Neurochem. Res.,* 25, 733, 2000.
15. Hong, Y. et al., The role of selenium-dependent and selenium-independent glutathione peroxidases in the formation of prostaglandin F2 alpha, *J. Biol. Chem.,* 264, 13793, 1989.
16. Cox, B. et al., Human colorectal cancer cells efficiently conjugate the cyclopentenone prostaglandin, prostaglandin J (2), to glutathione, *Biochim. Biophys. Acta,* 1584, 37, 2002.
17. Paumi, C.M. et al., Glutathione S-transferases (GSTs) inhibit transcriptional activation by the peroxisomal proliferators-activated receptor γ (PPAR γ) ligand, 15-deoxy-$\Delta^{12,14}$-prostaglandin J$_2$ (15-d-PGJ$_2$), *Biochemistry,* 43, 2345, 2004.
18. Jowsey, I.R., Smith, S.A., and Hayes, J.D., Expression of the murine glutathione S-transferase 3 (GSTA3) subunit is markedly induced during adipocyte differentiation: activation of the GSTA3 gene promoter by the proadipogenic eicosanoid 15-deoxy-$\Delta^{12,14}$-prostaglandin J$_2$, *Biochem. Biophys. Res. Commun.,* 312, 1226, 2003.
19. Itoh, K. et al., Transcription factor Nrf2 regulates inflammation by mediating the effect of 15-dexoy-$\Delta^{12,14}$-prostaglandin J$_2$, *Mol. Cell. Biol.,* 24, 36, 2004.
20. McMahon, M. et al., Keap1-dependent proteasomal degradation of transcription factor Nrf2 contributes to the negative regulation of antioxidant response element-driven gene expression, *J. Biol. Chem.,* 278, 21592, 2003.

21. Wakabayashi, N. et al., Protection against electrophile and oxidant stress by induction of the phase 2 response: fate of cysteines of the Keap1 sensor modified by inducers, *Proc. Natl. Acad. Sci. USA,* 101, 2040, 2004.

22. Rossi, A. et al., Anti-inflammatory cyclopentenone prostaglandins are direct inhibitors of IκB kinase, *Nature,* 403, 103, 2000.

23. Hayes, J.D., Flanagan, J.U., and Jowsey, I.R., Glutathione transferases, *Annu. Rev. Pharmacol. Toxicol.,* 45, 51, 2004.

24. Talalay, R., Mechanisms of induction of enzymes that protect against chemical carcinogenesis, *Adv. Enzyme Reg.,* 28, 237, 1989.

25. Uchida, K., Induction of glutathione S-transferases by prostaglandins, *Mech. Aging Develop.,* 116, 135, 2000.

26. Zhao, T. et al., The role of human glutathione S-transferase hGSTA1-1 and hGSTA2-2 in protection against oxidative stress, *Arch. Biochem. Biophys.,* 367, 216, 1999.

27. Samuelsson, B. et al., Leukotrienes and lipoxins: structures, biosynthesis, and biological effects, *Science,* 237, 1171, 1987.

28. Jakobsson, P.J. et al., Identification of and characterization of a novel microsomal enzyme with glutathione dependent transferase and peroxidase activities., *J. Biol. Chem.,* 272, 22934, 1997.

29. Hayes, J.D. and Pulford, D.J., The glutathione S-transferase supergene family: regulation of GST and the contribution of the isozymes to cancer chemoprotection and drug resistance, *Crit. Rev. Biochem. Mol. Biol.,* 30, 445, 1995.

30. Awasthi, Y.C., Sharma, R., and Singhal, S.S., Human glutathione S-transferases, *Int. J. Biochem.,* 26, 295, 1994.

31. Awasthi, Y.C., Dao, D.D., and Saneto, R.P., Interrelationship between anionic and cationic forms of glutathione S-transferases of human liver, *Biochem. J.,* 191, 1, 1980.

32. Little, C. and O'Brien, P.J., An intracellular GSH-peroxidase with a lipid peroxide substrate, *Biochem. Biophys. Res. Commun.,* 131, 45, 1968.

33. Brigelius-Flohe, R., Tissue-specific functions of individual glutathione peroxidases, *Free Radical Biol. Med.,* 27, 951, 1999.

34. Thomas, J.P. et al., Protective action of phospholipids hydroperoxide glutathione peroxidase against membrane-damaging lipid peroxidation. *In situ* reduction of phospholipids and cholesterol hydroperoxides, *J. Biol. Chem.,* 265, 454, 1990.

35. Mills, G.C., Hemoglobin catabolism. I. Glutathione peroxidase, an erythrocyte enzyme which protects hemoglobin from oxidative breakdown, *J. Biol. Chem.,* 229, 189, 1957.

36. Lawrence, R.A. and Burk, R.F., Glutathione peroxidase activity in selenium-deficient rat liver, *Biochem. Biophys. Res. Commun.,* 71, 952, 1976.

37. Tan, K.H. et al., Inhibition of microsomal lipid peroxidation by glutathione and glutathione transferases C and AA. Role of endogenous phospholipase A2, *Biochem. J.,* 220, 243, 1984.

38. Yang, Y. et al., Role of glutathione S-transferases in protection against lipid peroxidation. Overexpression of hGSTA2-2 in K562 cells protects against hydrogen peroxide–induced apoptosis and inhibits JNK and caspase 3 activation, *J. Biol. Chem.,* 276, 19220, 2001.

39. Singhal, S.S. et al., Glutathione S-transferases of human lung: characterization and evaluation of the protective role of the alpha-class isozymes against lipid peroxidation, *Arch. Biochem. Biophys.,* 299, 232, 1992.

40. Yang, Y. et al., Protection of HLE B-3 cells against hydrogen peroxide- and naphthalene-induced lipid peroxidation and apoptosis by transfection with hGSTA1 and hGSTA2, *Invest. Ophthalmol. Vis. Sci.,* 43, 434, 2002.

41. Hurst, R. et al., Phospholipids hydroperoxide glutathione peroxidase activity of human glutathione transferases, *Biochem, J.,* 332, 97, 1998.

42. Yang, Y. et al., Role of alpha class glutathione S-transferases as antioxidant enzymes in rodent tissues, *Toxicol. Appl. Pharmacol.,* 182, 105, 2002.

43. Alin, P., Danielson, U.H., and Mannervik, B., 4-Hydroxyalk-2-enals are substrates for glutathione transferase, *FEBS Lett.,* 179, 267, 1985.

44. Zimniak, P. et al., A subgroup of alpha-class glutathione S-transferases. Cloning of cDNA for mouse lung glutathione S-transferase GST 5.8, *FEBS Lett.,* 313, 173–176, 1992.

45. Stenberg, G. et al., Cloning and heterologous expression of cDNA encoding class alpha rat glutathione transferase 8-8, an enzyme with highly catalytic activity towards genotoxic alpha, beta-unsaturated carbonyl compounds, *Biochem. J.,* 284, 313, 1992.

46. Zimniak, P. et al., Estimation of genomic complexity, heterologous expression, and enzymatic characterization of mouse glutathione S-transferase mGSTA4-4 (GST 5.7), *J. Biol. Chem.,* 269, 992, 1994.

47. Singh, S.P. et al., Membrane association of glutathione S-transferase mGSTA4-4, an enzyme that metabolizes lipid peroxidation products, *J. Biol. Chem.,* 227, 4232, 2002.

48. Berhane, K. et al., Detoxification of base propenals and other alpha, beta-unsaturated aldehyde products of radical reactions and lipid peroxidation by human glutathione transferases, *Proc. Natl. Acad. Sci. USA,* 91, 1480, 1994.

49. Dieckhaus, C.M. et al., Role of glutathione S-transferases A-1-1, M1-1, and P1-1 in the detoxification of 2 phenylpropenal, a reactive felbamate metabolite, *Chem. Res. Toxicol.,* 14, 511, 2001.

50. Bull, A.W. et al., Conjugation of the linoleic acid oxidation product, 13-oxooctadeca-9-11-dienoic acid, a bioactive endogenous substrate for mammalian glutathione transferase, *Biochim. Biophys. Acta,* 1571, 77, 2002.

51. Zarkovic, N., 4-Hydroxynonenal as a bioactive marker of pathogenesis. *Mol. Aspects Med.,* 24, 281, 2003.

52. He, N.G. et al., Transfection of a 4-hydroxynonenal metabolizing glutathione S-transferase isozyme, mouse GSTA4-4, confers doxorubicin resistance to Chinese hamster ovary cells, *Arch. Biochem. Biophys.,* 333, 214, 1996.

53. Awasthi, S. et al., Curcumin protects against 4-hydroxy-2-trans-nonenal–induced cataract formation in rats' lenses, *Am. J. Clin. Nutr.,* 64, 761, 1996.

54. Srivastava, S.K. et al., Attenuation of 4-hydroxynonenal-induced cataractogenesis in rat lens by butylated hydroxytoluene, *Curr. Eye Res.,* 15, 749, 1996.

55. Cheng, J.Z. et al., Effects of mGSTA4 transfection on 4-hydroxynonenal-mediated apoptosis and differentiation of K562 human erythroleukemia cells, *Arch. Biochem. Biophys.,* 372, 29, 1999.

56. Cheng, J.Z. et al., Transfection of mGSTA4 in HL-60 cells protects against 4-hydroxynonenal-induced apoptosis by inhibiting JNK-mediated signaling, *Arch. Biochem. Biophys.,* 392, 197, 2001.

57. Zimniak, L. et al., Increased resistance to oxidative stress in transfected cultured cells overexpressing glutathione S-transferase mGSTA4-4, *Toxicol. Appl. Pharmacol.,* 143, 221, 1997.

58. Cheng, Z.J. et al., Accelerated metabolism and exclusion of 4-hydroxynonenal through induction of RLIP76 and hGST5.8 is an early adaptive response of cells to heat and oxidative stress, *J. Biol. Chem.,* 276, 41213, 2001.

59. Yang, Y. et al., Cells preconditioned with mild, transient UVA irradiation acquire resistance to oxidative stress and UVA-induced apoptosis: role of 4-hydroxynonenal in UVA-mediated signaling for apoptosis, *J. Biol. Chem.,* 278, 41380, 2003.

60. Pandya, U., et al., Attenuation of galactose cataract by low levels of dietary curcumin, *Nutr. Res.*, 20, 515, 2000.

61. Pandya, U. et al., Dietary curcumin prevents ocular toxicology of naphthalene in rats, *Toxicol. Lett.*, 115, 195, 2000.

62. Anandatheerthavarda, H.K. et al., Phosphorylation enhances mitochondrial targeting of GSTA4-4 through increased affinity for binding to cytoplasmic Hsp70, *J. Biol. Chem.*, 278, 18960, 2003.

63. Forman, H.J., Dickinson, D.A., and Iles, K.E., HNE-signaling pathways leading to its elimination, *Mol. Aspects Med.*, 24, 189, 2003.

64. Krzywanski, D.M., et al., Variable regulation of glutamate cysteine ligase subunit proteins affect glutathione biosynthesis in response to oxidative stress, *Arch. Biochem. Biophys.*, 23, 116, 2004.

65. Giblin, F.J., McCready, J., and Reddy, V.N., The role of glutathione metabolism in the detoxification of H_2O_2 in rabbit lens, *Invest. Ophthalmol. Vis. Sci.*, 22, 330, 1982.

66. Cheng, H.M. et al., Effect of glutathione deprivation on lens metabolism, *Expt. Eye Res.*, 39, 355, 1984.

67. Witztum, J.L. and Steiberg, D., Oxidative modification hypothesis of atherosclerosis. Does it hold for humans?, *Trends Cardiovasc. Med.*, 11, 93, 2001.

68. Srivastava, S. et al., Identification of biochemical pathways for the metabolism of oxidized low-density lipoprotein-derived aldehyde-4-hydroxy trans-2-nonenal in vascular smooth muscle cells, *Atherosclerosis*, 158, 339, 2001.

69. Yang, Y. et al., Glutathione S-transferase A4-4 modulates oxidative stress in endothelium: possible role in human atherosclerosis. *Atherosclerosis*, 173, 211, 2004.

70. Misra, P. et al., Glutathione S-transferase 8-8 is localized in smooth muscle cells of rat aorta and is induced in an experimental model of atherosclerosis, *Toxicol. Appl. Pharmacol.*, 133, 27, 1995.

71. Sayere, L.M. et al., 4-Hydroxynonenal-derived advanced lipid peroxidation products are increased in Alzheimer's disease, *J. Neurochem.*, 68, 2092, 1997.

72. Yoritaka, A. et al., Immunohistochemical detection of 4-hydroxynonenal protein adducts in Parkinson's disease, *Proc. Natl. Acad. Sci. USA*, 93, 2696, 1996.

73. Eckl, P.M., Ortner, A., and Esterbauer, H., Genotoxic properties of 4-hydroxyalkenals and analogous aldehydes, *Mut. Res.*, 290, 183, 1993.

74. Hu, W. et al., The major lipid peroxidation product, trans-4-hydroxy-2-nonenal preferentially forms D adducts at codon 249 of human p53 gene, a unique mutational hot spot in hepatocellular carcinoma, *Carcinogenesis*, 23, 1781, 2002.

75. Awasthi, Y.C., Singhal, S.S., and Awasthi, S., Mechanisms of anti-carcinogenic effects of antioxidant nutrients, in *Nutrition and Cancer*, Watson, R.R. and Mufti, S.I. (eds.), CRC Press, Boca Raton, FL, 139, 1995.

76. Litwack, G., Ketterer, B., and Arias, I.M., Ligandin: a hepatic protein which binds steroids, bilirubin, carcinogens and a number of exogenous organic anions, *Nature (London)*, 234, 446, 1971.

77. Tipping, E. and Ketterer, B., The role of intracellular proteins in the transport and metabolism of lipophilic compounds, in *Transport Proteins*, Blasuser, G. and Sund, H. (eds.), Walter de Gruyter, Berlin, 368, 1978.

78. Board, P., Ligandin revisited: resolution of the alpha class glutathione transferase gene family, *Pharmacogenetics*, 12, 275, 2002.

79. LeTrong, I. et al., 1.3-A resolution structure of human glutathione S-transferase with S-hexyl glutathione bound reveals a possible extended ligandin-binding site, *Protein: Structure, Funct. Gen.*, 48, 618, 2002.

80. Sayed, Y., et al., Themodynamics of the ligandin function of human class alpha glutathione transferase A1-1: energetics of organic anion ligand binding, *Biochem. J.*, 363, 341, 2002.

81. Ketterer, B., Ross-Mansell, P., and Whitehead, J.K., The isolation of carcinogen-binding protein from livers of rats given 4-dimethylaminobenzene, *Biochem. J.*, 102, 316, 1967.

82. Singh, S.V., Srivastava, S.K., and Awathi, Y.C., Binding of benzo(a)pyrene to rat lung glutathione S-transferases *in vivo*, *FEBS Lett.*, 111, 179, 1985.

83. Singh, S.V. et al., Glutathione S-transferases of mouse lung. Selective binding of benzo(a)pyrene metabolites by the subunits which are preferentially induced by t-butylated hydroxyanisole, *Biochem. J.*, 243, 351, 1987.

84. Kamisaka, K. et al., Multiple forms of human glutathione S-transferase and their affinity for bilirubin, *Eur. J. Biochem.*, 60, 153, 1975.

85. Wang, T. et al., Glutathione S-transferase P1-1 (GSTP1-1) inhibits c-Jun N-terminal kinase (JNK1) signaling through interactions with the C terminus, *J. Biol. Chem.*, 276, 20999, 2001.

86. Sharma, R. et al., Transfection with 4-hydroxynonenal-metabolizing glutathione S-transferase isozymes leads to phenotypic transformation and immortalization of adherent cells, *Eur. J. Biochem.*, 271, 1690, 2004.

87. Van Dorp, D.A. et al., Isolation and properties of enzymes involved in prostaglandin biosynthesis, *Acta Biol. Med. Geriat.*, 37, 691, 1978.

88. Agarwal, R. et al., Glutathione S-transferase–dependent conjugation of leukotrienes A4-methyl ester to leukotrienes C4-methyl ester in mammalian skin, *Biochem. Pharmacol.*, 44, 2047, 1992.

89. Meyer, D.J. and Ketterer, B., 5 Alpha, 6 alpha-epoxy-choletan-3 beta-ol (cholesterol alpha-oxide): a specific substrate for rat liver glutathione transferase B, *FEBS Lett.*, 150, 499, 1982.

90. Flohé, L., Günzler, W.A., and Schock, H.H., Glutathione peroxidase: a selenoenzyme. *FEBS Lett.*, 32, 132, 1973.

91. Awasthi, Y.C., Beutler, E., and Srivastava, S.K., Purification and properties of human erythrocyte glutathione peroxidase, *J. Biol. Chem.*, 250, 5144, 1975.

92. Chu, F.F., Doroshow, J.H., and Esworthy, R.S., Expression, characterization, and tissue distribution of a new selenium-dependent glutathione peroxidase, GSH-Px-GI, *J. Biol. Chem.*, 268, 2571, 1993.

93. Takahashi, K. et al., Purification and characterization of human plasma glutathione peroxidase: a selenoglycoprotein distinct from the known cellular enzyme, *Arch. Biochem. Biophys.*, 256, 677, 1987.

94. Ursini, F., Maiorino, M., and Gregolin, C., The selenoenzyme phospholipid hydro-peroxide glutathione peroxidase, *Biochim. Biophys. Acta*, 839, 62, 1985.

95. Brigelius-Flohé, R. et al., Phospholipid-hydroperoxide glutathione peroxidase. Genomic DNA, cDNA, and deduced amino acid sequence, *J. Biol. Chem.*, 269, 7342, 1994.

96. Hubatsch, I., Ridderstorm, M., and Mannervik, B., Human glutathione S-transferase A4-4: an alpha-class enzyme with high catalytic efficiency in conjugation of other xenotoxic products of lipid peroxidation, *Biochem. J.*, 15, 175, 1998.

10 Glutathione S-Transferases as Modulators of Signal Transduction

Yusong Yang and Yogesh C. Awasthi

CONTENTS

205

10.1   INTRODUCTION

The role of glutathione S-transferases (GSTs) in protecting cells from lipid peroxidation was discussed in the preceding chapter. In recent years, the physiological significance of this function of GSTs has begun to unfold, suggesting not only their importance as antioxidant enzymes but also their role in the regulation of stress-mediated signaling. Recent studies indicate that the α-class GSTs can modulate stress-mediated signaling for apoptosis by regulating the intracellular concentration of lipid peroxidation products, particularly the hydroperoxides and 4-hydroxynonenal (4-HNE).[1–3] Furthermore, some of these studies suggest that GSTs may affect cell growth, differentiation, transformation, and expression of key cell cycle genes by regulating the intracellular levels of 4-HNE.[4,5] Exposure to stress — particularly heat, oxidants, and UV irradiations — invariably leads to increased lipid peroxidation and 4-HNE formation[6,7] and there is credible evidence that the lipid peroxidation products, particularly 4-HNE, are involved in stress-mediated signaling.[1–11] The intracellular levels of lipid hydroperoxides and 4-HNE may be crucial for cell survival during stress conditions. Recent studies that show that at least some cell types undergo proliferation, differentiation, or apoptosis in culture depending on whether 4-HNE concentrations are below or above basal constitutive or "physiological" levels[4,5] are consistent with this idea. GSTs are the major determinants of the intracellular concentrations of lipid peroxidation products including 4-HNE because these enzymes not only limit the extent of lipid peroxidation through GSH-dependent reduction of hydroperoxide intermediates,[12] but also metabolize the majority of cellular 4-HNE via its conjugation to GSH.[13–15] This would suggest that GSTs could be involved in the mechanisms regulating stress-mediated signaling and perhaps the events of cell cycle signaling. In this chapter, these putative roles of GSTs are critically evaluated.

10.2   LIPID PEROXIDATION AND SIGNALING

Lipid peroxidation products in general have been associated with toxicity. The toxicity of many xenobiotics including CCl_4, paraquat, doxorubicin, and various quinonoids has been attributed to lipid peroxidation induced by these agents, which can lead to membrane dysfunction, necrosis, and ultimately cell death. Since a basal level of lipid peroxidation must occur in every aerobic organism due to the ever-present reactive oxygen species (ROS), trace metal ions, and unsaturated lipids in cell membranes, it has long been speculated that lipid peroxidation products could also have some physiological role. However, little or no information existed on the physiological roles of these compounds before the pioneering work by Esterbauer and his group,[10] who not only identified 4-HNE and its homologous α,β-unsaturated alkenals as stable end products of lipid peroxidation but also showed the relevance of 4-HNE in signaling mechanisms.[8–11,16] These studies opened a new area of research on potential physiological functions of lipid peroxidation products, and now there is overwhelming evidence that 4-HNE and its precursor hydroperoxides can affect a variety of signaling processes including activation of various signaling kinases, cell growth, differentiation, and apoptosis. Lipid peroxidation results in the formation of a vast multitude of lipoxy, alkoxy, and hydroperoxy radicals; hydroperoxides; and aldehydes, including the α,β-unsaturated 4-hydroxyalkenals (e.g., 4-HNE), which all could be considered potential modulators of oxidative stress-induced signaling. So far most studies have focused on the role of lipid hydroperoxides and 4-HNE in these processes. In the following sections evidence for the involvement of hydroperoxides and 4-HNE in signaling processes is evaluated and the role of GSTs in the regulation of these processes is discussed.

10.2.1 Lipid Hydroperoxide–Mediated Signaling

There is overwhelming evidence linking lipid hydroperoxides to signaling processes. For example, membrane phospholipid hydroperoxides (PL-OOH) have been shown to activate protein kinase C (PKC), which in turn promotes cytosolic phospholipase A_2 (cPLA$_2$) phosphorylation and its translocation to the plasma membrane through mitogen-activated protein kinases (MAPK). These findings suggest that PL-OOH can affect cPLA$_2$ activity independent of the intracellular concentration of free calcium.[17,18] Preparations isolated from oxidized low-density lipoprotein, presumably containing hydroperoxides, have been shown to contain platelet-activating factor-like activity which can stimulate neutrophil adhesion and smooth muscle cell proliferation.[19] Studies from our laboratory have shown that phosphatidylcholine hydroperoxide can induce apoptosis in human erythroleukemia (K562) and human lens epithelial (HLE B-3) cells through a sustained activation of stress-activated protein kinase/c-Jun N-terminal kinase (SAPK/JNK) and caspase 3.[20,21]

Hydroperoxides are involved in the regulation of biosynthesis of prostaglandins, and the requirement of low amounts of lipid hydroperoxides has been demonstrated for cyclooxygenase (COX) activity and prostaglandin biosynthesis.[22,23] Hydroperoxides may also play an important role in signaling for protective mechanisms against oxidative stress through activation of nitric oxide synthase (NOS). Nitric oxide is known to inhibit lipid peroxidation. 15-Hydroxyperoxy endoperoxide prostaglandin G2 (PGG2) and 15-hydroperoxyeicosatetraenoic acid (15-HPETE) stimulate NOS activity,[24] suggesting that during signaling for response to oxidative stress, lipid hydroperoxides can enhance nitric oxide formation and mitigate lipid peroxidation. Involvement of the hydroperoxy group of PGG2 and 15-HPETE in the stimulation of NOS activity is suggested by the findings that neither PGH2 nor 15-HETE (which are the corresponding hydroxyl compounds formed from the reduction of PGG2 and 15-HPETE, respectively) can stimulate NOS activity.[24] For lipooxygenase (LOX) activity, small amounts of lipid hydroperoxides are also required.[25,26] LOX activity in membranes can be suppressed by the preincubation of the membrane fractions with phospholipid hydroperoxide glutathione peroxidase (GPx4), which reduces the hydroperoxides required for LOX activation. However, LOX activity in these membrane fractions can be restored by the addition of 13S-hydroperoxy-9Z,11E-octadecadienoic acid indicating the requirement of hydroperoxides for LOX activity.[27] Thus, it appears that lipid hydroperoxide–mediated signaling is involved in the regulation of biosynthesis of prostaglandins and leukotrienes as well as in the response of cells to oxidative stress. Even though it is not known whether or not GSTs modulate this signaling, it is expected that by reducing hydroperoxides through their GPx activity, GSTs should be able to influence this process.

In studies adopting a different approach, the role of lipid hydroperoxides in signaling has been studied by altering the levels of lipid hydroperoxides in cells through their GSH-dependent reduction by GSTs and Se-GPxs. The α-class GST isozymes GSTA1-1 and GSTA2-2, which efficiently catalyze GSH-dependent reduction of PL-OOH and fatty acid hydroperoxides (FA-OOH), can inhibit H_2O_2- and naphthalene-induced apoptosis of HLE B-3 and K562 cells, suggesting the involvement of hydroperoxides in signaling for apoptosis.[20,21] Likewise, it has been shown that overexpression of PL-OOH glutathione peroxidase (GPx-4) in a human endothelial cell line results in the inhibition of interleukin-1–induced NF-κB activation, and this effect has been attributed to the reduction of PL-OOH by GPx-4.[28]

10.2.2 4-HNE AND SIGNALING

4-HNE is a potent alkylating agent that can react with a variety of nucleophilic sites of DNA, proteins, and lipids to form corresponding 4-HNE adducts. Since the discovery of 4-HNE as a stable end product of lipid peroxidation, a multitude of studies have suggested that it can influence various signaling processes.[1–11,16] Results from different laboratories using a variety of cell lines convincingly demonstrate that inclusion of 4-HNE in cell cultures can activate SAPK/JNK, a member of the MAPK family that is involved in apoptosis perhaps through alternate mechanisms in different cells. For example, it has been proposed that in hepatic stellate cells, 4-HNE activates JNK by direct binding and not through phosphorylation.[29] On the other hand, 4-HNE appears to activate JNK through the redox-sensitive MAPK kinase cascade in other cell lines.[30–32] It has been suggested that 4-HNE–induced JNK activation promotes its translocation in the nucleus where JNK-dependent phosphorylation of c-Jun and the transcription factor activator protein (AP-1) binding take place, leading to the transcription of a number of genes having AP-1 consensus sequences in their promoter regions.[33] Activation of JNK by 4-HNE is accompanied by the activation of caspase 3 and eventual apoptosis.[31,33] 4-HNE–induced activation of p38 MAPK has been attributed to the induction of COX-2 through stabilization of its mRNA.[34]

4-HNE has also been shown to modulate the expression of various genes, including PKC βII,[35] c-myc,[36] procollagen type I,[37] aldose reductase,[38] c-myb,[39] and transforming growth factor β1.[36] In HL-60 cells, physiological concentrations of 4-HNE inhibit proliferation and induce a granulocyte-like differentiation.[39] The "physiologic" concentrations of 4-HNE are, however, difficult to define and these vary from cell to cell depending on the extent of lipid peroxidation and levels of antioxidant enzymes including GSTs. Thus, these effects may vary from cell to cell. For example, studies in our laboratory have shown that while K562 cells rapidly undergo differentiation when exposed to 20 μM 4-HNE, under similar conditions other cells such as HLE B-3 remain unaffected. Tables 10.1 and 10.2 list some of the studies where lipid hydroperoxides and 4-HNE have been shown to be involved in signaling processes and that GSTs can modulate these processes.

TABLE 10.1
Signaling Roles of Lipid Hydroperoxides and 4-Hydroxynonenal*

Lipid Hydroperoxides	4-Hydroxynonenal
Activate cytosolic phospholipase A$_2$[16,17]	Apoptosis[4,6,7,29–33]
Platelet-activation and mitogenic for vascular smooth muscle cells[18]	Stimulate caspase[4,6,7]
	Increase AP-1 DNA binding[31]
Stimulate NF-κB activation[28]	Inhibit PKCβII at higher concentration[29,35,38]
Increase synaptic NOS activity[24]	Activate JNK [6,7,29–33]
Inhibit neuronal Na$^+$-K$^+$ ATPase activity[23]	
Activate 15-LOX and COX[22,23,25,26]	Stimulate TGFβ expression[40]andCOX-2[34]
Induce programmed cell death[19]	At low concentrations, 4-HNE stimulates cell
Activate caspase 3 and JNK[20,21]	proliferation[41,42] and activates PKC βII[41]

* Only a limited number of studies are cited. For numerous other studies on the role of lipid peroxidation products in signaling, see reviews in references 1–3, 8–11, and 16.

TABLE 10.2
Role of GSTs in Signaling

Effect of overexpression of 4-HNE–metabolizing enzymes GSTA4-4 and GST5.8 on signaling	Cells transfected with mGSTA4-4 have lower 4-HNE levels, grow faster,[4] and are resistant to apoptosis caused by oxidative stress.[4,31]
	Activation of JNK and increased AP-1 DNA binding induced by H$_2$O$_2$ or 4-HNE is blocked by mGSTA4-4 overexpression.[4]
	4-HNE metabolism is accelerated in cells exposed to mild heat, UVA, or H$_2$O$_2$ treatment. Cells preconditioned with these mild stress conditions acquire capacity to exclude cellular 4-HNE at a faster rate and are resistant to H$_2$O$_2$, UVA, superoxide anion, and 4-HNE-induced JNK and caspase activation, and apoptosis.[6,7]
Effect of overexpression of hGSTA1-1 and hGSTA2-2 on signaling	Cells transfected with hGSTA1-1/A2-2 have lower levels of lipid hydroperoxides even after exposure to oxidative stress and are resistant to H$_2$O$_2$ and naphthalene-induced apoptosis.[20,21]
	Overexpression of hGSTA1-1/hGSTA2-2 in cells blocks H$_2$O$_2$-induced JNK and caspase activation.[20,21]

10.2.3   CONCENTRATION-DEPENDENT EFFECT OF 4-HNE

Interestingly, 4-HNE appears to affect the signaling processes in a concentration-dependent manner. While numerous studies have shown that 4-HNE is pro-apoptotic (Table 10.1), it has been shown to stimulate proliferation of some cell types[41,42] at relatively lower concentrations. In K562 cells, 4-HNE induces differentiation at moderate concentrations[4] and apoptosis at high concentrations.[4,6] Likewise, 4-HNE activates PKCβII at lower concentrations, while at higher concentrations 4-HNE inhibits this enzyme.[43] These studies strongly suggest that 4-HNE differentially affects the signals for proliferation, differentiation, and apoptosis.

The concentration-dependent effect of 4-HNE on cell cycle events is further suggested by recent studies demonstrating that overexpression of GSTA4-4 in attached cells HLE B-3 and CCL-75 leads to their transformation and immortalization, and this activity depends on the depletion of intracellular 4-HNE by GSTA4-4.[5] Lowering 4-HNE levels in these cells seems to profoundly affect the expression of key cell cycle genes. Our unpublished studies with mGSTA4-null mice also show that cell cycle regulatory gene p53 is upregulated due to higher 4-HNE levels in the tissues of these mice. These studies are consistent with a significant physiological role for 4-HNE and suggest that GSTs are modulators of 4-HNE–mediated signaling processes and may regulate cell cycle events by affecting the homeostasis of 4-HNE in cells. The potential role of GSTs in modulating stress-mediated signaling by regulating lipid hydroperoxides and 4-HNE homeostasis in cells is outlined in Figure 10.1.

10.3 GSTS AS DETERMINANTS OF THE INTRACELLULAR LEVELS OF 4-HNE

10.3.1 GST ISOZYMES WITH SUBSTRATE PREFERENCE FOR 4-HNE

It has been shown that α-class GSTs A1-1 and A2-2 can effectively reduce cellular lipid peroxides and protect cells from toxicity of oxidants.[12,20] Reduction of PL-OOH also limits the generation of 4-HNE. GST-mediated conjugation of 4-HNE to GSH is the major metabolic detoxification pathway for 4-HNE.[14,15] 4-HNE can nonenzymatically form conjugates with GSH, but the conjugation of 4-HNE to GSH in cells is facilitated by the GSTs' catalyzed reaction.[13] GST isozymes with substrate preference for 4-HNE[44–49] have been identified in rat (rGSTA4-4), mouse (mGSTA4-4), bovine (bGST5.8), and human (hGST5.8 and hGSTA4-4) as well as in Drosophila (DmGSTD1-1). These enzymes, belonging to a subgroup of the α-class GSTs with a remarkably high catalytic efficiency for conjugating 4-HNE to GSH, appear to have evolved to negate the toxicity of 4-HNE and its homologues generated during lipid peroxidation in aerobic organisms.[46] In humans, two immunologically distinct 4-HNE-metabolizing enzymes with remarkably high catalytic efficiency (K_{cat}/K_m values of $>2,000 s^{-1}$ mM^{-1}) have been identified.[50] While hGSTA4-4 has been cloned, the primary structure of hGST5.8 is still unknown and its cDNA has not been cloned perhaps due to very low constitutive levels of the message for this isozyme. However, the kinetic properties of tissue-purified hGST5.8 have been studied and its immunological similarity with mouse enzyme mGSTA4-4 suggests structural similarities.

In humans, other GST isozymes also contribute to the metabolism and detoxification of 4-HNE. GSTA1-1 and GSTA2-2 show only barely detectable activity toward 4-HNE. But these isozymes compose up to 90% of GST protein in livers and could contribute significantly to the metabolism of 4-HNE in this organ despite their low catalytic efficiency. In extrahepatic tissues, the π- and μ-class GSTs are predominant and the contribution of these enzymes in the metabolism of 4-HNE may also be substantial despite their low catalytic efficiency toward 4-HNE. This may be the

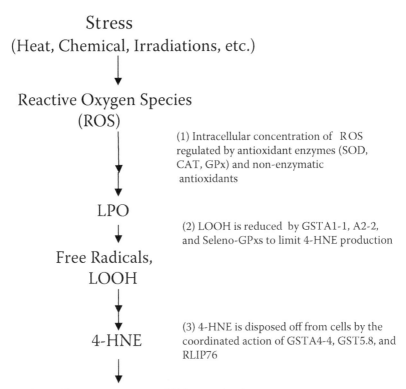

FIGURE 10.1 Regulation of stress-induced lipid peroxidation and signaling by GSTs. GSTs can attenuate LPO stress induced by reducing lipid hydroperoxides (LOOH) and by conjugating 4-HNE to GSH. LPO, lipid peroxidation; ROS, reactive oxygen species; SOD, superoxide dismutase; CAT, catalase; RLIP76, Ral-interacting protein synonymous with Ral BP1 (Ral-binding protein 1), which catalyzes ATP-dependent transport of GS-HNE and other GSH conjugates from cells.

reason that approximately 40% of the residual GST activity toward 4-HNE is retained in the tissues of mGSTA4 knockout mice generated in our laboratory.[51]

Whereas the majority of 4-HNE is metabolized through its GST-catalyzed conjugation to GSH,[14,15] other pathways are also operative in the metabolism of 4-HNE. Among these pathways, the roles of aldehyde dehydrogenases (ALDHs) and aldoketoreductase have been studied in detail.[52–56] Aldehyde dehydrogenase oxidizes the aldehyde group of 4-HNE and other fatty aldehydes to corresponding acids. 4-Hydroxy-2-nonenoic acid (4-HNA) formed from oxidation of 4-HNE can be utilized in the fatty acid β-oxidation pathway. Cytochrome P_{450} 4A–mediated ω-hydroxylation of 4-HNA has also been reported.[57] Aldoketoreductases (AKRs) catalyze NADPH-dependent reduction of 4-HNE to corresponding alcohol, 1, 4-dihydroxynonene (DHN). The role of aldose reductase, a member of this family, has been extensively studied in the metabolism of 4-HNE and GS-HNE.[14,52,57] ALDHs and AKRs can utilize GS-HNE as a substrate. For example, GS-HNE is reduced by aldose reductase to yield the corresponding alcohol GS-DHN, which has been shown to be involved in signaling mechanisms.[58] Alternatively, GS-HNE can be oxidized to the corresponding acid GS-HNA by ALDHs.[56] These pathways, however, appear to contribute only a minor fraction of GS-HNE metabolism because the majority of GS-HNE formed in cells is transported as such through ATP-dependent transport catalyzed by RLIP76.[6,59]

Because 4-HNE contains an α,β-unsaturated double bond, it is also a substrate for NADPH-dependent alkenal/one oxidoreductase (AOR), which is believed to be involved in the detoxification of α,β-unsaturated carbonyls.[60] The carbon–carbon double bond of 4-HNE is reduced by AOR to saturated aldehyde, which is less toxic than 4-HNE. It is interesting to note that enzymes including GSTs, AKRs, and ALDHs are enzyme families, members of which have a varying degree of substrate preference for 4-HNE, indicating redundancy among the 4-HNE metabolizing enzymes. The redundancy in the 4-HNE-metabolizing enzymes seems to be similar to that observed with GPxs, which are also crucial for the detoxification of endogenously generated toxicants, H_2O_2 and lipid hydroperoxides. The redundancy in 4-HNE-metabolizing enzymes provides backup defense against oxidative stress and 4-HNE toxicity, and it is perhaps important for a tight regulation of 4-HNE homeostasis because of its hormetic effect on signaling events. GSTs, being the main regulators of 4-HNE homeostasis, not only protect cells from its toxic effects but can also regulate 4-HNE-mediated signal transduction.

10.4 MODULATION OF SIGNALING BY GSTS WITH GPX ACTIVITY

10.4.1 Protection against Stress-Mediated Apoptosis

If stress-mediated signaling is conveyed through lipid peroxidation products even partially, then it should be possible to interrupt or inhibit this signaling by limiting the extent of lipid peroxidation. This prediction seems to be true at least in some cell lines as indicated by *in vitro* studies with cells overexpressing hGSTA1-1 or hGSTA2-2.[20,21] Transfection of K562 cells with *hGSTA1* or *hGSTA2* cDNA results in about a 10-fold higher expression of these enzymes with a corresponding increase in the GPx activity toward PL-OOH and FA-OOH. The activities of other antioxidant enzymes such as catalase, superoxide dismutase, and Se-dependent GPx activity toward H_2O_2 remain unaltered in the transfected cells. Both *hGSTA1*- and *hGSTA2*-transfected cells are remarkably well protected against lipid peroxidation. Even during oxidative stress (treatment with Fe^{2+}/H_2O_2), levels of malondialdehyde (MDA) as well as 4-HNE in *hGSTA1*- or *hGSTA2*-transfected cells are lower than those in the wild-type or empty vector–transfected cells.[20] When treated with H_2O_2, the wild-type and empty vector–transfected cells, which show a dramatic increase (>3-fold) in lipid peroxidation, undergo apoptosis (Figure 10.2) accompanied with a sustained activation of JNK (Figure 10.3A, B) and caspase 3 (Figure 10.3). By

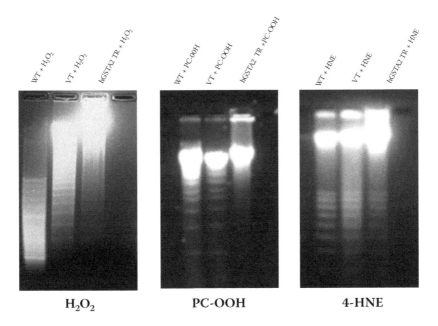

H₂O₂   **PC-OOH**   **4-HNE**

FIGURE 10.2 Overexpression of hGSTA2-2 in K562 cells blocks H₂O₂ and PC-OOH but not 4-HNE–induced apoptosis. The wild-type, vector-transfected, and hGSTA2-transfected K562 cells were treated with 30 μM H₂O₂ for 48 hours, 40 μM PC-OOH for eight hours, or 40 μM 4-HNE for eight hours in RPMI complete medium. After the incubations, genomic DNA was extracted and electrophoresced on 2% agarose gel. Lanes representing the wild-type (WT), vector-transfected (VT), and *hGSTA2*-transfected K562 (*hGSTA2* TR) cells are marked. Apoptosis is indicated by the characteristic DNA laddering.

contrast, *hGSTA1*- or *hGSTA2*-transfected cells upon identical treatment with H_2O_2 show only minimal lipid peroxidation, are resistant to apoptosis (Figure 10.2), and show no activation of JNK and caspase 3 (Figure 10.3).

The protective effect of hGSTA1-1 or hGSTA2-2 overexpression against H_2O_2-mediated lipid peroxidation and apoptosis must be exerted through their GPx activity toward PL-OOH/FA-OOH because these enzymes do not display any detectable GPx activity toward H_2O_2. This would strongly suggest that signaling for apoptosis by H_2O_2 is transduced through lipid hydroperoxides or their downstream products in the cascade of lipid peroxidation reactions. This is consistent with the results of studies showing that phosphatidylcholine hydroperoxide (PC-OOH) per se can cause apoptosis in the wild-type and vector-transfected K562 cells (Figure 10.2), but hGSTA1- and hGSTA2-transfected cells are resistant to PC-OOH–induced apoptosis. The transfected cells are, however, not protected against 4-HNE–induced apoptosis (Figure 10.2) because 4-HNE is downstream to PL-OOH in the cascade of lipid peroxidation reactions. Further support for the idea that lipid peroxidation products are involved in stress-mediated signaling for apoptosis and that these products can be regulated by GSTs is provided by unpublished studies in our laboratory that show that overexpression of hGSTA1-1 or hGSTA2-2 protects various cell types from UVA-induced apoptosis. Likewise, hGSTA1-1– and hGSTA2-

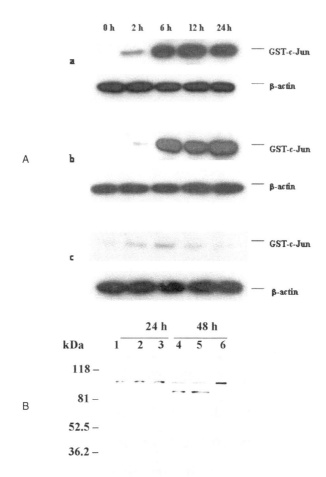

FIGURE 10.3 Inhibition of H$_2$O$_2$-induced SAPK/JNK and caspase 3 activation by hGSTA2-2 in K562 cells. Panel A: K562 cells were incubated with 30 μM H$_2$O$_2$ for the indicated times. Cell extracts containing 250 μg proteins from the wild-type (a), vector-transfected (b), and *hGSTA2*-transfected (c) cells were incubated overnight with 2 μg of GST-c-Jun (1-89) fusion protein. After extensive washing, the kinase reaction was performed in the presence of 100 μM of cold ATP and phosphorylation of c-Jun at Ser 63 was detected by Western blot analysis using Phospho-c-Jun (Ser63) antibody. β-actin expression was used to ascertain that similar amounts of protein were incubated with c-Jun. Panel B: Cells were incubated with 30 μM H$_2$O$_2$ in the medium for the indicated times. Cell lysates were subjected to Western blot analysis using the monoclonal antibody against caspase 3 substrate PARP, which recognizes the full length PARP (116 kDa) as well as its 89-kDa fragment. Lanes 1, 2, and 3 — lysates from the wild-type, vector-transfected, and hGSTA2-transfected cells, repectively — were treated with H$_2$O$_2$ for 24 hours. Lanes 4, 5 and 6 — lysates from the wild-type, vector-transfected, and *hGSTA2*-transfected cells, respectively — were treated with H$_2$O$_2$ for 48 hours.

2–overexpressing cells are resistant to apoptosis by oxidative stress-causing agents such as xanthine/xanthine oxidase, adriamycin, and naphthalene.

10.4.2 INHIBITION OF THE ACTIVATION OF JNK AND CASPASE 3

The protective effect of GSTA1-1 and GSTA2-2 against apoptosis caused by stress does not seem to be limited to a particular cell line. Unlike K562 cells that grow indefinitely in suspension, human lens epithelial cells (HLE B-3) have finite life spans and grow as attached cells.[61] Wild-type and empty vector–transfected HLE B-3 cells undergo apoptosis when oxidative stress-causing agents such as H_2O_2 or naphthalene are included in the medium. However, the cells overexpressing hGSTA2-2 are resistant to H_2O_2- or naphthalene-induced apoptosis under these conditions (Figure 10.4A). Overexpression of hGSTA2-2 suppresses lipid peroxidation caused by these oxidants, inhibits the activation of JNK, caspase 3 (Figure 10.4B), and provides protection against apoptosis. These studies provide further evidence for the involvement of PL-OOH or their downstream products such as 4-HNE in reactive oxygen species (ROS)–mediated signaling for apoptosis in cell types of differing origin and suggest that lipid peroxidation, a consequence of stress, may be a common factor in the mechanisms of the signaling for apoptosis by oxidative stress, chemical agents, and UV irradiation. Therefore, GST isozymes that limit lipid peroxidation through their Se-independent GPx activity may be considered as important regulators of stress-induced signaling for apoptosis. This novel physiological role of GSTs in regulation of signaling may also be relevant to their well-established pharmacological role in protective mechanisms against chemical carcinogenesis.

10.5 MODULATION OF 4-HNE–MEDIATED SIGNALING BY GSTS

As discussed in the preceding sections, there is ample evidence that 4-HNE is a key signaling molecule and is involved in the mechanisms of stress-mediated signaling for apoptosis. This would imply that cells overexpressing 4-HNE–metabolizing GSTs should be protected against stress-induced apoptosis. Likewise, differentiation of cells,[4] activation of various kinases,[1–3,8–11,15] and caspases[6,7] by 4-HNE and its concentration-dependent effects on signaling processes should also be affected by levels of the expression of 4-HNE–metabolizing GSTs. Recent *in vitro* studies with cell cultures discussed in the following sections are consistent with these predictions and provide strong evidence for the involvement of GSTs in the mechanisms for regulation of signal transduction.

10.5.1 OVEREXPRESSION OF 4-HNE–METABOLIZING GSTS PROMOTE PROLIFERATION

In studies by Cheng et al.,[4] K562 cells stably transfected with mGSTA4 cDNA were shown to have about 5-fold higher GST activity toward 4-HNE as compared to the controls. 4-HNE levels in *mGSTA4*-transfected cells were remarkably lower as

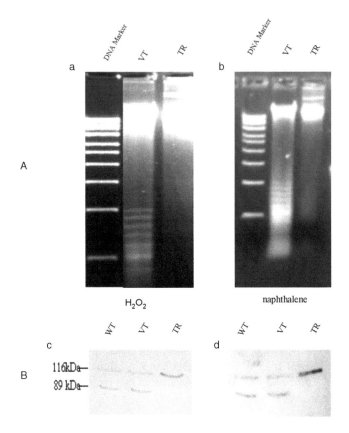

FIGURE 10.4 Inhibition of H_2O_2- and naphthalene-induced apoptosis and caspase activation by hGSTA1-1 in HLE B-3 cells. Panel A: The vector-transfected and *hGSTA1*-transfected HLE B-3 cells were treated with 100 µM H_2O_2 for six hours (a) or 200 µM naphthalene for 48 hours (b) in the culture medium and apoptosis was detected by DNA laddering assay. Lanes contained DNA markers; genomic DNA from vector-transfected cells (VT), and genomic DNA from *hGSTA1*-transfected cells (TR). Panel B: Effect of hGSTA1-1 overexpression on H_2O_2, and naphthalene-induced caspase 3 activation (a and b) and PARP cleavage (c and d). Cells (1×10^6) were incubated with 100 µM H_2O_2 (a and c) in the medium for six hours or with 200 µM naphthalene in the medium for 48 hours (b and d). Lanes containing lysates from the wild-type cells (WT), vector-transfected cells (VT), and *hGSTA1*-transfected cells (TR) are marked. Western blots were performed using antibodies against caspase 3 (a and b) and PARP (c and d).

compared to those observed in the empty vector–transfected or wild-type cells, suggesting that overexpression of this enzyme led to a significant depletion of cellular 4-HNE. *mGSTA4*-transfected cells did not show any noticeably compensatory effect on the expression of other GST isozymes or the conventional antioxidant enzymes such as CAT, SOD, and GPx. Surprisingly, *mGSTA4*-transfected cells showed a 50% higher growth rate as compared to their wild-type or empty vector–transfected counterparts, suggesting that lowering the intracellular 4-HNE levels promoted their proliferation. Promotion of proliferation of cells overexpressing 4-HNE-metabolizing GSTs has also been observed in studies with other cell lines. For example, HL-60 and HLE B-3 cells transfected with *mGSTA4*[31] and *hGSTA4*,[5] respectively, grow at a faster rate than the wild-type or vector-transfected cells due to the depletion of 4-HNE by GSTs. In earlier studies[30] proliferation of cells observed upon inclusion of low levels of 4-HNE in the culture medium is also consistent with the idea that lower intracellular levels of 4-HNE may lead to their proliferation. Thus the regulation of 4-HNE homeostasis by GSTs may be highly relevant to cell cycle signaling events.

10.5.2 INHIBITION OF DIFFERENTIATION AND APOPTOSIS

The significance of 4-HNE homeostasis in signaling is further suggested by studies showing that when wild-type K562 cells are exposed to 20 μM 4-HNE in the culture medium, the cells undergo a marked erythroid differentiation. However, the cells overexpressing mGSTA4-4 and consequently having lower intracellular concentrations of 4-HNE do not undergo such differentiation (Figure 10.5). These studies[4] not only indicate a role of 4-HNE in signaling for differentiation but also demonstrate that it can be regulated by GSTs through modulation of the intracellular concentrations of 4-HNE. Exposure of the wild-type or vector-transfected K562 cells for

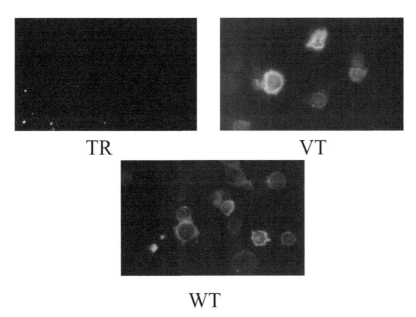

FIGURE 10.5 Erythroid differentiation of K562 cells by 4-HNE. Wild-type (WT), vector-transfected (VT), and *mGSTA4*-transfected (TR) K562 cells were treated with 20 μM 4-HNE in PBS for two hours followed by addition of growth medium and incubated for 16 hours. Erythroid cells were detected by immunochemical staining using anti-hemoglobin antibodies.

longer periods or a higher concentration of 4-HNE in the medium results in a marked increase in apoptotic cells. In contrast, the cells transfected with *mGSTA4* are relatively resistant to 4-HNE–induced apoptosis under identical conditions.[4] More important, the *mGSTA4*-transfected cells also show resistance to H_2O_2-induced apoptosis, which implies that the signaling for H_2O_2-induced apoptosis is in part conveyed through 4-HNE. Because the transfection with *mGSTA4* does not affect antioxidant enzymes such as CAT, GPx, and SOD, the apoptotic effect of H_2O_2 would be blocked by the overexpression of mGSTA4 only if 4-HNE were directly involved in signaling.

10.5.3 INHIBITION OF THE ACTIVATION OF JNK AND CASPASES

There is strong evidence that a sustained activation of JNK precedes apoptosis.[31] It has been shown that the prolongation of TNF-α–induced JNK activation by incubating cells with TNF-α in the presence of cycloheximide, actinomycin D, or orthovandadate leads to apoptosis.[62] A sustained activation of JNK achieved by inhibiting the expression of mitogen-activated protein kinase phosphatase-1 (MKP-1), a dual specific phosphatase that inactivates JNK, has been shown to potentiate TNF-α–induced apoptosis.[63] A sustained activation of JNK is also observed during H_2O_2-induced apoptosis in K562 and HL-60 cells. However, these cells can be protected from H_2O_2-induced apoptosis by overexpression of 4-HNE–metabolizing isozyme mGSTA4-4, which results in lower intracellular levels of 4-HNE and blockage of JNK activation.[31]

Studies from our laboratory indicate a role of 4-HNE in the activation of JNK and caspase 3 in several cell lines of diverse origin. We have shown that the wild-type or empty vector–transfected HL-60 cells undergo a substantial degree of apoptosis within two hours when 20 μM 4-HNE is included in the medium. In contrast, cells transfected with mGSTA4, having an enhanced capacity to metabolize 4-HNE and therefore lower intracellular levels of 4-HNE, do not undergo apoptosis even after a prolonged period of exposure to 4-HNE.[31] In the wild-type or vector-transfected cells, 4-HNE–induced apoptosis is preceded by a sustained activation of JNK, and an increase in AP-1–DNA binding is observed within two hours. These effects of 4-HNE are significantly delayed in *mGSTA4*-transfected cells. 4-HNE–treated wild-type cells show caspase 3 activation within two hours, while a barely detectable activation of caspase 3 is seen in mGSTA4-transfected cells only after eight hours.[31] Besides reaffirming a key role of 4-HNE in the signaling events upstream to the activation of JNK and caspases, these studies also indicate that GSTs, particularly those that either limit 4-HNE formation or accelerate its metabolism, can modulate oxidative stress-induced signaling. It is well known that different cells respond differentially to oxidative stress-induced apoptosis. For example, H_2O_2 treatment causes apoptosis in K562 cells, but under the similar conditions of H_2O_2 treatment RPE cells are resistant to apoptosis. It is possible that levels of antioxidant enzymes, particularly the α-class GSTs, may be at least partially responsible for the differential response of cells to stress agents.

10.5.4 ACCELERATED METABOLISM OF 4-HNE BY GSTs PROTECTS AGAINST STRESS-INDUCED APOPTOSIS

The majority of 4-HNE is metabolized through its GST-catalyzed conjugation to GSH. The two GST isozymes, hGST5.8 and hGSTA4-4, catalyzing the conjugation of 4-HNE to GSH, are present in humans.[50] The conjugate of 4-HNE and GSH (GS-HNE) must be transported out of the cell because it can inhibit GSTs and other GSH-linked enzymes and can impair a sustained detoxification of 4-HNE and other electrophiles. It has been suggested that the ATP-dependent transport of GS-HNE is mediated by multidrug resistance-associated protein 1 (MRP1). However, our recent studies have demonstrated that in a variety of cell lines of human origin, the majority (approximately 70%) of the ATP-dependent transport of GS-HNE is mediated by a previously described Ral-interacting GTPase-activating protein (RALBP1 or RLIP76) while MRP1 accounts only for less than 30% of this transport.[59,64,65] We have also demonstrated that GSTs and RLIP76 are the major determinants of the intracellular concentrations of 4-HNE, which is conjugated to GSH by GSTs and the resulting conjugate, GS-HNE, is subsequently transported by RLIP76.[6,7] Physiological significance of GSH conjugates of endogenous electrophiles and the transporters involved in their ATP-dependent efflux is covered in Chapter 11 of this volume.

Our studies[6,7] show that when human cells from varying origins are exposed to low levels of H_2O_2, heat (42°C), or mild UVA irradiation, a transient increase in lipid peroxidation and consequently the formation of 4-HNE are observed along with a rapid induction of hGST5.8 and RLIP76. Cells preexposed to these mild stress conditions and rested for two hours (stress-preconditioned cells) acquire the capability to exclude 4-HNE from their intracellular environment at an accelerated rate because of the increase in conjugation of 4-HNE by hGST5.8 and efflux of GS-HNE by RLIP76. Consequently, these stress-conditioned cells acquire resistance to 4-HNE toxicity. Studies with stress-preconditioned cells show that these cells acquire resistance not only to 4-HNE–induced apoptosis but also to H_2O_2 or O_2- and UVA-induced apoptosis, suggesting that 4-HNE is a common denominator in signaling for apoptosis induced by these agents (Figure 10.6). These findings further reaffirm the role of 4-HNE in signaling for stress-induced apoptosis and demonstrate that GSTs modulate this signaling by regulating 4-HNE homeostasis through its conjugation to GSH and subsequent efflux of GS-HNE by RLIP76.[6,7]

10.5.5 INHIBITION OF GS-HNE EFFLUX POTENTIATES STRESS-INDUCED APOPTOSIS

If stress-preconditioned cells do acquire resistance to stress-induced apoptosis by accelerating the exclusion of 4-HNE from the cells, then this resistance should be compromised upon blockage of the efflux of GS-HNE from these cells by anti-RLIP76 antibodies that have been shown to inhibit transport of GSH conjugates.[65] Consistent with this prediction, when stress-preconditioned cells are coated with anti-RLIP76 IgG, the efflux of GS-HNE is blocked and the cells become sensitive to 4-HNE, H_2O_2, or O_2- or UVA-induced apoptosis (Figure 10.6). Mild stress pre-

A   B

C   D

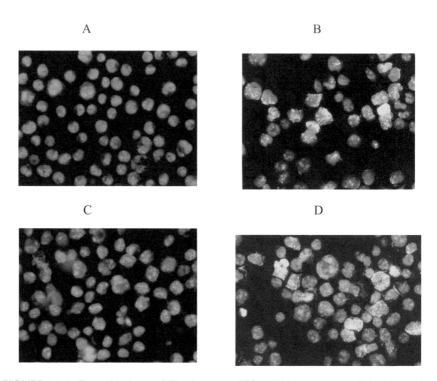

FIGURE 10.6 **(See color insert following page 178.)** Mild stress preconditioning confers resistance to 4-HNE-induced apoptosis, which is abrogated by anti-RLIP76 IgG. Cells were fixed onto poly-L-lysine-coated slides and the TUNEL apoptosis assay was performed to detect the apoptotic cells. The slides were analyzed by fluorescence microscope (photomicrographs at 400× magnification are presented). Apoptotic cells showed characteristic green fluorescence. A: control K562 cells pretreated with heat shock (42°C, 30 minutes) and allowed to recover for two hours at 37°C. B: control cells without heat shock pretreatment, incubated with 20 µM 4-HNE for two hours. C: cells pretreated with heat shock, allowed to recover for two hours at 37°C, followed by incubation in medium containing 20 µM 4-HNE for two hours at 37°C. D: heat-shock pretreated cells, allowed to recover for one hour at 37°C after which anti-RLIP76 IgG was added to medium (20 µg/ml final concentration) and incubated for an additional hour. Cells were then incubated for two hours at 37°C in medium containing 20 µM 4-HNE. (Reprinted with permission from *J. Biol. Chem.* 276, 41213, 2001.)

conditioning by a short exposure to UVA, heat shock, or H_2O_2 resulting in the induction of hGST5.8 and RLIP76 and acquisition of resistance against oxidative stress-mediated apoptosis is observed in a variety of cell lines of human origin (e.g., HL-60, H69, HLE B-3, RPE, H-226), which suggests that the involvement of 4-HNE in oxidative stress-mediated signaling for apoptosis is not limited to a specific cell type but appears to be a generalized phenomenon[6,7] and that GSTs in conjunction with GS-HNE transporter RLIP76 can modulate stress-mediated signaling in these cells by controlling 4-HNE homeostasis.

10.5.6 DEPLETION OF 4-HNE BY GSTs LEADS TO TRANSFORMATION AND IMMORTALIZATION OF SOME ADHERENT CELLS

Evidence for a pivotal role of GSTs in the modulation of cell cycle signaling is suggested by studies showing that the transfection of HLE B-3 cells with 4-HNE–metabolizing enzyme hGSTA4-4 results in the transformation, anchorage-independent rapid growth, and immortalization of these cells.[5] HLE B-3 cells are human lens epithelium cells transformed with SV-40/adenovirus, grow as adherent cells, and have finite life spans of up to 76 doublings.[61] When these cells are transfected with hGSTA4, as expected, the intracellular level of 4-HNE goes down due to its accelerated conjugation to GSH. Surprisingly, *hGSTA4*-transfected cells with lower 4-HNE levels become transformed as round, smaller cells that detach from the surface and grow indefinitely in suspension with an accelerated growth rate.[5] Cells transfected with the enzymatically inactive mutant of *hGSTA4* do not undergo this transformation and grow like the vector-transfected cells, which specifically link the transforming effects of hGSTA4-4 to its activity toward 4-HNE. That the depletion of intracellular 4-HNE by GSTs is essential for their transforming activity is further indicated by results of studies in which active hGSTP1-1, Drosophila GST DmGSTD1-1, and mouse mGSTA4-4 were microinjected in HLE B-3 cells. Both DmGSTD1-1[49] and mGSTA4-4[45] are functionally similar to hGSTA4-4 and have high catalytic efficiency to conjugate 4-HNE to GSH. On the other hand, hGSTP1-1 has only minimal activity toward 4-HNE. Incorporation of hGSTP1-1 in HLE B-3 cells did not cause transformation and the cells maintained their original phenotype. By contrast, incorporation of mGSTA4-4 or DmGSTD1-1 resulted in transformation of cells similar to that caused by transfection with hGSTA4 cDNA or microinjection with enzymatically active hGSTA4-4. Another adherent cell line, CCL-75, also underwent transformation similar to HLE B-3 cells when active hGSTA4-4 was microinjected in the cells.[5] Microinjection of hGSTA4-4 mutant Y212F, which has no significant activity toward 4-HNE but maintains its activity toward 1-chloro-2,4-dinitrobenzene, also did not transform the cells, which further confirmed that conjugating activity for 4-HNE was required for transformation.

Studies discussed above underscore the physiological importance of GSTs in the maintenance of cellular 4-HNE homeostasis, which is crucial to the mechanisms of cell cycle signaling. These results not only define an important physiologic role of GSTs in the mechanisms of signaling but they also interject some novel concepts in the mechanisms of stress-mediated signaling. It is known that oxidants, UVA exposure, starvation, heat, etc., can cause lipid peroxidation. Apoptosis caused by these stressors has been linked to ROS, but how ROS initiates the cascades for signaling is not clear. If we hypothesize that ROS-induced lipid peroxidation is the initial event in stress-induced apoptosis and that lipid peroxidation products, including 4-HNE and possibly other α,β-unsaturated alkenals having Michael acceptor groups, act as the signaling molecules, a much larger diversity can be envisaged in the mechanisms of signaling in response to ROS. Intracellular concentrations of 4-HNE or similar Michael acceptor molecules may be important in the initiation of the signaling mechanisms; GSTs, together with GSH, may perhaps globally affect the expression of relevant genes by regulating the concentration of these signaling molecules. In fact, studies discussed below suggest that there is a profound change in the expression of key cell cycle genes in HLE B-3 cells in which 4-HNE is depleted by hGSTA4-4.

10.5.7 OVEREXPRESSION OF HGSTA4-4 AFFECTS EXPRESSION OF A MULTITUDE OF GENES IN HLE B-3 CELLS

It is well documented that HNE can affect expression of key regulatory genes such as cyclins and genes involved in the pRb/E2F pathway.[8,9] Our studies[5] demonstrate that in HLE B-3 cells overexpressing active hGSTA4-4 and consequently having lower steady-state 4-HNE levels, the expression of transforming growth factor beta 1 (TGFβ1), cyclin-dependent kinase (CDK2), protein kinase C beta II (PKC βII), and extracellular signal-regulated kinase (ERK) is upregulated.[5] On the other hand, the expression of p53 is almost completely suppressed in these cells. These results suggest that some of the more prominent genes suggested to be involved in promoting proliferation are upregulated upon depletion of 4-HNE in HLE B-3 cells. This, along with an almost complete suppression of p53, may account for the observed 3-fold higher growth rate of the transformed HLE B-3 cells obtained by overexpression of 4-HNE–metabolizing GSTs. Studies on the effects of 4-HNE depletion using cDNA microarrays indicate a profound effect on somewhat global gene expression (Table 10.3) in HLE B-3 cells overexpressing hGSTA4-4, further indicating a role of GSTs in the mechanisms of signaling.[66]

10.5.8 GSTs CAN MODULATE SIGNALING BY INTERACTING WITH SAPK/JNK

It has been suggested that GSTs can interact with JNK and consequently affect JNK-mediated signaling for apoptosis. It has been shown that the monomeric form of GSTP 1-1 binds to JNK and inhibits its activation.[67] However, during UVA exposure or oxidative stress, GSTP1 subunits oligomerize and disassociate from JNK, resulting in its activation and triggering apoptosis. The role of the μ class of GST isozyme MGST 1-1 has also been suggested in the regulation of stress-mediated signaling for apoptosis.[68] Mouse enzyme mGSTM 1-1 can bind to ASK1 (apoptosis signal-regulating kinase, also identified as MAPKKK), which is known to activate JNK/SAPK and the p38 signaling cascade.[69] mGSTM 1-1 overexpression in cultured cells has been shown to suppress stress-induced activation of ASK-1, block oligo-merization of ASK-1, and protect against ASK-1–dependent apoptosis. It has been proposed that during stress, GSTM 1-1 may also be induced to provide protection against stress-induced apoptosis.[68] It is to be considered that if GSTM 1-1 does indeed regulate the stress-mediated apoptosis, then about 50% of the Caucasian population with deleted *GSTM1* may lack this regulatory mechanism. Alternatively, other GST isozymes may fill this void because the role of other GSTs besides the π and μ classes in the regulation of the various kinases is possible. Further studies

TABLE 10.3

Changes in Gene Expression in HLE B-3 Cells Transformed by Transfection with hGSTA4

Gene Name	Genebank ID Number	Fold Change	Function
Cyclin-dependent kinase inhibitor 2A (melanoma, p16)	NM_000077	−895	Inhibitor of CDK4 kinase; stabilizer of the tumor suppressor protein p53; regulate cell cycle G1 progression
Amyloid beta (A4) precursor protein (APP)	NM_000484	−717	Involved in aging, Alzheimer's disease, and cerebroarterial amyloidosis
Laminin, gamma 1 (LAMC1)	NM_002293	−598	Cell adhesion molecule activity; structural molecule activity
Gap junction protein, alpha 1, 43kDa (connexin 43)	NM_000165	−186	Major protein of gap junctions in the heart; ion transporter activity; connexion channel activity
Integrin, alpha 6 (ITGA6)	AV733308	−70	Cell adhesion and cell surface–mediated signaling
Fibronectin 1 (FN1)	AJ733308	−65	Involved in cell adhesion and migration processes including embryogenesis, wound healing, blood coagulation, host defense, and metastasis
Integrin, alpha 7 (ITGA7)	AF072132	−58	Homophilic cell adhesion; cell-matrix adhesion; integrin-mediated signaling pathway
Tumor protein p53 (Li-Fraumeni syndrome, TP53)	NM_000546	−53	Tumor suppressor protein; cell cycle checkpoint; DNA repair; apoptosis; induction of apoptosis by hormones; DNA damage response, signal transduction resulting in induction of apoptosis; negative regulation of cell cycle
Tumor necrosis factor receptor superfamily, member 6 (Fas antigen, CD95)	AA164751	−31	A central role in the physiological regulation of programmed cell death; this receptor has also activated NF-kappaB, MAPK3/ERK1, and MAPK8/JNK and is found to be involved in transducing the proliferating signals in normal diploid fibroblast and T cells
Protein kinase C, mu (PRKCM)	NM_002742	−22	Neurological process, cell cycle progression; tumorigenesis
Transforming growth factor, alpha (TGFA)	NM_003236	−9	40% sequence homology with epidermal growth factor (EGF); competes with EGF for binding to the EGF receptor, stimulating its phosphorylation and producing a mitogenic response
Cyclin-dependent kinase inhibitor 1A (p21, Cip1)	NM_00389	8	The expression is tightly controlled by the tumor suppressor protein p53, through which this protein mediates the p53-dependent cell cycle G1 phase arrest in response to a variety of stress stimuli
c-Jun (AP-1)	NM_002228	−5	Proliferation, colony survival, cell death, apoptosis, developmental process, morphology, transformation, motility, chemotaxis

TABLE 10.3 *(Continued)*
Changes in Gene Expression in HLE B-3 Cells Transformed by Transfection with hGSTA4

Gene Name	Genebank ID Number	Fold Change	Function
Integrin, beta 1 (fibronectin receptor)	NM_033669	−4	Homophilic cell adhesion; cell-matrix adhesion
BCL2-antagonist of cell death (BAD)	U66879	−4	Positively regulates cell apoptosis by forming heterodimers with BCL-xL and BCL-2, reversing their death repressor activity
Mitogen-activated protein kinase kinase kinase 8 (MAP3K8)	NM_005204	3	Oncogenic transforming activity in cells; activates both the MAP kinase and JNK kinase pathways; activates IkappaB kinases
Cyclin D3	NM_001760	4	Forms a complex with and functions as a regulatory subunit of CDK4 or CDK6, whose activity is required for cell cycle G1/S transition
c-MYC	NM_002467	4	Replication, apoptosis, cell cycle progression, transformation
Cyclin E1	AI671049	5	Forms a complex with and functions as a regulatory subunit of CDK2, whose activity is required for cell cycle G1/S transition; overexpression of this gene has been observed in many tumors
Pim-1 oncogene	M24779	15	Proliferation, developmental process, tumorigenesis
Spleen tyrosine kinase (SYK)	NM_003177	39	Organization, developmental process, tumorigenesis
C-MYB	NM_005375	66	Apoptosis, transformation, developmental process, tumorigenesis
BMX nonreceptor tyrosine kinase	NM_001721	78	Proliferation, developmental process, apoptosis, tumorigenesis

are required to investigate this potential role of GSTs in the regulation of signal transduction.

10.6 FUTURE DIRECTIONS

Studies discussed in this chapter strongly suggest that 4-HNE is not only involved in stress-mediated apoptosis but that it can also affect cell cycle signaling events in a concentration-dependent manner; that the intracellular levels of 4-HNE may be one of the determinants for leading cells toward pathways for transformation, differentiation, proliferation, or apoptosis; and that GSTs are involved in regulating these processes. Only a small increase (approximately 50%) in 4-HNE levels triggers the activation of JNK upon UVA or heat exposure.[6,7] Conversely, a decrease in 4-HNE levels of only approximately 50% in *mGSTA4*-transfected K562 or HL-60 cells causes the cells to proliferate.[4,30] Attached HLE B-3 or CCL-75 cells show even more dramatic effects of alterations in intracellular 4-HNE levels.[5] These cells undergo transformation and grow indefinitely in an anchorage-independent manner upon depletion of only about 50% of cellular 4-HNE. These findings strongly suggest that a very narrow window of physiological levels of 4-HNE tightly controlled by GSTs exists in cells and that alterations in the 4-HNE levels beyond this narrow range may profoundly affect the cell cycle events.

We propose a very simple yet testable hypothesis (Figure 10.7) for the role of lipid peroxidation products, particularly 4-HNE in stress-mediated signaling 4-HNE is a common denominator in the mechanisms of apoptosis caused by UV, oxidative

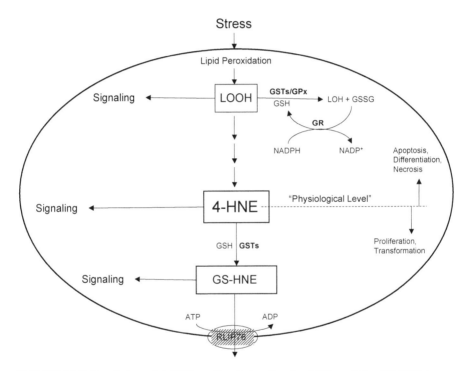

FIGURE 10.7 Regulation of 4-HNE-mediated signaling by GSTs and the GS-HNE transporter, RLIP76 (Ral BP-1). Cellular stressors (endogenous electrophiles, exogenous chemical oxidants, heat shock, UVA irradiation, etc.) induce the peroxidation of membrane lipids leading to the formation of lipid hydroperoxides (LOOH). LOOH can directly influence the signaling processes. Intracellular concentration of LOOH is regulated by seleno-glutathione peroxidase (Se-GPx) and GST A1-1/A2-2 through their GSH-dependent reduction to the corresponding alcohols. The primary α,β-alkenal product of LOOH degradation is 4-hydroxynonenal (4-HNE), which is known to have variegated effects on various signaling pathways and is conjugated to GSH primarily by the highly efficient enzymes hGST5.8 and hGSTA4-4. This conjugation forms GS-HNE, which inhibits GST activity and therefore is shuttled from the cell in an ATP-dependent manner by the membrane-associated transport protein RLIP76. Metabolites of 4-HNE such as GS-HNE and corresponding alcohol GS-DHN are also involved in signaling.

stress, and chemical agents and alterations in 4-HNE homeostasis cause profound changes in the expression of key cell cycle genes.[5–7] These studies suggest that when intracellular 4-HNE levels exceed the physiological range, a programmed cell death signal is triggered. On the other hand, proliferation is favored when 4-HNE levels drop below the normal physiological range. While this speculation does not have solid experimental evidence, it provides a platform for future studies to delineate the mechanistic basis for the intriguing hormetic effects of 4-HNE on cell cycle signaling, which seem to be tightly regulated by GSTs. It would be interesting to examine the role of other enzymes involved in the metabolism of 4-HNE and GS-HNE. The reported role of aldose reductase[58] in the modulation of signaling pathways strongly suggests that enzymes involved in the metabolism and transport of GS-HNE[6] may play a crucial role in the signaling mechanism.

The mechanisms through which 4-HNE affects signaling processes in a concentration-dependent manner are obscure and need to be investigated. Biological effects of HNE are likely to arise through its interaction with proteins. 4-HNE can react with nucleophilic groups of proteins, nucleic acids, and lipids.[70] It interacts with protein thiols and also with the nucleophilic nitrogen atoms of proteins and phospholipids, with varying affinity. It is possible that at low concentrations, 4-HNE selectively affects pathways favoring proliferation by interacting with nucleophilic groups having high affinity for this endogenous elecrophile. On the other hand, at higher concentrations of 4-HNE, the effects of these interactions may be overwhelmed by its reactions with low-affinity groups of cellular nucleophiles to trigger the pathways favoring apoptosis. There is no direct experimental evidence for this speculation but it is consistent with the studies discussed earlier in the chapter, which show that the βI and βII isoforms of phosphoinositide specific protein kinase C in several cell types are activated by submicromolar levels but are inhibited by higher levels of 4-HNE. Further studies on the possible chemical interaction of 4-HNE with cellular nucleophiles including proteins, nucleic acids, and lipids, and a possible correlation between these interactions and signaling cascades may provide clues to the mechanisms through which 4-HNE affects signaling events as well as the role of GSTs in regulating these events.

Besides the regulation of 4-HNE–mediated signaling, GSTs may also affect signaling pathways modulated by other lipid-derived ligands. As discussed in the previous chapter, GSTs are involved in the biosynthesis or metabolism of prostaglandins, leukotrienes, and steroid hormones. While prostaglandin PGD_2-, PGE_2-, and $PGF_2\alpha$-mediated signaling processes primarily involve G protein–coupled receptors, GSTs can still perhaps modulate these processes by affecting the intracellular concentrations of the prostanoids. This idea is consistent with recent studies showing that 15d-PGJ_2–induced signaling can be modulated by GSTs. The J series prostaglandin 15d-PGJ_2 can be conjugated to GSH through a reaction catalyzed by GST. This prostanoid is an activating ligand for the peroxisome proliferators-activator receptor γ (PPARγ). It has been shown that overexpression of GST inhibits transactivation of gene expression by 15d-PGJ_2 mediated by PPAR due to GST-catalyzed conjugation of 15d-PGJ_2 to GSH.[71] 15d-PGJ_2 can also stimulate Nrf2-mediated induction of gene expression[72,73] and inhibit NF-κB–dependent gene expression.[74] It is possible that GSTs may regulate these processes through conjugation of 15d-PGJ_2

to GSH. The role of GSTs in regulating these processes must be explored. Regulation of signaling processes by GSTs may perhaps be more important than realized at present because it is possible that these enzymes may be involved in the regulation of additional signaling cascades by regulating the intracellular concentrations of other potential signaling ligands having Michael acceptor groups similar to that present in 4-HNE.

ACKNOWLEDGMENT

Research for this chapter was supported in part by NIH grants EY04396 and ES012171.

REFERENCES

1. Awasthi, Y.C. et al., Regulation of 4-hydroxynonenal-mediated signaling by glutathione S-transferases, *Free Rad. Biol. Med.*, 37, 607, 2004.
2. Awasthi, Y.C. et al., Role of 4-hydroxynonenal in stress-mediated apoptosis signaling, *Mol. Aspects Med.*, 24, 219, 2003.
3. Yang, Y. et al., Lipid peroxidation and cell cycle signaling: 4-hydroxynonenal, a key molecule in stress-mediated signaling, *Acta Biochim. Polon.*, 50, 319, 2003.
4. Cheng, J.Z. et al., Effects of *mGST A4* transfection on 4-hydroxynonenal–mediated apoptosis and differentiation of K562 human erythroleukemia cells, *Arch. Biochem. Biophys.*, 372, 29, 1999.
5. Sharma, R. et al., Transfection with 4-hydroxynonenal-metabolizing glutathione S-transferase isozymes leads to phenotypic transformation and immortalization of adherent cells, *Eur. J. Biochem.*, 271, 1690, 2004.
6. Cheng, J.Z. et al., Accelerated metabolism and exclusion of 4-hydroxynonenal through induction of RLIP76 and hGST5.8 is an early adaptive response of cells to heat and oxidative stress, *J. Biol. Chem.*, 276, 41213, 2001.
7. Yang, Y. et al., Cells preconditioned with mild, transient UVA irradiation acquire resistance to oxidative stress and UVA-induced apoptosis: role of 4-hydroxynonenal in UVA–mediated signaling for apoptosis, *J. Biol. Chem.*, 278, 41380, 2003.
8. Barrera, G., Pizzimenti, S., and Dianzani, M.U., 4-Hydroxynonenal and regulation of cell cycle: effects on the pRb/E2F pathway, *Free Rad. Biol. Med.*, 37, 597, 2004.
9. Dianzani, M.U., 4-Hydroxynonenal from pathology to physiology, *Mol. Aspects Med.*, 24, 263, 2003.
10. Esterbauer, H., Schaur, R.J., and Zollner, H., Chemistry and biochemistry of 4-hydroxynonenal, malonaldehyde and related aldehydes, *Free Rad. Biol. Med.*, 11, 81, 1991.
11. Poli, G. and Schaur, R.J., 4-Hydroxynonenal in the pathomechanism of oxidative stress, *IUBMB Life,* 50, 315, 2000.
12. Zhao, T. et al., The role of human glutathione S-transferases hGSTA1-1 and hGSTA2-2 in protection against oxidative stress, *Arch. Biochem. Biophys.*, 367, 216, 1999.
13. Alin, P., Danielson, U.H., and Mannervik, B., 4-Hydroxyalk-2-enals are substrates for glutathione transferase, *FEBS Lett.*, 179, 267, 1985.
14. Hartley, D.P., Ruth, J.A., and Petersen, D., The hepatocellular metabolism of 4-hydroxynonenal by alcohol dehydrogenase, aldehyde dehydrogenase, and glutathione S-transferase, *Arch. Biochem. Biophys.*, 316, 197, 1995.

15. Srivastava, S. et al., Metabolism of lipid peroxidation product, 4-hydroxy-trans-2-nonenal, in isolated perfused rat heart, *J. Biol. Chem.*, 273, 10893, 1998.

16. Dianzani, M.U., Barrera, G., and Parola, M., 4-Hydroxy-2, 3-nonenal as a signal for cell function and differentiation, *Acta Biochim. Polon.*, 46, 61, 1999.

17. Rashba-Step, J. et al., Phospholipid peroxidation induces cytosolic phospholipase A2 activity: membrane effects versus enzyme phosphorylation, *Arch. Biochem. Biophys.*, 343, 44, 1997.

18. Suzuki, Y.J., Forman, H.J., and Sevanian, A., Oxidants as stimulators of signal transduction, *Free Rad. Biol Med.*, 22, 269, 1997.

19. Heery, J.M. et al., Oxidatively modified LDL contains phospholipids with platelet-activating factor-like activity and stimulates the growth of smooth muscle cells, *J. Clin. Invest.*, 96, 2322, 1995.

20. Yang, Y. et al., Role of glutathione S-transferases in protection against lipid peroxidation. Overexpression of hGSTA2-2 in K562 cells protects against hydrogen peroxide–induced apoptosis and inhibits JNK and caspase 3 activation, *J. Biol. Chem.*, 276, 19220, 2001.

21. Yang, Y. et al., Protection of HLE B-3 cells against hydrogen peroxide- and naphthalene-induced lipid peroxidation and apoptosis by transfection with hGSTA1 and hGSTA2, *Invest. Ophthalmol. Vis. Sci.*, 43, 434, 2002.

22. Hemler, M.E. and Lands, W.E., Evidence for a peroxide-initiated free radical mechanism for prostaglandin biosynthesis, *J. Biol. Chem.*, 255, 6253, 1980.

23. Kulmacz, R.J. and Land, W.E., Requirements for hydroperoxide by the cyclooxygenase and peroxidase activation of prostaglandin H synthase, *Prostaglandins*, 25, 531, 1983.

24. Foley, T.D., The cyclooxygenase hydroperoxide product PGG(2) activates synaptic nitric oxide synthase: a possible antioxidant response to membrane lipid peroxidation, *Biochem. Biophy. Res. Commun.*, 286, 235, 2001.

25. Ludwig, P. et al., A kinetic model for lipooxygenases based on experimental data with the lipooxygenase reticulocytes, *Eur. J. Biochem.*, 168, 325, 1987.

26. Schilstra, M.J., Veldink, G.A., and Vilegenthart, J.F.G., Kinetic analysis of the induction period in lipooxygenase cat-aylsis, *Biochemistry*, 32, 7686, 1993.

27. Schnurr, K. et al., The selenoenzyme phospholipids hydroperoxide glutathione peroxidase controls the activity of the 15-lipoxygenase with complex substrates and preserves the specificity of the oxygenation products, *J. Biol. Chem.*, 271, 4653, 1996.

28. Brigelius-Flohe, R. et al., Interleukin-1-induced nuclear factor kappa B activation is inhibited by overexpression of phospholipids hydroperoxide glutathione peroxidae in a human endothelial cell line, *Biochem. J.*, 328, 199, 1997.

29. Parola, M. et al., HNE interacts directly with JNK isoforms in human hepatic stellate cells, *J. Clin. Invest.*, 102, 1942, 1998.

30. Uchida, K. et al., Activation of stress signaling pathways by the end product of lipid peroxidation, *J. Biol. Chem.*, 274, 2234, 1999.

31. Cheng, J.Z. et al., Transfection of mGSTA4 in HL-60 cells protects against 4-hydroxynonenal–induced apoptosis by inhibiting JNK-mediated signaling, *Arch. Biochem. Biophys.*, 392, 197, 2001.

32. Yang, Y. et al., Glutathione-S-transferase A4-4 modulates oxidative stress in endothelium: possible role in human atherosclerosis, *Atherosclerosis*, 173, 211, 2004.

33. Camandola, S., Poli, G., and Mattson, M.P., The lipid peroxidation product 4-hydroxy-2,3-nonenal increases AP-1 binding activity through caspase activation in neurons, *J. Neurochem.*, 74, 159, 2000.

34. Uchida, K. and Kumagai, T., 4-Hydroxy-2-nonenal as a COX-2 inducer, *Mol. Aspects Med.*, 24, 213, 2003.

35. Chiarpotto, E. et al., Regulation of rat hepatocyte protein kinase C beta isoenzymes by the lipid peroxidation product 4-hydroxy-2,3-nonenal: a signaling pathway to modulate vesicular transport of glycoproteins, *Hepatology*, 29, 1565, 1999.

36. Fazio, V.M. et al., 4-Hydroxynonenal, a product of cellular lipid peroxidation, which modulates c-myc and globin gene expression in K5622 erythroleukemic cells, *Cancer Res.*, 52, 4866, 1992.

37. Parola, M. et al., Stimulation of lipid peroxidation or 4-hydroxynonenal treatment increases procollagen alpha 1 (I) gene expression in human liver fat-storing cells, *Biochem. Biophys. Res. Commun.*, 194, 1044, 1993.

38. Spycher, S. et al., 4-Hydroxy-2,3-trans-nonenal induces transcription and expression of aldose reductase, *Biochem. Biophys. Res. Commun.*, 226, 512, 1996.

39. Barrera, G. et al., 4-Hydroxynonenal specifically inhibits c-myb but does not affect c-fos expressions in HL-60 cells, *Biochem. Biophys. Res. Commun.*, 227, 589, 1996.

40. Leonarduzzi, G. et al., The lipid peroxidation end product 4-hydroxy-2,3-nonenal up-regulates transforming growth factor beta1 expression in the macrophage lineage: a link between oxidative injury and fibrosclerosis, *FASEB J.*, 11, 851, 1997.

41. Barrera, G. et al., Effects of 4-hydroxynonenal, a product of lipid peroxidation, on cell proliferation and ornithine decarboxylase activity, *Free Rad. Res. Commun.*, 14, 81, 1991.

42. Ruef, J. et al., Induction of rat aortic smooth muscle cell growth by the lipid peroxidation product 4-hydroxy-2-nonenal, *Circulation*, 97, 1071, 1998.

43. Marionari, U.M. et al., Role of PKC-dependent pathways in HNE–induced cell protein transport and secretion, *Mol. Aspects Med.*, 24, 205, 2003.

44. Stenberg, G. et al., Cloning and heterologous expression of cDNA encoding class alpha rat glutathione transferase 8-8, an enzyme with high catalytic activity towards genotoxic alpha, beta-unsaturated carbonyl compounds, *Biochem. J.*, 284, 313, 1992.

45. Zimniak, P. et al., Estimation of genomic complexity, heterologous expression, and enzymatic characterization of mouse glutathione S-transferase mGSTA4-4 (GST5.7), *J. Biol. Chem.*, 269, 992, 1994.

46. Zimniak, P. et al., A subgroup of class alpha glutathione S-transferases. Cloning of cDNA for mouse lung glutathione S-transferase GST 5.7, *FEBS Lett.*, 313, 173, 1992.

47. Hubatsch, I., Ridderstorm, M., and Mannervik, B., Human glutathione transferase A4-4: an alpha class enzyme with high catalytic efficiency in the conjugation of 4-hydroxynonenal and other genotoxic products of lipid peroxidation, *Biochem. J.*, 330, 175, 1998.

48. Singhal, S.S. et al., Several closely related glutathione S-transferase isozymes catalyzing conjugation of 4-hydroxynonenal are differentially expressed in human tissues, *Arch. Biochem. Biophys.*, 311, 242, 1994.

49. Sawicki, R. et al., Cloning expression and biochemical characterization of one epsilon-class (GST3) and ten delta-class (GST1) glutathione S-transferases from drosophilia melanogaster, and identification of additional nine members of the epsilon class, *Biochem. J.*, 370, 661, 2003.

50. Cheng, J.Z. et al., Two distinct 4-hydroxynonenal metabolizing glutathione S-transferase isozymes are differentially expressed in human tissues, *Biochem. Biophys. Res. Commun.*, 282, 1268, 2001.

51. Engle, M.R. et al., Physiological role of mGSTA4-4, a glutathione S-transferase metabolizing 4-hydroxynonenal: generation and analysis of mGsta4 null mouse, *Toxicol. Appl. Pharmacol.*, 194, 296, 2004.

52. Luckey. S.W. and Petersen, D.R., Metabolism of 4-hydroxynonenal by rat kupffer cells, *Arch. Biochem. Biophys.*, 389, 77, 2001.

53. Droon, J., Srivastava, S.K., and Petersen, D.R., Aldose reductase catalyzes reduction of the lipid peroxidation product 4-oxonon-2-enal, *Chem. Res. Tox.*, 16, 1418, 2003.

54. Rechard, J.F., Vasilioa, V., and Petersen, D.R., Characterization of 4-hydroxy-2-nonenal metabolism in stellate cell lines derived from normal and cirrhotic rat liver, *Biochem. Biophys. Acta,* 1487, 222, 2000.

55. Srivastava, S. et al., Metabolism of lipid peroxidation product, 4-hydroxynonenal (HNE) in rat erythrocytes: role of aldose reductase, *Free Rad. Biol. Med.,* 273, 1089, 2000.

56. Alary, J., Gueraud, F., and Cravedi. J.P., Fate of 4-hydroxynonenal *in vivo*: disposition and metabolic pathways, *Mol. Aspects Med.,* 24, 177, 2003.

57. Guerand, F., *In vivo* involvement of cytochrome P450 4A family in the oxidative metabolism of the lipid peroxidation product trans-4-hydroxy-2-nonenal, using PPA Rα-deficient mice, *Lipid Res.,* 40, 152, 1999.

58. Ramana, K.V. et al., Aldose reductase mediates mitogenic signaling in vascular smooth muscle cells, *J. Biol. Chem.,* 277, 32063, 2002.

59. Awasthi, S. et al., Novel function of human RLIP76: ATP-dependent transport of glutathione conjugates and doxorubicin, *Biochemistry,* 39, 9327, 2000.

60. Dick, R.A. and Kenster, T.W., The catalytic and kinetic mechanisms of NADPH-dependent alkenal/one oxidoreductase, *J. Biol Chem.,* 279, 17269, 2004.

61. Andley, U.P. et al., Propagation and immortalization of human lens epithelial cells in culture, *Invest. Ophthal. Visual Sci.,* 25, 3094, 1994.

62. Baker, S.J. and Reddy, E.P., Modulation of life and death by the TNF receptor superfamily, *Oncogene,* 17, 3261, 1998.

63. Guo, Y.L., Kang, B., and Williamson, J.R., Inhibition of the expression of mitogen-activated protein phosphatase-1 potentiates apoptosis induced by tumor necrosis factor-alpha in rat mesangial cells, *J. Biol. Chem.,* 273, 10362, 1998.

64. Awasthi, S. et al., RLIP76, a novel transporter catalyzing ATP-dependent efflux of xenobiotics, *Drug Metab. Dispos.,* 30, 1300, 2002.

65. Sharma, R., RLIP76 is a major ATP-dependent transporter of glutathione-conjugates and doxorubicin in human erythrocytes, *Arch. Biochem. Biophys.,* 391, 71, 2001.

66. Patrick, B. et al. Depletion of 4-hydroxynonenal in hGSTA4-transfected HLE B-3 cells results in profound changes in gene expression, *Biochem. Biophys. Res. Commun.,* 334, 425, 2005.

67. Adler, V. et al., Regulation of JNK signaling by GSTp, *Embo. J.,* 18, 1321, 1999.

68. Cho, S.G. et al., Glutathione S-transferase mu modulates the stress-activated signals by suppressing apoptosis signal-regulating kinase 1, *J. Biol. Chem.,* 276, 12749, 2001.

69. Ichijo, H. et al., Induction of apoptosis by ASK1, a mammalian MAPKKK activates SAPK/JNK and p38 signaling pathways, *Science,* 275, 90, 1998.

70. Schaur, R.J., Basic aspects of the biochemical reactivity of 4-hydroxynonenal, *Mol. Aspects Med.,* 24, 149, 2003.

71. Paumi, C.M. et al., Glutathione S-transferases (GSTs) inhibit transcriptional activation by the peroxisomal proliferators-activated receptor gamma (PPAR gamma) ligand, 15 deoxy-delta 12, 14 prostaglandin J2 (15-d-PGJ2), *Biochemistry,* 43, 2345, 2004.

72. Jowsky, I.R., Smith, S.A., and Hayes, J.D., Expression of the murine glutathione S-transferase alpha3 (GSTA3) subunit is markedly induced during adipocyte differentiation: activation of the GSTA3 gene promoter by the pro-adipogenic eicosanoid 15-deoxy-delta 12,14-prostaglandin J2, *Biochem. Biophys. Res. Commun.,* 312, 1226, 2003.

73. Itoh, K. et al., Transcription factor Nrf2 regulates inflammation by mediating the effect of 15-deoxy-delta (12, 14)-prostaglandin j (2), *Mol. Cell Biol.,* 24, 35, 2004.

74. Rossi, A. et al., Anti-inflammatory cyclopentenone prostaglandins are direct inhibitors of IKappaB kinase, *Nature,* 403, 103, 2000.

11 Glutathione-Conjugate Transport and Stress-Response Signaling: Role of RLIP76

Sharad S. Singhal and Sanjay Awasthi

CONTENTS

11.1 INTRODUCTION

The central role of stress-defense mechanisms in age-related and toxin exposure–related diseases dictates that rational preventative and therapeutic strategies
must be based on a specific and mechanistic theory that unites observations at
chemical, biochemical, and signaling levels. Glutathione (GSH), a sulfhydryl-containing ubiquitous tripeptide, is the chief cellular chemical antioxidant and nucleophile.[1] In concert with GSH-linked enzymes such as glutathione S-transferases
(GSTs) and glutathione peroxidases (GPx) GSH functions as a biochemical defense
to protect cellular macromolecules, particularly DNA, from oxidant and free-radical
damage and to guard against programmed cell death caused by various acute stresses
including radiation stress, mechanical/sheer stress, and chemical stress from oxidants, heavy metals, polycyclic-aromatic hydrocarbons, and alkylating anti-neoplas-
tic agents. GSTs and GPx have also been shown recently to directly modulate stress-
mediated signaling through effects on a wide array of stress-inducible signaling
pathways in a cell-, tissue-, organ-, and organism-specific manner. These roles of
GSH-linked enzymes, particularly GSTs, in chemical, biochemical, and signaling
defense mechanisms have been covered individually in this volume, but a unifying
model is lacking. The key role of GSTs as the rate-limiting step in metabolism of
electrophilic xenobiotics and endobiotics by catalyzing the formation of GS-elec-
trophile thioethers (GS-E), and the known potent inhibition of GSTs and glutathione
reductase by GS-E, indicate that the disposition of GS-E is crucial for proper
functioning of the GSH-linked biochemical pathways for xenobiotic and endobiotic
stress defense. We discovered that RLIP76 (a known Ral-binding and Rho-GAP
protein crucial for receptor–ligand endocytosis) is the predominant cellular GS-E
transporter, and we have found a novel link between GSH metabolism and a number
of signaling pathways. Taken together with results of studies from other investigators
on the roles of RLIP76 in a wide variety of stress-signaling pathways, our findings
of the activity of RLIP76 as a GS-E transporter have led us to propose a unifying
model for stress defense in which GS-E transport plays a central regulatory role as
a biochemical modulator of phase I and II biotransformation pathways, as a modulator of expression of stress-defense proteins, as a regulator of apoptosis though
GAP activity exhibited toward cdc42, as a mediator of chemotaxis through leukotriene effux, and as a signal-termination mechanism for a number of membrane
receptor–ligand signaling pathways (Figure 11.1).

Recent studies show that RLIP76 is a stress-responsive, multispecific membrane
transport protein, which accounts for the majority of GS-E transport in mammalian
cells including humans.[2–12] RLIP76 was previously characterized as a human
GTPase-activating protein, Ral-binding protein-1 (RALBP1).[13–15] In this chapter, to
avoid confusion we uniformly refer to this protein and to its mouse counterpart (RIP-
1)[14] or rat counterpart (RalBP1)[15] collectively as RLIP76.[2,16] There is evidence that
RLIP76 represents a link between stress-inducible protein signaling, receptor
tyrosine-kinase signaling, insulin signaling, endocytosis, heat-shock and stress-
defense pathways, and transport-mediated drug resistance.[5,17–23] Based on our studies
and other research on the biochemical and physiological functions of RLIP76 and
its links to signaling pathway components and GSH metabolism, we propose a model

for stress defenses that places RLIP76 at a strategic location as a Ral effector ATPase in these processes. In this model the rate of GS-E efflux mediated by RLIP76 is the primary determinant of the rate of clathrin-coated pit-mediated receptor–ligand endocytosis. The balance between the transport activity of RLIP76 and its GAP- and protein-binding activities directly regulates apoptosis and stress responses. In parallel, RLIP76-mediated transport of GS-E serves to limit cellular accumulation of pro-apoptotic lipid-peroxidation products such as HNE, which forms GS-HNE through a reaction catalyzed by GSTs and transported by RLIP76.[9,24] Therefore the inhibition of RLIP76 either specifically or through any xenobiotics or metabolites that share this efflux mechanism should competitively inhibit GS-E efflux, promote apoptosis, and inhibit endocytosis. Conversely, augmenting RLIP76 should inhibit apoptosis and promote endocytosis. Available evidence to support this hypothesis generated through biochemical, cellular, and *in vivo* studies is critically evaluated in this chapter against the backdrop of an abbreviated account of known proteins catalyzing the active transport of GS-E.

11.2 TRANSPORT OF GS-E

Cell membrane is impermeable to GS-E, which are amphiphilic compounds and must be extruded by active transport to avoid their toxic effects such as the inhibition of various enzymes including GSTs.[25] ATP-dependent transport of oxidized glutathione, GSSG (a molecule in which two glutathionyl residues are conjugated) from human erythrocytes, was first demonstrated in the late 1960s.[26,27] In later studies, the requirement of ATP for the efflux of the GSH-conjugate of an electrophilic xenobiotic, 1-chloro-2,4-dinitrobenzene (DNP-SG), was shown.[28] The transport of DNP-SG into inside-out vesicles (IOVs) prepared from erythrocytes was shown to be a primary active, ATP-dependent process that was saturable with respect to both ATP and DNP-SG.[29] The presence of membrane proteins having intrinsic ATPase activity that was differentially stimulated by GSSG[30] and GS-E[31,32] suggested heterogeneity in GS-E transporters present in cell membranes.

11.2.1 DNP-SG ATPᴀꜱᴇ

Identification and characterization of a transporter having ATPase activity, which was stimulated not only by GS-E such as DNP-SG but also by other xenobiotics such as doxorubicin (DOX), for the first time suggested the presence of a versatile transporter in the erythrocyte membranes, which besides GS-E could also transport cationic chemotherapeutic drugs such as DOX and daunomycin.[29,33–36] This transporter, designated as dinitrophenyl S-glutathione (DNP-SG) ATPase, was purified and the ATP-dependent transport of various GS-E and amphiphilic drugs in proteoliposomes reconstituted with purified DNP-SG ATPase was demonstrated.[33–36] These and other studies established that DNP-SG ATPase was expressed ubiquitously in various human tissues and the derived cell lines[4,5,33,37–39] and that a similar protein was present in rats.[40,41] DNP-SG ATPase was distinct from the previously reported GSSG transporter[30] and also from MRP1, p-glycoprotein (Pgp), and other known ATP-binding cassette (ABC) proteins.[42–45] The basal ATPase activity of DNP-SG

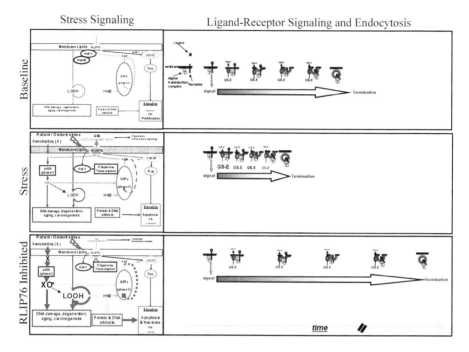

FIGURE 11.1 (See color insert following page 178.)

ATPase was stimulated in the presence of not only organic anions including GS-E (e.g., DNP-SG, leukotrienes, glucuronides, and sulfates) but also weakly cationic drugs such as DOX, vinblastine, daunomycin, and their metabolites, suggesting a promiscuous nature of this transporter.[33,35–39] Transport measurement studies in proteoliposomes reconstituted with purified DNP-SG ATPase showed that it catalyzed ATP-dependent transport of GS-E (e.g., DNP-SG) as well as of DOX and colchicine.[35,36] The transport of these substrates or allocrites by DNP-SG ATPase against a concentration gradient was saturable, temperature-dependent, sensitive to the osmolarity of the assay medium, and ATP-dependent. Transport activity of DNP-SG ATPase was not observed in the presence of methylene-adenosine triphosphate (Met-ATP), a nonhydrolyzable analogue of ATP, indicating that the energy for transport was provided by ATP hydrolysis.[9] Polyclonal antibodies raised against DNP-SG ATPase inhibited transport of both DOX and DNP-SG in IOVs prepared from erythrocyte membranes, indicating that the transport was specifically catalyzed by DNP-SG ATPase.[24,33] Immunological studies also demonstrated that DNP-SG ATPase was distinct from the known ABC proteins MRP1 or Pgp.[34]

Attempts to purify DNP-SG ATPase consistently showed that this protein was prone to degradation; depending on the conditions of purification and the nature of detergents used to extract it from membranes, peptides of varying chain lengths were observed in denaturing SDS gels.[2,3] Thus for over a decade the molecular identity of DNP-SG ATPase could not be established due to the lack of purified protein. The purified preparations of DNP-SG ATPase consistently showed a 38-kDa peptide

FIGURE 11.1 (See figure, facing page.) The role of RLIP76 in stress signaling and recep-tor–ligand signaling and endocytosis under baseline, stressed, and RLIP76-inhibited conditions. In the absence of significantly increased oxidative or radiant stress, a baseline level of lipid-hydroperoxide generation from membrane lipids gives rise to gentoxic lipid peroxy radicals through the chain reaction of lipid peroxidation. 4-Hydroxynonenal (HNE) and other α,β-unsaturated lipid aldehydes as well as peroxides and epoxides are metabolized by GSTs to GS-E, which are effluxed by RLIP76 (and other ABC-family GS-E transporters). The concen-tration of GS-E is maintained low, minimizing product inhibition, and permitting optimal GST activity. Because of low ambient GS-E formation, the activities of RLIP76 are optimally balanced between transport and GAP (toward cdc42), thus inhibiting this pro-apoptotic path-way. Under conditions of low HNE, DNA adducts formed from reactive lipid-peroxidation products, as well as protein adducts formed by the α,β-unsaturated aldehydes, is kept low and proliferation is favored. Under these conditions, signaling through ligand–receptor interactions (i.e., EGF/EGF-R, insulin-insulin-R) proceeds normally and initiation of signal termination by endocytosis is carried out at a baseline rate, determined by the level of RLIP76 transport activity toward GS-E. Under conditions of stress (oxidant, radiant, etc.), lipid peroxidation is increased, resulting in a greater amount of genotoxicity and a greater amount of HNE formation, which is catalyzed to GS-E by GST. Because ambient GS-E rise, some product inhibition of GSTs is expected initially until RLIP76 activity is increased (within minutes, likely as a consequence of PKC-mediated RLIP76 phosphorylation). If xenobiotic toxins (i.e., polycyclic aromatic hydrocarbons, chemotherapy drugs, carcinogens) are also present under these conditions, because of competition between lipid-metabolites and activated P450 metabolites (XO*) for GST-mediated conjugation with GSH, the concentration of XO* rises more rapidly and geno-toxicity is enhanced, particularly because XO* can themselves propagate lipid peroxidation. Because of increased transport substrate, GS-E, the transport activity of RLIP76 is favored over its GAP activity toward cdc42, partially releasing that inhibition. The net effect, depending on the severity of stress, would result in decreased proliferative signals and increased tendency to apoptosis, as a result of loss of cdc42 inhibition and as a result of increased electrophilic lipid or xenobiotic oxidation products. Because of increased GS-E transport, endocytosis proceeds more rapidly, thus allowing more rapid signal termination from ligands such as EGF, TGF, insulin, and others. If RLIP76 is inhibited under conditions of stress, not only is there rapid accumulation of GS-E but also inhibition of GSTs by GS-E results in accumulation of XO*, which are also formed more rapidly because of increased substrate xenobiotic due to lack of the xenobiotic-efflux activity represented by RLIP76. Under these conditions, geno-toxicity and apoptosis signaling are greatly increased and proliferation is diminished. The consequence of RLIP76 inhibition of ligand–receptor signaling is a delay in endocytotic-signal termination, thus more prolonged signaling.

fragment that displayed ouabain, and EGTA-insensitive ATPase activity that was stimulated by DNP-SG. Surprisingly, a purified preparation of this fragment cata-lyzed DNP-SG and DOX transport in reconstituted proteoliposomes, which led us to erroneously conclude that this peptide was the intact DNP-SG ATPase protein.[34,35]

11.2.2 ABC TRANSPORTERS

Studies of the transport mechanisms responsible for the drug resistance of cancer cells led to the discovery of Pgp as the first reported multispecific transporter of cationic drugs in cancer cells.[44,45] It was demonstrated that this transporter conferred a drug accumulation defect phenotype to cells and that the cells overexpressing Pgp acquired multidrug resistance (MDR). Subsequently, several members of the MDR family were characterized in humans and rodents and their transport properties toward various cationic substrates were studied using the *in vitro* models as well as the knockout animals.[46–48] These studies suggested that Pgp did not catalyze the transport of GS-E. The fact that drug accumulation defect phenotype and MDR were exhibited by a number of cell lines not overexpressing Pgp suggested the presence of other transporters besides the MDR family transporters. Using clonal genetics and novel proteomic approaches, another multispecific transporter, multidrug resistance–associated protein (MRP) was identified.[49] At least nine members of the MRP family of transporters have since been identified.[42,43] Structural studies with MDR and MRP nucleotide-binding domains led to the identification of a superfamily of transporters called ABC proteins as putative transporters in various organisms, which have characteristic Walker domain ATP-binding sites.[44,50,51] Overexpression of ABC transporters has been linked to the drug resistance of certain bacteria, parasites, and human cancer cells.[44,52,53] In the human genome at least 48 sequences have been identified that correspond to the typical ABC proteins.[54] However, only two major subfamilies of the transporters MRP and MDR have been extensively studied and their involvement in the mechanisms of multidrug resistance have been established. Details of ABC transporters and their possible role in drug resistance to chemotherapeutic agents have been covered extensively in many excellent reviews during the past decade.[42–46,50,51] Table 11.1 gives the outline of some of these transporters that are relevant to humans and rodents; these transporters are often used as suitable models to investigate the mechanisms of drug resistance.

Although the first demonstration of the ability of a single multispecific transporter to catalyze the transport of organic cations as well as anions was demonstrated for DNP-SG ATPase,[33] later studies demonstrated that MRP1 could also catalyze the transport of both cations and anions.[42,43,51] Thus the transport of GS-E in reconstituted proteoliposomes has been demonstrated only for DNP-SG and MRP1. The relative contributions of DNP-SG ATPase and MRP1 remain debatable but studies with MRP1[-/-] mice[55] suggest that they account for a relatively minor fraction of GS-E transport. Our recent studies identify DNP-SG ATPase with RLIP76 and show that RLIP76 is the predominant transporter of GS-E.[16]

11.2.3 IDENTITY OF DNP-SG ATPASE WITH A KNOWN GTPASE-ACTIVATING PROTEIN, RLIP76 (RALBP1)

The polyclonal antibodies against the 38-kDa peptide referred to as DNP-SG ATPase in our initial studies were eventually used to clone the transporter. Immunoscreening of a human bone marrow cDNA library using these antibodies consistently yielded RLIP76,[2] a previously known Ral-binding and GTPase-activating protein (GAP), which was suggested to bridge the Ral, Rac, and cdc42 pathways.[13] Similar to DNP-SG ATPase, bacterially expressed recombinant RLIP76 could be purified by DNP-SG affinity chromatography, and during the purification it underwent proteolytic degradation, yielding peptide patterns in SDS gels similar to those observed during purification of DNP-SG ATPase.[2] Extensive structural, biochemical, and functional

TABLE 11.1
Structural and Functional Characteristics of ABC Transporters Involved in Mechanisms of Multidrug Resistance

	Chromosome Localization	Family Members Identified in Human Tissues	Structural Attributes	Transport Substrates/Allocrites	Expression in Human Tissues	References
MDR family	7p21; 7p21, 1	MDR1–MDR3	Two TMDs* and two NBDs containing Walker motifs	Chemotherapeutic agents like anthracyclines, vinca alkaloids, and taxanes	Expressed in adrenal gland, kidney, brain, pancreas, colon, and small intestine	44, 46, 50, 56
MRP superfamily	16p13.1; 10q24; 17q21.3; 13q27; 16p13.1; 6p21; 16q	MRP1–MRP9	Two TMDs and two NBDs with an additional TMD0 connected with an L0 loop; these additional features are absent in MRP4, MRP5, MRP8, and MRP9; NBDs contain Walker motifs	Chemotherapeutic amphiphilic and cationic drugs like anthracyclines, vinca alkaloids; organic anions such as conjugates of glutathione, glucuronate, and sulfates, etc.; transport of chemotherapeutic agents requires GSH co-transport	Widely expressed in human tissues including liver, lung, muscle, brain, heart, and peripheral blood cells; expression ranges from low to high	42, 43, 46, 57, 68
Breast cancer resistance protein (BCRP/MXR/ABCG2)	4q22	Unknown	One TMD and one NBD with Walker motif	Chemotherapeutic drugs like mitoxantrone, anthracyclines	Placenta, intestine	69–73

* TMDs: transmembrane domains; NBDs: nucleotide binding domains.

characterization of rec-RLIP76 established that DNP-SG ATPase and RLIP76 were identical. In these studies, bacterially expressed RLIP76 showed structural and catalytic properties identical to those of DNP-SG ATPase purified from human tissues. It showed constitutive basal ATPase activity similar to that of tissue-purified DNP-SG ATPase activity that was stimulated by not only GS-E (e.g., DNP-SG), but also by other anionic as well as cationic (e.g., DOX) ligands. The K_m for the ATPase activity of RLIP76 for ATP, DNP-SG, GS-HNE, DOX, and colchicine were similar to that of DNP-SG ATPase.[2–12] Recombinant RLIP76 and tissue-purified DNP-SG ATPase showed identical profiles in SDS-PAGE. The antibodies against recombinant RLIP76 specifically recognized DNP-SG ATPase and vice versa (Figure 11.2). More important, RLIP76 when reconstituted in proteoliposomes catalyzed ATP-dependent primary active transport of DNP-SG, DOX, colchicine, daunomycin, LTC$_4$, and other allocrites with similar kinetics to that of DNP-SG ATPase.[2–5,9,12,24,36]

11.3   THE MAJOR MULTISPECIFIC TRANSPORTER OF GS-E IN MAMMALIAN TISSUES

The substrate profile of RLIP76 is clearly distinct from either Pgp or MRP1, the two most-studied ABC proteins of human tissues. While the transport of DOX and other amphiphilic cationic drugs has been demonstrated in proteoliposomes reconstituted with Pgp,[45,56] Pgp does not catalyze the transport of anionic conjugates. On the other hand, MRP1 mediates the transport of GS-E including DNP-SG, leukotrienes, GS-HNE, and other organic anions such as glucuronides.[57] MRP1 also catalyzes transport of vincristine and daunomycin, but it requires GSH cotransport.[53,58] The direct evidence for MRP1-mediated transport of DOX is lacking even though it was originally cloned from cells selected for resistance against this drug.[49] We have provided indirect evidence for MRP as a transporter of DOX in studies showing that about one-third of total DOX transport in lung cancer cells is inhibited by anti-MRP1 antibodies.[7] However, cross-reactivity of these antibodies with the half-transporter BCRP suggests that both of these transporters play a role in DOX transport. In light of our findings with RLIP76, it is also likely that previous studies of DOX and GS-E transport in crude-membrane vesicles for MRP1 transport[53,58] were in fact confounded by the presence of RLIP76, which is ubiquitously expressed in tissues. Transport studies with purified RLIP76 preparations free from any contaminated protein have clearly demonstrated that its substrate spectrum is wider than that of either Pgp or MRP1, because it can catalyze the transport of organic anions as well as cations without GSH cotransport.[35] RLIP76-overexpressing cells acquire resistance to both DOX and 4-HNE–induced cytotoxicity by accelerating the efflux of these compounds.[2] These studies, together with the ability of RLIP76 to transport leukotrienes,[9] suggest that transport catalyzed by RLIP76 has toxicological as well as physiological significance and is a modulator of drug resistance. RLIP76 is a low-affinity, high-capacity, broad substrate specificity transporter of amphiphilic small molecules that include chemotherapy drugs such as DOX, daunorubicin, vincristine, vinblastine, vinorelbine, and colchicine.[2–12,33–36] Studies with a panel of nonsmall-cell lung cancer (NSCLC) and small-cell lung cancer (SCLC) lines show

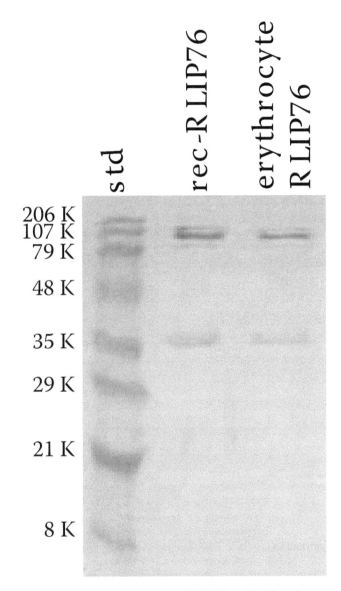

FIGURE 11.2 Western blots of rec-RLIP76 and RLIP76 purified from human erythrocytes. DNP-SG affinity purified RLIP76 (5 μg) from recombinant source and human erythrocytes were subjected to SDS-PAGE and Western blot analyses using rabbit anti-human RLIP76 IgG as a primary antibody. Western blots were developed using horseradish peroxidase-conjugated goat anti-rabbit secondary antibody with 4-chloro-1-naphthol as a substrate.

that RLIP76 accounted for more than 70% of the transport activity toward GS-E in these cells.[6–8] Likewise, in human erythrocytes, more than two-thirds of the transport activity is contributed by RLIP76 while MRP1 accounted only for a small fraction (17–30%) of this activity.[7,24]

More convincing evidence for RLIP76 being the major GS-E pump in mammalian tissues is provided by *in vivo* studies. In agreement with cell studies, RLIP76[-/-] mice have only 20% residual GS-E and DOX transport activity.[16] Western blot analysis shows the complete loss of RLIP76 cross-reacting protein in RLIP76[-/-] mice tissue indicating complete loss of RLIP76 expression in the knockout mice (Figure 11.3A). When DNP-SG and DOX transport activities were compared in the membrane vesicles prepared from the livers and hearts of the knockout and wild-type mice it was found that as compared to RLIP76[+/+] mice, the RLIP76[-/-] mice had only about 20% transport activity for GS-E (DNP-SG) as well as DOX (Figures 11.3B and 11.3C). This was consistent with the results of studies showing that RLIP76 was the major transporter of these substrates in human erythrocytes[24] and various human cell lines.[7,12] Thus in human and rodent tissues, RLIP76, rather than MRP1, is the major efflux pump for not only GS-E but also for some of the unconjugated xenobiotics such as anthracyclines used in cancer chemotherapy.

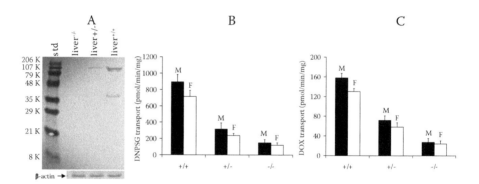

FIGURE 11.3 Effect of RLIP76 disruption on GS-E and DOX transport. RLIP76 knockout mice were generated from an embryonic stem cell clone in which the RLIP76 gene is disrupted by insertion of a retroviral vector, using Cre-Lox technology by Lexicon Genetics.[74,75] Genotyping was performed on tail tissue DNA using PCR primers designed based on the sequence around the insertion site of LTR trapping cassette.[16] Twelve-week-old animals were sacrificed and Western blot analyses (100 μg crude membrane fraction), using anti-RLIP76 IgG as primary antibodies, were performed on tissues from RLIP76[+/+], RLIP76[+/-], and RLIP76[-/-] mice. Results for liver tissues are shown. The blots were developed using 4-chloro-1-naphthol. β-actin was used as an internal control (panel A). Total transport activity for dinitrophenyl S-glutathione (DNP-SG) (panel B) and doxorubicin (DOX) (panel C) was measured as described previously[2] in inside-out vesicles (IOVs) prepared from membrane fractions. Representative results for heart tissue from RLIP76[+/+], RLIP76[+/-], and RLIP76[-/-] are presented. Statistical analyses by ANOVA were significant at $p < 0.01$ for RLIP76[+/+] vs. RLIP76[+/-], RLIP76[+/+] vs. RLIP76[-/-], and RLIP76[+/-] vs. RLIP76[-/-]. M, male; F, female.

TABLE 11.2
Effect of RLIP76 Knockout on Brain and Serum Phenytoin Levels

Genotype	Phenytoin Dose (mg/kg)	Tissue Levels (µg/mg protein)		Ratio (Serum to Brain)
		Brain	Serum	
RLIP76[+/+]	33	0.10	0.14	1.41
	83	0.10	0.35	3.53
	4166	0.05	1.12	21.53
RLIP76[-/-]	33	0.12	1.22	10.63
	83	0.29	1.24	4.25

Note: RLIP76[+/+] and RLIP76[-/-] C57B mice were administered phenytoin phosphate intraperitoneally and sacrificed two hours later. Blood and brains were collected and measurements of phenytoin and protein were performed using a multichannel analyzer.

11.4 PHARMACOLOGICAL AND PHYSIOLOGICAL SIGNIFICANCE

11.4.1 DRUG RESISTANCE

The role of RLIP76 in the drug resistance of cancer cells is suggested by studies showing that its overexpression confers resistance and drug-accumulation defects by increasing the rate of drug efflux in various types of cancer cells.[12] By contrast, specific inhibition of RLIP76 potentiates drug effects synergistically.[7,8] Likewise, depletion of RLIP76 in lung cancer cells by siRNA sensitizes these cells to DOX toxicity and apoptosis.[12] The relevance of the transport functions of RLIP76 to drug resistance in other diseases besides cancer is suggested by our recent studies showing that RLIP76 is expressed in the endothelium of the blood–brain barrier, that increased expression of RLIP76 is found in drug-resistant epilepsy, and that antiepileptic agents such as phenytoin and carbamazepine are substrates for transport by RLIP76.[59]

In the RLIP76[-/-] mice, administration of diphenylhydantoin phosphate caused status epilepticus. The brain and serum levels of phenytoin were elevated significantly in RLIP76[-/-] animals as compared with the wild-type RLIP76[+/+] animals (Table 11.2). To achieve similar serum and brain levels, a 125× greater dose of phenytoin was required in the RLIP76[+/+] animals as compared with the RLIP76[-/-] animals. These studies indicate that drugs other than chemotherapy agents are also substrates for transport by RLIP76 and that the physiological functions of RLIP76 include exclusion of drugs from the brain, as well as renal drug excretion. The ability of RLIP76 to transport xenobiotics as well as their metabolites suggests that it is one of the major components of the phase III detoxification systems.

11.4.2 INFLAMMATION

Among the inflammation recruitment signals released by stressed cells, GS-E such as leukotrienes are among the most potent. We have shown that leukotrienes are

TABLE 11.3
Genes Upregulated in the Heart Tissue of RLIP76 Knockout Mouse

Protein	(+/–) vs. (+/+)	(–/–) vs. (+/–)	(–/–) vs. (+/+)
Heat shock protein	1.09	1.53	2
Heat shock protein 1, alpha	1.38	1.36	2.19
Heat shock protein Hsp40	1.08	2.09	2.27
105-kDa heat shock protein	1.12	2.57	2.35
25-kDa mammalian stress protein 1	1.41	1.54	2.21
Stress-induced phosphoprotein 1	1.56	1.37	2.08
Insulin-like growth factor binding protein 5	2	1.72	3.62

Note: An Affymetrix murine Genome U74v2 array was used to compare RLIP76$^{+/+}$ vs. RLIP76$^{+/-}$, RLIP76$^{+/+}$ vs. RLIP76$^{-/-}$, and RLIP76$^{+/-}$ vs. RLIP76$^{-/-}$, each in duplicate and analyzed using IOBION software. Significant effects were selected by stipulating ≥ 2-fold increase and by stipulating stepwise effects defined such that the upregulation fold between RLIP76$^{+/+}$ vs. RLIP76$^{-/-}$ is within 20% of the product of the upregulation folds of RLIP76$^{+/+}$ vs. RLIP76$^{+/-}$ and RLIP76$^{+/-}$ vs. RLIP76$^{-/-}$. The seven upregulated genes satisfying these criteria are presented.

excellent physiological substrates for transport by RLIP76.[9] These studies show that RLIP76 is the predominant low affinity–high capacity transport mechanism for leukotrienes in a variety of cancer cells.[9] Likewise in human erythrocytes about 70% of LTC$_4$ transport is mediated by RLIP76.[24] In RLIP76$^{-/-}$ mice, only 20% of transport activity for GS-E is observed.[16] Microarray studies with the tissues of these mice show that in RLIP76$^{-/-}$ mice, five of the seven highest upregulated proteins are heat shock proteins (Table 11.3). This finding is consistent with the idea of a central role of RLIP76 in stress-response signaling and also with previous studies showing that RLIP76 sequesters Hsf-1, a transcription factor for chaperones.[17]

11.4.3 PROTECTION FROM HEAT SHOCK AND OXIDATIVE STRESS

Oxidative stress usually leads to enhanced generation of endogenous electrophiles, which are eventually metabolized to GS-E by GSTs. Besides being toxic, the products of lipid peroxidation, particularly 4-HNE and its GS-E metabolites, are known to be important signaling molecules.[60,61] Competitive inhibition of GS-E transport by xenobiotics[33–35] would imply that cellular signaling pathways involving oxidative stress-induced signaling may be directly interfered with by such xenobiotics. Alternatively, the disposition and pharmacological effects of such xenobiotics could be modulated by oxidative stress. A major role of RLIP76 in defense against the deleterious effects of oxidative stress is suggested by studies showing that a transient exposure to mild heat shock, oxidative stress, or UVA onto human cell lines derived from diverse origins leads to induction of RLIP76 as an early stress response even prior to the appearance of the previously known heat shock proteins or the antioxidant enzymes.[62,63] Under these stress conditions, enhanced formation of electrophilic products of lipid peroxidation is observed, and consequently GS-E levels rise. For example, when cells are exposed to a mild heat shock (42°C for 30 minutes), oxidative stress (50 μM H_2O_2 for 20 minutes), or UVA exposure and allowed to recover for two hours, enhanced levels of 4-HNE and GS-HNE, a rapid induction of hGST 5.8 (which catalyzes the formation of GS-HNE), and RLIP76 are observed. With induced RLIP76, the stress-preconditioned cells acquire at least a 3-fold higher capacity to transport GS-HNE. This increased efflux of GS-HNE can be blocked by coating the cells with antibodies against RLIP76, indicating that GS-HNE is specifically transported by RLIP76.[62] The stress-preconditioned cells with an enhanced ability to exclude GS-HNE (or GS-E in general), due to induced RLIP76, acquire resistance to 4-HNE, H_2O_2, or O_2^- and UVA-mediated toxicity. Also these cells acquire resistance to apoptosis by suppressing a sustained activation of c-Jun N-terminal kinase and caspase 3. The protective effect of stress preconditioning caused by these stress agents can be abrogated by inhibiting GS-E transport. These studies suggest that the intracellular concentrations of endogenous GS-E such as GS-HNE can affect stress-mediated signaling and that RLIP76 and GSTs are the major determinants of the intracellular levels of 4-HNE and GS-HNE. Induction of RLIP76 by mild, transient stress and the resulting resistance of stress-preconditioned cells to apoptosis appears to be a general phenomenon because it is not limited to a specific cell type but is observed in cells from diverse origins such as lung cancer cells, human leukemia cells (HL60), human retinal pigmented epithelial cells, and human lens epithelial cell lines, indicating that the transport activity of RLIP76 toward GS-E regulates the intracellular levels of 4-HNE, a lipid peroxidation product that is known to be involved in various signaling pathways.[60,61]

11.4.4 PROTECTION FROM RADIATION TOXICITY

The generation of reactive oxygen species is one of the common effects of high-energy radiation as well as from heat shock and oxidant exposure. This would imply that RLIP76 should protect against radiation toxicity. This hypothesis was tested by examining the effect of increased intracellular RLIP76 on radiation sensitivity. It has been demonstrated that RLIP76 containing proteoliposomes are avidly taken up by cultured cells, resulting in increased cellular RLIP76 levels.[2] SCLC cell lines (H182 and H1618) in culture were treated with liposomes reconstituted in the presence of purified RLIP76 prior to exposure to a single dose of radiation at 500 cGy. After radiation, cells were serially cultured by reinoculation into the medium each day at a density of 5×10^5 cells/ml and cell density was determined at 24-hour intervals. Results showed that the cells enriched with RLIP76 were significantly more resistant to radiation when compared with the control cells, suggesting that the efflux of GS-E formed from reactive intermediates of lipid peroxidation contributed significantly to radiation toxicity and that an increased efflux of GS-E from RLIP76-enriched cells rendered these cells resistant to radiation.[4]

The significance of GS efflux and the protective role of RLIP76 against X-radiation has been confirmed by *in vivo* studies. Diminished capacity to transport the GS-E in RLIP76[-/-] mice should result in increased cellular levels of GS-E and their precursors including alkenals and hydroperoxides. Consequently, RLIP76[-/-] mice should be more sensitive to radiation toxicity as compared to their RLIP76[+/+] counterparts. Consistent with these predictions, our studies have shown that

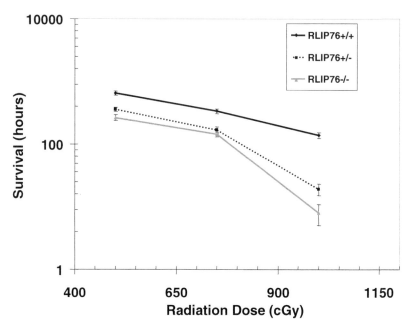

FIGURE 11.4 Radiation sensitivity conferred by loss of RLIP76 in mice. C57B mice were genotyped using tail tissue and segregated according to RLIP76$^{+/+}$, RLIP76$^{+/-}$, and RLIP76$^{-/-}$. Genotypes were randomized (six per group) to be treated with either 500, 750, or 1000 cGy whole body radiation using a Varian 2100C Linear Accelerator. Animals were placed individually in cages and monitored for survival by observations at eight-hour intervals.

RLIP76$^{-/-}$ animals have up to 2–7-fold increased tissue levels of electrophilic alkenals and hydroperoxides.[16] Upon exposure to whole body X-irradiation, the RLIP76$^{-/-}$ animals' survival is significantly shorter when compared with RLIP76$^{+/+}$ mice (Figure 11.4). Intraperitoneal administration of RLIP76 proteoliposomes leads to higher intracellular levels of RLIP76 in mouse tissues.[16] Treatment of RLIP76$^{-/-}$ animals with RLIP76 liposomes resulted in replenishment of RLIP76 levels in the tissues, including the brains of these animals, and conferred a marked survival advantage, when compared with RLIP76$^{-/-}$ mice given only blank liposomes.[16] Treatment with RLIP76 proteoliposomes even after 72 hours' exposure to X-radiation resulted in prolongation of survival.[16] Most remarkably, the RLIP76 liposomes were found to traverse the blood–brain barrier. These findings further affirm the protective role of RLIP76 against X-irradiation and disposition of GS-E and strongly suggest a potential for the use of RLIP76 liposomes for the treatment of radiation poisoning.

The ability of RLIP76 to protect against stress from chemical sources (anthracyclines, hydrogen peroxide), heat, and radiation, as well as from amphiphilic xenobiotic toxins such as vinca alkaloids used in chemotherapy, indicates that RLIP76 is a central component of cellular defense mechanisms against oxidative and xenobiotic stresses and its protective effect can be attributed to its multispecific transport activity. Induction of apoptosis on RLIP76 inhibition,[62] the known role of physiologic GS-E as signaling molecules, and the known participation of RLIP76 in the Ras/Ral/Rho/Rac G-protein signaling pathways[13–15,18–22] (which are involved in regulating cell growth, apoptosis, and membrane plasticity) point to a novel paradigm for the role of GS-E in regulating these physiologic events. This idea is supported primarily by chemical and biochemical evidence and is reinforced by independent evidence from cell biological studies on the role of RLIP76 as a Ral-binding GAP that regulates the Rho/Rac class of G-proteins, which are downstream-signaling proteins in the Ral pathway. In order to assess the role of GS-E and RLIP76-mediated transport of GS-E in cellular signaling pathways, it is important to discuss the role of Ral, RLIP76, and other downstream-signaling proteins in these processes.

11.5 A MULTIFUNCTIONAL KEY REGULATORY PROTEIN

Besides its newly discovered transport function, RLIP76 has multiple activities that affect various signaling cascades. RLIP76 was initially cloned as a Ral-effector and Ral-GAP that bridged the Ras and Ral pathways, and it displayed GAP activity toward Rho/Rac G-proteins, particularly cdc42, which is a key G-protein linked to Ras involved in regulating cell movement and apoptosis.[13–15,18–22] Recent studies indicate that RLIP76 plays a crucial role in clathrin-coated pit-mediated receptor–ligand pair endocytosis, particularly as related to TGF-α, EGF, and insulin.[13–15,17–23] Binding cdc2 to RLIP76 is essential to shut off endocytosis during mitosis, and the overexpression of POB1 (the partner of RalBP1, the first described RLIP76-binding protein) triggers apoptosis in prostate cancer cells[64,65] (Figure 11.5).

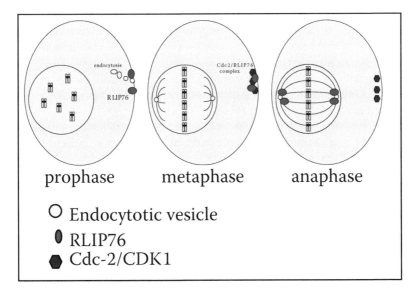

FIGURE 11.5 (See color insert.) Role of RLIP76 in the mitotic spindle. Before anaphase, cdc2 binding to RLIP76 terminates endocytosis and RLIP76 is translocated from a membrane location to the mitotic spindle in the centrosomes. Blockade of this event results in a mitotic cell death.

Drug resistance is strongly suggested because of the seminal roles of each of the above RLIP76-interacting pathways in carcinogenesis and cancer pharmacology. Furthermore, because of the relatively promiscuous nature of substrate affinity of RLIP76 with respect to transport activity, numerous amphiphilic chemotherapy drugs can function as its substrates or as competitive inhibitors of its physiological function as a transporter of GS-E. A direct implication of these findings is that the execution of endocytosis, mitosis, and apoptosis may be regulated by signaling pathways such as Ral, Ras, Rho/Rac, and CDK1 through the rate of efflux of the physiological, pro-apoptotic GS-E (e.g., GS-HNE) by RLIP76, which couples ATP hydrolysis with GS-E transport.[5] A novel integrated signaling model can be envisaged in which RLIP76 plays a key role in different signaling pathways through its GS-E transport activity. Dissociation of RLIP76 from membrane upon binding with cdc2 results in translocation to the mitotic spindle where it is purported to function as a motor for spindle movement.[23] These studies are consistent with a more general hypothesis that RLIP76 functions as a modular ATPase that provides energy to different cellular proteins by selectively binding to different adaptor proteins.

11.6   AN INTEGRATED MODEL FOR SIGNALING

The convergence of evidence from the different lines of investigation outlined above suggests that RLIP76 through as yet incompletely defined mechanisms plays a central role in signaling at several different levels:

1. GAP activity toward Rho-subfamily proteins involved in signaling for xenobiotic stress responses and apoptosis
2. Inhibiting the Ras-signaling pathway through its GAP activity toward cdc42
3. Regulating the termination of ligand–receptor signaling by clathrin-coated pit-mediated internalization
4. Regulating neurotransmitter signaling through effects in the exocyst complex
5. Regulating the termination of leukotriene and GS-HNE–mediated intra-cellular signaling through PPAR-γ
6. Transporting these molecules out of the cell to regulate intercellular signaling and chemotaxis by functioning as a conduit for release of leukotrienes from cells

The involvement of RLIP76 in these diverse processes (Figure 11.1) could be attributed to different individual functions associated with its multiple domains. However, it is possible that the newly discovered functionality of RLIP76 as an ATPase as well as a transporter of GS-E provides a common link to these diverse cellular processes by functioning as a molecular motor. Evidence in support of this idea is provided by our recent studies and by other investigators.

The considerable basal ATPase activity of RLIP76, which is upregulated by xenobiotics, may imply that RLIP76 can function as a molecular motor capable of transducing energy from nucleotide hydrolysis to facilitate diverse processes including association and dissociation of the clathrin complex, conformational changes during endocytosis, trafficking of vesicles during exocytosis, movement of the mitotic spindle, or transmembrane movement of GS-E and other xenobiotic and endobiotic substrates. The transport of endobiotics and xenobiotics may be an obligate requisite, inextricably linked to and necessary for some of these membrane events such as endocytosis and exocytosis that are associated with RLIP76 function. On the other hand, the ATP hydrolysis alone could be coupled to some of the other RALBP1-linked functions such as mitotic spindle movement. The ATPase activity of RALBP1 alone, and not its transport function, may be sufficient for its function in receptor endocytosis. However, some of the available experimental evidence argues against this possibility. The C-terminal domain of RLIP76 has been shown to inhibit both EGF and insulin-receptor internalization, and the N-terminal RLIP76 inhibits insulin-receptor internalization; however, full-length RALBP1 does not inhibit either. We have shown that both C- and N-terminals have ATP-binding sites and display ATPase activity, but neither the C- nor the N-terminal domains alone can mediate transport.[3] On the other hand, these domains combined constitute a transport function equivalent to that of intact full-length RLIP76.[3] These observations would be compatible with the hypothesis that transport function of RLIP76 is necessary for EGF/EGF-R or insulin/insulin-R downregulation through internalization, and that the observed inhibition of receptor internalization by either the N- or C-terminal RLIP76 alone could be due to a deviation from the optimal ratio of the N- and C-terminal fragments required for transport.

The hypothesis presented above predicts that proper functioning of the endocytosis pathways to terminate receptor-ligand signaling or of exocytosis to release neurotransmitters (or other molecules) would be coupled with efflux of physiological ligands (i.e., GS-E) (Figure 11.1). If so, the actions that affect the transport function of RLIP76 would affect endocytosis and exocytosis, or conversely that agents or actions that affect endocytosis or exocytosis would also affect RLIP76-mediated transport. This idea is consistent with the known effects of certain drugs such as vincristine, vinblastine, colchicine, and DOX, which are known not only to competitively inhibit RLIP76-mediated GS-E transport but also to interfere with many aspects of membrane trafficking. The finding that RLIP76 is a GS-E drug transporter fundamentally changes the paradigm regarding possible links between drug transport and resistance and stress-signaling membrane trafficking. Recent studies in our laboratory showing that alterations in the GS-E transport activity of RLIP76 and its intracellular levels affect various signaling processes are consistent with this hypothesis. These effects of RLIP76 transport activity and its intracellular levels on signaling events have been observed in cell culture systems where RLIP76 has been induced or supplemented through transfection, by liposomal delivery, or suppressed through siRNA and inhibition of its transport function. RLIP76 binds to various partner proteins during its multiple cellular functions. Binding of RLIP76 to these proteins may affect its transport functions, thus RLIP76-binding proteins may also modulate signaling processes by interfering with its transport properties. We have shown that POB1 binds to and inhibits the transport activity of RLIP76 and confers increased sensitivity to DOX through increased drug accumulation. These studies in lung cancer cells confirmed previous findings of other investigators that overexpression of POB1 causes apoptosis.[66] A retrospective evaluation of earlier studies linking drug transport and resistance with signaling, and our studies on the role of RLIP76-mediated GS-E transport supporting this link are discussed in the following sections.

11.6.1 POB1 OVEREXPRESSION INDUCES APOPTOSIS BY INHIBITING RLIP76 TRANSPORT ACTIVITY

POB1 (also referred to as REPS2) is a human RLIP76-binding protein that has been shown to cause apoptosis in prostate cancer cells upon overexpression.[64,65] POB1 contains a coiled-coil C-terminal region that binds the C-terminal region of RLIP76.[21,64–67] We have purified POB1 and its deletion mutant POB1[1-512] (lacking the RLIP76-binding domain) and examined their effects on the transport activity of RLIP76. Transport activities of RLIP76 for GS-E were inhibited by POB1 in a concentration-dependent manner but not by the deletion mutant of POB1 (POB1[1-512]), which lacks the RLIP76-binding domain. These studies demonstrate that POB1 binding of RLIP76 inhibits GS-E transport. Liposomal delivery of recombinant wild-type POB1 to H358 cancer cells caused apoptosis, due to inhibition of GS-E efflux, but liposomal delivery of the POB1 mutant deficient in the RLIP76-binding site had no effect on RLIP76-mediated transport of GS-E and did not induce apoptosis. Augmentation of cellular POB1 also results in increased intracellular DOX accumulation as well as a decreased rate of efflux from cells, further suggesting a role of RLIP76 in resistance of cancer cells to anthracyclines.[66] More important, these studies demonstrate that POB1 can regulate the transport function of RLIP76 and induces apoptosis in cancer cells through the accumulation of endogenously formed GS-E. Consistent with this finding, depletion of RLIP76 in a panel of SCLC and NSCLC cell lines by siRNA induces apoptosis.[12]

11.6.2 GS-E TRANSPORT: REGULATION OF RECEPTOR-TYROSINE KINASE SIGNALING AND ENDOCYTOSIS

During endocytosis, a complex of RLIP76/POB1/Epsin/Grb2/Nck/Src is formed in proximity to receptor-tyrosine kinases, the activation of which phosphorylates Src and other proteins.[5,13–15,18–23,67] Signaling through receptor–ligand couples of receptor-tyrosine kinases such as EGF-R and TGFβ-R is terminated though complex mechanisms beginning with clathrin-coated pit-mediated endocytosis. RLIP76 has been demonstrated by various investigators to be an integral component of endocytosis.[18–22] The multiprotein complex containing at least RALBP1, POB1, EH-domain proteins, and grb/nck or src, forms an arm that links clathrin via an AP2-adaptor protein to an EGF-receptor kinase. RLIP76 as a GS-E transporter could be a determinant of the rate of endocytosis and thus signal termination for these signaling pathways. Evidence in support of this hypothesis was provided by *in vivo* studies in our laboratory showing marked insulin sensitivity in RLIP76[-/-] mice. The ratio of plasma insulin to glucose (a marker of insulin sensitivity) was found to be lowest in RLIP76[-/-] mice, followed by a stepwise increase in RLIP76[+/-] and RLIP76[+/+] animals. Intraperitoneal administration of increasing doses of purified recombinant RLIP76 in proteoliposomes to RLIP76[+/+] animals resulted in a rise in the intracellular levels of RLIP76 and a further increase in the insulin to glucose ratio (Figure 11.6).

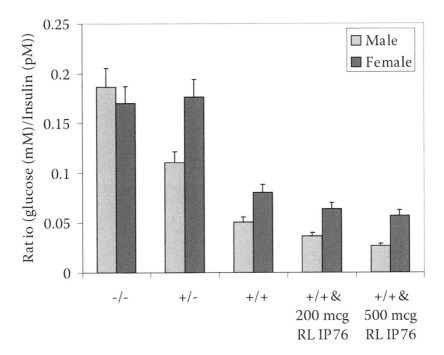

FIGURE 11.6 Glucose to insulin ratio in RLIP76-deficient and -supplemented mice. Insulin and glucose measurements were performed in serum 24 hours after intraperitoneal injection of either control liposomes to RLIP76$^{+/+}$, RLIP76$^{+/-}$, and RLIP76$^{-/-}$ animals, or RLIP-76 liposomes (equivalent to either 200 or 500 μg RLIP76 protein) to RLIP76$^{+/+}$ animals (to increase tissue levels of RLIP76 above that found in wild-type animals). Tissue delivery by this method was confirmed by Western blot analyses and immunohistochemistry. The measurements were performed at Michigan State University by Dr. Kent R. Refsal. The values are presented as mean ± SD from three separate determinations with three replicates (n = 9). The results presented show that the glucose to insulin ratio, a surrogate for insulin sensitivity, declines with increasing RLIP76 from RLIP76$^{-/-}$ to the RLIP76$^{+/+}$ animals supplemented with additional RLIP76.

The results of these *in vivo* studies are corroborated by *in vitro* studies with SCLC and NSCLC, which also showed a link between the GS-E transport activity of RLIP76 and the rate of endocytosis. Our studies showed that insulin-independent glucose uptake was significantly greater in NSCLC, which have about 2-fold GS-E transport activity mediated by RLIP76 when compared with SCLC. FITC-insulin endocytosis was increased in RLIP76-overexpressing cells that could be blocked by anti-RLIP76 IgG. Treatment with either 50 μM H_2O_2 (oxidant stress) or 25 mM glucose (glycemic stress) results in an increased endocytosis of insulin-receptor, along with marked induction of RLIP76 and enhanced GS-E transport. A stress-induced increase in endocytosis could be blocked by anti-RLIP76 IgG, which inhibits GS-E transport (Figure 11.7). The rate of endocytosis of rhodamine-tagged EGF was increased with RLIP76 overexpression in H358 cells; endocytosis was nearly abrogated upon inhibition of RLIP76 to mediate transport by antibodies in control

Pre-immune IgG   anti-RLIP76 IgG

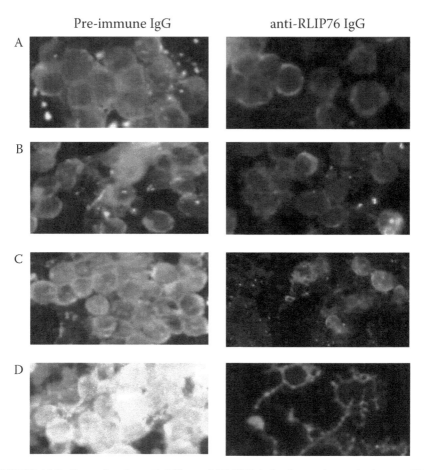

FIGURE 11.7 (See color insert.) Effect of RLIP76 induction and transfection on I/IR endocytosis. H358 control (panel A) and RLIP76-transfected (panel D) cells (0.5×10^6 cells/ml) were grown on sterilized glass cover slips (18 mm size) in RPMI 1640 medium in tissue culture treated 12 well plates overnight, followed by incubation of control cells with either 50 μM H_2O_2 (panel B) or 25 mM glucose (panel C) for 20 minutes at 37°C. Subsequently, cells were washed with PBS four to five times and resuspended in the RPMI 1640 medium and allowed to recover for two hours at 37°C in a CO_2 incubator. Cells were washed three to four times with PBS and treated with 10% goat serum for 30 minutes at room temperature and incubated with 100 ng/ml FITC-conjugated insulin for 60 minutes on ice. Cells were then incubated at 37°C in a humidified chamber for 10 minutes followed by fixation with 4% paraformaldehyde. Slides were analyzed using confocal laser scanning microscopy with the Zeiss 510 meta system, with excitation at 494 nm and emission at 518 nm.

and transfected cells (Figure 11.8). These findings are consistent with the hypothesis that RLIP76-mediated GS-E transport is crucial to the regulation of endocytosis.

11.7 OTHER ACTIVITIES AND POSSIBLE LINKS TO GS-E TRANSPORT

11.7.1 ROLES OF CALCIUM AND CDC2 IN REGULATING RLIP76 ACTIVITY

Cdc2 (CDK1) is a cell-cycle checkpoint control kinase that binds to RLIP76 during mitosis, such that endocytosis is inhibited.[18] Binding of proteins such as RLIP76 to POB1 inhibits its transport activity.[66] Our unpublished studies with recombinant His-tagged cdc2 show a significant inhibition of RLIP76 transport activity by cdc2 in a concentration-dependent manner. Interestingly, although calcium had no effect on the transport activity of RLIP76 by itself, its presence reversed cdc2-mediated inhibition. These findings suggest that cdc2 interactions with RLIP76 may modulate RLIP76 activity, perhaps differentially depending on calcium concentration. Further studies are required to study the significance of GS-E efflux to cdc2-mediated processes in relation to the role of calcium.

11.7.2 MITOSIS

At the G2-M transition, endocytosis is shut off. Recent studies have shown that the binding of cdc2 (CDK1) to RLIP76 is necessary for this to occur[18] (Figure 11.5). At the same time, the splice variant of RLIP76, cytocentrin, is found attached to the mitotic spindle at a site suggesting that it functions as an ATPase to provide energy for spindle formation and movement.[6,23] This observation is consistent with the idea that RLIP76 is a modular multifunctional protein incorporating an ATPase activity and adapter regions, which allow it to perform distinct motor functions in different contexts of protein binding and subcellular distribution.

11.7.3 G-PROTEIN SIGNALING

As described earlier, RLIP76 was initially cloned as a Ral-binding effector protein and proposed to bridge the Ras-Ral pathways.[13] It is regulated by Ral and displays GAP (inhibitory) activity toward Rho proteins, most important, cdc42, a pro-apop-totic G-protein that exerts an inhibitory effect on Ras. It is unknown whether this GAP activity is a competing or complementary activity to its ATPase and transport activities. Investigators who initially cloned RLIP76[13-15] predicted its major function to be an effector protein linking the Ras and Ral pathways.

Studies discussed in this chapter provide strong evidence for a link among various RLIP76-mediated processes to its GS-E transport activity and a novel frame-work for an integrated view of GSH-linked stress-defense mechanisms and signaling pathways. Considerable additional experimental evidence is, however, needed to elucidate the mechanisms of stress signaling and the role of the GAP, ATPase, transport, motor, and scaffold roles of RLIP76. Such studies should lead toward a

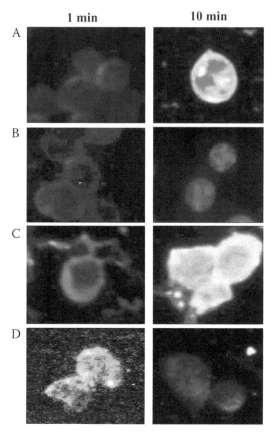

FIGURE 11.8 (See color insert.) Effect of RLIP76 on EGF internalization. EGF binding and internalization are performed in control (panels A and B) and RLIP76-transfected (panels C and D) H358 cells grown on the sterilized cover slips and treated with either pre-immune IgG (panels A and C) or anti-RLIP76 IgG (panels B and D) overnight, followed by washing with PBS four to five times. Cells were treated with 10% goat serum for 30 minutes at room temperature and were incubated with 40 ng/ml EGF-rhodamine for 60 minutes in ice (4°C). Cells were then incubated at 37°C in a humidified chamber for 1 or 10 minutes followed by fixation with 4% paraformaldehyde at the respective time. Slides were analyzed using confocal laser scanning microscopy with the Zeiss 510 meta system, with excitation at 555 nm and emission at 580 nm.

better understanding not only of the regulation of signaling but also of the mechanisms of drug resistance and stress diabetes.

ACKNOWLEDGMENT

Research for this chapter was supported in part by USPHS grants CA 77495 and CA 104661.

REFERENCES

1. Beutler, E., Nutritional and metabolic aspects of glutathione, *Annu. Rev. Nutr.*, 9, 287, 1989.
2. Awasthi, S. et al., Novel function of human RLIP76: ATP-dependent transport of glutathione conjugates and doxorubicin, *Biochemistry,* 39, 9327, 2000.
3. Awasthi, S. et al., Functional reassembly of ATP-dependent xenobiotics transport by the N- and C-terminal domains of RLIP76 and identification of ATP binding sequence, *Biochemistry,* 40, 4159, 2001.
4. Awasthi, S. et al., RLIP76, a novel transporter catalyzing ATP-dependent efflux of xenobiotics, *Drug Metab. Disp.*, 30, 1300, 2002.
5. Awasthi, S. et al., Transport of glutathione-conjugates and chemotherapeutic drugs by RLIP76 (RALBP1): a novel link between G-protein and tyrosine kinase signaling and drug resistance, *Int. J. Cancer,* 106, 635, 2003.
6. Singhal, S.S. et al., Role of RLIP76 in lung cancer doxorubicin resistance. I. The ATPase activity of RLIP76 correlates with doxorubicin and 4-hydroxynonenal resistance in lung cancer cells, *Int. J. Oncol.*, 22, 365, 2003.
7. Awasthi, S. et al., Role of RLIP76 in lung cancer doxorubicin resistance. II. Doxorubicin transport in lung cancer by RLIP76, *Int. J. Oncol.*, 22, 713, 2003.
8. Awasthi, S. et al., Role of RLIP76 in lung cancer doxorubicin resistance. III. Anti-RLIP76 antibodies trigger apoptosis in lung cancer cells and synergistically increase doxorubicin cytotoxicity, *Int. J. Oncol.*, 22, 721, 2003.
9. Sharma, R. et al., RLIP76 (RALBP1)-mediated transport of leukotriene C4 (LTC$_4$) in cancer cells: implications in drug resistance, *Int. J. Cancer*, 112, 934, 2004.
10. Yadav, S. et al., Identification of membrane anchoring domains of RLIP76 using deletion mutants analyses, *Biochemistry*, 43, 16243, 2004.
11. Stuckler, D. et al., RLIP76 transports vinorelbine and mediates drug resistance in non-small cell lung cancer, *Cancer Res.,* 65, 991, 2005.
12. Singhal, S. et al., Depletion of RLIP76 sensitizes lung cancer cells to doxorubicin, *Biochem. Pharmacol.*, 70, 481, 2005.
13. Jullien-Flores, V. et al., Bridging Ral GTPase to Rho pathways. RLIP76, a Ral effector with CDC42/Rac GTPase activating protein activity, *J. Biol. Chem.*, 270, 22473, 1995.
14. Park, S.H. and Weinberg, R.A., A putative effector of Ral has homology to Rho/Rac GTPase activating proteins, *Oncogene*, 11, 2349, 1995.
15. Cantor, S.B., Urano, T., and Feig, L.A., Identification and characterization of Ral-binding protein 1, a potential downstream target of Ral GTPases, *Mol. Cell. Biol.*, 15, 4578, 1995.
16. Awasthi, S. et al., RLIP76 is a major determinant of radiation sensitivity, *Cancer Res.*, 65, 6022, 2005.
17. Hu, Y. and Mivechi, N.F., HSF-1 interacts with Ral-binding protein 1 in a stress-responsive, multiprotein complex with HSP90 *in vivo*, *J. Biol. Chem.*, 278, 17299, 2003.
18. Rosse, C. et al., RLIP, an effector of the Ral GTPases, is a platform for cdk1 to phosphorylate epsin during the switch off of endocytosis in mitosis, *J. Biol. Chem.*, 278, 30597, 2003.
19. Moskalenko, S. et al., The exocyst is a Ral effector complex, *Nat. Cell Biol.*, 4, 66, 2002.
20. Nakashima, S. et al., Small G protein Ral and its downstream molecules regulate endocytosis of EGF and insulin receptors, *Embo. J.,* 18, 3629, 1999.

21. Ikeda, M. et al., Identification and characterization of a novel protein interacting with Ral-binding protein 1, a putative effector protein of Ral, *J. Biol. Chem.*, 273, 814, 1998.

22. Jullien-Flores, V. et al., RLIP76, an effector of the GTPase Ral, interacts with AP2 complex: involvement of the Ral pathway in receptor endocytosis, *J. Cell Sci.*, 113, 2837, 2000.

23. Quaroni, A. and Paul, E.C., Cytocentrin is a Ral-binding protein involved in the assembly and function of the mitotic apparatus, *J. Cell Sci.*, 112, 707, 1999.

24. Sharma, R. et al., RLIP76 is the major ATP-dependent transporter of glutathione-conjugates and doxorubicin in human erythrocytes, *Arch. Biochem. Biophys.*, 391, 171, 2001.

25. Awasthi, S. et al., Interactions of glutathione S-transferase π with ethacrynic acid and its glutathione conjugate, *Biochim. Biophys. Acta*, 1164, 173, 1993.

26. Srivastava, S.K. and Beutler, E., The transport of oxidized glutathione from the erythrocytes of various species in the presence of chromate, *Biochem. J.*, 114, 833, 1969.

27. Srivastava, S.K. and Beutler, E., The transport of oxidized glutathione from human erythrocytes, *J. Biol. Chem.*, 244, 9, 1969.

28. Board, P.G., Transport of glutathione S-conjugate from human erythrocytes, *FEBS Lett.*, 124, 163, 1981.

29. LaBelle, E.F. et al., Dinitrophenyl glutathione efflux from human erythrocytes is primary active ATP-dependent transport, *Biochem. J.*, 238, 443, 1986.

30. Kondo, T. et al., Glutathione disulfide-stimulated Mg2+-ATPase of human erythrocyte membranes, *Proc. Natl. Acad. Sci. USA*, 84, 7373, 1987.

31. LaBelle, E.F. et al., A novel dinitrophenylglutathione-stimulated ATPase is present in human erythrocyte membranes, *FEBS Lett.*, 228, 53, 1988.

32. Singhal, S.S. et al., The anionic conjugates of bilirubin and bile acids stimulate ATP hydrolysis by S-(dinitrophenyl) glutathione ATPase of human erythrocyte, *FEBS Lett.*, 281, 255, 1991.

33. Awasthi, S. et al., Adenosine triphosphate-dependent transport of doxorubicin, dauno-mycin, and vinblastine in human tissues by a mechanism distinct from the P-glyco-protein, *J. Clin. Invest.*, 93, 958, 1994.

34. Awasthi, S. et al., ATP-dependent human erythrocyte glutathione-conjugate trans-porter. I. Purification, photoaffinity labeling, and kinetics characteristics of ATPase activity, *Biochemistry*, 37, 5231, 1998.

35. Awasthi, S. et al., ATP-dependent human erythrocyte glutathione-conjugate trans-porter. II. Functional reconstitution of transport activity, *Biochemistry*, 37, 5239, 1998.

36. Awasthi, S. et al., ATP-dependent colchicine transport by human erythrocyte glu-tathione conjugate transporter, *Toxicol. Appl. Pharmacol.*, 155, 215, 1999.

37. Sharma, R. et al., Purification and characterization of dinitrophenylglutathione ATPase of human erythrocytes and its expression in other tissues, *Biochem. Biophys. Res. Commun.*, 171, 155, 1990.

38. Saxena, M. et al., Dinitrophenyl S-glutathione ATPase purified from human muscle catalyzes ATP hydrolysis in the presence of leukotrienes, *Arch. Biochem. Biophys.*, 298, 231, 1992.

39. Awasthi, Y.C. et al., Purification and characterization of an ATPase from human liver which catalyzes ATP hydrolysis in the presence of the conjugates of bilirubin bile acids and glutathione, *Biochem. Biophys. Res. Commun.*, 175, 1090, 1991.

40. Pikula, S. et al, Organic anion-transporting ATPase of rat liver. I. Purification, photoaffinity labeling, and regulation by phosphorylation, *J. Biol. Chem.,* 269, 27566, 1994.

41. Pikula, S. et al, Organic anion-transporting ATPase of rat liver. II. Functional reconstitution of active transport and regulation by phosphorylation, *J. Biol. Chem.,* 269, 27574, 1994.

42. Borst, P. et al., The multidrug resistance protein family, *Biochim. Biophys. Acta,* 1461, 347, 1999.

43. Tammur, J. et al., Two new genes from the human ATP-binding cassette transporter superfamily, ABCC11 and ABCC12, tandemly duplicated on chromosome 16q12, *Gene,* 273, 89, 2001.

44. Gottesman, M.M. and Pastan, I., Biochemistry of multidrug resistance mediated by the multidrug transporter, *Annu. Rev. Biochem.,* 62, 385, 1993.

45. Shapiro, A.B. and Ling, V., Reconstitution of drug transport by purified P-glycoprotein, *J. Biol. Chem.,* 270, 16167, 1995.

46. Dean, M., Hamon, Y., and Chimini, G., The human ATP-binding cassette (ABC) transporter superfamily, *J. Lipid Res.,* 42, 1007, 2001.

47. Johnson, D.R. et al., The pharmacological phenotype of combined multidrug-resistance mdr1a/1b- and mrp1-deficient mice, *Cancer Res.,* 61, 1469.

48. Rupniak, N.M. et al., P-Glycoprotein efflux reduces the brain concentration of the substance P (NK1 receptor) antagonists SR140333 and GR205171: a comparative study using mdr1a-/- and mdr1a+/+ mice, *Behav. Pharmacol.,* 14, 457, 2003.

49. Cole, S.P. et al., Over-expression of a transporter gene in a multidrug-resistant human lung cancer cell line, *Science,* 258, 1650, 1992.

50. Higgins, C.F., ABC transporters: from microorganisms to man, *Annu. Rev. Cell Biol.,* 8, 67, 1992.

51. Leslie, E.M., Deeley, R.G., and Cole, S.P., Multidrug resistance proteins: role of P-glycoprotein, MRP1, MRP2, and BCRP (ABCG2) in tissue defense, *Toxicol. Appl. Pharmacol.,* 204, 216, 2005.

52. Saier, M.H. and Paulsen, I.T., Phylogeny of multidrug transporters, *Semin. Cell Dev. Biol.,* 12, 205, 2001.

53. Renes, J. et al., The (patho)physiological functions of the MRP family, *Drug Resist. Update,* 3, 289, 2000.

54. Higgins, C. and Linton, K.J., The XYZ of ABC transporters, *Science,* 293, 1782, 2001.

55. Wijnholds, J. et al., Increased sensitivity to anticancer drugs and decreased inflammatory response in mice lacking the multidrug resistance–associated protein, *Nat. Med.,* 3, 1275, 1997.

56. Ambudkar, S.V. et al., Biochemical, cellular, and pharmacological aspects of the multidrug transporter, *Annu. Rev. Pharm. Toxicol.,* 39, 361, 1999.

57. Leslie, E.M., Deeley, R.G., and Cole, S.P., Toxicological relevance of the multidrug resistance protein 1, MRP1 (ABC c1) and related transporters, *Toxicology,* 1673, 2001.

58. Loe, D.W., Deeley, R.G., and Cole, S.P., Characterization of vincristine transport by the M(r) 190,000 multidrug resistance protein (MRP): evidence for cotransport with reduced glutathione, *Cancer Res.,* 58, 5130, 1998.

59. Awasthi, S. et al., RALBP1: a novel mechanism of drug resistance in epilepsy, *BMC. Neurosci.,* in press.

60. Awasthi, Y.C. et al., Regulation of 4-hydroxynonenal-mediated signaling by glutathione S-transferases, *Free Radic. Biol. Med.,* 37, 607, 2004.

61. Peterson, D.R. and Doorn, J.A., Reactions of 4-hydroxynonenal with proteins and cellular targets, *Free Radic. Biol. Med.*, 37, 937, 2004.

62. Cheng, J.Z. et al., Accelerated metabolism and exclusion of 4-hydroxynonenal through induction of RLIP76 and hGST5.8 is an early adaptive response of cells to heat and oxidative stress, *J. Biol. Chem.*, 276, 41213, 2001.

63. Yang, Y. et al., Cells preconditioned with mild, transient UVA irradiation acquire resistance to oxidative stress and UVA-induced apoptosis: role of 4-hydroxynonenal in UVA-mediated signaling for apoptosis, *J. Biol. Chem.*, 278, 41380, 2003.

64. Oosterhoff, J.K. et al., REPS2/POB1 is downregulated during human prostate cancer progression and inhibits growth factor signalling in prostate cancer cells, *Oncogene,* 22, 2920, 2003.

65. Oosterhoff, J.K. et al., EGF signalling in prostate cancer cell lines is inhibited by a high expression level of the endocytosis protein REPS2, *Int. J. Cancer,* 113, 561, 2005.

66. Yadav, S. et al., POB1 over-expression inhibits RLIP76-mediated transport of glutathione-conjugates, drugs and promotes apoptosis, *Biochem. Biophys. Res. Commun.,* 328, 1003, 2005.

67. Yamaguchi, A. et al., An Eps homology (EH) domain protein that binds to the Ral-GTPase target, RalBP1, *J. Biol. Chem.,* 272, 31230, 1997.

68. Adachi, M., Reid, G., and Schuetz, J.D., Therapeutic and biological importance of getting nucleotides out of cells: a case for the ABC transporters, MRP4 and 5, *Adv. Drug Deliv. Rev.*, 54, 1333, 2002.

69. Childs, S. et al., Identification of a sister gene to P-glycoprotein, *Cancer Res.,* 55, 2029, 1995.

70. Doyle, L.A. et al., A multidrug resistance transporter from human mcf-7 breast cancer cells. *Proc. Natl. Acad. Sci. USA*, 95, 15665, 1998.

71. Lage, H. and Dietel, M., Effect of the breast cancer resistance protein on atypical multidrug resistance, *Lancet Oncol.*, 1, 169, 2000.

72. Litman, T. et al., The multidrug resistant phenotype associated with over-expression of the new ABC half-transporter, Mxr (ABC g2), *J. Cell. Sci.*, 113, 2011, 2000.

73. Rocchi, E. et al., The product of the ABC half-transporter gene ABC g2 (BCRP/MXR/ABC p) is expressed in the plasma membrane, *Biochem. Biophys. Res. Commun.*, 271, 42, 2000.

74. Sauer, B., Inducible gene targeting in mice using the Cre/*lox* system, *Methods Enzymol.*, 14, 381, 1998.

75. Joyner, A.L., *Gene Targeting: A Practical Approach*, Oxford University Press, Oxford, England, 2000.

12 Glutathione S-Transferases and Oxidative Injury of Cardiovascular Tissues

Yongzhen Yang and Paul J. Boor

CONTENTS

12.1 INTRODUCTION

Oxidative stress is thought to be associated with the toxicity of numerous chemicals, with the pathogenesis of many disease processes in several different organ systems, and with diseases of aging. To protect against the deleterious effects of oxyradicals and to prevent lipid peroxidation processes, both nonenzymatic and enzymatic antioxidant defense systems have evolved. Nonenzymatic cellular defenses include low molecular weight compounds such as Vitamins A and E, ascorbate, urate, and reduced glutathione (GSH). A second line of defense includes enzymes such as catalase, superoxide dismutase, and the GSH-dependent enzymatic system, where the key protecting enzymes are glutathione peroxidase (GPx) and the glutathione S-transferases (GSTs).

The discovery of the ubiquitous tripeptide GSH has greatly advanced biomedical and nutritional sciences over the past 125 years.[1] The near ubiquitous cellular content of GSH has been known for years and has been described in cardiovascular tissues since at least 1979.[2] GSH is an important water-phase antioxidant and essential cofactor for antioxidant enzymes; it is found mainly in the cell cytosol and other aqueous phases of cells where it provides protection for intracellular components (especially mitochondria) against endogenously generated oxygen radicals.

In cells, glutathione can exist in two forms: the antioxidant-reduced glutathione tripeptide is generally referred to as GSH; the oxidized form is a sulfur–sulfur linked dimer glutathione disulfide or GSSG. The GSSG to GSH ratio is a sensitive indicator of oxidative stress. GSH status is tightly controlled, being continually self-adjusting with respect to the balance between GSH synthesis (by GSH synthetase enzymes), its recycling from GSSG (by GSH reductase), and its utilization by enzymes including peroxidases, transferases, transhydrogenases, and transpeptidases.[3]

GSH participates in many cellular reactions. The basic mechanism of antioxidant defense is through effectively scavenging free radicals and other reactive oxygen species (e.g., hydroxyl radical, lipid peroxyl radical, peroxynitrite, and H_2O_2) directly, and indirectly through enzymatic reactions. In such reactions, GSH is oxidized to form GSSG, which is then reduced to GSH by the NADPH-dependent glutathione reductase (hence, the usefulness of the GSSG to GSH ratio as an indicator of intracellular oxidative stress). GSH can react with various electrophiles and in cells these reactions are catalyzed by GSTs. A detailed description of GST isozymes, both cytosolic and microsomal, is extensively covered in various chapters of this volume.

GSTs are frequently attributed with the role of preventing oxidation of biomolecules, such as conjugating with NO to form an S-nitrosglutathione adduct, which is cleaved by the thioredoxin system to release GSH and NO.[4] A group of immunologically related α-class mammalian GSTs, which utilize 4-hydroxynonenal as their preferred substrate, has been proposed to be a major cellular defense system against oxidative injury by products of lipid peroxidation. These isozymes, clustered as a subgroup of the α-class GSTs, use 4-hydroxynonenal as their preferred substrates and are found in several species, including humans. This has been covered in Chapters 9 and 10 of this volume. This group of GSTs has been of special interest to our laboratory because of their possible role in protecting against oxidative stress-induced injury of cardiovascular tissues.[5,6–11] GSH also may serve as a substrate for formaldehyde dehydrogenase, which converts formaldehyde and GSH to S-formyl-glutathione.[1] Thus many endogenous and exogenous or xenobiotic substances are believed to be detoxified through GSH and GST pathways.

GSH deficiency contributes to oxidative stress in many tissues and therefore may play a key role in aging and the pathogenesis of diseases that are believed to be linked to intracellular oxidative stress. Diseases that have been attributed at least partially to imbalances of GSH and associated pathways include kwashiorkor, seizure disorders, Alzheimer's disease, Parkinson's disease, fibrotic liver diseases (cirrhosis), cystic fibrosis, sickle cell anemia, HIV, AIDS, cancer, heart disease, and diabetes.[6] Besides maintaining the oxidative/reductive state of the cell, myriad other functions have also been known for years to involve GSH. For example, cell growth and division, protein synthesis, cellular reducing reactions, regulation of sulfhydryl-containing enzymes, and leukotreine synthesis have all been linked to the oxidative state of the cell regulated by GSH.[3] Many of these functions, however, vary according to the organ, tissue, or cells being examined. In this review therefore we concentrate on what is presently known about the role played by GSH, GSTs, and other related GSH enzymes in cardiovascular tissues (i.e., in the heart and blood vessels).

12.2   GSH AND GST ISOZYMES IN CARDIOVASCULAR TISSUES

Although GSH was first described in hepatic and other tissues, it was soon realized that cardiac muscle or myocardium contains similarly high levels of GSH.[7] The activity of the GSH-associated enzymes, including the GSTs, has been described in myocardium, and GST isozymes belonging to various classes have been described in hearts.[8] Several closely related GSTs with clearly distinguishable catalytic properties and high specific activity for short chain alkenal products of lipid peroxidation such as 4-hydroxynon-2-enal (4-HNE) were described in human tissues, including myocardium.[8] These isozymes, initially characterized in mice,[9] were also described in pulmonary endothelial cells[10] and in rat vascular tissues, specifically the aorta.[11] More recently, significant alterations of GSH during oxidative stress have been described and studied in vascular tissues, especially the more readily studied large- to medium-sized muscular or elastic arteries, such as the aorta.[12] Because blood vessels are essentially composed of two major cell types — the endothelial cell and a variable component of vascular smooth muscle cells — many of the studies relevant to blood vessels have been performed in these isolated cell systems.

12.3   THE ROLE OF GSTS IN DOXORUBICIN CARDIOTOXICITY

12.3.1   HISTORICAL PERSPECTIVE

The well-studied and long-recognized cardiotoxicity of the anticancer drug doxorubicin serves as an excellent example of the protective role played by GSH and GSTs in cardiovascular tissues. Doxorubicin belongs to a wide family of compounds, the anthracyclines, that were originally produced by cultivating *Streptomyces* organisms. These compounds are glycosides whose aglycone is a tetracyclic anthraquinone chromophore. The most widely used drug of this class of compounds — doxorubicin (Adriamycin) — was originally approved for use in the United States in 1973 and has been one of the most successful and extensively used anticancer drugs against a variety of leukemias and solid tumors.[13] Soon after their extensive clinical usage began, the anthracyclines doxorubicin and the related drug daunorubicin were both recognized to cause a severe, dose-limiting cardiotoxicity characterized by dose-dependent contractile dysfunction that results in severe, intractable heart failure and the pathologic picture of a dilated type of cardiomyopathy.

12.3.2 OXIDATIVE STRESS AND DOXORUBICIN CARDIOTOXICITY

Extensive early studies in both humans[14] and a variety of animal models showed that subcellular sites of damage included the nucleus, mitochondria, and — in chronic animal models — the contractile apparatus.[15] Also early in the experimental work on doxorubicin cardiomyopathy, it was demonstrated by our and other laboratories[7,16] that a marked decrease in GSH content in myocardium implicated oxidative stress as an underlying mechanism of cardiac myocyte injury. This concept gained support from the extensive studies of Doroshow, Myers, Olson, and others,[16–21] which defined the metabolism of the anthracycline drugs to a highly reactive semiquinone intermediate that generates reactive oxygen species, thus inducing lipid peroxidation in cellular components of the myocyte, especially the sarcoplasmic reticulum, mitochondria,[22] and contractile apparatus. Transfection of a mouse GST isozyme, mGSTA4-4, which has homology to an α-class GST found in several species, including humans, was found to confer doxorubicin resistance, supporting a GSH and GST mechanism for cellular defense against anthracycline-induced peroxidative injury.[23]

12.3.3 DOXORUBICIN METABOLISM AND TOXICITY

More recent studies of anthracycline cardiotoxicity have attempted to formulate a unifying concept that involves oxidative stress (with generation of H_2O_2), iron, and a role for the metabolite doxorubicinol in causing cardiac myocyte apoptosis.[24] Furthermore, a role for NF-κB in pro-apoptotic signaling in cardiac myocytes has been implicated.[25]

Free radical oxidative stress during anthracycline metabolism has been found to be dependent on Fe and to target mitochondria.[26] Studies with iron-chelating agents have been undertaken to develop drugs that would counteract the cardiotoxicity of doxorubicin in a variety of experimental animals,[27,28] and these successful types of agents have achieved some limited clinical usage.[25,29] It is likely that these initial iron-dependent metabolic events result in lipid peroxidation downstream, and the protection afforded by GST (as noted above) is related to effective metabolism of the highly reactive products, such as aldhydes, of lipid peroxidation.

The most recent approaches taken to ameliorate anthracycline cardiotoxicity include the development of third-generation anthracycline derivatives, including fluoro-derivatives, disaccharide analogs, and other epimers.[30] Furthermore, synthetic derivatives of doxorubicin and liposomal encapsulation formulations have been employed with clinical success in some specific types of tumors, including relapsed ovarian cancer and Kaposi's sarcoma.[25] Thus the historic search for the mechanisms of myocardial toxicity of this important class of antitumor drugs has led to a better understanding of how metabolism generates metabolite- and radical-based oxidant stress, involves GSH defense mechanisms, and critically injures subcellular organelles in the heart. Further, the understanding of the mechanisms of anthracycline cardiotoxicity has resulted in a variety of successful therapeutic modalities for minimizing its toxicity to the heart.

12.4 GSH AND GSTS DURING MYOCARDIAL AND VASCULAR INJURY

12.4.1 THE ISCHEMIC HEART

Although coronary artery atherosclerosis and resultant ischemic heart disease is the most common cause of death in the Western world, there is surprisingly little evidence that GSH plays a major role in potentially protecting the heart against acute myocardial ischemic, or anoxic, injury. This is contradictory to the clear-cut role demonstrated for GSH in many toxic or xenobiotic forms of myocardial injury, such as doxorubicin toxicity discussed above, or in its role as a protective mechanism during oxidative injury of other tissues, such as liver.

Myocardial hypoxic injury occurs in humans either in areas of the heart undergoing an injurious lack of blood flow (hypoxia) or total ischemia (anoxia), and also when perfusion to an area of the myocardium is reestablished following a short but lethal period of ischemia ("reperfusion injury").[31,32] In all such situations, it is well established that the major cellular injury is due to lack of oxygen with consequent failure of mitochondrial generation of high-energy phosphates in the myocardium, which is an obligate aerobic tissue. In the case of reperfusion injury, the reestablishment of an oxygen source results in an aberrant burst of oxygen radical production, with accelerated structural cellular injury.

It appears that in the myocardium, damage to mitochondria and the consequent loss of mitochondrial oxidative function, with a major redox imbalance, is largely behind the loss of myocardial contractility and eventual cell death. The role played by GSH and GSH-linked enzymes including GSTs in this process, however, is not clear and needs to be investigated further. Earlier studies have suggested that GSSG is increased in ischemic myocardium while GSH does not change drastically.[33] It may be that ischemic or ischemia-reperfusion conditions have lesser effects on GSH of the myocardium than has been demonstrated in other tissues such as brain, liver, or the endothelium. It should also be noted, however, that studies in patients undergoing ischemia-reperfusion injury associated with cross-clamping of the aorta during cardiac surgery demonstrated changes in plasma GPx and alterations consistent with oxidative stress and GSH depletion in the myocardium, at least under the drastic conditions of cardiac surgery.[34] Other experimental work with a model of ischemia-reperfusion injury in rat hearts suggests that the alkenals generated from lipid peroxidation (most notably 4-HNE), in conjunction with the depletion of mitochondrial GSH, play an important role in the mitochondrial dysfunction that is a hallmark of ischemic myocardial injury.[35] Since conjugation of endogenous electrophilic toxicants including 4-HNE is catalyzed by GSTs, these enzymes should be involved in these mechanisms.

The work of Kehrer and Lund, and others[36,37] indicates that in intact heart tissue, despite major depletion of ATP, little oxidation of GSH occurs, contrasting distinctly with findings in hepatocytes. Hypoxic hearts retain a normal reductive capacity toward GSSG even after major ATP depletion.[38] With regard to the myocardium, tissue differences and comparisons that are beyond the scope of this review must exist and are likely to be worthy of further study in view of the importance of ischemic heart disease as a major cause of morbidity and mortality. In particular the role of GSTs in both their conjugating activity toward endogenous electrophiles generated during oxidative stress and peroxidase activity toward endogenous hydroperoxides is of interest because both these activities can modulate GSH homeostasis.

12.4.2 BLOOD VESSELS

Our laboratory's interest in vascular GSH and GSH-linked enzymes stems from earlier studies of the cardiovascular toxin, allylamine, or 3-aminopropene. Allylamine has been known for many years to cause relatively specific vascular and heart lesions in a wide variety of species when administered by many different routes. Acutely, a single dose allylamine causes subendocardial myocardial necrosis that bears many similarities to acute myocardial infarction.[39] Given chronically, allylamine results in progressive loss and scarring of the myocardium, a morphologic picture that is analogous to chronic ischemic heart disease.[40,41]

Recent evidence indicates that this picture of ischemic or anoxic damage is secondary to pathophysiologic effects on coronary vessels known as vasospasm or vasomotion.[42,43] In large- to medium-sized arteries, such as coronary arteries or the aorta, chronic administration of allylamine induces intimal lesions that markedly resemble the early fibrous plaque of atherosclerosis. Early studies from this laboratory[44,45] and others indicated that depletion of GSH accompanied the acute toxic changes in both myocardial tissue and vascular tissues *in vivo*. Extensive subsequent study of aortic tissue as an indicator of vascular changes following allylamine administration showed acute depletion and subsequent rebound of GSH levels, similar to those shown in livers following a variety of toxins, inducing oxidative stress.[12]

A body of data from several laboratories has indicated that the toxic effects of allylamine on vascular tissue and the myocardium, both *in vivo* and *in vitro,* are related to the deamination of allylamine to the highly reactive aldehyde acrolein.[12,46–50] Among the α,β-unsaturated aldehydes, acrolein is by far the strongest electrophile and shows high reactivity toward nucleophilic groups.[51,52] The toxicity of allylamine bears many similarities to the hepatotoxic effects of allylalcohol because allylalcohol is also metabolized to acrolein. The metabolism of allylamine was shown to be catalyzed by a little-studied tissue and plasma amine oxidase (known as semicarbazide-sensitive amine oxidase), which has highest activities in connective tissues, especially vascular smooth muscle cells.[53] Subsequent metabolism of the generated α,β-unsaturated aldehyde to the mercapturic acid, which was excreted as the major urinary metabolite. GSTs catalyze the first rate-limiting step in the detoxification pathway.[12,49,54] Thus the metabolism and toxicity of allylamine bear many similarities to the hepatotoxic effects of allylalcohol.

12.4.3 INDUCTION OF GST BY ALLYLAMINE

Earlier studies from this laboratory showed that GSH and a specific α-class GST known in rats as GST8-8 (in revised nomenclature used in this volume this enzyme is designated as rGSTA4-4) were likely to be involved in the detoxification of the acrolein generated at the vascular smooth muscle level within large- and medium-sized arteries such as the aorta and coronary artery,[55] and also possibly in endothelial cells where similar GSTs are expressed.[10] rGSTA4-4, which has the highest specific activity of any GSTs toward 4-HNE, was found to be markedly induced in vascular smooth muscle cells following chronic dosing with allylamine, and inhibitors of GSTs markedly exacerbated cytotoxicity in cultured vascular smooth muscle cells in a synergistic fashion. Conversely, transfection with *mGSTA4* (the mouse homolog of rat GST8-8) resulted in increased GST activity toward alkenals such as acrolein and 4-hydroxynonenal and was associated with decreased cytotoxicity.[23] Although GSH and the GSTs appear to have little well-established roles during the oxidative injury seen in the myocardium following ischemia or anoxia, a significant role for specific α-class GSTs in vascular tissue in defending against xenobiotic injury has been proposed. Such a role may also be implicated in chronic, aging-associated diseases of blood vessels such as atherosclerosis and aneurysm formation as discussed below.

12.5 STUDIES WITH ENDOTHELIAL CELLS

12.5.1 THE ENDOTHELIUM AND ATHEROSCLEROSIS

There is considerable evidence that reactive oxygen and nitrogen species are intimately associated with the development of atherosclerosis.[5,56] It would be of great clinical interest to be able to manipulate the antioxidative capacity of cells undergoing oxidative stress during such processes. To this end, the antioxidants GSH and α-class GSTs that catalyze the conjugation of GSH with 4-HNE have attracted considerable clinical interest as both markers of oxidative stress and potential preventative/therapeutic agents. Endothelial cells that line the vascular lumen provide a selective barrier to the uptake of circulatory substances, such as regulatory molecules and metabolic precursors for antioxidant defense in the vascular wall. Endothelial cells also are intimately involved in regulating vascular tone and blood flow. The "endothelial injury" hypothesis of atherosclerosis is based on the assumption that subtle injury of endothelial cells is a seminal event in the initiation of the atherosclerotic plaque.[32] Thus the endothelial cell may act as the first line of defense against oxidative stress in the vessel wall. In support of this concept, Voskoboinik et al.[57] demonstrated that endothelial cells restrict cysteine or GSH-stimulated synthesis in smooth muscle cells when metabolic precursors are supplied from the luminal or endothelial side.

12.5.2 THE ROLE OF SPECIFIC GST ISOZYMES

In experiments involving isolated endothelial cells and in human atherosclerotic tissues, we attempted to elucidate the potential defensive role of GSH and GSTs in endothelial cells. In *in vitro* experiments, we transfected cDNA of mGSTA4-4 into mouse pancreatic islet endothelial cells (MS1).[5] Our data showed that experimental overexpression of mGSTA4-4 in endothelial cells significantly increased resistance to the cytotoxicity of several α,β-unsaturated aldehydes, such as acrolein and 4-HNE, and direct oxidative stress through H_2O_2 (Figure 12.1), and protected cells against oxidative stress-induced apoptosis (Figure 12.2). Our results showed that the protective effect of mGSTA4-4 for endothelial cells was perhaps due to inhibition of JNK (c-Jun N-terminal kinase), Bax, and a p53-activated apoptosis pathway (Figure 12.3). Our studies of human hGSTA4-4 in early human atherosclerotic plaques demonstrated the expression of GSTs including hGSTA4-4 in normal human vascular walls and showed the appearance of intense cytoplasmic staining in endothelial cells overlying atherosclerotic plaque (Figure 12.4), thus suggesting that endothelial induction of hGSTA4-4 occurs early as a protective mechanism adopted by endothelial cells exposed to oxidative stress during atherogenesis. These studies are consistent with the idea of a protective role of GSTs against atherogenesis and suggest that detoxification of electrophilic lipid peroxidation products by GSH/GSTs is a major mechanism for this protection.

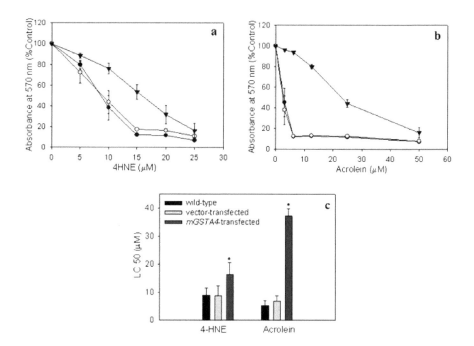

FIGURE 12.1 Effect of *mGSTA4* transfection on cytotoxicity of 4-HNE and acrolein toward MS1 endothelial cells. Cytotoxicity of 4-HNE$_2$ (a), acrolein (b) toward MS1 wild-type (●), vector-transfected (○) and *mGSTA4*-transfected (▼) endothelial cells were measured for MTT absorbance at 570 nm, and values are the means ±SE of five determinations and are representative of four separate experiments. LC50s were derived and shown (c). Reprinted from Reference 5. With permission from Elsevier.

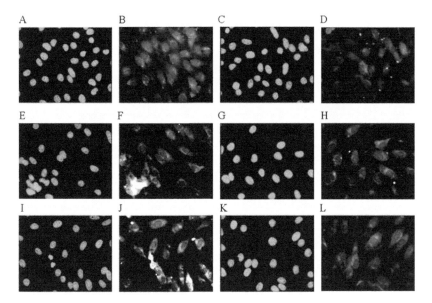

FIGURE 12.2 (See color insert following page 178.) Effect of *mGSTA4* transfection on 4-HNE–mediated caspase activation. Cells 1×10^5 were cultured into poly-L-lysine-coated slides and the CaspACE TITC-VAD-FMK (Promega, Madison, WI) *in situ* marker was used to detect caspase activation of cells. The slides were viewed by fluorescence microscopy (Nikon Eclips 600, Japan) using a standard fluorescein filter set to view green fluorescence at 520 ± 20 nm; blue (DAPI) at 460 nm. Photomicrographs at 400× magnification are presented. Total cells fluoresce blue (A, E, I, C, G, K); caspase-activated cells fluoresce green (B, F, J, D, H, L). The upper (A, B, C, D), middle (E, F, G, H), and lower rows (I, J, K, L) are control (no 4-HNE treatment), with 15 minutes and 30 minutes treatment with 4-HNE 40 µM. The left two columns are vector-transfected MS1 endothelial cells and the right two columns are *mGSTA4*-transfected MS1 endothelial cells. Reprinted from Reference 5. With permission from Elsevier.

LDL oxidation by arterial wall macrophages is also an important factor in early atherosclerosis. The macrophage oxidative state depends on the balance between cellular NADPH-oxidase and the glutathione system.[58] Rosenblat and Aviram[59] have shown that enhancing the macrophage GSH–GPx status results in inhibition of LDL oxidation, which may thus contribute to the attenuation of the atherosclerotic process. Because a major portion of the GPx activity in human tissues toward lipid peroxides is contributed by GSTs, primarily those belonging to the class of GSTs, a protective role against atherosclerotic process is implied by these studies.

Another potentially important atherogenic factor is the reactive nitrogen species peroxynitrite ($ONOO^-$). Reaction with GSH was proposed to be a major detoxification pathway of $ONOO^-$ in the biological system. Cao and Li[56] recently demonstrated that upregulating endogenous GSH by 3H-1,2-dithiol-3-thione protects vascular smooth muscle cells against $ONOO^-$ cytotoxicity. Their studies further point to the feasibility of protecting against $ONOO^-$-mediated vascular cell injury via upregulating endogenous GSH biosynthesis by chemical inducers. The role of GST isozymes also needs to be studied in these protective mechanisms.

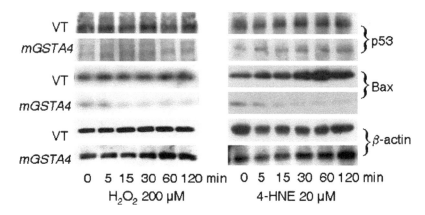

FIGURE 12.3 The effect of *mGSTA4* transfection on regulation of JNK-activated signal transduction pathway. Cell lysates (50 μg) were loaded into 4 to 20% SDS-PAGE and probed with different antibodies. Anti-p53 antibodies were diluted at 1:100. Anti-Bax antibodies were diluted at 1:4,000. Anti-β-actin antibodies (1:2,000) were used to reprobe the membrane to confirm equivalent amounts of protein. VT: vector-transfected endothelial cells; *mGSTA4*: mGSTA4-transfected endothelial cells. Reprinted from Reference 5. With permission from Elsevier.

12.6   GSH/GST AND NITRATE TOLERANCE

Recent research has raised clinical interest in the potential role of GSH and GSTs in the metabolism of important commonly used medical treatments for cardiovascular diseases. Specifically, GSH and GSTs are believed to be involved in the metabolism of organic nitrates, which are some of the most common and effective therapeutic agents in relieving the symptoms of angina pectoris and for preventing such symptoms.[60–65] These compounds, the most well known of which is nitroglycerin (NTG) or glyceryl trinitrate, are deemed to be safe and rapidly acting drugs, but tolerance to their actions has been recognized since their first use over a hundred years ago.[64] Biochemically, nitrates are thought to exert their action through the liberation of NO in the vasculature, but the metabolic pathways by which this is accomplished, particularly in relation to tolerance development, are still incompletely defined. It is thought that NTG is metabolized either by metabolism to NO or alternatively by a clearance-based pathway that yields the less potent metabolite NO_2.[64] The details of these two pathways, however, are not well defined.

Numerous candidate enzymes for NTG metabolism, as well as a multiplicity of tolerance mechanisms, have been proposed in the literature, but a unifying hypothesis that links these phenomena together has not appeared. Earlier studies indicated that tolerance to NTG was not related to decreased intracellular thiols and sulfhydryl supplementation does not alter tolerance. Similarly, the study by Matsuzaki et al.[65] found that physiologic vasorelaxation was decreased by inhibition of GSTs but the increase in vascular cGMP content produced by NTG was not significantly changed. This result suggested that the enzymatic process for resolving NTG using GSH or GSTs was not important in the NO production process or in the development of tolerance.

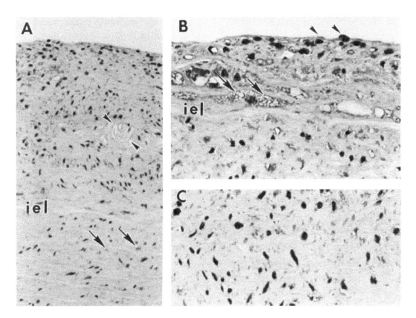

FIGURE 12.4 (See color insert.) Immunocytochemical localization of hGSTA4-4 in human vessel and human atherosclerosis. Findings are representative of 15 aortic samples from five humans, ages 16 to 25 (see methods for detail). (a) A small fibrous aortic plaque in a 25-year old; staining for human GST A4-4 (hGSTA4-4) is found in normal aortic media (arrows) and appears increased in proliferating vascular smooth muscle cells within the plaque, suggesting upregulation of the isozyme during atherogenesis. Note focal, small collections of cholesterol clefts (arrowheads) consistent with very early Stage 4 plaque. (b) Higher power of an even earlier, very small (less than 1 mm) Stage 3 plaque also shows increased staining in proliferating vascular smooth muscle cells and foam cells (arrows). The endothelial cells overlying the early plaque (arrowheads) exhibit increased cytoplasmic staining whereas normal endothelium away from plaque showed no staining. (c) High power of vascular smooth muscle cells [from plaque shown in (a)] demonstrates intense staining and marked proliferation. Immunohistochemical staining with hGSTA4-4 as primary antibody; ×350 (a) and ×490 (b, c). Reprinted from Reference 5. With permission from Elsevier.

Later studies, however, suggest that GSTs are important in the bioactivation of NTG and hence are responsible for its pharmacologic effects. The μ-class of GSTs was implicated by studies of Kenkare et al.[61] in which GST inhibitors were used to dissect the individual activity of GST subtypes toward NTG in rabbit aortas. Subsequent work by Singhal et al.,[60] utilizing a variety of GST inhibitors and GSTs purified from rabbit aortas and characterized as μ- or α-class GSTs, concluded that the higher abundance of the aortic α-class GSTs suggests their greater importance in the metabolism of NTG to denitration products. These authors suggested that because of various assays being employed, including physiologic as well as metabolic, it may prove difficult to extrapolate to the human situation with regard to NTG bioactivation and tolerance.

There is abundant evidence that decreased GSH content and inhibition of GST activities are related to vasorelaxation and the development of tolerance by nitrates, at least in experimental situations.[62,66,67] Lee and Fung[63] recently examined the partial chemical inactivation of GST by NTG. In their study, they observed that tetranitromethane (TNM, a NO_2^+/NO_2 donor) inactivated GST through oxidation of the protein sulfhydryl groups, as well as through nitration of tyrosine residues whereas NTG inactivated GST primarily through protein sulfhydryl oxidation. Their proposed mechanism for the partial inactivation of GST by NTG involves the formation of S-oxidized GST dimers. Further studies are still needed to clarify the role of GSH and GSTs in these important clinical drugs and in the development of tolerance to these drugs' beneficial effects.

12.7   GST POLYMORPHISMS AND HUMAN ATHEROSCLEROSIS

From human epidemiologic studies that have spanned many decades, risk factors that predict atherosclerotic diseases such as ischemic heart disease (also known as coronary heart disease, CHD, or coronary artery disease, CAD) and stroke have now been extensively defined. However, it is estimated that the known risk factors, such as smoking, hypertension, and diabetes, only account for approximately two-thirds of the incidence of CHD. Thus it is likely that genetic "familial" predisposition plays a major role in the underlying pathogenesis of atherosclerotic disease.

Recent epidemiologic studies have begun to examine the possible interaction between GST polymorphism and CHD or atherosclerosis. The μ (*GSTM1*) and θ (*GSTT1*) classes of GSTs, which are known to exhibit polymorphisms, are the isozymes most frequently implicated in human diseases. Although these classes of polymorphic GSTs have been associated with an increased risk of atherosclerosis or CAD in smokers, some major controversies still exist.[68–79] Generally speaking, *GSTM1* and *GSTT1* are polymorphic in humans and the homozygous null genotype for each of these genes has a high prevalence in all human populations (10–65%).[46] Further details on GST polymorphism are discussed in Chapter 7. In recent years, extensive epidemiologic studies have analyzed the potential effects of these polymorphisms on atherosclerotic-related diseases such as CHD. In the vast majority of the studies to date, however, these statistical analyses of GST polymorphisms seem to indicate a role in the development of CHD and atherosclerosis only in those who have previously smoked.[69–71,73–76]

Li et al.[79] were among the first investigators to provide evidence that the risk of CHD was affected by polymorphisms of *GSTM1* and *GSTT1* in humans. These investigators found that smokers who also carried the *GSTM1* null genotype were approximately 1.5 times more likely to suffer from CHD when compared to smokers with a functional allele. Interaction between *GSTT1* and smokers for the risk of CHD, however, showed the opposite relationship: there was an approximately three times higher risk of CHD in individuals who had smoked (≥20 pack-years) and also carried a functional *GSTT1* allele. In a separate study, Li et al.[77] found that a functional *GSTT1* allele was associated with an increased risk of lower extremity arterial disease in smokers, where no effect of *GSTM1* was found. Subsequent studies somewhat contradicted these earlier results. Specifically, Tamer et al.[74] (in a case-controlled study of 247 healthy controls and 148 consecutive patients who had undergone coronary angiography for suspicion of CHD) found that patients who had previously smoked and also had the null genotypes of *GSTM1* and *GSTT1*[80] were at a higher risk for developing CHD. Their results were confirmed by more recent studies of Masetti et al.[69] who evaluated the distribution of GST genotypes in 430 angiographically defined patients. The frequencies of GST null genotypes did not differ significantly between patients with and without CAD. However, smokers with *GSTM1* and *GSTT1* null genotypes had a significantly higher risk of CAD than those who never smoked with these same genotypes. Furthermore, in smokers who concurrently lacked both *GSTM1* and *GSTT1* genes, the risk of CAD was further increased. In that study, smokers lacking both GST genes had a greater extent and severity of stenosis in coronary vessels (evaluated by several methods) when compared to smokers expressing the genes.

Still other studies have shown somewhat more controversial results. For instance, Habdous et al.,[68] through their meta-analysis, confirmed the previously observed association of the *GSTM1* null allele with increased risk for CHD in smokers, but, surprisingly, for *GSTT1*, the functional allele, not the null allele, there was an interaction with smoking on increasing the risk of CHD. In the latter study, the *GSTT1* functional allele was also significantly associated with an increased risk of lower extremity arterial disease. Similar findings were observed in earlier research that used cross-sectional data from a subset of participants in the Atherosclerosis Risk in Communities (ARIC) study. The ARIC cohort is a probability sample of 15,792 predominantly white or African-American (AA) men and women aged 45–64 years at the baseline survey (1987–1989), randomly selected from four U.S. communities. Olshan et al.[71] found the interaction between the functional *GSTT1* genotype and smoking on the risk of carotid artery atherosclerosis by the analysis of a stratified random sample of 1,394 individuals from the ARIC study but found no interaction between smoking and *GSTM1*.

Other studies that evaluated the interaction of the *GSTM1* genotype with atherosclerosis or CAD[76,81,82] showed controversial results. De Waart et al.[81] suggested that smokers lacking the detoxifying enzyme *GSTM1* develop progression of atherosclerosis at an increased rate. However, Wang et al.[82] showed that *GSTM1* deficiency alone is not sufficient to cause CAD. Wilson et al.[76] analyzed 398 patients admitted for angiographic investigation of chest pain and 196 age- and sex-matched controls. The patients were subdivided into those with and without previous acute myocardial infarction (AMI). The *GSTM1* null genotype occurred at a significantly lower frequency in the AMI patient group (48%) compared both to patients with no history of AMI (59%) and to the control group (57.2%). When subjects were stratified for smoking status, a significant association was observed only in smokers.

Data obtained in populations moving from South Asia to Europe and North America indicated that the *GSTM1* null genotype protected against both CAD and AMI, but no significant associations were observed between the *GSTT1* genotype and cardiovascular disease.[75] The association was independent of smoking history with both nonsmokers and smokers showing a similar pattern of genotype distribution. It is possible that the mechanism of action of *GSTM1* in disease pathogenesis is different between these two groups of subjects or that a lifestyle factor correlated with cigarette smoking is causing the enhanced risk in Caucasians. Corresponding to these results, another interesting study showed increased risk to carotid atherosclerosis in Korean postmenopausal rheumatoid arthritic (RA) women without histories of smoking.[72] This study suggests a weak association between the *GSTT1* null genotype and atherosclerosis. However, an independent relationship between the *GSTT1* genotype and carotid atherosclerosis was not observed. Because the studies from different races or group sizes showed largely inconsistent results of GST polymorphism genotype in the development of atherosclerosis and CHD or CAD, more extensive research is needed.

It is likely that the conflicting results from these various studies may be related to the many different types of end points being studied (e.g., CAD, carotid thickening, peripheral vascular disease) as well as the various methods of analysis (clinical measurements, physiologic measurements, clinical and historical data). Another possibility is that variation in the populations being studied as well as study group sizes may affect the data obtained and conclusions drawn.

Another subpopulation that has been studied is that of patients with systemic sclerosis, an inflammatory disease known to be associated with increased oxidative stress and prevalence of atherosclerosis. In a study of 54 patients with systemic sclerosis, Palmer et al.[78] found a highly significant increase in the degree of carotid artery disease in persons with systemic sclerosis who had a specific genotype involving three GSTs (*GSTT1*-null, *GSTM1*-null, and either heterozygous or homozygous carriers of the *GSTP1* Val105 allele). In view of this and other studies of different races or group sizes that show somewhat variable correlations of GST polymorphism genotypes with the development of atherosclerosis and CHD or CAD, more extensive research would definitely seem to be indicated.

With regard to the underlying mechanisms by which *GSTM1* and *GSTT1* might modify the risk of atherosclerosis in smokers, at least two studies have addressed the relation of GST polymorphism with some common biomarkers. Miller et al.[70] provided some limited evidence that *GSTM1* and *GSTT1* polymorphisms modify the effect of smoking on inflammation, hemostasis, and endothelial function in combination with smoking and diabetes. Dušinská et al.,[80] on the other hand, found a link only between smoking and oxidized pyrimidines in DNA in the GSTT1 null group. These studies suggest that polymorphisms in GSTs may be important determinants of commonly measured biomarkers.

Many diverse studies have implicated specific subtypes of GST alleles as factors in the incidence of atherosclerosis-associated human diseases. However, the current evidence does little to elucidate the mechanisms underlying this association. Furthermore, various studies often present somewhat conflicting results with regard to this association, perhaps due to the difference in the populations studied or the methods of investigation. Therefore, it seems likely that continued studies of this topic will be forthcoming, given the provocative data suggesting the importance of GST isoforms in the complex disease process of atherosclerosis.

12.8   GPX DEFICIENCY AND VASCULAR DISEASE; OTHER FACTORS

A few recent studies have suggested that variability in glutathione peroxidase activity may result in endothelial dysfunction and oxidative stress in the vasculature, possibly associated with vascular pathology. Forgione et al.[83] found that a murine model of heterozygous deficiency of GPx-1 was associated with altered vascular responses in mesenteric arterioles, increased plasma and aortic levels of an isoprostane marker of oxidant stress, and pathologic vascular changes including coronary intimal thickening and perivascular fibrosis. These structural and biochemical alterations were accompanied by augmented diastolic cardiac dysfunction following an ischemia-reperfusion myocardial injury protocol. Such changes were deemed to be related to generalized endothelial dysfunction caused by increased oxidative stress in mice with GPx-1 deficiency.

Similarly, in a single human prospective study of patients with suspected CAD,[84] erythrocyte GPx-1 levels were found to be inversely associated with risk of cardiovascular events, including death from cardiovascular causes or nonfatal myocardial infarction. Both smoking status and sex also proved to be strong predictors of the level of GPx-1. The authors concluded that GPx-1 activity in humans may have prognostic value in addition to the more traditional risk factors such as blood lipids, smoking, hypertension, and family history.

The fact that the glutathione peroxidases are selenoproteins may be important from the viewpoint of deficiencies, or variations in selenium intake or metabolism.[85] It has long been known that dietary selenium deficiency is associated with an Asian endemic cardiomyopathy and with long-term parenteral nutrition. Although most epidemiologic studies have not shown a clear association of selenium intake and risk of cardiovascular diseases or atherosclerosis, theories still prevail regarding the connection of selenium's protective role on inflammatory and immune processes, the oxidation of blood lipoproteins, eicosanoid or homocysteine metabolism, and modulation of oxidative stress through the action of GPx. It is likely that the interactions of nutritional selenium, especially with other factors such as the lipid-soluble vitamins, will be the subject of future research.

The GSH synthetic pathways have also received some experimental attention, specifically in regard to a potential role for glutamate-cysteine ligase (GCL) in human cardiovascular diseases. The recent study by Nakamura et al.[86] in over 950 patients found a significant correlation between a T polymorphism of GCL (-588C/T; the T allele shows lower promoter activity in response to oxidants) and an increased incidence of myocardial infarction. Hence, another possible human risk factor has been suggested by this study.

12.9 CONCLUSIONS

The roles of GSH, GSTs, and GSH-linked enzymes in heart and vascular tissue have been well established. Earlier studies of doxorubicin cardiotoxicity served as a model of how basic science investigations of a wide variety eventually led to an understanding of the underlying mechanisms of xenobiotic metabolism and oxidative damage that are hallmarks of that form of toxic cellular injury. These studies also produced important clinical modalities for treatment that continue to be developed today. The importance of GSH-related mechanisms in oxidative stress-related injury of the myocardium and vascular tissues is established, but the exact subcellular mechanisms involved in such important diseases as atherosclerosis, cardiomyopathy, or vascular aneurysm are yet to be thoroughly defined.

With regard to human disease, many epidemiologic and basic science studies have implicated specific polymorphisms of GST and potentially other GSH enzymes and alleles as factors in the incidence of atherosclerosis-associated diseases. However, many of these studies have somewhat conflicting conclusions, and the current evidence does little to elucidate the mechanisms underlying this association. Although much information has been gained about the cardiovascular system and GSH in health and disease, vast areas remain to be explored.

ACKNOWLEDGMENT

This work was supported by NIH grants HL 65416 and ES 013038.

REFERENCES

1. Wu, G., Fang, Y.Z., Yang, S., & Lupton, J.R. Glutathione metabolism and its implications for health. *J. Nutr.* 134, 489–492, 2004.
2. Boor, P.J. Cardiac glutathione: diurnal rhythm and variation in drug-induced cardiomyopathy. *Res. Comm. Chem. Pathol. Pharmacol.* 24, 27–36, 1979.
3. Kidd, P.M. Glutathione: systemic protectant against oxidative and free radical damage. *Alt. Med. Rev.* 2, 155–176, 1997.
4. Fang, Y.Z., Yang, S., & Wu, G. Free radicals, antioxidants, and nutrition. *Nutrition* 18, 872–879, 2002.
5. Yang, Y. et al. Glutathione S-transferase A4-4 modulates oxidative stress in endothelium: possible role in human atherosclerosis. *Atherosclerosis* 173, 211–221, 2004.
6. Townsend, D.M., Tew, K.D., & Tapiero, H. The importance of glutathione in human disease. *Biomed. Pharmacother.* 57, 145–155, 2003.
7. Boor, P.J. Cardiac glutathione: diurnal rhythm and variation in drug-induced cardiomyopathy. *Res. Commun. Chem. Pathol. Pharmacol.* 24, 27–36, 1979.
8. Singhal, S.S. et al. Several closely related glutathione S-transferase isozymes catalyzing conjugation of 4-hydroxynonenal are differentially expressed in human tissues. *Arch. Biochem. Biophys.* 311, 242–250, 1994.
9. Medh, R.D. et al. Characterization of a novel glutathione S-transferase isoenzyme from mouse lung and liver structural similarity expressed in human tissues. *Arch. Biochem. J.* 278(Pt. 3), 793–799, 1991.
10. He, N.G. et al. Purification and characterization of a 4-hydroxynonenal metabolizing glutathione S-transferase isozyme from bovine pulmonary microvessel endothelial cells. *Biochim. Biophys. Acta* 1291, 182–188, 1996.
11. Misra, P. et al. Glutathione S-transferase 8-8 is localized in smooth muscle cells of rat aorta and is induced in an experimental model of atherosclerosis. *Toxicol. Appl. Pharmacol.* 133, 27–33, 1995.
12. Awasthi, S. & Boor, P.J. Lipid peroxidation and oxidative stress during acute allylamine-induced cardiovascular toxicity. *J. Vasc. Res.* 31, 33–41, 1994.
13. Arcamone, F.M. From the pigments of the actinomycetes to third generation antitumor anthracyclines. *Biochimie* 80, 201–206, 1998.
14. Buja, L.M., Ferrans, V.J., Mayer, R.J., Roberts, W.C., & Henderson, E.S. Cardiac ultrastructural changes induced by daunorubicin therapy. *Cancer* 32, 771–788, 1973.
15. Singal, P.K., Deally, C.M., & Weinberg, L.E. Subcellular effects of adriamycin in the heart: a concise review. *J. Mol. Cell Cardiol.* 19, 817–828, 1987.

16. Doroshow, J.H., Locker, G.Y., Baldinger, J., & Myers, C.E. The effect of doxorubicin on hepatic and cardiac glutathione. *Res. Commun. Chem. Pathol. Pharmacol.* 26, 285–295, 1979.

17. Doroshow, J.H., Locker, G.Y., & Myers, C.E. Enzymatic defenses of the mouse heart against reactive oxygen metabolites: alterations produced by doxorubicin. *J. Clin. Invest.* 65, 128–135, 1980.

18. Doroshow, J.H., Locker, G.Y., & Myers, C.E. Experimental animal models of adriamycin cardiotoxicity. *Cancer Treat. Rep.* 63, 855–860, 1979.

19. Batist, G. et al. Cardiac and red blood cell glutathione peroxidase: results of a prospective randomized trial in patients on total parenteral nutrition. *Cancer Res.* 45, 5900–5903, 1985.

20. Olson, R.D. et al. Regulatory role of glutathione and soluble sulfhydryl groups in the toxicity of adriamycin. *J. Pharmacol. Exp. Ther.* 215, 450–454, 1980.

21. Doroshow, J.H. Doxorubicin-induced cardiac toxicity. *N. Engl. J. Med.* 324, 843–845, 1991.

22. Aversano, R.C. & Boor, P.J. Histochemical alterations of acute and chronic doxorubicin cardiotoxicity. *J. Mol. Cell Cardiol.* 15, 543–553, 1983.

23. He, N.G. et al. Transfection of a 4-hydroxynonenal metabolizing glutathione S-transferase isozyme, mouse GSTA4-4, confers doxorubicin resistance to Chinese hamster ovary cells. *Arch. Biochem. Biophys.* 333, 214–220, 1996.

24. Hasinoff, B.B. Inhibition and inactivation of NADH-cytochrome c reductase activity of bovine heart submitochondrial particles by the ironIII-adriamycin complex. *Biochem. J.* 265, 865–870, 1990.

25. Herman, E.H. & Ferrans, V.J. Amelioration of chronic anthracycline cardiotoxicity by ICRF-187 and other compounds. *Cancer Treat. Rev.* 14, 225–229, 1987.

26. Herman, E.H. & Ferrans, V.J. Influence of vitamin E and ICRF-187 on chronic doxorubicin cardiotoxicity in miniature swine. *Lab Invest.* 49, 69–77, 1983.

27. Speyer, J.L. et al. ICRF-187 permits longer treatment with doxorubicin in women with breast cancer. *J. Clin. Oncol.* 10, 117–127, 1992.

28. Gianni, L. et al. Anthracyclines. *Cancer Chemother. Biol. Response Modif.* 21, 29–40, 2003.

29. Konorev, E.A., Kennedy, M.C., & Kalyanaraman, B. Cell-permeable superoxide dismutase and glutathione peroxidase mimetics afford superior protection against doxorubicin-induced cardiotoxicity: the role of reactive oxygen and nitrogen intermediates. *Arch. Biochem. Biophys.* 368, 421–428, 1999.

30. Bertazzoli, C. et al. Experimental systemic toxicology of 4′-epidoxorubicin, a new, less cardiotoxic anthracycline antitumor agent. *Toxicol. Appl. Pharmacol.* 79, 412–422, 1985.

31. Opie, L. In *Braunwald's Heart Disease: A Textbook of Cardiovascular Medicine*, ed. D.P. Zipes, 482–485, W.B. Saunders, Philadelphia, 2005.

32. Libby, P. In *Braunwald's Heart Disease: A Textbook of Cardiovascular Medicine*, ed. D.P. Zipes, 921–935, W.B. Saunders, Philadelphia, 2005.

33. Hoshida, S. et al. Brief myocardial ischemia affects free radical generating and scavenging systems in dogs. *Heart Vessels* 8, 115–120, 1993.

34. Carlucci, F. et al. Cardiac surgery: myocardial energy balance, antioxidant status, and endothelial function after ischemia-reperfusion. *Biomed. Pharmacother.* 56, 483–491, 2002.

35. Jefferies, H. et al. Glutathione. *ANZ. J. Surg.* 73, 517–522, 2003.

36. Kehrer, J.P. & Lund, L.G. Cellular reducing equivalents and oxidative stress. *Free Radic. Biol. Med.* 17, 65–75, 1994.

37. Kehrer, J.P., Paraidathathu, T., & Lund, L.G. Effects of oxygen deprivation on cardiac redox systems. *Proc. West Pharmacol. Soc.* 36, 45–52, 1993.

38. Lund, L.G., Paraidathathu, T., & Kehrer, J.P. Reduction of glutathione disulfide and the maintenance of reducing equivalents in hypoxic hearts after the infusion of diamide. *Toxicology* 93, 249–262, 1994.

39. Boor, P.J., Moslen, M.T., & Reynolds, E.S. Allylamine cardiotoxicity: I. Sequence of pathologic events. *Toxicol. Appl. Pharmacol.* 50, 581–592, 1979.

40. Boor, P.J., Nelson, T.J., & Chieco, P. Allylamine cardiotoxicity: II. Histopathology and histochemistry. *Am. J. Pathol.* 100, 739–764, 1980.

41. Boor, P.J. & Ferrans, V.J. Ultrastructural alterations in allylamine cardiovascular toxicity. Late myocardial and vascular lesions. *Am. J. Pathol.* 121, 39–54, 1985.

42. Conklin, D.J., Boyce, C.L., Trent, M.B., & Boor, P.J. Amine metabolism: a novel path to coronary artery vasospasm. *Toxicol. Appl. Pharmacol.* 175, 149–159, 2001.

43. Conklin, D.J., Langford, S.D., & Boor, P.J. Contribution of serum and cellular semi-carbazide-sensitive amine oxidase to amine metabolism and cardiovascular toxicity. *Toxicol. Sci.* 46, 386–392, 1998.

44. Awasthi, S. & Boor, P.J. Semicarbazide protection from *in vivo* oxidant injury of vascular tissue by allylamine. *Toxicol. Lett.* 66, 157–163, 1993.

45. Boor, P.J. & Nelson, T.J. Allylamine cardiotoxicity: III. Protection by semicarbazide and *in vivo* derangements of monoamine oxidase. *Toxicology* 18, 87–102, 1980.

46. Nelson, T.J. & Boor, P.J. Allylamine cardiotoxicity: IV. Metabolism to acrolein by cardiovascular tissues. *Biochem. Pharmacol.* 31, 509–514, 1982.

47. Boor, P.J. & Nelson, T.J. Biotransformation of the cardiovascular toxin, allylamine, by rat and human cardiovascular tissue. *J. Mol. Cell Cardiol.* 14, 679–682, 1982.

48. Boor, P.J. & Hysmith, R.M. Allylamine cardiovascular toxicity. *Toxicology* 44, 129–145, 1987.

49. Boor, P.J., Sanduja, R., Nelson, T.J., & Ansari, G.A.S. *In vivo* metabolism of the cardiovascular toxin, allylamine. *Biochem. Pharmacol.* 36, 4347–4353, 1987.

50. Boor, P.J., Hysmith, R.M., & Sanduja, R. A role for a new vascular enzyme in the metabolism of xenobiotic amines. *Circ. Res.* 66, 249–252, 1990.

51. Beauchamp, R.O., Jr., Andjelkovich, D.A., Kligerman, A.D., Morgan, K.T., & Heck, H.D. A critical review of the literature on acrolein toxicity. *Crit. Rev. Toxicol.* 14, 309–380, 1985.

52. Witz, G. Biological interactions of alpha,beta-unsaturated aldehydes. *Free Radic. Biol. Med.* 7, 333–349, 1989.

53. Lyles, G.A. & Singh, I. Vascular smooth muscle cells: a major source of the semi-carbazide-sensitive amine oxidase of the rat aorta. *J. Pharm. Pharmacol.* 37, 637–643, 1985.

54. Sanduja, R., Ansari, G.A.S., & Boor, P.J. 3-Hydroxypropylmercapturic acid: a biologic marker of exposure to allylic and related compounds. *J. Appl. Toxicol.* 9, 235–238, 1989.

55. He, N., Singhal, S.S., Awasthi, S., Zhao, T., & Boor, P.J. Role of glutathione S-transferase 8-8 in allylamine resistance of vascular smooth muscle cells *in vitro*. *Toxicol. Appl. Pharmacol.* 158, 177–185, 1999.

56. Cao, Z. & Li, Y. Protecting against peroxynitrite-mediated cytotoxicity in vascular smooth muscle cells via upregulating endogenous glutathione biosynthesis by 3H-1,2-dithiole-3-thione. *Cardiovasc. Toxicol.* 4, 339–354, 2004.

57. Voskoboinik, I., Soderholm, K., & Cotgreave, I.A. Ascorbate and glutathione homeostasis in vascular smooth muscle cells: cooperation with endothelial cells. *Am. J. Physiol.* 275, C1031–C1039, 1998.

58. Aviram, M. & Fuhrman, B. LDL oxidation by arterial wall macrophages depends on the oxidative status in the lipoprotein and in the cells: role of prooxidants vs. antioxidants. *Mol. Cell Biochem.* 188, 149–159, 1998.

59. Rosenblat, M. & Aviram, M. Macrophage glutathione content and glutathione peroxidase activity are inversely related to cell-mediated oxidation of LDL: *in vitro* and *in vivo* studies. *Free Radic. Biol. Med* 24, 305–317, 1998.

60. Singhal, S.S. et al. Rabbit aorta glutathione S-transferases and their role in bioactivation of trinitroglycerin. *Toxicol. Appl. Pharmacol.* 140, 378–386, 1996.

61. Kenkare, S.R., Han, C., & Benet, L.Z. Correlation of the response to nitroglycerin in rabbit aorta with the activity of the mu class glutathione S-transferase. *Biochem. Pharmacol.* 48, 2231–2235, 1994.

62. Kenkare, S.R. & Benet, L.Z. Tolerance to nitroglycerin in rabbit aorta. Investigating the involvement of the mu isozyme of glutathione S-transferases. *Biochem. Pharmacol.* 51, 1357–1363, 1996.

63. Lee, W.I. & Fung, H.L. Mechanism-based partial inactivation of glutathione S-transferases by nitroglycerin: tyrosine nitration vs sulfhydryl oxidation. *Nitric. Oxide* 8, 103–110, 2003.

64. Fung, H.L. Biochemical mechanism of nitroglycerin action and tolerance: is this old mystery solved? *Annu. Rev. Pharmacol. Toxicol.* 44, 67–85, 2004.

65. Matsuzaki, T. et al. Effects of glutathione S-transferase inhibitors on nitroglycerin action in pig isolated coronary arteries. *Clin. Exp. Pharmacol. Physiol.* 29, 1091–1095, 2002.

66. Hill, K.E., Hunt, R.W., Jr., Jones, R., Hoover, R.L., & Burk, R.F. Metabolism of nitroglycerin by smooth muscle cells. Involvement of glutathione and glutathione S-transferase. *Biochem. Pharmacol.* 43, 561–566, 1992.

67. Lau, D.T. & Benet, L.Z. Effects of sulfobromophthalein and ethacrynic acid on glyceryl trinitrate relaxation. *Biochem. Pharmacol.* 43, 2247–2254, 1992.

68. Habdous, M., Siest, G., Herbeth, B., Vincent-Viry, M., & Visvikis, S. Glutathione S-transferases genetic polymorphisms and human diseases: overview of epidemiological studies. *Ann. Biol. Clin. Paris* 62, 15–24, 2004.

69. Masetti, S. et al. Interactive effect of the glutathione S-transferase genes and cigarette smoking on occurrence and severity of coronary artery risk. *J. Mol. Med.* 81, 488–494, 2003.

70. Miller, E.A. et al. Glutathione-S-transferase genotypes, smoking, and their association with markers of inflammation, hemostasis, and endothelial function: the atherosclerosis risk in communities, ARIC study. *Atherosclerosis* 171, 265–272, 2003.

71. Olshan, A.F. et al. Risk of atherosclerosis: interaction of smoking and glutathione S-transferase genes. *Epidemiology* 14, 321–327, 2003.

72. Park, J.H. et al. Glutathione S-transferase M1, T1, and P1 gene polymorphisms and carotid atherosclerosis in Korean patients with rheumatoid arthritis. *Rheumatol. Int.* 24, 157–163, 2004.

73. Salama, S.A. et al. Polymorphic metabolizing genes and susceptibility to atherosclerosis among cigarette smokers. *Environ. Mol. Mutagen.* 40, 153–160, 2002.

74. Tamer, L. et al. Glutathione S-transferase gene polymorphism as a susceptibility factor in smoking-related coronary artery disease. *Basic Res. Cardiol.* 99, 223–229, 2004.

75. Wilson, M.H., Grant, P.J., Kain, K., Warner, D.P., & Wild, C.P. Association between the risk of coronary artery disease in South Asians and a deletion polymorphism in glutathione S-transferase M1. *Biomarkers* 8, 43–50, 2003.

76. Wilson, M.H., Grant, P.J., Hardie, L.J., & Wild, C.P. Glutathione S-transferase M1 null genotype is associated with a decreased risk of myocardial infarction. *FASEB J.* 14, 791–796, 2000.

77. Li, R. et al. Interaction of the glutathione S-transferase genes and cigarette smoking on risk of lower extremity arterial disease: the Atherosclerosis Risk in Communities, ARIC study. *Atherosclerosis* 154, 729–738, 2001.

78. Palmer, C.N., Young, V., Ho, M., Doney, A., & Belch, J.J. Association of common variation in glutathione S-transferase genes with premature development of cardiovascular disease in patients with systemic sclerosis. *Arthritis Rheum.* 48, 854–855, 2003.

79. Li, R. et al. Glutathione S-transferase genotype as a susceptibility factor in smoking-related coronary heart disease. *Atherosclerosis* 149, 451–462, 2000.

80. Dušinská, M. et al. Glutathione S-transferase polymorphisms influence the level of oxidative DNA damage and antioxidant protection in humans. *Mutat. Res.* 482, 47–55, 2001.

81. de Waart, F.G. et al. Effect of glutathione S-transferase M1 genotype on progression of atherosclerosis in lifelong male smokers. *Atherosclerosis* 158, 227–231, 2001.

82. Wang, X.L. et al. Glutathione S-transferase mu1 deficiency, cigarette smoking and coronary artery disease. *J. Cardiovasc. Risk* 9, 25–31, 2002.

83. Forgione, M.A. et al. Heterozygous cellular glutathione peroxidase deficiency in the mouse: abnormalities in vascular and cardiac function and structure. *Circulation* 106, 1154–1158, 2002.

84. Blankenberg, S. et al. Glutathione peroxidase 1 activity and cardiovascular events in patients with coronary artery disease. *N. Engl. J. Med.* 349, 1605–1613, 2003.

85. Alissa, E.M., Bahijri, S.M., & Ferns, G.A. The controversy surrounding selenium and cardiovascular disease: a review of the evidence. *Med. Sci. Monit.* 9, RA9–18, 2003.

86. Nakamura, S. et al. Polymorphism in the 5'-flanking region of human glutamate-cysteine ligase modifier subunit gene is associated with myocardial infarction. *Circulation* 105, 2968–2973, 2002.

13 Mitochondrial Glutathione S-Transferase Pool in Health and Disease

Haider Raza and Narayan G. Avadhani

CONTENTS

13.1 INTRODUCTION

Mitochondria play an essential role in harnessing chemical energy in addition to carrying out key reaction steps for a number of metabolic pathways. These organelles not only help maintain the cellular metabolic homeostasis but also play an important role in the regulation of pathways leading to cell death.[1–3] Over 95% of the molecular oxygen consumed by the cells undergoes two electron reductions as part of mito-chondrial respiration coupled with oxidative phosphorylation to produce water. The consequence of this physiological process is the generation of reactive oxygen species (ROS), mainly O_2^-, which then converts to other molecular forms including H_2O_2, hydroxyl radical, and lipid peroxides.[1,2,4] It is estimated that under normal physiological conditions nearly 1 to 2% of the respiratory O_2 in different tissues is converted to O_2^-, and this rate sharply increases under various disease conditions.[5] It is also now widely accepted that mitochondria contain nitric oxide synthetase (NOS), although the precise molecular and genetic characteristics of mitochondrial NOS may vary depending on the cells and tissues.[6] Mitochondria generate substantial amounts of NO, which functions either as part of the physiological signaling process or as reactive molecular species that eventually contribute to cellular and tissue damage through the formation of peroxide radicals or protein nitrosylation.[7] NO has also been shown to modulate the activity of cytochrome c oxidase, the terminal oxidase of the mitochondrial electron transport chain, by direct interaction with the Fe^{2+} and Cu^{2+} at the active site of the enzyme.[8] Currently there is compelling evidence that mitochondria in mammalian cells are the predominant source of cellular ROS both under normal and pathological conditions, which not only contribute to mitochondrial dysfunction but also affect diverse types of cellular processes.[9]

Mammalian cell mitochondria are also the direct targets of oxidative and chemical stresses that lead to mitochondrial mutations and altered metabolic and genetic stresses.[9–11] A persistent oxidative stress has been implicated in the pathogenesis of diseases, including atherosclerosis, cancer, diabetes and ischemia/reoxygenation injury, calcium imbalance, apoptosis, and alcohol- and drug-induced toxicity.[1,12,13] Mitochondria are susceptible to the damaging effects of ROS leading to both accidental cell death and programmed cell death (necrosis and apoptosis). ROS-induced membrane permeability changes and release of cytochrome c are critical steps in the propagation of apoptotic signals.[3,14,15] Mitochondrially generated ROS are also reported to induce mitochondrial membrane permeability transition leading to the disruption of mitochondrial transmembrane potential, $\Delta\Psi m$.[5] Additionally, mitochondria from different mammalian tissues contain both xenobiotic inducible and constitutively expressed cytochrome P450-type monooxygenases, which actively contribute to substrate metabolism, and phase II drug-metabolizing enzymes for the detoxification of metabolic products by conjugation.[16,17] In this way mitochondria are also implicated in chemically induced stress.

Maintenance of the mitochondrial GSH (glutathione) pool presents an important yardstick that allows cells to acclimate against ROS-induced cell signaling and toxicity. The mitochondrial antioxidant system is composed of various thiols, SOD, NADP transhydrogenases, NADH, Vitamin C, Vitamin E, and thiol peroxidase.[18] The mitochondrial GSH system consists of GSH/GSSG, selenium glutathione peroxidase (GPX), glutathione S-transferases (GSTs), and glutathione reductase, which can sustain GSH-dependent free radical scavenging activity in mitochondria.[19,20] In addition, the mitochondrial thioredoxin system (thioredoxin, thioredoxin reductase, and thioredoxin peroxidase) is also very active in ROS removal, mitochondrial DNA repair, and cell signaling.[21] This chapter, however, mostly covers topics related to the mitochondrial GSH and GST pools in physiological and pathophysiolagal situations and their involvement in respiratory function and ROS production.

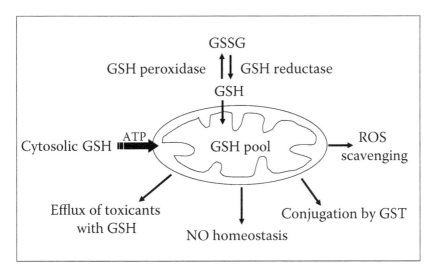

FIGURE 13.1 Maintenance of the mitochondrial GSH pool. The figure shows schematic representation of dynamic mitochondrial GSH homeostasis in mitochondria.

13.2 MITOCHONDRIAL GSH, ROS, AND RESPIRATORY FUNCTION

13.2.1 SOURCE AND SIZE OF MITOCHONDRIAL GSH POOL

The mitochondrial membrane compartment contains about 15% of the total cellular GSH pool, and more than 90% of it in the reduced form under normal physiological conditions.[22,23] Mitochondria lack some of the essential enzymes for the synthesis of GSH from amino acid components.[24] Currently there is general agreement that a sizable fraction of the mitochondrial pool is derived by importation from the cytosolic compartment[22,25] and a negligible part, if any, synthesized inside the mitochondrial compartment. Initial studies[24] reported the existence of an ATP-dependent GSH transporter protein on the inner mitochondrial membrane. More recently this system has been characterized as an exchanger of dicarboxylate and 2-oxogluterate in the inner mitochondrial membranes of kidneys and livers.[22] The lipid-protein composition and fluidity of the inner membrane are highly critical for the proper functioning of this transporter. A number of studies also suggest that mitochondrial uncoupler UCP2, which is induced by cellular stress conditions, may downregulate GSH transport into mitochondria.[26] An outline of mitochondrial GSH usage and its sources is presented in Figure 13.1.

GSH is highly critical for mitochondrial integrity and functions such as membrane permeability transition-induced proton potential, activities of respiratory chain complexes, protection against mitochondrial DNA and membrane damage, and oxidative phosphorylation. Despite its exclusive synthesis in the cytosol, GSH is distributed in subcellular organelles such as the nucleus and endoplasmic reticulum (ER). In most of the subcellular compartments, GSH is found mainly in the reduced form. The exceptions appear to be the ER where the mostly oxidized form (GSSG) is predominant mainly due to the lack of significant redox recycling.[22]

13.2.2 THE ROLE OF MITOCHONDRIAL GSH AND ROS IN RESPIRATION

The cellular GSH pool and ROS production are markedly altered in a number of physiological conditions.[27] The superoxide anion (O_2^-) and H_2O_2 are the normal products of aerobic metabolism in mitochondria due to "leakage" of electrons from the inner membrane. The physiological rate of ROS production depends on the mitochondrial metabolic state, the ADP/O_2 ratio, the membrane proton gradient, and the presence of the antioxidant defense system including GSH and GSH-metaboliz-ing enzymes.[28–30] At the physiological levels ROS are also involved in different cellular functions — as mediators of signal transduction pathways, mitosis, apoptosis, activation of tyrosine kinase and MAP kinase pathways, and NFκB-, ARE-, AP1-, and JNK-dependent gene expressions.[31–33] Recent studies also suggest that mitochondrial ROS and redox signaling may activate the Hif pathway, which regulates many cellular functions through activation of specific nuclear genes.[34] In addition to mitochondrial metabolic stress due to ROS, mitochondrial DNA breaks and mutations are also induced when the mitochondrial GSH pool is depleted.[35] Mitochondrial DNA depletion also induces hypoxia-like GSH pool shifts by inhibiting GSH reductase and GSH recycling.[36] GSH peroxidase was markedly higher in mitochondrial DNA-depleted cells. This suggests that the lack of a respiratory chain prompts the cells to reduce the need for antioxidant defenses, exposing them to a major risk of oxidative injury. Oxidative modification (mainly nitrosylation and carbonylation) of mitochondrial proteins, which become preferred substrates for proteasome-based degradation, is also reported under depleted mitochondrial GSH and increased ROS production as seen in hypoxia.[37,38] A shift in the mitochondrial GSH/GSSG pool leads to increased glutathionylation of mitochondrial respiratory chain complexes and inactivation of critical mitochondrial enzymes. Increased glutathionylation also renders proteins more resistant to ROS-induced proteolytic degradation, suggesting a protective role of glutathionylation.[39] A complex interrelationship between the mitochondrial GSH pool, GSH redox status, NADP/NADPH status, respiratory function, and H_2O_2 production seems to exist in mitochondria and when mitochondrial GSH is depleted by about 50%, H_2O_2 diffuses from mitochondria to an extramitochondrial compartment under oxidative stress conditions.[40,41]

13.3 MITOCHONDRIAL GSH POOL UNDER OXIDATIVE AND CHEMICAL STRESS

13.3.1 GENERAL ASPECTS OF MITOCHONDRIAL GSH

Glutathione (GSH) homeostasis and cellular compartmentalization have been implicated in toxicity, cell signaling, survival, and cell death.[41–43] The cellular GSH pool is markedly altered in a number of physiological conditions including aging, starvation, pregnancy, obesity, and the diurnal cycle.[44] Various kinds of chemicals and pathological insults are also known to cause cellular GSH stress. Two major cellular compartments — the cytosol and mitochondria — are known to contain about 5–15 mM GSH pools.[22] A critical function of GSH is to participate in cellular redox activities that are essential for survival. An imbalance in the redox cycling can cause oxidative stress. Thus GSH homeostasis is essential for countering oxidative stress-induced alterations in a number of physiological and pathological conditions.

13.3.2 THE ROLE OF MITOCHONDRIAL GSH IN DETOXIFICATION OF ROS

Mitochondrial redox activities are essential for oxygen utilization, ATP synthesis, membrane proton gradient, cell signaling, cell survival, and cell death.[45,46] It is known that a small percentage (1 to 2%) of O_2 is partially reduced by a single electron that forms ROS and other oxidant species (O_2^-, H_2O_2, hydroxyl radicals, etc.). Antioxidant defense systems inside the mitochondrial membrane compartment protects the organelles against deleterious effects of ROS. The mitochondrial superoxide dismutase (MnSOD) is the major antioxidant defense, which converts superoxide to H_2O_2. Because of the abundance of Cu_2^+-containing and heme/non-heme Fe_2^+-containing enzymes in the mitochondrial compartment, H_2O_2 is converted to a more reactive hydroxyl radical through the Fenton reaction. GSH reacts covalently with ROS and activated chemicals and also serves as a substrate for glutathione peroxidase and GSTs, which besides the transferase activity also have peroxidase activity. Conjugation of 4-hydroxynonenal (4-HNE) and other toxic lipid alkenals to GSH and reduction of lipid peroxides represent major cellular defense mechanisms against oxidative stress-induced toxicity.[1,47–49]

With the exception of the heart, most tissues do not contain detectable catalase activity in the mitochondrial compartment. Thus, the GSH defense system is the main route of detoxification of H_2O_2 in conjunction with GSH peroxidase and GSSG reductase activities. Depletion of GSH below a critical level would therefore compromise detoxification of H_2O_2, leading to build-up of O_2^- and hydroxyl radicals. Recent reports show that selective depletion of mitochondrial GSH by (R,S)-3-hydroxy-4-pentenoate leads to increased ROS generation from complex I and III of the respiratory chain,[40,50,51] which can be effectively reversed by restoring the GSH levels in the mitochondria. Mitochondrial thiols and NO together play important roles not only in maintaining the mitochondrial GSH homeostasis but also in the reduction of both H_2O_2 and organic hydroperoxides.[52–54] An important but relatively less understood role of mitochondrial GSH is modulation of nutrient transport such as glucose transport by inducing altered membrane permeability.[55] Recent results in our laboratory show that depletion of mtGSH causing increased ROS production initiates a mitochondria-to-nucleus stress signaling that alters the nuclear gene expression pattern.[10,11,56]

13.3.3 THE ROLE OF MITOCHONDRIAL GSH IN NITRIC OXIDE METABOLISM

Another fast emerging area of research involves the role of mitochondrial GSH in the detoxification of mitochondrially generated NO as part of physiological or pathological processes. Results show that NO induces oxidative stress in mitochondria, which simultaneously increases protein modification (carbonylation and nitrosylation).[8] Other studies show that mitochondrial GSH may have a direct or indirect protective role in NO-mediated inhibition of both complex I and complex IV.[7] Some models propose that mitochondrial NO reacts with O_2^- produced in the mitochondrial respiratory chain to form highly reactive $ONOO^-$, which in turn causes irreversible inhibition of respiratory complex activities.[50,53] Production of NO at physiological or pathological levels also differentially alters the cytosolic and mitochondrial GSH pools.[57] NO-induced inhibition of complex I and II activities following the depletion of the mitochondrial GSH pool and GSTs may contribute to many neurodegenerative disorders, including Parkinson's disease.[42,50] GSH conjugation of NO and members of GST are also considered as a natural reservoir and carriers of NO.[58] Thus it is clear that NO and GSH homeostasis are tightly integrated in mitochondrial signaling and oxidative stress.

13.4 MULTIPLE FORMS OF GLUTATHIONE S-TRANSFERASES IN MITOCHONDRIA

The canonical GST family is thought to have evolved from a thioredoxin/glutaredoxin progenitor.[59] The induction of GST appears to be an evolutionarily conserved response of cells to oxidative stress. The mitochondrial GSTK1-1, however, appears to be a parallel evolutionary pathway not from the other canonical GST family.[60,61] Three major families of GST proteins that are widely distributed in nature exhibit GST activity. Two of the cytosolic and mitochondrial GSTs comprise soluble enzymes that are only distantly related. The third family is composed of microsomal GST, which is now referred to as membrane-associated protein in eicosanoid and GSH (MAPEG) metabolism.[62] One membrane-bound microsomal GST isoform was also detected in the outer membrane of mitochondria.[63] Cytosolic and mitochondrial GSTs share some similarities in their 3-D folds[59–61] but do not share a structural resemblance to the MAPEG family. The cytosolic GSTs and mitochondrial GSTs are catalytically active as dimers, with the dimer interphase providing a noncatalytic ligand-binding site. Based on amino acid sequence similarities, seven classes of cytosolic GSTs are recognized in mammalian species, designated as alpha, mu, pi, sigma, theta, omega, and zeta.[59] Human and rodent cytosolic GST isoenzymes within a class typically share more than 40% identity and those between classes share less than 25% identity. The majority of cytosolic GST isoenzymes are found in the cytoplasm of the cell. However, mouse and human alpha classes can be translocated into mitochondria and other cytoplasmic membrane compartments.[64–67]

13.4.1 MITOCHONDRIAL GSTK1-1

Harris et al.[68] had initially characterized a novel rat liver mitochondrial GST, designated as GST13-13. This enzyme was later classified as GSTK1-1 and a human homolog hGSTK1-1 was also identified.[69] The mitochondrial GSTK1-1 is more closely related to the protein disulfide bond isomerase from *E. coli* than it is to other members of the GST superfamily. GSTK1-1 is a dimeric form of mitochondrial GST, comprising 226 amino acids.[59,60,70] GSTK1-1 exhibits high activity for CDNB and can also reduce cumene hydroperoxide and (S)-15 hydroperoxy-5,8,11,13-eicosatetraenoic acid.[70] Electron microscopy and other molecular biological approaches have been used to confirm the occurrence of GST kappa in mitochondria and peroxisomes of different tissues.[71,72]

13.4.2 OTHER MITOCHONDRIAL GST FORMS

We previously reported the existence of the alpha-class GSTs in mouse liver mitochondria, which are structurally and catalytically similar to their cytosolic counterparts[16] and exhibit high activity for both CDNB and cumene hydroperoxide. Recently, we presented evidence for the existence of multiple forms of cytosolic GSTs in mouse liver mitochondria. We purified three isoforms of cytosolic GST alpha and mu from mouse liver mitochondria.[66] Based on N-terminal MALDI/TOF analysis, catalytic activities, and immunochemical properties, we characterized them as GSTA1-1, GSTA4-4, and GSTM1-1. All these GSTs showed peroxidase activity but the maximum activity was associated with GSTA4-4. We also showed that mitochondrial GSTs efficiently catalyzed the conjugation of GSH to 4-HNE, an end product of lipid peroxidation. The expression of mitochondrial GSTs, particularly GSTA4-4, was increased in COS cells when treated with 4-HNE[66] or inducers of protein kinases, cAMP, and phorbol ester (Figure 13.2). GSH pools in the cytosol and mitochondria were found to be differentially affected under oxidative stress conditions. The recovery of the mitochondrial GSH pool was delayed compared to the cytsolic GSH when oxidants were withdrawn from the medium. This delay is presumably related to the delayed recycling of mitochondrial GSH by GSH reductase and GSH peroxidases.[66]

Our previous *in vivo* studies with nicotine-treated rats have also shown increased expression of mitochondrial GSTA4-4 at mRNA and protein levels in the brain.[73] Gardner and Gallagher[65] showed the presence of GSTA4-4 only in the mitochondrial compartment but not in the cytosol and that the mitochondrial form was about 4 kDa longer than the cytosolic form. By contrast, we and others have shown the presence of GSTA4-4 in both the cytosol and mitochondria.[47,66] Furthermore, we did not observe any difference in size or amino acid sequences of the mitochondrial and cytosolic forms. We also showed that GSTA4-4 both in the cytosol and mitochondria are induced under oxidative stress conditions. Using the *in vivo* streptozotocin-induced diabetes model in rats, we recently demonstrated an increased expression of GSTA4-4 in mitochondria, which was accompanied with increases in the expression on the stress marker protein HSP70 and CYP2E1.[74] In this study we also showed increased ROS production and membrane lipid peroxidation in the mitochondria of tissues in diabetic rats. The increased expression of GSTA4-4 along with other stress marker proteins suggests its role in the protection of mitochondrial functions against deleterious effects of ROS and 4-HNE. 4-HNE is also believed to act as an intracellular signaling molecule[47] and therefore its conjugation to GSH should influence cell survival and cell death signaling.[46,75] Currently we are studying the characterization of other GST forms in mouse and rat liver mitochondria.

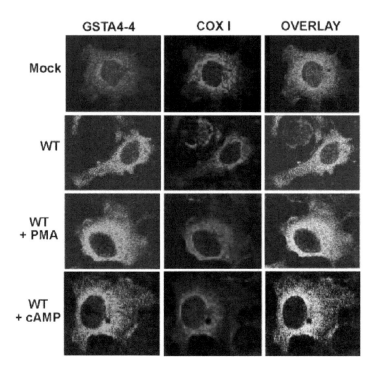

FIGURE 13.2 (See color insert following page 178.) Confocal immunofluorescence micro-scopic studies on mitochondrial accumulation of GSTA4-4 in COS cells. COS cells were mock transfected with empty vector or with cDNA for GSTA4-4 cloned in pCMV. Cells were treated with cAMP or phorbol myristyl acetate (PMA) and mitochondrial translocated GSTA4-4 was visualized by using an antibody against GSTA4-4. A bonafide mitochondrial protein, cytochrome c oxidase (COX), was also visualized under similar conditions. The overlapping nature of the expression of these proteins suggests increased mitochondrial accumulation of GSTA4-4 by kinases.

13.5   MECHANISMS OF MITOCHONDRIAL IMPORT OF CYTOSOLIC GSTS

Dual targeting of mitochondrial GST kappa in mitochondria and peroxisomes has been reported recently.[70] There is only a single GSTK1-1 in both organelles pre-sumably involved in beta-oxidation of fatty acids. The exact mechanism of targeting of this protein has not been studied. A possible N-terminal cleavage and involvement of chaperone HSP60 has been postulated by Morel et al.[70] A peroxisomal sequence ala-arg-leu at the C-terminus end of GSTK1 was also identified.

13.5.1   MECHANISM OF MITOCHONDRIAL TARGETING OF GSTA4-4

We have reported multiple forms of GSTs in the mitochondrial matrix that are structurally and functionally similar to their cytosolic counterparts. Because there was no detectable processing of N-terminal or C-terminal ends of mitochondrial GSTs purified from mouse liver mitochondria,[16,66] we reasoned that mitochondrial importation may involve the activation of cryptic signals similar to that established for some of the CYP proteins.[76] In our recent study, using *in vitro* mitochondrial import assay and transient transfection in COS cells, we demonstrated that essentially intact cytosolic GSTA4-4 is imported into mitochondria.[77] The *in vitro* import assay was performed using ^{35}S-methionine-labeled *in vitro* translation product of mouse GSTA4-4 cDNA in rabbit reticulocyte system. Import of the translation product was shown using freshly prepared mitochondria into a protease-resistant membrane compartment. Interestingly, mitochondrial import of GSTA4-4, both in the *in vitro* and intact cell systems, required phosphorylation of a tandem PKA (protein kinase A)/PKC (protein kinase C) site located close to the C-terminus. Mutations at Ser-189 and Thr-193, the residues that are phosphorylated by PKA and PKC, respectively, show that phosphorylation is critical for mitochondrial import and that increased phosphorylation enhanced the mitochondrial translocation of GSTA4-4. A similar observation was made in COS cells transfected with GSTA4-4 cDNA, where an increased mitochondrial GSTA 4-4 was detected in the presence of added cAMP or phorbol ester (Figure 13.2). Accordingly, we also observed that mitochondria-targeted GSTA4-4 was phosphorylated at a higher level compared to the cytosolic GSTA4-4.[77]

Our results also showed that cytosolic HSP70 plays an important role in mitochondrial translocation of GSTA4-4. Phosphorylated GSTA4-4 bound to HSP70 seems to be preferentially translocated as indicated by co-immunoprecipitation experiments. Adding HSP70 protein to an *in vitro* import reaction mixture enhanced the mitochondrial import of GSTA4-4, while immunodepletion of HSP70 from rabbit reticulocyte lysate or the inhibition of PKA/PKC-mediated phosphorylation greatly reduced mitochondrial translocation.[77] Figure 13.3 shows a proposed model for mitochondrial translocation of GSTA4-4 and the role of different kinases in oxidative stress conditions.

13.6   MITOCHONDRIAL GSTS IN THE PATHOGENESIS OF DISEASE

GST isoforms play an important role in maintaining GSH pools in different cellular compartments. Selective overexpression of isoforms is considered to be an adaptation to chemical or oxidative stress and presents a defense mechanism against these stress conditions. GSTs also play critical roles in drug-induced resistance and efflux of toxins as GSH conjugates. Increased ROS production and lipid peroxidation of mitochondrial membrane lipids may result in increased 4-HNE production, which also affects numerous mitochondrial functions and induces proteolytic degradation of critical proteins. Cytosolic GSTs of the alpha family and particularly GSTA4-4 have been implicated in the metabolism of 4-HNE and in protection against oxidative stress.[47,78,79] GSTA4-4 is one of the main cytosolic isoforms identified to be translocated into mitochondria or the plasma membrane under oxidative stress conditions.[64,65,77]

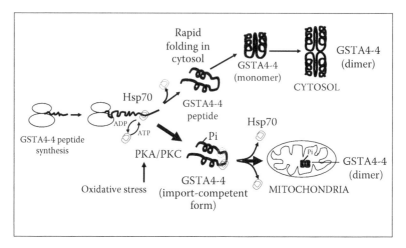

FIGURE 13.3 Schematic model of GSTA4-4 translocation into mitochondria. This figure shows the process of GSTA4-4 synthesis, energy-dependent interaction with molecular chaperone HSP70, and enhanced import of GSTA4-4 into mitochondria after PKA/PKC-dependent phosphorylation under increased oxidative stress conditions.[77]

A number of endogenous and exogenous substrates regulate the expression of various GSTs in normal and disease conditions, thus influencing the levels of mitochondrial GSTs.[80] An important function of mitochondrial GSTs is the metabolism of endogenous 4-HNE. Increased 4-HNE is known to affect the activity of signaling pathways such as that involving the Keap I/Nrf2 pathway,[71,75,78,81] thus ultimately affecting the patterns of nuclear gene expression. The level of 4-HNE is increased in order of magnitude in diabetes and other oxidative stress-related conditions, which may alter nuclear gene expression in addition to affecting mitochondrial functions and membrane-bound proteins.[74,82] As noted above, the mitochondrial electron transport chain is the main source of ROS during normal metabolism, and the rate of ROS production from mitochondria is increased in a variety of pathologies. Diseases such as long-term ethanol intake, hepatitis, cirrhosis, and cerebral ischemia are associated with a decrease in the mitochondrial GSH pool.[27,37,39,40] A hypoxic environment is known to promote tumor growth and survival by several mechanisms, all of which point to ROS as the critical effectors and GSH/GST as the scavengers.[42,45] Nitric oxide is another regulator of ROS production, GSH homeostasis, and mitochondrial respiratory functions, which are involved in aging and neurodegenerative disorders such as Parkinson's disease and Alzheimer's disease.[7,41,57,83] In Figure 13.4 we highlighted various pathophysiological conditions that are known to be associated with the mitochondrial GSH pool, ROS, and NO metabolism.

13.7 CONCLUSIONS

The mitochondrial GSH/GST redox system is directly or indirectly related to mitochondrial respiratory chain functions, and a downward shift in this pool would affect mitochondrial structural and functional integrity. The mitochondrial GSH/GST pool

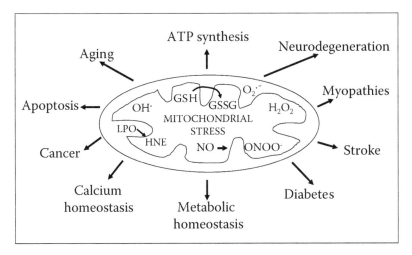

FIGURE 13.4 Mitochondrial stress-related abnormalities. This figure shows numerous pathophysiological conditions where the mitochondrial GSH pool, ROS, and NO have been implicated.

shifts in disease conditions and in drug- or alcohol-induced toxicity. Alterations in respiratory complex activities increase ROS production exponentially. We observed an increase in mitochondrial ROS formation in increased oxidative stress conditions and have provided evidence that ROS production in experimental diabetes is linked to an increased mitochondrial translocation of CYP2E1 and GSTs. Our findings on the PKA/PKC-mediated translocation of cytosolic GSTs to mitochondria have important implications in modulating the mitochondrial GST levels under various pathologies. It is becoming increasingly apparent that mitochondrial ROS and associated membrane and DNA damage may play critical roles in various diseases including cancer, diabetes, neurodegeneration, and aging. In fact, a major effort is being undertaken to find ways of manipulating mitochondrial antioxidant defense under these pathological conditions. Selective modulation of mitochondrial GSH/GSTs using chemicals or drugs such as esterified GSH and ectopic expression of modified GSTs similar to that used in our study may be of great importance for future strategies of "mitochondrial medicine" for treatment of diseases.

ACKNOWLEDGMENTS

We are thankful to Drs. Subbuswamy K. Prabu and Marie-Anne Robin for their help and valuable contributions to our ongoing research. We also acknowledge National Institutes of Health grants GM-49683 and GM34881 and Terry Fox Cancer Research Fund, FMHS, UAE University for supporting this research.

REFERENCES

1. Genova, M.L. et al. The mitochondrial production of reactive oxygen species in relation to aging and pathology, *Ann. N.Y. Acad. Sci.*, 1011, 86, 2004.

2. Davies, K.J. Oxidative stress: the paradox of aerobic life, *Biochem. Soc. Symp.* 61, 1, 1995.
3. Green, D.R. and Reed, J.C. Mitochondria and apoptosis, *Science,* 281, 1309, 1998.
4. Garcia-Ruiz, C. et al. Role of oxidative stress generated from the mitochondrial electron transport chain and mitochondrial glutathione status in loss of mitochondrial function and activation of transcription factor nuclear factor-kappa B: studies with isolated mitochondria and rat hepatocytes, *Mol. Pharmacol.,* 48, 825, 1995.
5. Zamzami, N. et al. Sequential reduction of mitochondrial transmembrane potential and generation of reactive oxygen species in early programmed cell death, *J. Exp. Med.,* 182, 367, 1995.
6. Ghafourifar, P. and Cadenas, E. Mitochondrial nitric oxide synthase, *Trends Pharmacol. Sci.,* 26, 190, 2005.
7. Shiva, S. et al. Nitroxia: the pathological consequence of dysfunction in the nitric oxide-cytochrome c oxidase signaling pathway, *Free Rad. Biol. Med.,* 38, 297, 2005.
8. Antunes, F., Boveris, A., and Cadenas, E. On the mechanism and biology of cytochrome oxidase inhibition by nitric oxide, *Proc. Natl. Acad. Sci. USA,* 101, 16774, 2004.
9. Poderoso, J.J., Boveris, A., and Cadenas, E. Mitochondrial oxidative stress: a self-propagating process with implications for signaling cascades, *Biofactors,* 11, 43, 2000.
10. Amuthan, G. et al. Mitochondrial stress-induced calcium signaling, phenotypic changes and invasive behavior in human lung carcinoma A549 cells, *Oncogene,* 21, 7839, 2002.
11. Biswas, G., Anandatheerthavarada, H.K., and Avadhani, N.G. Mechanism of mitochondrial stress-induced resistance to apoptosis in mitochondrial DNA-depleted C2C12 myocytes, *Cell Death Differ.,* 12, 266, 2005.
12. Cadenas, E. Mitochondrial free radical production and cell signaling, *Mol. Asp. Med.,* 25, 17, 2004.
13. Cadenas, E. and Davies, K.J. Mitochondrial free radical generation, oxidative stress, and aging, *Free Rad. Biol. Med.,* 29, 222, 2000.
14. Zamzami, N. et al. Mitochondrial implication in accidental and programmed cell death: apoptosis and necrosis, *J. Bioener. Biomem.,* 29, 185, 1997.
15. Wang, X. The expanding role of mitochondria in apoptosis, *Genes Dev.,* 15, 2922, 2001.
16. Addya, S. et al. Purification and characterization of a hepatic mitochondrial glutathione S-transferase exhibiting immunochemical relationship to the alpha-class of cytosolic isoenzymes, *Arch. Biochem. Biophys.,* 310, 82, 1994.
17. Anandatheerthavarada, H.K. et al. Localization of multiple forms of inducible cytochromes P450 in rat liver mitochondria: immunological characteristics and patterns of xenobiotic substrate metabolism, *Arch. Biochem. Biophys.,* 339, 136, 1997.
18. Halliwell, B. and Gutteridge, J.M.C. *Free Radicals in Biology and Medicine,* Clarendon Press, Oxford, 1989.
19. Marchetti, P. et al. Redox regulation of apoptosis: impact of thiol oxidation status on mitochondrial function, *Eur. J. Immunol.,* 27, 296, 1997.
20. Meister, A. Mitochondrial changes associated with glutathione deficiency, *Biochim. Biophys. Acta,* 1271, 35, 1995.
21. Arner, E.S. and Holmgren, A. Physiological functions of thioredoxin and thioredoxin reductase, *Eur. J. Biochem.,* 267, 6102, 2000.
22. Fernandez-Checa, J.C. and Kaplowitz, N. Hepatic mitochondrial glutathione: transport and role in disease and toxicity, *Toxicol. Appl. Pharmacol.,* 204, 263, 2005.

23. Filomeni, G., Rotilio, G., and Ciriolo, M.R. Cell signalling and the glutathione redox system, *Biochem. Pharmacol.*, 64, 1057, 2002.

24. Griffith, O.W. and Meister, A. Origin and turnover of mitochondrial glutathione, *Proc. Natl. Acad. Sci. USA*, 82, 4668, 1985.

25. Fernandez-Checa, J.C. et al. Mitochondrial glutathione: importance and transport, *Sem. Liver Dis.*, 18, 389, 1998.

26. Arsenijevic, D. et al. Disruption of the uncoupling protein-2 gene in mice reveals a role in immunity and reactive oxygen species production, *Nature Genet.*, 26, 435, 2000.

27. Giordano, F.J. Oxygen, oxidative stress, hypoxia, and heart failure, *J. Clin. Invest.*, 115, 500, 2005.

28. Inarrea, P. et al. Redox activation of mitochondrial intermembrane space Cu,Zn-superoxide dismutase, *Biochem. J.*, 387, 203, 2005.

29. Chandel, N.S. and Schumacker, P.T. Cellular oxygen sensing by mitochondria: old questions, new insight, *J. Appl. Physiol.*, 88, 1880, 2000.

30. Kowaltowski, A.J. et al. Catalases and thioredoxin peroxidase protect *Saccharomyces cerevisiae* against Ca(2+)-induced mitochondrial membrane permeabilization and cell death, *FEBS Lett.*, 473, 177, 2000.

31. Pinkus, R. et al. Role of oxidants and antioxidants in the induction of AP-1, NF-kappa B, and glutathione s-transferase gene expression, *J. Biol. Chem.*, 271, 13422, 1996.

32. Lundberg, A.S. et al. Genes involved in senescence and immortalization, *Curr. Opin. Cell Biol.*, 12, 705, 2000.

33. Clement, M.V. and Pervaiz, S. Reactive oxygen intermediates regulate cellular response to apoptotic stimuli: an hypothesis, *Free Rad. Res.*, 30, 247, 1999.

34. Chandel, N.S. et al. Reactive oxygen species generated at mitochondrial complex III stabilize hypoxia-inducible factor-1alpha during hypoxia: a mechanism of O2 sensing, *J. Biol. Chem.*, 275, 25130, 2000.

35. Hauptmann, N. et al. The metabolism of tyramine by monoamine oxidase A/B causes oxidative damage to mitochondrial DNA, *Arch. Biochem. Biophys.*, 335, 295, 1996.

36. Beer, S.M. et al. Glutaredoxin 2 catalyzes the reversible oxidation and glutathiony-lation of mitochondrial membrane thiol proteins: implications for mitochondrial redox regulation and antioxidant DEFENSE, *J. Biol. Chem.*, 279, 47939, 2004.

37. Lluis, J.M. et al. Critical role of mitochondrial glutathione in the survival of hepato-cytes during hypoxia, *J. Biol. Chem.*, 280, 3224, 2005.

38. Avila, M.A. et al. Regulation by hypoxia of methionine adenosyltransferase activity and gene expression in rat hepatocytes, *Gastroenterology*, 114, 364, 1998.

39. Kil, I.S. and Park, J.W. Regulation of mitochondrial NADP+-dependent isocitrate dehydrogenase activity by glutathionylation, *J. Biol. Chem.*, 280, 10846, 2005.

40. Han, D. et al. Effect of glutathione depletion on sites and topology of superoxide and hydrogen peroxide production in mitochondria, *Mol. Pharmacol.*, 64, 1136, 2003.

41. Shen, D. et al. Glutathione redox state regulates mitochondrial reactive oxygen production, *J. Biol. Chem.*, 280, 25305, 2005.

42. Sims, N.R., Nilsson, M., and Muyderman, H. Mitochondrial glutathione: a modulator of brain cell death, *J. Bioener. Biomem.*, 36, 329, 2004.

43. Sauer, H., Wartenberg, M., and Hescheler, J. Reactive oxygen species as intracellular messengers during cell growth and differentiation, *Cell. Physiol. Biochem.*, 11, 173, 2001.

44. Thannickal, V.J. and Fanburg, B.L. Reactive oxygen species in cell signaling, *Am. J. Physiol. — Lung Cell. Mol. Physiol.*, 279, 11005, 2000.

45. Wenger, R.H. Cellular adaptation to hypoxia: O2-sensing protein hydroxylases, hypoxia-inducible transcription factors, and O2-regulated gene expression, *FASEB J.*, 16, 1151, 2002.

46. Levonen, A.L. et al. Cellular mechanisms of redox cell signalling: role of cysteine modification in controlling antioxidant defences in response to electrophilic lipid oxidation products, *Biochem. J.,* 378, 373, 2004.

47. Awasthi, Y.C. et al. Regulation of 4-hydroxynonenal-mediated signaling by glutathione S-transferases, *Free Rad. Biol. Med.,* 37, 607, 2004.

48. LeBras, B.M. et al. Reactive oxygen species and the mitochondrial signaling pathway of cell death, *Histol. Histopathol.,* 205, 1920.

49. Dickinson, D.A. et al. Cytoprotection against oxidative stress and the regulation of glutathione synthesis, *Biol. Chem.,* 384, 527, 2003.

50. Hsu, M. et al. Glutathione depletion resulting in selective mitochondrial complex I inhibition in dopaminergic cells is via an NO-mediated pathway not involving peroxynitrite: implications for Parkinson's disease, *J. Neurochem.,* 92, 1091, 2005.

51. Chen, Q. et al. Production of reactive oxygen species by mitochondria: central role of complex III, *J. Biol. Chem.,* 278, 36027, 2003.

52. Boveris, A. and Cadenas, E. Mitochondrial production of hydrogen peroxide regulation by nitric oxide and the role of ubisemiquinone, *IUBMB Life,* 50, 245, 2000.

53. Brookes, P.S. et al. Mitochondria: regulators of signal transduction by reactive oxygen and nitrogen species, *Free Rad. Biol. Med.,* 33, 755, 2002.

54. Netto, L.E. et al. Thiol enzymes protecting mitochondria against oxidative damage, *Methods Enzymol.,* 348, 260, 2002.

55. Hammond, C.L., Lee, T.K., and Ballatori, N. Novel roles for glutathione in gene expression, cell death, and membrane transport of organic solutes, *J. Hepatol.,* 34, 946, 2001.

56. Srinivasan, S. and Avadhani, N. Unpublished research, 2005.

57. Roychowdhury, S. et al. Cytosolic and mitochondrial glutathione in microglial cells are differentially affected by oxidative/nitrosative stress, *Nitric Oxide,* 8, 39, 2003.

58. De, M.F. et al. The specific interaction of dinitrosyl-diglutathionyl-iron complex, a natural NO carrier, with the glutathione transferase superfamily: suggestion for an evolutionary pressure in the direction of the storage of nitric oxide, *J. Biol. Chem.,* 278, 42283, 2003.

59. Armstrong, R.N. Structure, catalytic mechanism, and evolution of the glutathione transferases, *Chem. Res. Toxicol.,* 10, 2, 1997.

60. Ladner, J.E. et al. Parallel evolutionary pathways for glutathione transferases: structure and mechanism of the mitochondrial class kappa enzyme rGSTK1-1, *Biochemistry,* 43, 352, 2004.

61. Robinson, A. et al. Modelling and bioinformatics studies of the human kappa-class glutathione transferase predict a novel third glutathione transferase family with similarity to prokaryotic 2-hydroxychromene-2-carboxylate isomerases, *Biochem. J.,* 379, 541, 2004.

62. Bresell, A. et al. Bioinformatic and enzymatic characterization of the MAPEG superfamily, *FEBS J.,* 272, 1688, 2005.

63. Morgenstern, R. et al. The distribution of microsomal glutathione transferase among different organelles, different organs, and different organisms, *Biochem. Pharmacol.,* 33, 3609, 1984.

64. Singh, S.P. et al. Membrane association of glutathione S-transferase mGSTA4-4, an enzyme that metabolizes lipid peroxidation products, *J. Biol. Chem.,* 277, 4232, 2002.

65. Gardner, J.L. and Gallagher, E.P. Development of a peptide antibody specific to human glutathione S-transferase alpha 4-4 (hGSTA4-4) reveals preferential localization in human liver mitochondria, *Arch. Biochem. Biophys.,* 390, 19, 2001.

66. Raza, H. et al. Multiple isoforms of mitochondrial glutathione S-transferases and their differential induction under oxidative stress, *Biochem. J.,* 366, 45, 2002.

67. Prabhu, K.S. et al. Microsomal glutathione S-transferase A1-1 with glutathione peroxidase activity from sheep liver: molecular cloning, expression and characterization, *Biochem. J.,* 360, 345, 2001.

68. Harris, J.M. et al. A novel glutathione transferase (13-13) isolated from the matrix of rat liver mitochondria having structural similarity to class theta enzymes, *Biochem. J.,* 278, 137, 1991.

69. Pemble, S.E., Wardle, A.F., and Taylor, J.B. Glutathione S-transferase class kappa: characterization by the cloning of rat mitochondrial GST and identification of a human homologue, *Biochem. J.,* 319, 749, 1996.

70. Morel, F. et al. Gene and protein characterization of the human glutathione S-transferase kappa and evidence for a peroxisomal localization, *J. Biol. Chem.,* 279, 16246, 2004.

71. Jowsey, I.R. et al. Biochemical and genetic characterization of a murine class kappa glutathione S-transferase, *Biochem. J.,* 373, 559, 2003.

72. Thomson, R.E. et al. Tissue-specific expression and subcellular distribution of murine glutathione S-transferase class kappa, *J. Histochem. Cytochem.,* 52, 653, 2004.

73. Bhagwat, S.V. et al. Preferential effects of nicotine and 4-(N-methyl-N-nitrosamine)-1-(3-pyridyl)-1-butanone on mitochondrial glutathione S-transferase A4-4 induction and increased oxidative stress in the rat brain, *Biochem. Pharmacol.,* 56, 831, 1998.

74. Raza, H. et al. Elevated mitochondrial cytochrome P450 2E1 and glutathione S-transferase A4-4 in streptozotocin-induced diabetic rats: tissue-specific variations and roles in oxidative stress, *Diabetes,* 53, 185, 2004.

75. Ishii, T. et al. Role of Nrf2 in the regulation of CD36 and stress protein expression in murine macrophages: activation by oxidatively modified LDL and 4-hydroxynonenal, *Circ. Res.,* 94, 609, 2004.

76. Robin, M.A. et al. Bimodal targeting of microsomal CYP2E1 to mitochondria through activation of an N-terminal chimeric signal by cAMP-mediated phosphorylation, *J. Biol. Chem.,* 277, 40583, 2002.

77. Robin, M.A. et al. Phosphorylation enhances mitochondrial targeting of GSTA4-4 through increased affinity for binding to cytoplasmic Hsp70, *J. Biol. Chem.,* 278, 18960, 2003.

78. Okada, K. et al. 4-Hydroxy-2-nonenal-mediated impairment of intracellular proteolysis during oxidative stress. Identification of proteasomes as target molecules, *J. Biol. Chem.,* 274, 23787, 1999.

79. Cheng, J.Z. et al. Two distinct 4-hydroxynonenal metabolizing glutathione S-transferase isozymes are differentially expressed in human tissues, *Biochem. Biophys. Res. Commun.,* 282, 1268, 2001.

80. Hayes, J.D., Flanagan, J.U., and Jowsey, I.R. Glutathione transferases, *Annu. Rev. Pharmacol. Toxicol.,* 45, 51, 2005.

81. Nguyen, T., Sherratt, P.J., and Pickett, C.B. Regulatory mechanisms controlling gene expression mediated by the antioxidant response element, *Annu. Rev. Pharmacol. Toxicol.,* 43, 233, 2003.

82. Eaton, P. et al. Formation of 4-hydroxy-2-nonenal-modified proteins in ischemic rat heart, *Am. J. Physiol.,* 276, H935, 1999.

83. Zoccarato, F., Toscano, P., and Alexandre, A. Dopamine-derived dopaminochrome promotes H(2)O(2) release at mitochondrial complex I: stimulation by rotenone, control by Ca(2+), and relevance to Parkinson disease, *J. Biol. Chem.,* 280, 15587, 2005.

14 Activation of Microsomal Glutathione Transferase 1 in Toxicology

*Miyuki Shimoji, Yoko Aniya,
and Ralf Morgenstern*

CONTENTS

14.1 INTRODUCTION

Glutathione transferases (GSTs, EC 2.5.1.18) play an important role in cellular biotransformation and detoxification of toxic/genotoxic electrophiles of xenobiotic as well as endogenous origin. The existence of both soluble and membrane-bound enzymes that share substrate specificity[1,2] underscores the broad and overlapping functional role of these enzymes. Model substrates for the glutathione transferases (GSTs) are 1-chloro-2,4-dinitrobenzene (CDNB), epoxides, organic nitrite esters, and ethacrynic acid among many others, while endogenous substrates include lipid hydroperoxides and hydroxyalkenals.[2–4] Several cytosolic enzymes and microsomal glutathione transferase 1 (MGST1) display glutathione (GSH)-dependent peroxidase (GPx) activity toward organic hydroperoxides and phospholipid hydroperoxides.[2,5,6] The influence of several reagents and stress conditions on the activity of cytosolic GSTs and MGST1 has been studied. Among the GSTs, only MGST1 is activated by sulfhydryl reagents (e.g., GST activity toward CDNB[7–9] and GPx activity toward cumene hydroperoxide[10,11]).

As a result of the progress in genome sequence analysis and bioinformatics, six human proteins including MGST1 could be designated as MAPEG (membrane-associated proteins involved in eicosanoid and glutathione metabolism) members.[12,13] The superfamily contains MGST1, MGST2, MGST3, microsomal prostaglandin E synthase 1 (mPGES-1), leukotriene C_4 synthase (LTC$_4$S), and 5-lipoxygenase activating protein (FLAP). MGST1, MGST2, MGST3, LTC$_4$S, and mPGES-1 have GSH-dependent transferase activities whereas human MGST3 and LTC$_4$S do not display catalytic activity toward CDNB.[14–18] All MGSTs and mPGES-1 display GPx activity.[14–18] FLAP is hypothesized to act as a substrate provider for 5-lipoxygenase in leukotriene A_4 production and is necessary for its efficient activity.[19–21] The main role of mPGES-1 is efficient PGE production, with other activities being very low.[16] The information to date points to a general role in detoxication for MGST1, MGST2, and MGST3 where the latter also could take part in leukotriene production in certain cell types. mPGES-1, LTC$_4$S, and FLAP on the other hand have specific roles in the production of prostaglandins and leukotrienes. They commonly convert reactive lipophilic substances (such as those formed during lipid peroxidation or drug metabolism or by the specific conversion of arachidonic acid to epoxide or endoperoxide) to nonreactive water-soluble products targeted for excretion (e.g., MGST1-3) or to specific and powerful mediators in fever, pain, and inflammation relief (e.g., mPGES-1, LTC$_4$S, and FLAP). All MAPEG proteins display a similar structure containing about 150 amino acids harboring four hydrophobic segments that traverse the membrane.[12,22] Members investigated so far (MGST1, mPGES-1, and LTC$_4$S) are homotrimers[16,23,24] where the active site faces the cytosol.[25] Amino acids that are necessary for activity have been identified.[26,27] Structural work is progressing[28] that will hopefully put these into context and reveal the detailed mechanism of this group of enzymes.

14.1.1 Microsomal GST (MGST1)

MGST1 was originally purified in its activated form from rat liver microsomes that had been incubated with N-ethyl maleimide (NEM).[9] Rat liver microsomes and the unactivated purified enzyme show a maximal 8- and 20-fold activation by NEM, respectively.[2] The microsomal fraction actually contains cytosolic GSTs, and several cytosolic isoforms (e.g., cytosolic GST alpha and zeta[5,6]) appear to localize also in the endoplasmic reticulum (ER) membrane.[29-31] However, cytosolic GSTs in the ER membrane are not activated by sulfhydryl reagents.[2,32,33] MGST1 contains one cysteine residue per subunit in the homotrimer, which causes activation following covalent modification.[2,8,9] Mixed-disulfide formation also results in activation[34] and in addition inter-subunit cross-linking of cysteines, leading to oligomerization, has been suggested as an activation mechanism.[7] In mammals, MGST1 is expressed ubiquitously in the ER and outer mitochondrial membranes.[35-41] MGST is broadly expressed in nonvertebrate and mammalian species but not in nematode and yeast.[12,13,42,43]

As opposed to MGST1, neither cytosolic GST nor other MAPEG members display activation by NEM as summarized below. In addition, NEM activation appears to be a late phenomenon in evolution as activation has so far only been observed in mammals; in fact amphibian and fish MGSTs can be inactivated by NEM.[42] This is also supported by unpublished observations from the authors' laboratories. The capacity of MGST1 to become activated is logically mirrored in its general unresponsiveness to common inducers of drug metabolism.[44] Several investigations over the years have shown that MGST1 can be activated by a plethora of treatments *in vivo* and *in vitro* (Figure 14.1). Although the detailed mechanisms in many cases have not been investigated, it is clear that chemical and proteolytic modification as well as ligand interactions can activate the enzyme in an irreversible or reversible fashion. Several protein modification sites, including the Cysteine-49 sulfhydryl and cleavage after Lysine-41, are indicated (Figure 14.2). The activated MGST1 appears to afford increased protection from lipid peroxidation,[45-47] and it

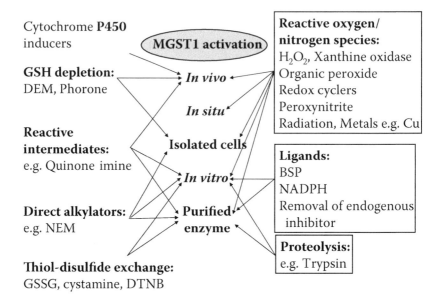

FIGURE 14.1 Summary of MGST1 activation mechanisms in systems ranging from the purified enzyme to whole organisms.

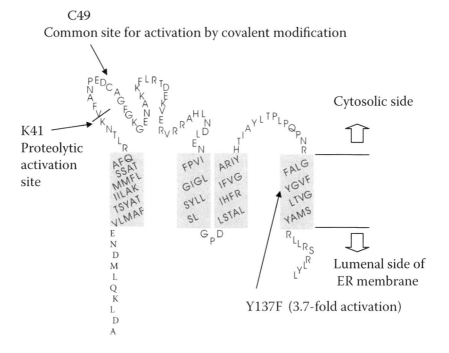

FIGURE 14.2 Target sites for MGST1 activation by chemical modification, proteolysis, or mutation (modified from reference 28).

remains to be demonstrated whether the activated enzyme yields an increased level of protection in more complex systems.

It is a peculiar fact that activation of MGST1 could not be observed in most extrahepatic microsomes except for rat testes microsomes,[30] although it is known that the enzyme is present in most tissues.[36] Also, in freshly prepared human liver microsomes the enzyme could not be activated until further solubilized and purified.[48] Hydrogen peroxide was found to activate the enzyme only under specific experimental conditions.[49] Thus activation is enigmatic by being curiously absent in some systems. This chapter summarizes and critically examines data on the activation of MGST1, with the aim to provide a broad collective perspective on mechanisms of activation and to suggest avenues for future research.

14.2 CONDITIONS LEADING TO ACTIVATION OF MGST1

14.2.1 IN VIVO ACTIVATION

In general, when *in vivo* experiments are performed it is important to be able to discriminate between induction and activation. By using Western blot analysis it is possible to determine whether protein levels were increased. This has been performed in a few cases. It might appear difficult to obtain the level of precision required to analyze activation levels of the 1.3- to 2-fold often encountered. However, as cytosolic GSTs account for some 70% of the activity in untreated rat liver microsomes[50] an increase in activity of 1.3-fold actually corresponds to at least doubling of the specific activity of MGST1. Overall, as induction of MGST1 is rarely observed, the increase in activity resulting from short-term *in vivo* treatments reviewed below and summarized in Table 14.1 is most likely explained by activation, although this has not been rigorously determined. The *in vivo* models examined include oxidative stress, disease, and reactive intermediate formation. Activation of MGST1 by various agents and plausible mechanisms of activation are discussed in the following sections.

14.2.1.1 Cytochrome P450 (P450) Inducers as Activators of MGST1

Inducers of P450s can be good activators of MGST1 under certain conditions. Thus phenobarbital (PB), 3-methylcholanthrene (3-MC), and acetone have activated MGST1 by 1.2- to 1.5-fold.[51,52] Both GST and GPx activities of MGST1 are enhanced by PB and 3-MC treatment. In addition, acetone-treated rats showed an increase of GST activity in liver microsomes.[51] It has been suggested that reactive oxygen species (ROS) formed by P450 result in the activated forms of MGST1[52] as ROS can activate MGST1 *in vitro*.[7,53] The increase of MGST1 activity by several different P450 inducers (Table 14.1) indicates that ROS generated from various P450 isoforms can activate MGST1. Recently electron spin resonance (ESR) and chemiluminescence measurements show that production of ROS occurs and it corresponds with the activation of MGST1 (Aniya, unpublished data). The P450 system also produces reactive intermediates by a variety of reactions. It has been shown that

TABLE 14.1

Activation of MGST (GST and GPx) in Microsomes of Rat Livers after Different *In Vivo* Treatments

Treatment	% of Control Microsomes	
Phenobarbital[52]	147	187
3-Methylcholanthrene[52]	121	109
Acetone[51]	125	
CoCl$_2$[61]	141	54
Di-ethyl maleate[62]	177	
Di-ethyl maleate[63]	214	157
Phorone[62]	246	
Alpha-naphtylisothiocyanate[73]	182	
1,2-dibromo-ethane[75]	128	
CCl$_4$[75]	159	
CCl$_4$[77]	139	
Acetaminophen[80]	176	126
Allyl alcohol[83]	190	
Galactosamine[67]	135	70
Galactosamine + LPS[67]	160	85
Streptozotocin[72]	75	
Streptozotocin + Insulin[72]	98	

Note: GST and GPx activity were measured with CDNB and CuOOH, respectively. Percentage change from control in the references indicated was calculated.

reactive intermediates — such as quinones and quinone imines — can cause direct activation of MGST1.[54–57] In the treatment of rats with PB, other factors can possibly activate MGST1, as for example the altered phospholipid content during remodeling of the proliferating ER membrane.[58] Although ROS and reactive intermediates are certainly attractive candidates, the exact mechanism of activation via P450 inducers *in vivo* remains to be clarified.

14.2.1.2 Activation of MGST1 during Hypoxia

Cobalt chloride is a well-known inducer of heme oxygenase, following P450 suppression,[59] and it causes hypoxia experimentally in cells and *in vivo*.[60] Treatment of rats for 24 hours with cobalt chloride increased GST activity in liver microsomes 1.4-fold; however, it inhibited the GPx activity[61] (Table 14.1). Since actinomycin D prevented the increase, MGST1 was most likely induced in this case. It is possible that the GST and GPx activities of MGST1 are modulated differentially under conditions of hypoxia. Alternatively, the increased activity may originate from a

GST other than MGST1 that does not support GPx activity or the GPx activity of other cytosolic GSTs could be comparatively more or less sensitive to hypoxia.

14.2.1.3 Activation of MGST1 by GSH Depletors

Di-ethyl maleate (DEM) and phorone treatment results in augmented MGST1 activity and it has been suggested that an increase of oxidized glutathione (GSSG) results in the activation of MGST1 *in vivo*.[62] Liver microsomes of DEM-treated rats show an increase of the transferase and GPx activities concomitant with GSH depletion, and the mechanism has been suggested to be a covalent binding of DEM after its metabolism *in vivo*,[63] in analogy to *in vitro* incubations where NEM binds MGST1 covalently and both GST and GPx activities are markedly increased.[8,9,11] Only GST activity remains increased after GSH levels are recovered again, suggesting the possibility of differential activation. As dithiothreitol (DTT) does not reverse the activation with DEM *in vitro* incubations,[63] the involvement of thiol-disulfide interchange forming a protein GSH-mixed disulfide could be ruled out. In contrast, MGST activity is not increased in isolated rat hepatocytes after DEM treatment.[64] Two activation mechanisms could explain the *in vivo* results. One is an activation mechanism that depends on the suggested GSSG formation *in vivo*,[62,63] which is less likely because DEM and phorone react with GSH to form a conjugate. Also a considerable amount of GSSG relative to GSH needs to be formed to cause activation because it has been shown that a 20-fold excess is required to yield half the maximal activation of MGST1 via thiol-disulfide interchange.[65] In other words the Cys-49 sulfhydryl in MGST1 is not easily oxidized to a mixed disulfide. The other mechanism might entail activation by hydrogen peroxide (that causes *in vitro* activation, vide infra) with expected increased concentrations of hydrogen peroxide when GSH is depleted. By contrast, phorone, another GSH depletor, activates MGST1 both in *in vivo* and in isolated rat hepatocytes. DEM and phorone thus display discrete activation characteristics and it should be underscored that phorone does not activate MGST1 directly. Phorone activation, which is very pronounced (3-fold) in isolated hepatocytes,[64] may offer a convenient experimental approach for understanding *in vivo* activation.

14.2.1.4 Lipopolysaccharide (LPS) and D-Galactosamine (GalN) as Activators of MGST1

LPS produces nitrosative stress. NO, ONOO$^-$ formation, induces cytokines (e.g., interleukin [IL]-1 and IL-6) and promotes nuclear transcription factor kappa B (NF-κB)–related signaling.[66] GalN induces liver damage similar to viral hepatitis and cytosolic GST activity has been shown to be decreased by GalN due to release into serum.[67] Contrary to the cytosolic GSTs, MGST1 is activated by treatment with GalN plus LPS *in vivo*.[67] The fact that hepatic MGST1 mRNA is downregulated by LPS[68] further supports that its activation does indeed occur. As oxidative stress is believed to be the main cause of GalN/LPS-induced liver injury, ROS thus generated may contribute to the activation of MGST1. S-nitrosylation also causes activation *in vitro*[69] and is a possible *in vivo* activator. However, MGST1 is not S-nitrosylated during endotoxin-mediated shock[70] while other liver microsomal proteins are modified.

14.2.1.5   MGST1 and Diabetes

During diabetes ROS are produced at higher levels.[71] MGST1 activation in rat livers after chemically induced diabetes has been examined and a decrease in liver cytosolic and microsomal GST activities has been observed in long-term streptozotocin-induced diabetic rats. Only microsomal GST activity is restored by insulin treatment.[72]

14.2.2   EFFECT OF HEPATOTOXINS ON MGST1

14.2.2.1   Alpha-Naphtylisothiocyanate (ANIT)

ANIT induces cholestatic liver injury that is mediated by a biotransformation of ANIT via the P450 system. ANIT can activate MGST1 up to 1.8-fold[73] and phenobarbital does not show any additive activation effect. Covalent binding of reactive metabolites and removal of inhibitors by increased bile acids has been suggested as activation mechanisms. It is possible that increased ligand levels, such as bile acids, could activate the enzyme directly.

14.2.2.2   Carbon Tetrachloride and 1,2-Dibromo-Ethane

CCl_4 represents a classical hepatotoxin where the reactive intermediates *in vivo* are radicals that can also lead to ROS production.[74] MGST1 activity was enhanced by CCl_4. The mechanism was suggested to involve covalent binding of a reactive intermediate to Cys-49,[75] oxidative modification by radical compounds, and subsequent reactions.[76,77] The activation of GST could offer enhanced protection against damage by lipid peroxidation in this case.[76,77] In addition 1,2-dibromo-ethane causes a slight activation possibly by the known reactive intermediate GSH conjugate.[78]

14.2.2.3   Acetaminophen (AAP)

AAP is an analgesic/antipyretic drug that can cause necrotic liver injury.[79] The reactive intermediate that is formed via P450 biotransformation, N-acetyl-*p*-benzo-quinone imine, binds to MGST1,[54,56,57] resulting in its activation. Upon treatment of rats with AAP, both microsomal GST and GPx activities increase up to 3-fold and the liver GSH content is depleted markedly.[80] During AAP toxicity inducible NO synthase (iNOS) is increased along with the superoxide levels.[79,81] Peroxynitrite $(ONOO^-)$ is formed from the reaction between NO and superoxide and subsequently nitrotyrosine is formed upon reaction of protein tyrosine residues with $ONOO^-$ in the presence of carbon dioxide. Nitrotyrosine formation is increased during AAP-induced liver injury. As it is known that $ONOO^-$ can activate MGST1 *in vitro*,[46,82] this could be an additional mechanism of MGST1 activation during liver injury. More important, it forms an attractive activation mechanism following liver injury involving iNOS induction. The extent of MGST1 nitration under *in vivo* conditions clearly is an interesting area for exploration.

14.2.2.4   Effects of Herbal Extracts on MGST1

Several herbs and plant constituents have been reported to display antioxidant effects, protection against liver toxicity in rodents, and chemopreventive effects in rodent cancer models[67,84–87] (Aniya, unpublished data). It has been observed that herbal extracts with antioxidant action cause a decrease in MGST1 activity *in vivo*[76] and that activation of MGST1 resulting from the pretreatment of GalN plus LPS or CCl$_4$ is suppressed by *Monascus anka* (mold) extracts[77] or by herbal extract[76] (*Thonningia sanguinea*) corresponding to protection from hepatotoxicity and oxidative stress. In addition, antioxidant constituents involved in some herbs including quercetin and gallic acid can differentially affect MGST1 activity *in vitro,* leading to its activation or inhibition (Aniya, unpublished data). Thus these compounds as well as the herbal extracts containing these compounds can be evaluated both as potential activators of MGST1 or preventive modulators of MGST1 activation in toxicity models.

14.2.3   IN SITU ACTIVATION OF MGST1: MODELS OF OXIDATIVE STRESS AND LIPID PEROXIDATION

14.2.3.1   Ischemia/Reperfusion

Ischemia/reperfusion in isolated rat livers did activate MGST1 but not cytosolic GSTs. An activation mechanism was suggested involving Cys-49 modification by the stress treatment.[88] Ischemia reflow followed by perfusion with the nitrovasodilator Nicorandil (SG-75) in isolated rat livers inhibited activation of MGST1. Nicorandil has known free radical–scavenging properties and it has been suggested to inhibit radical-induced activation of MGST1 in the ischemia/reperfusion experiment.[89]

14.2.3.2   Organic Hydroperoxide

Tert-butyl hydroperoxide perfusion of isolated rat livers activated MGST1. Dimerization was suggested as the activation mechanism, because a correlation between dimer formation in SDS-polyacrylamide electrophoresis gels and an increase of transferase activity was observed.[47,90] In addition, almost quantitative mixed disulfide formation with GSH could be demonstrated by Western blot analysis after perfusion with a high concentration of cumene hydroperoxide.[65]

14.2.4   IN VITRO ACTIVATION OF MGST1

To date numerous agents have been investigated for their effects on MGST1 activity. These agents are listed in Table 14.2 with citations, and possible mechanisms of activation of MGST1 by some of these agents are discussed in the following sections. In general these mechanisms point toward covalent binding and mixed disulfide formation.

14.2.4.1   N-Ethyl Maleimide

The "classical" activator NEM covalently binds Cys-49 of MGST1, resulting in the activation of both GST and GPx activities.[8,9,11] Hydrogen/deuterium exchange exper-

TABLE 14.2
Activation of MGST Activity *In Vitro*

| | | Activation Fold | | |
| | | Liver | Purified | |
Treatment	Reference	Microsomes	Enzyme	Mechanism
N-ethyl maleimide	8, 10	7.8	15	Covalent binding to Cys-49
	69		13	
	83	7.0		
	34	5.7	20	
	11	2.4		
Diamide	34, 92	1.8	2.8	Protein disulfide formation
Diamide + GSH	34	5.4	9.6	Mixed disulfide formation
Cystamine	34	5.4	3.3	Mixed disulfide formation
Cystamine + GSH	34		3.2	Mixed disulfide formation
GSSG	34	3.2		Mixed disulfide formation
	62, 65	2.0		
		4–5		
Cystine	62	2.6		Mixed disulfide formation
Homocystine	62	2.5		Mixed disulfide formation
GSNO	69	2		S-nitrosylation/mixed disulfide formation
DEA (1,1-diethyl-2-hydroxy-2-nitrosohydrazine)/NO	69	2		S-nitrosylation
ONOO⁻	46	3	4.3	Tyrosine nitration
Acrolein (generated from allyl alcohol)	83	2.2		Modification of Cys-49
N-phenylthiourea, N-alpha-methylbenzyl-thiourea, N,N′-diethylthiourea, N,N′-diphenylthiourea, thioperamide	102	1.7–2.7		Modification of Cys-49
Te-phenyl-L-tellurcysteine, Se-phenyl-L-selenocysteine	104	1.5		Modification of Cys-49
1,2,4-benzene triol	103	2	6	Modification of Cys-49 by autoxidation product
Benzoquinone	103	2	8	Modification of Cys-49
Chlorambucil	105		2.5	Modification of Cys-49
Methyldopa + NADPH	55	2.3		Arylation of Cys-49
4-methyl-ortho-benzoquinone	55	2.9		Arylation of Cys-49
Noradrenaline	101	1.6		ROS-generated oxidative
	53	1.8		modification of Cys-49 and covalent modification by ortho-quinone
Noradrenaline + NADPH	53	4.0		ROS-generated oxidative modification of Cys-49
Hydrogen peroxide	7	1.5	15	Oxidation of Cys-49

TABLE 14.2 *(Continued)*
Activation of MGST Activity *In Vitro*

Treatment	Reference	Activation Fold		Mechanism
		Liver Microsomes	Purified Enzyme	
Linoleic acid hydroperoxide + GSH	90	1.4		Oxidation of Cys-49/mixed disulfide formation
T-butyl hydroperoxide	90	2.2		Oxidation of Cys-49
Cumene hydroperoxide	90	2.2		Oxidation of Cys-49
NADPH-generating system	49, 53	2.0	1.4	Oxidation of Cys-49 by ROS generation or ligand-induced autoxidation
NADPH	49		1.2	Ligand-induced autoxidation
ATP	49		1.3	Ligand-induced autoxidation
Xanthine oxidase + xanthine	53, 101	1.1–1.9		Oxidation of Cys-49 by ROS or proteolysis by contaminating protease
Radiation	115	2		Oxidation by ROS
Proteoliposome preparation	4		2.2–4.2	Oxidation during prolonged dialysis
Xanthine oxidase	101	1.8		Proteolysis by contaminating protease
Trypsin	110	5	10	Proteolysis at Lys-41
Di-ethyl maleate	63	2.4		
Heating 50°C	112	2.6		
Small unilamellar vesicles	111	5.0		Removal of endogenous inhibitor
Bromosulphophtalein	108, 109		2.5	Reversible ligand interaction during turnover

iments demonstrated that the dynamics of the activated enzyme resemble those of the GSH-bound state, suggesting that activation facilitates GSH interactions.[91]

14.2.4.2 Diamide, Cystamine, and Glutathione Disulfide

Diamide, which oxidizes thiols to disulfides and cystamine, increased MGST1 activity 3- and 2-fold, respectively. The former compound increased MGST1 activity by 10-fold when GSH was present, probably through an efficient mixed disulfide formation and dimer formation of MGST1.[53,92,93] Activation with glutathione disulfide alone has also been observed by several investigators.[34,94–96] This activation occurs through reversible thiol/disulfide exchange[34] and is thus a potential mechanism for *in vivo* regulation. However, it has been shown that the equilibrium for mixed disulfide formation on MGST1 is very unfavorable[94] and half maximal activation requires a large (20-fold) excess of oxidized versus reduced GSH as mentioned above. In all cases mixed disulfide formation could be confirmed by reversal of activation by the reductant DTT. This reagent thus forms an important tool to investigate the possibility of mixed disulfide-dependent activation of MGST1 but has not been employed in some of the *in vivo* activation experiments where this type of activation was suggested.[96] Although thermodynamics do not favor activation by mixed disulfide formation[65] it is possible that slow kinetics of reduction favor a transient activated state.

14.2.4.3   S-Nitrosoglutathione (GSNO) and ONOO⁻

Nitric oxide (NO) is an important signaling molecule.[97] S-nitrosoglutathione (GSNO) may function as a storage or transport form of NO *in vivo* and may participate in transnitrosation reactions.[69] Interestingly, GSNO is formed catalytically from organic nitrite esters by GSTs including MGST1.[98] GSNO activates MGST1 by S-nitrosylation.[69] Peroxynitrite also activates MGST1.[46,82] Oxidative modification of Cys-49 and nitration of tyrosine residues of MGST1 have been suggested as possible activation mechanisms.[46] In general, the modification of thiols by ROS and reactive nitrogen species (RNS) is potentially important and a variety of modified forms have been defined.[99] It is likely that different chemical modifications result from hydrogen peroxide and ONOO⁻ (ONOO⁻ is a stronger oxidant). Furthermore, xenobiotics can cause nitrosative stress as mentioned above,[66,100] and ONOO⁻ is also produced by activated phagocytes and endothelial cells.[97] We are presently pursuing the molecular mechanism of MGST1 activation by ONOO⁻, described earlier,[46,82] using mass spectrometry.

14.2.4.4   Alpha-Methyldopa and Noradrenaline

Reactive products from alpha-methyldopa formed by the P450 system are able to stimulate MGST1.[55] It has been shown that the quinone of alpha-methyldopa is responsible for the stimulation of MGST1 via arylation of Cys-49. Noradrenaline also activates MGST1 but different activation mechanisms have been suggested.[53,101] One possibility is the generation of reactive intermediates (or autoxidation to an ortho-quinone) generating covalent modification similar to the activation by alpha-methyldopa.[55] Modification of Cys-49 could also occur via ROS that are generated during autoxidation of catechols.[53]

14.2.4.5   Acrolein

Allyl alcohol is metabolized to acrolein that reacts readily with sulfhydryl groups. When rats were treated with allyl alcohol, liver microsomal GST activity increased 1.9-fold.[83] This result was confirmed by *in vitro* incubations of acrolein with rat liver microsomes where the same degree of activation was also detected. Acrolein is suggested to be the causative agent for liver necrosis. Interestingly, acrolein activation has no effect on GSH-dependent protection against lipid peroxidation in rat liver microsomes, which is thought to be one function of MGST1. It is thus enigmatic whether MGST1 can offer better protection in its activated form. It is conceivable that acrolein-dependent sulfhydryl modification also results in extensive sulfhydryl modification of microsomal proteins. As sulfhydryls are contributors to membrane antioxidant capacity, this would counteract protection and lead to a status quo in this particular case. This suggestion is consistent with the observation that cystamine, a charged sulfhydryl reagent that would have access to fewer membrane sulfhydryls, does activate protection.[45] In addition, we suggested a protection by activated MGST1 in different combinations of CCl_4 and herbal extracts.[76,77,86]

14.2.4.6 Other Activators

It has been reported that MGST1 is activated by reactive thionosulfur intermediates formed from N-phenylthioureas during metabolism via microsomal flavin-containing monooxygenase (FMO).[102] MGST1 is localized in the same membrane as FMOs and P450s. The close spatial interaction between these drug-metabolizing enzymes could thus be a very important factor in activating MGST1. The anticancer drug chlorambucil increases MGST1 activity toward the standard substrate CDNB in a time- and dose-dependent manner.[105] As activation also increases the catalytic capacity toward chlorambucil, the activation of MGST1 has been suggested to contribute to drug resistance toward chlorambucil in cancer cells.

Tellurium, selenium, and sulfur compounds are known as effective antioxidants and chemoprotectors. Te-Phenyl-L-tellurocysteine and Se-Phenyl-L-selenocysteine show strong, time-dependent MGST1 activation in the presence of NADPH and it has been suggested that an oxidation-prone selenolate/tellurolate intermediate forms a mixed diseleno/telluro-thiol disulfide with Cys-49.[104] NADPH-dependent activation with metabolites of phenol has been detected in rat liver microsomes.[103] The reactive metabolites, benzoquinone and 1,2,4-benzene triol, increased GST activity, and the incubation of rat liver microsomes with ^{14}C-labeled phenol showed covalent binding of the metabolites.

14.3 ACTIVATION OF MGST1 BY REACTIVE OXYGEN SPECIES

14.3.1 HYDROGEN PEROXIDE AND ORGANIC HYDROPEROXIDES

Activation of MGST1 by hydrogen peroxide has been found to vary in reports from different laboratories. Liver microsomes prepared without a liver perfusion step did not yield activation of MGST1 by hydrogen peroxide. Membranes from the BL21 *E. coli* strain expressing MGST1 did not show activation with hydrogen peroxide in spite of an observed activation with NEM (maximum 2-fold). By contrast, potassium chloride–perfused liver and the purified enzyme showed activation with hydrogen peroxide (independent of GSH content).[7,49,92] Furthermore, the microsomal fraction of cultured cells did not show marked activation with hydrogen peroxide. In a side-by-side set of experiments we showed that the microsome preparation procedure influences the rate at which hydrogen peroxide is decomposed in the incubations.[49] Thus varying (or lack of) activation of MGST1 by hydrogen peroxide after different preparation procedures, in different sources and different organs, can now be rationalized.[7,46,49,55] The suggested mechanisms of activation involve sulfhydryl oxidation and dimer formation.[7,92] Linoleic acid hydroperoxide can also activate MGST1,[47] wherefore one might speculate that lipid hydroperoxides formed in the membrane could also be effective.

14.3.2 MGST1 ACTIVATION BY AN NADPH-GENERATING SYSTEM OR XANTHINE OXIDASE (XO)

Microsomal incubations containing an NADPH-generating system produce ROS (e.g., superoxide anion). MGST1 is activated up to 1.5-fold[49,53] in microsomes incubated with NADPH, although lack of activation has also been reported.[55,102] It has been suggested that ROS formation contributes to oxidative modification of Cys-49. However, NADPH could have an alternate activation mechanism because NADPH (and ATP) can activate MGST1 directly, as observed with the purified enzyme.[49] Thus the mechanism might be complex. Activation can be diminished simply by reducing the NADPH concentration.[103]

The enzyme XO has been used to produce oxidative stress *in vitro* because XO generates ROS during turnover.[106] Our previous ischemia/reperfusion study in rat livers showed activation of MGST1[88] could be mediated by ROS formed in part by XO. Recently experiments with rat liver microsomes revealed that hydroxyl radicals generated by the Fenton reaction could activate MGST1 (Aniya, unpublished data). However, it has been demonstrated that the increase of GST activity can also result from a contaminating protease in addition to radical-induced activation.[101] Therefore, although it is clear that one of the products of XO (hydrogen peroxide) can activate MGST1, controls for proteolysis during XO treatment have to be included.

14.3.3 ANTIOXIDANTS INCLUDING SUPEROXIDE DISMUTASE, CATALASE, AND POLYPHENOLS

Although antioxidants would not be expected to activate MGST1, intermediates in their action could possibly contribute to activation and thus confound their role as experimental tools (for instance, by acting as pro-oxidants such as ascorbate or tocopherol or in enzymatic side reactions).

Incubation of these antioxidant enzymes with rat liver microsomes did not activate MGST1 as expected.[53] Catalase under some conditions, however, can activate MGST1. Incubation of rat liver microsomes with catalase and sodium azide (a catalase inhibitor) showed an additive activation to radical-induced activation of MGST1 (Aniya, unpublished data). A mechanism involving an azide-bound heme that generates reactive ROS or formation of an azidyl radical resulting in activation of MGST1 is possible. Superoxide dismutase increases the rate of hydrogen peroxide production in systems where superoxide is generated.

The antioxidative effect of quercetin has been studied by Bast's group, and thiol modification was reported,[107] raising the possibility that the compound could function as a covalent MGST1 activator. Our pilot experiments, however, showed that quercetin inhibited activation via hydrogen peroxide. Quercetin also strongly inhibits MGST1 activity (Aniya, unpublished data). Gallic acid, on the other hand, shows an additive activation of MGST1 in cultured cells treated with hydrogen peroxide (Aniya, unpublished data). Gallic acid is a structural unit of tannin and the structure resembles that of benzene triol, which is a known activator of MGST1 and a metabolite of phenol.[103] Thus certain antioxidants have the potential to form reactive intermediates or contribute to redox cycling with the capacity to yield various modifications of Cys-49 of MGST1 (e.g., disulfide, thioether, or oxidized forms of Cys-49) as described in detail.[99]

14.3.4 METAL IONS AND LIGAND BINDING

Metal ions have been found to increase MGST1 transferase activity (Aniya, unpublished data). MGST1 activity of microsomes and the purified enzyme was increased markedly by Cu compounds and formation of dimers was observed. The known radical formation capability of these activators has been confirmed by ESR. Bromo-sulphophthalein (BSP) activates and inhibits MGST1 directly in the assay (in the presence of a large excess of GSH).[108] The activation depends on the concentration of BSP and the nature of the substrate. BSP inhibited NEM-activated MGST1.[109] Thus BSP can work both as an activator and inhibitor, depending on concentration. Incubation of purified MGST1 with NADPH or ATP at room temperature results in a significant increase of the activity. The increase of MGST1 activity can be reversed by the reducing reagent DTT or inclusion of GSH. This implies that Cys-49 is involved in the stimulation of MGST1 activity with these compounds and that the interaction might involve facilitated oxidation.[49]

14.3.5 ACTIVATION OF MGST1 BY LIMITED PROTEOLYSIS

Trypsin activates MGST1 by limited proteolysis, resulting in fragmentation at Lys-41.[110] Cleavage at Lys-4 was also observed in rat MGST1. As human MGST1 does not have Lys-4 and can be activated by trypsin, any contribution of proteolysis at Lys-4 can be ruled out. Oxidative stress could promote limited proteolysis since pretreatment of microsomes with hydrogen peroxide and diamide resulted in more efficient trypsin activation.[93] Furthermore, BSP enhances proteolysis of MGST1 dramatically.[110] Putative microsomal proteases that degrade and activate MGST1 have been purified, and characterization is ongoing (Aniya, unpublished data). Cold stress by incubation of a rat liver microsomal suspension at 4°C promotes MGST1 fragmentation with a slight increase in MGST1 activity. Our pilot experiments suggest that serine proteases can activate MGST1 via limited proteolysis. It is thus reasonable to examine whether MGST1 is activated via limited proteolysis *in vivo*. This issue has been studied by Aniya et al. and at least PB, 3-MC, acetone, $CoCl_2$, and DEM treatment did not result in any proteolytic fragmentation of MGST1.[51,52,61,63]

14.3.6 REMOVAL OF ENDOGENOUS INHIBITORS

The activity of MGST1 was increased 5-fold after treatment of rat liver microsomes with small unilamellar vesicles made from phosphatidylcholine.[111] The activation mechanism is thought to involve removal of an inhibitor of MGST1 from the microsomal membrane. The inhibitor was suggested to be a physiological regulator of MGST1 in the ER. The activation observed when human liver microsomes are solubilized might be an example of the same phenomenon.[48]

14.4   MISCELLANEOUS MECHANISMS

Heating activates MGST1 in rat liver microsomes, but the mechanism is unknown.[112] A dimer formation but no fragmentation of MGST1 is observed after incubation of rat liver microsomes at 50°C as detected by Western blot analysis. Calreticulin, one of the ER chaperones, shows heat-induced oligomerization that correlates with increased binding to denatured proteins and peptides.[113] By analogy, a possible activation mechanism of MGST1 after heating could involve aggregation. In general a MGST1 dimer has been observed after incubation of liver microsomes with hydrogen peroxide,[34] ONOO$^-$,[82] or CuCl$_2$ (Aniya, unpublished data). Whether this dimer contributes significantly to the activation or is a minor tell-tale sign of these treatments remains to be determined.

During the incorporation of MGST1 in proteoliposomes the enzyme revealed a lack of a distinct preference for specific phospholipids and also became activated to a low degree.[4] Activation capacity (NEM) was lost correspondingly. Because the dialysis of proteoliposomes required seven days (for the removal of Triton X-100) it is conceivable that oxidation could take place. It is therefore difficult to discern specific activation of MGST1 by the lipid environment. The production of two-dimensional crystals[114] has revealed that the crystalline enzyme can be stable for months at room temperature in the presence of GSH. Therefore crystalline MGST1 appears to be at least partly protected from oxidation. Radiation activates or inhibits MGST1 depending on the radiation procedure.[115] It is well known that radicals are produced during radiation, and it is plausible that these modify the enzyme in various ways to account for both activation and inhibition.

Phosphorylation plays an important role in signaling pathways and it is involved in many cellular phenomena.[116] Matrix-assisted laser desorption/ionization time of flight (MALDI-TOF) analysis of purified MGST1 from control rats did not reveal evidence for post-translational modifications other than partial N-terminal acetylation.[117,118] Cytosolic GST (GSTM1-1) is regulated by p38MAPK and apoptosis signal-regulating kinase-1 (Ask1) that are involved in oxidative stress, heating, and phosphorylation of heat shock proteins.[119] Other phosphorylations of cytosolic GST (GSTA4-4) regulate mitochondrial targeting that is related to Hsp70 binding.[120] It would be important to study whether oxidative stress or heating leads to phosphorylation of MGST1 and whether this could have an impact on the known multiple cellular localization and distribution.

A typical inducer of ER stress, tunicamycin, did not activate MGST1 in Fao rat hepatoma cells (Morgenstern et al., unpublished data) although a slight induction was noted. ER stress by tunicamycin is induced by inhibition of N-glycosylation, which results in the accumulation of unfolded proteins. ER stress can also be induced by calcium efflux changes (thapsigargin) and reducing reagents (DTT).[121] Thapsigargin treatment actually resulted in slight activation (Aniya et al., unpublished observation). As multiple functions are being discovered for many enzymes, including cytosolic GSTs, a role for MGST1 activation as a sensing and signaling in stress-protection mechanisms remains an interesting possibility.

14.5 GENERAL DISCUSSION

14.5.1 INDUCTION VERSUS ACTIVATION

In general, MGST1 is rarely induced; however, some inducers appear to be able to activate the enzyme. It thus becomes very important to discriminate between activation and induction in future experiments. Clearly, Western blot and activity measurements need to be performed in parallel. Also, NEM activation of the activity in microsomes gives a clue to whether more protein was actually produced because the already activated enzyme is refractory to NEM treatment (most likely regardless of the activation mechanism because the enzyme activated by proteolysis can no longer be activated by NEM). Oxidative stress can enhance the transcriptional activity of the human MGST1 promoter;[122] therefore, induction upon longer treatments is a distinct possibility, as has been observed for butylated hydroxyanisole[123] and after overexpression of CYP2E1 in cells.[124]

14.5.2 DIFFERENT ACTIVATION CONDITIONS

Our previous work, together with that of several groups, reveals that different methodologies used in different laboratories result in either sensitive detection of activation during oxidative stress or a complete lack of activation. Since cellular subfractions (i.e., microsomes) will contain other organelles and also extracellular contaminants it appears extremely important to monitor the breakdown rate of added ROS when these are examined.[49] These considerations also apply when conditions that generate oxidative stress are examined. In our experience fresh microsomes, or nonwashed microsomes rapidly frozen at $-80°C$ as pellets, maintain a high activation capacity. The lack of activation in microsomes from most extrahepatic organs can have several explanations: endogenous inhibitors could be present at higher levels, the extrahepatic enzyme is perhaps already modified in a way that leads to suboptimal activation, or it could be a combination of both. On a more speculative note the enzyme could be involved in protein interactions that affect activity and the ability to become activated. Clearly this is an issue that needs to be resolved and that might give additional clues to the functions of MGST1. Activation by NEM is obtained in microsomes prepared from several cell lines.[125]

14.5.3 DIFFERENT ENZYMATIC ACTIVITIES OF MGST1 IN RELATION TO ACTIVATION

It has been observed that GST and GPx activation do not always occur in parallel *in vivo*, which is counterintuitive (Table 14.1). In general, activities toward CDNB and CuOOH (cumene hydroperoxide) are both increased upon activation *in vitro*, albeit the latter to a lower extent. Due to the lack of sensitivity of the GPx assay perhaps the reason is simply methodological. An added complexity results from the fact that other GSTs present have GPx activity that might be altered negatively. A safe recommendation at this stage is to base the conclusions on positive data and always aim for the most sensitive assay. It should be borne in mind that a variety of less reactive substrates are not turned over more rapidly by the activated enzyme.[126]

14.5.4 MECHANISM OF THE ACTIVATED ENZYME

Although it is known that different activation conditions result in different degrees of activation[8,127] it is reasonable to assume that the rate-limiting steps affected are similar. It has been demonstrated that the chemical step catalyzed by MGST1 is not altered in the activated enzyme;[128,129] instead the GSH-thiolate formation rate is enhanced 30-fold. This explains why activity to only certain electrophilic substrates can be activated, namely those that are more reactive, where the rate of the chemical conjugation step exceeds that of the slow rate of thiolate formation in the unactivated enzyme. We also have indications that a step after product formation could be affected. Interestingly, when GSH becomes rate limiting (as during depletion under toxic conditions) the activated enzyme is more efficient toward all classes of second substrates (reactive and less reactive[130]). Why then produce an enzyme in the unactivated form in the first place? The answer might lie in the fact that the unactivated enzyme is more stable than the activated form (Morgenstern and Aniya et al., unpublished observations). In addition, if the activated enzyme also functions in signaling, the switch could have an additional role to play as a stress sensor.

14.5.5 PHYSIOLOGICAL/TOXICOLOGICAL RELEVANCE

From the above it is evident that MGST1 is activated by a plethora of relevant toxic conditions. Chemical reactivity is certainly the most prominent common denominator, but other possibilities include ligand interactions and proteolysis. Although MGST1 can act as a chemical stress sensor (Figure 14.2), the physiological role of the activated enzyme is enigmatic. What then supports activation as a protective mechanism? The following observations are relevant:

1. The enzyme is present in the organelle where reactive intermediates are produced.[37]
2. Activation does occur *in vivo* during a variety of toxic conditions and activation results in an enzyme that is more efficient during severe toxicity (either when the reactive intermediate displays pronounced reactivity or against all substrates when GSH is low).[63]
3. The activated enzyme displays increased protection from *in vitro* lipid peroxidation.[46,90]
4. Cys-49 appears to be present in a hydrophobic binding pocket.[131]
5. Activation is a very rapid response in comparison to induction.[52]

Activation can be a useful tool to study potential drug toxicity because MGST1 itself has been successfully employed to sense the formation of reactive intermediates. The reader is referred to our recent review on this subject.[132] Interestingly, cytosolic GSTP is an inhibitor of c-Jun N-terminal protein kinase, which plays an important role in phosphorylation and a redox signaling pathway.[133] An MCF-7 cell line that stably overexpresses MGST1 is more resistant to oxidative stress than control cells (Morgenstern et al., unpublished observation). As the enzyme can detoxify (and become activated by) the cytostatic drug chlorambucil,[105] the role of MGST1 in drug resistance needs to be defined, and studies are ongoing in our laboratory. We have developed a new fluorescent substrate for GSTs that is very sensitive to activation as a tool to study potential MGST1 activation-driven drug resistance in cells.[134]

In an alternate hypothesis MGST1 as a chemical stress sensor could contribute to signaling mechanisms involving only a small population of the enzyme.[133,135–138] Future research should improve our understanding of whether MGST1 plays a key role as a chemical stress sensor. On a speculative note we propose three putative physiological functions of MGST1. First, as a channel (transporter) based on the structure that, at the current resolution, is consistent with a central pore;[22,28] second, in induction of microsome-triggered apoptosis by sensing reactive intermediates; and third, as a role for the enzyme in chemoprevention via activation of MGST1 by anticancer drugs (such as chlorambucil[105]) because it is known that several cancer cell lines express MGST1 with maintained NEM activation.[125]

In conclusion, a bewildering variety of toxicologically relevant phenomena activate MGST1, which is the most abundant microsomal glutathione transferase in the liver. The activated enzyme is becoming well understood at the molecular level. It remains a fascinating challenge to understand the role of activation in physiology and to harness the phenomenon in the study of reactive intermediates.

ACKNOWLEDGMENTS

Professor Kamataki is gratefully acknowledged for his support. We wish to thank Dr. Par L. Pettersson and Dr. Staffan Thoren (MBB, Karolinska Institutet) for critical reading of the manuscript. Studies from the authors' laboratories were supported by the Swedish Cancer Society, the National Board for Laboratory Animals, Carl Tryggers Foundation, funds from Karolinska Institutet, and University of the Ryukyus in Japan.

REFERENCES

1. Hayes, J.D., Flanagan, J.U., and Jowsey, I.R., Glutathione transferases, *Annu. Rev. Pharmacol. Toxicol.,* 45, 51–88, 2005.
2. Andersson, C. et al., Enzymology of microsomal glutathione S-transferase. In *Conjugation-Dependent Carcinogenicity and Toxicity of Foreign Compounds,* by Anders, M.W. and Dekant, W. Academic Press, San Diego, 1994, 19–35.
3. Mannervik, B. and Widersten, M., Human glutathione transferases: classification, tissue distribution, structure, and functional properties. In *Advances in Drug Metabolism in Man,* by Pacifici, G.M. and Fracchia, G.N. European Commission, Luxembourg, 1995, 407–460.
4. Mosialou, E. et. al., Microsomal glutathione transferase — lipid-derived substrates and lipid dependence, *Arch. Biochem. Biophys.,* 320 (2), 210–216, 1995.

5. Yang, Y. et al., Role of glutathione S-transferases in protection against lipid peroxidation. Overexpression of hGSTA2-2 in K562 cells protects against hydrogen peroxide–induced apoptosis and inhibits JNK and caspase 3 activation, *J. Biol. Chem.*, 276 (22), 19220–19230, 2001.

6. Cheng, J.Z. et al., Accelerated metabolism and exclusion of 4-hydroxynonenal through induction of RLIP76 and hGST5.8 is an early adaptive response of cells to heat and oxidative stress, *J. Biol. Chem.*, 276 (44), 41213–41223, 2001.

7. Aniya, Y. and Anders, M.W., Activation of rat liver microsomal glutathione S-transferase by hydrogen peroxide: role for protein dimer formation, *Arch. Biochem. Biophys.*, 296 (2), 611–616, 1992.

8. Morgenstern, R., DePierre, J.W., and Ernster, L., Activation of microsomal glutathione transferase activity by sulfhydryl reagents, *Biochem. Biophys. Res. Commun.*, 87, 657–663, 1979.

9. Morgenstern, R., Guthenberg, C., and DePierre, J.W., Microsomal glutathione transferase. Purification, initial characterization, and demonstration that it is not identical to the cytosolic glutathione transferases A, B, and C, *Eur. J. Biochem.*, 128, 243–248, 1982.

10. Morgenstern, R. and DePierre, J.W., Microsomal glutathione transferase, purification in unactivated form and further characterization of the activation process, substrate specificity, and amino acid composition, *Eur. J. Biochem.*, 134, 591–597, 1983.

11. Reddy, C.C. et al., Evidence for the occurrence of selenium-independent glutathione peroxidase activity in rat liver microsomes, *Biochem. Biophys. Res. Commun.*, 101 (3), 970–978, 1981.

12. Jakobsson, P.J. et al., Common structural features of MAPEG: a widespread superfamily of membrane-associated proteins with highly divergent functions in eicosanoid and glutathione metabolism, *Protein Sci.*, 8 (3), 689–692, 1999.

13. Jakobsson, P.J. et al., Membrane-associated proteins in eicosanoid and glutathione metabolism (MAPEG). A widespread protein superfamily, *Am. J. Respir. Crit. Care Med.*, 161 (Pt. 2), S20–S24, 2000.

14. Nicholson, D.W. et al., Purification to homogeneity and the N-terminal sequence of human leukotriene-C4 synthase: a homodimeric glutathione S-transferase composed of 18-kDa subunits, *Proc. Natl. Acad. Sci. USA*, 90 (5), 2015–2019, 1993.

15. Penrose, J.F. et al., Purification of human leukotriene C4 synthase, *Proc. Natl. Acad. Sci. USA*, 89 (23), 11603–11606, 1992.

16. Thoren, S. et al., Human microsomal prostaglandin E synthase-1: purification, functional characterization, and projection structure determination, *J. Biol. Chem.*, 278 (25), 22199–22209, 2003.

17. Sjostrom, M. et al., Human umbilical vein endothelial cells generate leukotriene C4 via microsomal glutathione S-transferase type 2 and express the CysLT(1) receptor, *Eur. J. Biochem.*, 268 (9), 2578–2586, 2001.

18. Jakobsson, P.-J. et al., Identification and characterization of a novel microsomal enzyme with glutathione-dependent transferase and peroxidase activities, *J. Biol. Chem.*, 272, 22934–22939, 1997.

19. Mancini, J.A. et al., 5-Lipoxygenase-activating protein is an arachidonate binding protein, *FEBS Lett.*, 318 (3), 277–281, 1993.

20. Miller, D.K. et al., Identification and isolation of a membrane protein necessary for leukotriene production, *Nature*, 343 (6255), 278–281, 1990.

21. Dixon, R.A. et al., Requirement of a 5-lipoxygenase-activating protein for leukotriene synthesis, *Nature*, 343 (6255), 282–284, 1990.

22. Schmidt-Krey, I. et al., The three-dimensional map of microsomal glutathione transferase 1 at 6 angstrom resolution, *Embo. J.,* 19 (23), 6311–6316, 2000.

23. Lundqvist, G., Yucel-Lindberg, T., and Morgenstern, R., The oligomeric structure of rat liver microsomal glutathione transferase studied by chemical cross-linking, *Biochim. Biophys. Acta,* 1159 (1), 103–108, 1992.

24. Schmidt-Krey, I. et al., Human leukotriene C(4) synthase at 4.5 A resolution in projection, *Structure,* 12 (11), 2009–2014, 2004.

25. Andersson, C. et al., Functional and structural membrane topology of rat liver microsomal glutathione transferase, *Biochim. Biophys. Acta,* 1204 (2), 298–304, 1994.

26. Lam, B.K. et al., Site-directed mutagenesis of human leukotriene C-4 synthase, *J. Biol. Chem.,* 272 (21), 13923–13928, 1997.

27. Murakami, M. et al., Regulation of prostaglandin E2 biosynthesis by inducible membrane-associated prostaglandin E2 synthase that acts in concert with cyclooxygenase-2, *J. Biol. Chem.,* 275 (42), 32783–32792, 2000.

28. Holm, P.J., Morgenstern, R., and Hebert, H., The 3-D structure of microsomal glutathione transferase 1 at 6 Å resolution as determined by electron crystallography of p22(1)2(1) crystals, *Biochim. Biophys. Acta,* 1594 (2), 276–285, 2002.

29. Prabhu, K.S. et al., Microsomal glutathione S-transferase A1-1 with glutathione peroxidase activity from sheep liver: molecular cloning, expression and characterization, *Biochem. J.,* 360 (Pt. 2), 345–354, 2001.

30. Shimoji, M. and Aniya, Y., Glutathione S-transferases in rat testis microsomes: comparison with liver transferase, *J. Biochem. (Tokyo),* 115 (6), 1128–1134, 1994.

31. Friedberg, T. et al., Identification, solubilization, and characterization of microsome associated glutathione S-transferases, *J. Biol. Chem.,* 254, 12028–12033, 1979.

32. Hayes, J.D. and Strange, R.C., Potential contribution of the glutathione S-transferase supergene family to resistance to oxidative stress, *Free Rad. Res.,* 22 (3), 193–207, 1995.

33. Sheehan, D. et al., Structure, function, and evolution of glutathione transferases: implications for classification of non-mammalian members of an ancient enzyme superfamily, *Biochem. J.,* 360 (Pt. 1), 1–16, 2001.

34. Aniya, Y. and Anders, M.W., Regulation of rat liver microsomal glutathione S-transferase activity by thiol/disulfide exchange, *Arch. Biochem. Biophys.,* 270 (1), 330–334, 1989.

35. Nishino, H. and Ito, A., Purification and properties of glutathione S-transferase from outer mitochondrial membrane of rat liver, *Biochem. Int.,* 20 (6), 1059–1066, 1990.

36. Morgenstern, R. et al., The distribution of microsomal glutathione transferase among different organelles, different organs, and different organisms, *Biochem. Pharmacol.,* 33, 3609–3614, 1984.

37. Estonius, M. et al., Distribution of microsomal glutathione transferase 1 in mammalian tissues. A predominant alternate first exon in human tissues, *Eur. J. Biochem.,* 260 (2), 409–413, 1999.

38. Otieno, M.A. et al., Immunolocalization of microsomal glutathione S-transferase in rat tissues, *Drug Metab. Dispos.,* 25 (1), 12–20, 1997.

39. Johnson, J.A. et al., Glutathione S-transferase isoenzymes in rat brain neurons and glia, *J. Neurosci.,* 13 (5), 2013–2023, 1993.

40. Rahilly, M. et al., Distribution of glutathione S-transferase activity in human ovary, *J. Reprod. Fert.,* 93, 303–311, 1991.

41. Kelner, M.J., Bagnell, R.D., and Morgenstern, R., Structural organization of the murine microsomal glutathione S-transferase gene (MGST1) from the 129/SvJ strain: identification of the promoter region and a comprehensive examination of tissue expression, *Biochim. Biophys. Acta,* 1678 (2–3), 163–169, 2004.

42. Sun, T.H. et al., A highly active microsomal glutathione transferase from frog (*Xenopus laevis*) liver that is not activated by N-ethylmaleimide, *Biochem. Biophys. Res. Commun.,* 246 (2), 466–469, 1998.

43. Toba, G. and Aigaki, T., Disruption of the microsomal glutathione S-transferase-like gene reduces life span of *Drosophila melanogaster, Gene,* 253 (2), 179–187, 2000.

44. Morgenstern, R. et al., Characterization of rat liver microsomal glutathione transferase activity, *Eur. J. Biochem.,* 104, 167–174, 1980.

45. Mosialou, E. et al., Evidence that rat liver microsomal glutathione transferase is responsible for glutathione-dependent protection against lipid peroxidation, *Biochem. Pharmacol.,* 45 (8), 1645–1651, 1993.

46. Ji, Y. and Bennett, B.M., Activation of microsomal glutathione S-transferase by peroxynitrite, *Mol. Pharmacol.,* 63 (1), 136–146, 2003.

47. Aniya, Y. and Daido, A., Organic hydroperoxide-induced activation of liver microsomal glutathione S-transferase of rats *in vitro, Jpn. J. Pharmacol.,* 62 (1), 9–14, 1993.

48. Mosialou, E. et al., Human liver microsomal glutathione transferase — substrate specificity and important protein sites, *FEBS Lett.,* 315 (1), 77–80, 1993.

49. Rinaldi, R. et al., NADPH-dependent activation of microsomal glutathione transferase 1, *Chem. Biol. Interact.,* 147 (2), 163–172, 2004.

50. Morgenstern, R. et al., The amount and nature of glutathione transferases in rat liver microsomes determined by immunochemical methods, *FEBS Lett.,* 160, 264–268, 1983.

51. Rinaldi, R. et al., Microsomal glutathione S-transferase 1 activation in relation to the cytochrome P450 system, *Chem. Biol. Interact.,* 133, 155–158, 2001.

52. Aniya, Y., Shimoji, M., and Naito, A., Increase in liver microsomal glutathione S-transferase activity by phenobarbital treatment of rats. Possible involvement of oxidative activation via cytochrome P450, *Biochem. Pharmacol.,* 46 (10), 1741–1747, 1993.

53. Aniya, Y. and Anders, M.W., Activation of rat liver microsomal glutathione S-transferase by reduced oxygen species, *J. Biol. Chem.,* 264, 1998–2002, 1989.

54. Chan, T.S. et al., Hydrogen peroxide supports hepatocyte P450 catalysed xenobiotic/drug metabolic activation to form cytotoxic reactive intermediates, *Adv. Exp. Med. Biol.,* 500, 233–236, 2001.

55. Haenen, G.R.M.M. et al., Activation of the microsomal glutathione S-transferase by metabolites of alpha-methyldopa, *Arch. Biochem. Biophys.,* 287 (1), 48–52, 1991.

56. Soucek, P., Ivan, G., and Pavel, S., Effect of the microsomal system on interconversions between hydroquinone, benzoquinone, oxygen activation, and lipid peroxidation, *Chem. Biol. Interact.,* 126 (1), 45–61, 2000.

57. Weis, M. et al., N-acetyl-p-benzoquinone imine-induced protein thiol modification in isolated rat hepatocytes, *Biochem. Pharmacol.,* 43 (7), 1493–1505, 1992.

58. Amatsu, T. et al., Endoplasmic reticulum proliferates without an increase in cytochrome P-450 in hepatocytes of mice treated with phenobarbital and cobalt chloride, *Eur. J. Cell Biol.,* 68 (3), 256–262, 1995.

59. Llesuy, S.F. and Tomaro, M.L., Heme oxygenase and oxidative stress. Evidence of involvement of bilirubin as physiological protector against oxidative damage, *Biochim. Biophys. Acta,* 1223 (1), 9–14, 1994.

60. Piret, J.P. et al., $CoCl_2$, a chemical inducer of hypoxia-inducible factor-1, and hypoxia reduce apoptotic cell death in hepatoma cell line HepG2, *Ann. NY Acad. Sci.,* 973, 443–447, 2002.

61. Daido, A. and Aniya, Y., Alteration of liver glutathione S-transferase and protease activities by cobalt chloride treatment of rats, *Jpn. J. Pharmacol.,* 66 (3), 357–362, 1994.

62. Masukawa, T. and Iwata, H., Possible regulation mechanism of microsomal glutathione S-transferase activity in rat liver, *Biochem. Pharmacol.*, 35, 435–438, 1986.

63. Aniya, Y. and Teruya, M., Activation of hepatic microsomal glutathione S-transferase of rats by a glutathione depletor, diethylmaleate, *J. Pharmacobio-Dyn.*, 15 (9), 473–479, 1992.

64. Lundqvist, G. and Morgenstern, R., Studies on the activation of rat liver microsomal glutathione transferase in isolated hepatocytes, *Biochem. Pharmacol.*, 43 (2), 131–135, 1992.

65. Dafre, A.L., Sies, H., and Akerboom, T., Protein S-thiolation and regulation of microsomal glutathione transferase activity by the glutathione redox couple, *Arch. Biochem. Biophys.*, 332 (2), 288–294, 1996.

66. Cadenas, S. and Cadenas, A.M., Fighting the stranger-antioxidant protection against endotoxin toxicity, *Toxicology*, 180 (1), 45–63, 2002.

67. Aniya, Y. et al., Screening of antioxidant action of various molds and protection of *Monascus anka* against experimentally induced liver injuries of rats, *Gen. Pharmacol.*, 32 (2), 225–231, 1999.

68. Fang, C. et al., Hepatic expression of multiple acute phase proteins and down-regulation of nuclear receptors after acute endotoxin exposure, *Biochem. Pharmacol.*, 67 (7), 1389–1397, 2004.

69. Ji, Y., Toader, V., and Bennett, B.M., Regulation of microsomal and cytosolic glutathione S-transferase activities by S-nitrosylation, *Biochem. Pharmacol.*, 63 (8), 1397–1404, 2002.

70. Shi, Q. and Lou, Y.J., Microsomal glutathione transferase 1 is not S-nitrosylated in rat liver microsomes or in endotoxin-challenged rats, *Pharmacol. Res.*, 51 (4), 303–310, 2005.

71. Caimi, G., Carollo, C., and Lo Presti, R., Diabetes mellitus: oxidative stress and wine, *Curr. Med. Res. Opin.*, 19 (7), 581–586, 2003.

72. Aniya, Y. et al., Glutathione S-transferases and chloroform toxicity in streptozotocin-induced diabetic rats, *Jpn. J. Pharmacol.*, 50 (3), 263–269, 1989.

73. Aniya, Y. et al., Alteration of glutathione S-transferase activity by alpha-naphthyl-isothiocyanate in hepatic microsomes of rats, *J. Biopharm. Sci.*, 1, 371–380, 1990.

74. Osawa, Y., Highet, R.J., and Pohl, L.R., The use of stable isotopes to identify reactive metabolites and target macromolecules associated with toxicities of halogenated hydrocarbon compounds, *Xenobiotica*, 22 (9–10), 1147–1156, 1992.

75. Botti, B. et al., Transient decrease of liver cytosolic glutathione S-transferase activities in rats given 1,2-dibromoethane or CC14, *Chem. Biol. Interact.*, 42, 259–270, 1982.

76. Gyamfi, M.A., Yonamine, M., and Aniya, Y., Free-radical scavenging action of medicinal herbs from Ghana: *Thonningia sanguinea* on experimentally-induced liver injuries, *Gen. Pharmacol.*, 32 (6), 661–667, 1999.

77. Aniya, Y. et al., Dimerumic acid as an antioxidant of the mold, *Monascus anka*, *Free Radic. Biol. Med.*, 28 (6), 999–1004, 2000.

78. Shimada, T. et al., Activation and inactivation of carcinogenic dihaloalkanes and other compounds by glutathione S-transferase 5-5 in *Salmonella typhimurium* tester strain NM5004, *Chem. Res. Toxicol.*, 9 (1), 333–340, 1996.

79. Jaeschke, H., Knight, T.R., and Bajt, M.L., The role of oxidant stress and reactive nitrogen species in acetaminophen hepatotoxicity, *Toxicol. Lett.*, 144 (3), 279–288, 2003.

80. Yonamine, M. et al., Acetaminophen-derived activation of liver microsomal glutathione S-transferase of rats, *Jpn. J. Pharmacol.*, 72 (2), 175–181, 1996.

81. James, L.P., Mayeux, P.R., and Hinson, J.A., Acetaminophen-induced hepatotoxicity, *Drug Metab. Dispos.,* 31 (12), 1499–1506, 2003.

82. Aniya, Y., Kunii, D., and Yamazaki, K., Oxidative and proteolytic activation of liver microsomal glutathione S-transferase (GSTm) of rats, *Chem. Biol. Interact.,* 133, 144–147, 2001.

83. Haenen, G.R.M.M. et al., Activation of the microsomal glutathione S-transferase and reduction of the glutathione dependent protection against lipid peroxidation by acrolein, *Biochem. Pharmacol.,* 37, 1933–1938, 1988.

84. Bast, A. et al., Antioxidant effects of carotenoids, *Int. J. Vitam. Nutr. Res.,* 68 (6), 399–403, 1998.

85. Lim, H.K. et al., Effects of bergenin, the major constituent of *Mallotus japonicus* against D-galactosamine-induced hepatotoxicity in rats, *Pharmacology,* 63 (2), 71–75, 2001.

86. Aniya, Y. et al., Free radical scavenging action of the medicinal herb *Limonium wrightii* from the Okinawa islands, *Phytomedicine,* 9 (3), 239–244, 2002.

87. Morioka, T. et al., The modifying effect of *Peucedanum japonicum,* an herb in the Ryukyu Islands, on azoxymethane-induced colon preneoplastic lesions in male F344 rats, *Cancer Lett.,* 205 (2), 133–141, 2004.

88. Aniya, Y. and Naito, A., Oxidative stress-induced activation of microsomal glutathione S-transferase in isolated rat liver, *Biochem. Pharmacol.,* 45 (1), 37–42, 1993.

89. Naito, A., Aniya, Y., and Sakanashi, M., Antioxidative action of the nitrovasodilator nicorandil: inhibition of oxidative activation of liver microsomal glutathione S-transferase and lipid peroxidation, *Jpn. J. Pharmacol.,* 65 (3), 209–213, 1994.

90. Aniya, Y. and Daido, A., Activation of microsomal glutathione S-transferase in tert-butyl hydroperoxide–induced oxidative stress of isolated rat liver, *Jpn. J. Pharmacol.,* 66 (1), 123–130, 1994.

91. Busenlehner, L. et al., Stress sensor triggers conformational response of the integral membrane protein microsomal glutathione transferase 1, *Biochemistry,* 43 (35), 11145–11152, 2004.

92. Aniya, Y. and Anders, M.W., Activation of hepatic microsomal glutathione S-transferase by hydrogen peroxide and diamide, in *Oxygen Radicals,* Yagi, K., Kondo, M., Niki, E., and Yoshikawa, T., Eds., Elsevier Science Publishers, New York, 1992, 737–740.

93. Shimoji, M., Aniya, Y., and Anders, M.W., Preferential proteolysis and activation of oxidatively modified liver microsomal glutathione S-transferase of rat, *Biol. Pharm. Bull.,* 19 (2), 209–213, 1996.

94. Sies, H., Dafre, A.L., Ji, Y., and Akerboom, T.P.M., Protein S-thiolation andredoz regulation membrane-bound glutathione transferase, *Chem. Biol. Interact.,* 111–112, 177–185, 1998.

95. Morgenstern, R. and DePierre, J.W., Microsomal glutathione transferases. In *Revs. in Biochem. Toxicol.,* by Hodgson, E., Bend, J.R., and Philpot, R.M. Elsevier Scientific Publishing, New York, 1985, 67–84.

96. Nishino, H. and Ito, A., Increase in glutathione disulfide level regulates the activity of microsomal glutathione S-transferase in rat liver, *Biochem. Int.,* 19 (4), 731–735, 1989.

97. Levonen, A.L. et al., Mechanisms of cell signaling by nitric oxide and peroxynitrite: from mitochondria to MAP kinases, *Antioxid. Redox. Signal.,* 3 (2), 215–229, 2001.

98. Ji, Y., Akerboom, T.P., and Sies, H., Microsomal formation of S-nitrosoglutathione from organic nitrites: possible role of membrane-bound glutathione transferase, *Biochem. J.,* 313 (Pt. 2), 377–380, 1996.

99. Cooper, C.E. et al., Nanotransducers in cellular redox signaling: modification of thiols by reactive oxygen and nitrogen species, *Trends Biochem. Sci.*, 27 (10), 489–492, 2002.

100. Moore, P.K. and Marshall, M., Nitric oxide releasing acetaminophen (nitroacetaminophen), *Dig. Liver Dis.*, 35 Suppl. 2, S49–S60, 2003.

101. Lundqvist, G. and Morgenstern, R., Mechanism of activation of rat liver microsomal glutathione transferase by noradrenaline and xanthine oxidase, *Biochem. Pharmacol.*, 43 (8), 1725–1728, 1992.

102. Onderwater, R.C. et al., Activation of microsomal glutathione S-transferase and inhibition of cytochrome P450 1A1 activity as a model system for detecting protein alkylation by thiourea-containing compounds in rat liver microsomes, *Chem. Res. Toxicol.*, 12 (5), 396–402, 1999.

103. Wallin, H. and Morgenstern, R., Activation of microsomal glutathione transferase activity by reactive intermediates formed during the metabolism of phenol, *Chem-Biol. Inter.*, 75, 185–199, 1990.

104. Rooseboom, M. et al., Comparative study on the bioactivation mechanisms and cytotoxicity of Te-phenyl-L-tellurocysteine, Se-phenyl-L-selenocysteine, and S-phenyl-L-cysteine, *Chem. Res. Toxicol.*, 15 (12), 1610–1618, 2002.

105. Zhang, J. and Lou, Y.J., Relationship between activation of microsomal glutathione S-transferase and metabolism behavior of chlorambucil, *Pharmacol. Res.*, 48 (6), 623–630, 2003.

106. Jaeschke, H., Xanthine oxidase–induced oxidant stress during hepatic ischemia-reperfusion: are we coming full circle after 20 years? *Hepatology*, 36 (3), 761–763, 2002.

107. Boots, A.W. et al., Oxidized quercetin reacts with thiols rather than with ascorbate: implication for quercetin supplementation, *Biochem. Biophys. Res. Commun.*, 308 (3), 560–565, 2003.

108. Andersson, C., Söderström, M., and Mannervik, B., Activation and inhibition of microsomal glutathione transferase from mouse liver, *Biochem. J.*, 249, 819–823, 1988.

109. Mosialou, E. and Morgenstern, R., Inhibition studies on rat liver microsomal glutathione transferase, *Chem. Biol. Interact.*, 74, 275–280, 1990.

110. Morgenstern, R. et al., Activation of rat liver microsomal glutathione transferase by limited proteolysis, *Biochem. J.*, 260, 577–582, 1989.

111. Boyer, T.D., Zakim, D., and Vessey, D.A., Studies of endogenous inhibitors of microsomal glutathione S-transferase, *Biochem. J.*, 207, 57–64, 1982.

112. Aniya, Y., Activation of liver microsomal glutathione S-transferase by heating, *J. Pharmacobio-Dyn.*, 12, 235–240, 1989.

113. Jorgensen, C.S. et al., Dimerization and oligomerization of the chaperone calreticulin, *Eur. J. Biochem.*, 270 (20), 4140–4148, 2003.

114. Schmidt-Krey, I. et al., Parameters for the two-dimensional crystallization of the membrane protein microsomal glutathione transferase, *J. Struct. Biol.*, 123 (2), 87–96, 1998.

115. Boyer, T.D., Vessey, D.A., and Kempner, E., Radiation inactivation of microsomal glutathione transferase, *J. Biol. Chem.*, 261, 16963–16968, 1986.

116. Brookes, P.S. et al., Mitochondria: regulators of signal transduction by reactive oxygen and nitrogen species, *Free Radic. Biol. Med.*, 33 (6), 755–764, 2002.

117. Shore, L.J., Odell, G.B., and Fenselau, C., Identification of an n-acetylated microsomal glutathione S-transferase by mass spectrometry, *Biochem. Pharmacol.*, 49 (2), 181–186, 1995.

118. Lengqvist, J. et al., Observation of an intact noncovalent homotrimer of detergent-solubilized rat microsomal glutathione transferase-1 by electrospray mass spectrometry, *J. Biol. Chem.*, 279 (14), 13311–13116, 2004.

119. Dorion, S., Lambert, H., and Landry, J., Activation of the p38 signaling pathway by heat shock involves the dissociation of glutathione S-transferase mu from Ask1, *J. Biol. Chem.*, 277 (34), 30792–30797, 2002.

120. Robin, M.A. et al., Phosphorylation enhances mitochondrial targeting of GSTA4-4 through increased affinity for binding to cytoplasmic Hsp70, *J. Biol. Chem.*, 278 (21), 18960–18970, 2003.

121. Breckenridge, D.G. et al., Regulation of apoptosis by endoplasmic reticulum pathways, *Oncogene*, 22 (53), 8608–8618, 2003.

122. Kelner, M.J. et al., Structural organization of the microsomal glutathione S-transferase gene (MGST1) on chromosome 12p13.1–13.2. Identification of the correct promoter region and demonstration of transcriptional regulation in response to oxidative stress, *J. Biol. Chem.*, 275 (17), 13000–13006, 2000.

123. Morgenstern, R. and Dock, L., A comparison of induction of microsomal glutathione S-transferase activity in the liver of the mouse and rat by dietary 2(3)-tert-butyl-4-hydroxyanisole (BHA), *Acta Chem. Scand. B*, 36, 255–256, 1982.

124. Mari, M. and Cederbaum, A.I., Induction of catalase, alpha, and microsomal glutathione S-transferase in CYP2E1 overexpressing HepG2 cells and protection against short-term oxidative stress, *Hepatology*, 33 (3), 652–661, 2001.

125. Weinander, R. et al., Heterologous expression of rat liver microsomal glutathione transferase in simian cos cells and *escherichia coli*, *Biochem. J.*, 311, 861–866, 1995.

126. Morgenstern, R. et al., Studies on the activity and activation of rat liver microsomal glutathione transferase, in particular with a substrate analogue series, *J. Biol. Chem.*, 263, 6671–6675, 1988.

127. Morgenstern, R., DePierre, J.W., and Ernster, L., Reversible activation of microsomal glutathione S-transferase activity by 5,5'-dithiobis(2-nitrobenzoic acid) and 2,2'-dipyridyl disulfide, *Acta Chem. Scand. B*, 34, 229–230, 1980.

128. Svensson, R. et al., Kinetic characterization of thiolate anion formation and chemical catalysis of activated microsomal glutathione transferase 1, *Biochemistry*, 43, 8869–8877, 2004.

129. Morgenstern, R. et al., Kinetic analysis of the slow ionization of glutathione by microsomal glutathione transferase MGST1, *Biochemistry*, 40 (11), 3378–3384, 2001.

130. Andersson, C. et al., Kinetic studies on rat liver microsomal glutathione transferase, consequences of activation, *Biochim. Biophys. Acta*, 1247, 277–283, 1995.

131. Svensson, R. et al., Reactivity of cysteine-49 and its influence on the activation of microsomal glutathione transferase 1: evidence for subunit interaction, *Biochemistry*, 39 (49), 15144–15149, 2000.

132. Rinaldi, R. et al., Reactive intermediates and the dynamics of glutathione transferases, *Drug. Metab. Dispos.*, 30 (10), 1053–1058, 2002.

133. Adler, V. et al., Regulation of JNK signaling by GSTp, *Embo. J.*, 18 (5), 1321–1334, 1999.

134. Svensson, R. et al., Synthesis and characterization of 6-chloroacetyl-2-dimethylaminonaphthalene as a fluorogenic substrate and a mechanistic probe for glutathione transferases, *Anal. Biochem.*, 311 (2), 171–178, 2002.

135. Wood, Z.A. et al., Structure, mechanism and regulation of peroxiredoxins, *Trends Biochem. Sci.*, 28 (1), 32–40, 2003.

136. Rundlof, A.K. and Arner, E.S., Regulation of the mammalian selenoprotein thioredoxin reductase 1 in relation to cellular phenotype, growth, and signaling events,
Antioxid. Redox. Signal., 6 (1), 41–52, 2004.

137. Kim, S.O. et al., OxyR: a molecular code for redox-related signaling, *Cell,* 109 (3),
383–396, 2002.

138. Thannickal, V.J. and Fanburg, B.L., Reactive oxygen species in cell signaling, *Am.
J. Physiol. Lung Cell Mol. Physiol.,* 279 (6), L1005–1028, 2000.

15 Glutathione S-Transferase Isozyme Composition of Human Tissues

Shaheen Dhanani and Yogesh C. Awasthi

CONTENTS

15.1 INTRODUCTION

From the description of GSTs in various chapters of this volume, it is clear that GST isozymes are expressed in a tissue-specific manner and the isozyme compositions of different tissues vary remarkably. More important, the tissue-specific expression of isozymes in various mammalian species studied so far is also significantly different. Toxicological and pharmacological studies to define the protective role of GSTs are usually performed in animal models and cell lines in order to extrapolate findings to humans. To understand the toxicological and pharmacological relevance of these studies to humans, it is important to know the comparative GST isozyme patterns in the tissues of humans and also of the animal model systems (e.g., in rats and mice) that are most frequently used in toxicological studies. There are significant differences in the isozyme patterns of mice, rats, and humans. For example, GSTP1-1, which is one of the major extrahepatic GST isoforms in humans, is less abundant in the rat extrahepatic tissues, and unlike rats and mice, GSTP1-1 is almost absent in human livers. Such differences also exist in the expression of other classes of GSTs between human and rodent tissues.

Since the first report on the GST isozyme composition of the human liver,[1] there have been extensive studies on human GST isozyme patterns of various tissues. These studies not only revealed tissue-specific differences, but also pointed out gender-specific differences[2–5] in the expression of GST isozymes in humans. In earlier studies, some of these human GSTs were not fully characterized and due to lack of information on their primary structures could not be identified with the different classes of GSTs as known today. Also because of the confusion caused by different nomenclatures used in earlier studies,[4] it was difficult to associate various human GST isozymes described in these studies with currently used classification and nomenclature.[6] With the current status of knowledge on the structure of GSTs, it is now possible to identify these earlier reported isozymes and to associate them with various classes of human GSTs. This chapter gives updated information on the composition and relative abundance of various GST isozymes in human tissues and provides comparative isozyme patterns of human, rat, and mouse tissues.

15.2 MAJOR GST ISOZYMES OF HUMAN TISSUES

More than 90% of the GST protein of most of the human tissues is accounted for by the isoenzymes belonging to the Alpha, Mu, and Pi classes of GSTs. The relative abundance of each of these three classes of GSTs in the human tissues compiled from various studies is arbitrarily given in Table 15.1. Because some of the studies referred to in the table were conducted before the currently accepted classification of GSTs, the isoenzymes in the table have been tentatively identified with respective GST classes on the basis of their reported properties or on the basis of published and unpublished Western blot data from the authors' laboratories. The assigned numbers of plus signs arbitrarily denote the relative abundance of these isozymes in the respective tissues. The numbers within parentheses correspond to the number of charge isoforms separated by isoelectric focusing/electrophoresis from the corresponding tissues, which have been partially or completely characterized. Complete characterization of some of these isozymes has not been conducted and it is not known which of the subunits constitute these dimeric isozymes.

The Alpha-class isozymes constitute a major portion of the GST protein of human livers and kidneys.[1,7–9] This class of GSTs is also present in substantial amounts in the adrenal gland,[10] cornea,[11] gastric mucosa,[12] pancreas,[13] retina,[14] small intestine,[15,16] testes,[17] and uterus[18] along with GST Pi and Mu. In individuals with *GSTM1* deletion, the Alpha-class GSTs constitute more than 90% of protein and activity of liver GSTs.[7] Isoenzyme patterns of liver Alpha-class GSTs as determined by isoelectric focusing,

TABLE 15.1

Expression and Relative Abundance of Major and Minor Classes of GSTs in Human Tissues

Tissues	Alpha	Mu	Pi	Minor GSTs
Adrenal gland[10,103]	++++	+	++	Kappa[51] Zeta[50]
Aorta[59,104]	+	+ (three)	++++	GST5.8
Bladder[105]	+	+	++++	
Biliary epithelium	NK	NK	++	Omega[44]
Brain[17,20]	+ (one)	++ (two)	++++	Theta[32] Zeta[50]
Breast[106]	ND	ND	++++	Zeta[50]
Colon[44,84]	+ (one)	++ (two)	++++	Theta[32] Omega[44]
Cornea[11]	++ (two)	ND	++++	
Diaphragm[107]	+	++	++++	
Duodenum[10]	++	+	++++	
Erythrocytes[108]	-	-	++++	Theta[32]
Fetal fibroblasts[109]	++++	ND	++	
Fetal liver[63]	+++	+	+++	Zeta[50]
Gastric mucosa[15]	+++	+	++++	Theta[32]
Heart[104,110,111]	++ (three)	++ (five)	++++	Theta[32]
Ileum[10,12]	+	+	++++	Theta[32]
Kidney[9,112,113]	++++ (six)	++ (two)	++ (two)	Theta[32] Kappa[51]
Lens[14,114]	+++	ND	+++	
Liver[1,7,8,115,116]	+++++ (at least six)	+++ (three)	+	Sigma[31] Theta[32,33] Zeta[50] Kappa[51] Omega[44]
Lung[19,21,71,117]	++ (three)	+ (two)	++++	Sigma[31] Theta[32] Zeta[50]
Muscle[25,118]	ND	++++ (four)	++++ (two)	Zeta[50]
Pancreas[13]	+++ (two)	++	+++	Sigma[31] Omega[44]
Placenta[119]	ND	ND	++++	Sigma[31] Omega[44]
Platelets[120]	NK	NK	++++	
Prostate gland[121,122]	++ (luminal secretary cells)	+	++++ (basal cells)	
Retina[14]	+++	ND	+	
Salivary gland[10]	+	++	++++	
Skin[26,27]	+	ND	++++	Theta[32] Sigma[31]

TABLE 15.1 *(Continued)*
Expression and Relative Abundance of Major and Minor Classes of GSTs in Human Tissues

Tissues	Alpha	Mu	Pi	Minor GSTs
Small intestine[12,16]	+++	+	++++	*Theta[32] Sigma[31]
Spleen[10]	+	++	++++	Theta[32]
Stomach[10]	+	+	++++	*Theta[32]
Testes[17]	+++	++	+	
Thyroid gland[123]	NK	NK	++++	
Uterus[18]	+ (two)	+ (two)	++++	

ND, not detected; NK, not known.

* In these tissues, expression of Theta is comparable to that of Alpha.
+ The assigned numbers of plus signs arbitrarily denote the relative abundance of GST isozymes in the respective tissues.

or HPLC, show a wide range of inter-individual variations. The reasons for these variations could be attributed to polymorphism. Because GSTs are inducible enzymes, these variations could also be due to the differential exposure of drugs.

The Mu-class GSTs are present in the livers of only about half of the Caucasian adults where the major Mu-class subunit M1 is deleted. GST Mu polymorphism and its toxicological relevance have been extensively discussed elsewhere in this volume. In GSTM1-positive individuals, Mu-class GSTs account for nearly half of the total GST activity of livers toward 1-chloro-2,4-dinitrobenzene (CDNB). Studies in our laboratory show that the specific activities of the Mu-class GSTs of human livers toward CDNB are generally higher than those of the Alpha-class GSTs.[8] Thus even though these two classes of isozymes may account for comparable amounts of activity toward CDNB, the Mu-class GSTs constitute a much smaller portion of total GST protein even in the livers of GSTM1-positive individuals as compared to the Alpha-class GSTs. GST Pi, which is the major isozyme in most of the extrahepatic tissues in humans, is only minimally expressed in the liver.[7,8]

A major portion of GST protein and activity of kidneys is accounted for by the Alpha-class GSTs, and several Alpha-class isoenzymes differing in their PI values have been separated from this organ.[9] The exact subunit composition of these isozymes is not known. The Pi- and Mu-class GSTs are also present in substantial amounts in the kidneys. In the lungs and brain, the Pi-class isoenzymes account for most of the GST activity toward CDNB.[17,19,20] However, it is interesting to note that even though Alpha-class GSTs constitute only about 5% of the total GSH-conjugating activity of lungs toward CDNB, about 30% of the total GST protein of this organ is accounted for by this class of GSTs.[19] This may suggest that the Alpha-class GSTs are involved in other functions besides their GSH-conjugating activity toward electrophilic xenobiotics. These isozymes may contribute primarily to the reduction of lipid hydroperoxides through their GSH-Px activity for protecting the lungs, which are constantly exposed to oxidative stress. This idea is consistent with the findings that in lungs, several Alpha-class GST isoenzymes are present. Each isoenzyme shows remarkably high activity toward fatty acid hydroperoxides and phospholipid hydroperoxides.[21] While the exact subunit structure of these isozymes has not been determined, subunits A1, A2, and A4 have been detected in lungs, and the role of these subunits in protection against lipid peroxidation is well known.[21–24]

GSTs of skeletal muscle are also heterogeneous and several Mu forms have been described in this tissue.[25] Alpha-class GSTs are minimally expressed or undetectable in skeletal muscle. Inter-individual differences in the pattern of the Mu-class GSTs are observed in skeletal muscle and in testes where a Mu-class isozyme with a blocked and extended amino-terminus with three additional residues (Pro-Val-Cys) at the carboxy terminus has been reported.[17] This isoenzyme is also present in the brain and is expressed in both GSTM1-positive and GSTM1-null individuals. In both these organs, several Mu-class GST isozymes are present. In human skin, the Mu-class GSTs were not detected either in males or females.[26,27] Of the 40 adult human skin (leg and breast) samples from both GSTM1-null and GSTM1-positive individuals analyzed in our laboratory, none showed detectable expression of Mu-class GSTs.[26] GST Pi is the major isoform of skin. In general, GST Pi is the most abundant isozyme in extrahepatic tissues and is reported to be the major form in most of the organs except in the adrenal gland, fetal fibroblasts, kidneys, retina, and testes. This pattern is somewhat similar to that of mice tissues where Pi is present in substantial amounts. However, in rat and mouse tissue there are significant differences in the expression of GST Pi. Inter-individual differences in the Pi-class GSTs in humans are primarily accounted for by GSTP1 polymorphisms discussed in detail elsewhere in this volume.

Not much information is available on the localization of these isoenzymes in specific cell types. Limited studies showing the immunohistochemical localization of GST isozymes in specific cell types are referred to in Table 15.1. Association of some of the cytosolic GSTs of the Alpha class with membranes has been suggested,[28] which is consistent with their role in protecting membranes from lipid peroxidation. The presence of multiple GST isozymes belonging to the Alpha and Mu classes in mouse liver mitochondria has been reported, and mitochondrial GSTA4-4 has been implicated in the defense against the reactive oxygen species in this organelle.[29] Thus it seems that GST isoenzymes are selectively expressed in specific organs, cell types, or organelles as dictated by the need for protection against exogenous and endogenous toxicants.

15.3 MINOR GST ISOZYMES

Besides the major cytosolic GSTs, a number of GSTs belonging to both cytosolic and MAPEG (membrane-associated proteins involved in eicosanoid and glutathione metabolism) families have been characterized in recent years.[5] These isoenzymes belong to distinct classes designated as Sigma, Theta, Omega, Zeta, and Kappa. The relative abundance of these isozymes is, however, much lower when compared with the Alpha, Mu, and Pi classes. However, the Theta-class GSTs are relatively more abundant in some human tissues and are expressed in amounts comparable to that of the major GSTs.[30]

15.3.1 GST SIGMA

The Sigma-class GST is also referred to as human Prostaglandin D Synthase (PGDS) as it catalyzes the isomerization of prostaglandin H_2 (PGH$_2$) to prostaglandin D_2 (PGD$_2$).[31] Besides PGDS activity, the enzymes from rats and humans express transferase and peroxidase activity. PGD$_2$ is synthesized in the central nervous system as well as in the peripheral tissues, and it is known to maintain the body temperature, prevent clot formation, relax smooth muscles, regulate the nervous system, and possibly mediate in asthma and tactile pain. The physiological significance of GST Sigma (PGDS) relative to GSH-independent PGDS is not completely understood. Studies on the comparative expression of PGDS indicate marked differences in the expression of this enzyme in rat and human tissues. For example, in rats, a high level of PGDS expression was found in the spleen and bone marrow along with low levels of expression in the small intestine, colon, liver, pancreas, and skin. By contrast, in the human bone marrow and spleen, mRNA for PGDS was undetectable. On the other hand, PGDS expression was detected in human lungs and was undetectable in rat lungs. Human tissues showing high expression of PGDS include adipose tissue, macrophages, and the placenta.[31]

15.3.2 GST THETA

Two Theta-class isozymes, GSTT1-1 and GSTT2-2,[32–36] have been characterized in humans. These isozymes show undetectable activity for 1-chloro-2,4-dinitrobenzene (CDNB) and utilize compounds such as methyl halides and sulfate esters as the preferred substrates.[32,33] The presence of GST Theta has been demonstrated in adult livers but is undetectable in fetal livers. Because it shows no activity toward CDNB and unlike other GSTs does not bind to the GSH-affinity column that is routinely used to purify GSTs, it may be missed during routine analysis of GSTs. Its expression in erythrocytes, lungs, kidneys, brains, skin, hearts, stomachs, small intestines, spleens, and colon mucosa has been demonstrated.[32] In human gastric and colonic mucosa, GST Theta isozymes are present in comparatively higher amounts, and their expression in the stomach has been shown to be comparable to that of the Alpha-class GSTs.[32] This suggests that these isozymes may play an important role in the detoxification of electrophilic toxicants including the chemical carcinogens present in the diet. Higher levels of expression of GST Theta in mice when compared with human tissues has been reported.[37] In humans, GST Theta expression was higher in kidneys when compared with livers, whereas in rats and mice its expression in the liver was higher than that in the kidney.[37] Inter-individual differences in the expression of GST Theta in human populations are attributed to the genetic polymorphism occurring due to complete or partial deletion of the GSTT1 gene. Ethnic differences have been observed in the frequency of GSTT1 polymorphism; for example, in the Chinese population the frequency of 70% has been observed as opposed to in the United States, which has a frequency of 15–20%.[30]

Association between the GSTT1 polymorphism with possible differential susceptibility toward chemical carcinogenesis has been extensively studied and presented elsewhere in this volume. Interestingly the lack of this enzyme has been suggested to increase the risk of cancer of the gastrointestinal tract, bladder, and oral cavity. However, the expression of GST Theta was found to be either unaffected or downregulated in the event of colon and gastric tumors.[32] On the other hand, increased risk of kidney and liver tumors in the human population exposed to halogenated solvents has been suggested in GSTT1-positive individuals.[30] This could be due to GST Theta-mediated bioactivation of these xenobiotics. In mice three GST Theta subunits are known[38] while in rats four GST Theta-class isozymes have been characterized.[39–41] GST Theta is present in lower organisms including prokaryotes, and the gene for the Theta-class subunit has been suggested to be the parent gene from which various classes of GSTs have evolved.[42]

15.3.3 GST Omega

GST Omega (GSTO1-1) shows thiol transferase and dehydroascorbate reductase activities[43,44] and because of these activities, the enzyme may play a role in the defense mechanism against oxidative stress. A mouse GST with 72% sequence identity with GSTO1 has been identified in a radiation-resistant cell line and has been linked to stress response mechanisms,[45] which is consistent with the idea of GSTO1-1 being an antioxidant enzyme. GSTO1-1 is expressed in many cell types as indicated by both cytoplasmic and nuclear staining. Cytoplasmic staining is seen in endocrine cells, pancreas, leydig cells of the testes, and macrophages. It is present in the nuclei and the nuclear membranes of cells including glial cells, myoepithelial cells in the breast, neuroendocrine cells of the colon, fetal myocytes, and many other cells.[44] Even though the physiological significance of GSTO1-1 in humans is not clear, widespread expression of GSTO1-1 suggests that it may have functional significance.

15.3.4 GST Zeta

GST Zeta catalyzes isomerization of maleylacetoacetate (MAA) to fumarylacetoacetate (FAA) and is also referred to as maleylacetoacetate isomerase (MAAI).[46–49] This enzyme is believed to contribute to the catabolism of tyrosine by catalyzing isomerization of MAA to FAA.[46] Its expression has been reported in various human tissues including lungs, livers, adrenal gland, skeletal muscle, and the brain.[50] It has been suggested that MAAI may be essential for survival under some nutritional conditions.[50] Deficiency of GSTZ1/MAAI is predicted to result in the accumulation of MAA, thus leading to tyrosinemia-like syndrome[49] and also to tubular dysfunction or Fanconi's syndrome.[50] However, no definitive evidence for the existence of human patients with MAAI deficiency is available.[50] Knockout mice models of GSTZ1 are now available. Studies with these mice show no phenotype on GSTZ1/MAAI gene deletion. In knockout mice, GST Mu and Pi are significantly induced. The significance of GSTZ1 in the detoxification of environmental pollutants is suggested by its ability to biotransform dichloroacetic acid (DCA), which is a contaminant of chlorinated water supplies, a metabolite of certain drugs such as chloral hydrate, and a metabolite of industrial solvents such as trichloroethylene and tetrachloro-ethylene.[48] DCA is also used in the management of congenital lactic acidosis and may cause pharmacologically induced MAAI deficiency.

15.3.5 GST KAPPA

It has been shown that even though the human GST κ (hGSTK1-1) isozyme is similar to the other GSTs in some aspects, its novel structure makes it significantly different from the other classes of soluble cytosolic GSTs.[51] Human, rat, and mouse Kappa-class isoenzymes are all active toward halogenated aromatics (e.g., chlo-robenzene, 1-2 dichlorobenzene, etc.).[51] In humans Kappa-class GSTs have been found in the mitochondrial matrix.[51]

15.3.6 GST COMPOSITION OF CELL LINES IN CULTURE

In most of the human cancer cell lines derived from various tissues, GST Pi is the predominant isozyme. In our laboratory cell lines have been extensively investigated for their GST isozyme composition. All these cell lines are rich in GST Pi and often do not show detectable amounts of the Alpha-class GSTs and minimally express Mu-class GSTs. Because of the altered expression of GSTs in cell lines presumably due to differential culture conditions, selection for drug resistance, passage numbers, and other unknown factors, it may not be valid to make quantitative comparisons between the compositions of GSTs in cell lines and their tissues of origin. GST Pi has been implicated in mechanisms of drug resistance and its overexpression in a number of drug-resistant cell lines is well documented.[52–54] In a chlorambucil-resistant cell line overexpression of Mu-class GSTs has also been reported.[55]

15.3.7 UNCHARACTERIZED GSTs

While the majority of human skeletal muscle GST protein is accounted for by the Mu and Pi classes of GSTs, a small but significant portion of GSH-conjugating activity of GSTs in this tissue is represented by an uncharacterized GST isoenzyme that has a blocked N-terminus and is immunologically distinct from the Alpha, Mu, and Pi classes.[25] This enzyme may be a member of the MAPEG family. Western blot studies using antibodies raised against mGSTA4-4 specifically recognize a human GST isozyme designated as GST 5.8. This isozyme is expressed in most of the human tissues including the liver, brain, heart, pancreas, and bladder and is recognized by antibodies raised against rGSTA4-4 and mGSTA4-4.[56–59] Although the constitutive expression of this enzyme in most of the tissue is minimal, studies with human cell lines in culture show that during oxidative stress it is rapidly but transiently induced.[60,61] The amino acid sequences of the CNBr peptide fragments of this isoenzyme show homology with rat and mouse GSTA4-4.[56] This enzyme has substrate preference for 4-hydroxynonenal (4-HNE) and is immunologically distinct from human hGSTA4-4, suggesting that in humans at least two isozymes with substrate preference for 4-HNE are present.[62] Information on the structure of hGSTA5.8 is unavailable and needs to be investigated.

15.3.8 Developmental Patterns of GSTs

Isoenzyme profiles of GSTs in human tissues and organs show both qualitative and quantitative differences and changes in their developmental patterns.[63–65] In fetal livers, GST Pi as well as GST Alpha are expressed, but only a minimal expression of GST Pi is observed postnatally, which persists in the adult liver.

15.3.9 Major GST Isozymes of Mice and Rats

Alpha, Mu, and Pi classes are also predominant in most of the mouse and rat tissues examined so far. These enzymes account for most of the GST protein present in the livers, lungs, kidneys, hearts, spleens, skin, muscles, etc., of rats and mice.[66–78] The intracellular concentration of these isozymes is, however, higher than that in humans. For example, the specific activity of GSTs in mouse lung homogenates is up to 10-fold higher than in human lung homogenates.[69] Significant compositional differences in liver GSTs exist between humans and rodents. In adult human livers only a barely detectable expression of GST Pi is observed but it is present in substantial amounts in mouse and rat livers. Similar to humans, GST Pi is one of the major isozymes of extrahepatic tissues in mice but not in rats. Noticeable differences also exist in the isozyme composition of human and rodent extrahepatic tissues; for example, in mouse and rat lungs, the Alpha- and Mu-class GSTs together constitute the majority of GST protein.[67,69] By contrast GST Pi is the predominant isozyme in human lungs.[19] Likewise in rat and mouse brains the Alpha- and Mu-class GSTs are more abundant as compared with the human brain.[20,77] In general, the available data suggest that the GST isozyme composition of human tissues is more comparable to that of mouse than of rat tissues, and it has been suggested that the mouse may be a more suitable model for toxicological studies.[3]

15.4 GENDER-RELATED DIFFERENCES

Similar to the Phase I detoxification enzymes,[79] gender-related differences also exist in the expression of GSTs in rodents and humans. For example, the expression of GST Pi is much less in the livers of female mice.[80] Studies in our laboratory have demonstrated sex-related qualitative and quantitative differences in the composition of GST isoenzymes in various mouse tissues.[81–83] In humans sex-related differences in the constitutive GST isozymes for composition of colon,[84] skin,[26] and erythrocytes[85] have been observed. Interestingly, these studies showed that GST Pi isolated from female skin had higher V_{max} toward CDNB when compared with that isolated from male skin. GST Pi isolated from male and female skin also differed in heat stability and kinetics of inhibition by several inhibitors.[26] These studies indicate qualitative as well as quantitative gender-related differences. The mechanisms responsible for gender-related differences in kinetic properties of GSTs are not known, but these may likely arise from hormone-dependent post-translational modifications or differential bindings of ligands. Differential carcinogenicity of benzo(a)pyrene to male and female mice has been attributed to gender-related differences in the expression of GST Pi.[86] Such a difference in humans may also be relevant to the protective mechanisms against chemical carcinogens and acquired drug resistance of tumor cells during cancer chemotherapy. The significance and potential toxicological implications of gender-related differences in human GSTs need to be investigated further.

15.4.1 Gender-Related Difference in Carcinogenicity of Benzo(a)pyrene (B[a]P) in Mice

The model of B(a)P-induced carcinogenesis of lungs and forestomachs in mice has been extensively used to evaluate the efficacy of tumor suppressors, and it has been demonstrated that a number of dietary antioxidants including butyalted hydroxytoulene (BHT), butylatedhydroxy anisole (BHA), ethoxyquin, and oltipraz can retard B(a)P-induced carcinogenesis in mice.[87–95] The chemoprotective effect of these agents has been attributed at least partly to the induction of GSTs.[96] Available evidence suggests that in mice gender-related differences in the constitutive GSTs may be a determinant of the differential carcinogenicity of B(a)P in male and female mice. In one of these studies, it has been shown that CD-1 female mice are more prone to develop lung tumors upon exposure to B(a)P.[86] At low doses B(a)P does not cause carcinogenicity in male mice but female mice develop lung tumors. At higher levels of B(a)P exposure, tumors in both male and female mice are observed. Interestingly the carcinogenic effect of B(a)P only at low doses is attenuated by BHA in the female mice. BHA does not provide any protection against the B(a)P-induced lung carcinogenesis in males, which are inherently more resistant to the carcinogenic effect of B(a)P than females, perhaps because of the high constitutive expression of GST Pi. Gender-related differences in the carcinogenicity of B(a)P in the livers of newborn CheB/JFXAJ mice have also been reported.[97]

Although the mechanistic basis of gender-related differential carcinogenicity of B(a)P to CD-1 mice is not completely understood, a tentative explanation could be the differential expression of GST Pi in males and females, which may be relevant to the human population. While the Alpha- and Mu-class GST isozymes are expressed in comparable amounts in the livers of male and female CD-1 mice,[82] the expression of GST Pi in the livers of female mice is less than half when compared to that of male mice.[70,80,82,83] Such gender-related differences in the expression of GST Pi in CD-1 mice should explain the observed higher carcinogenicity of B(a)P in female mice when compared to the males. Among the activated electrophilic metabolites of B(a)P, B(a)P diolepoxide (BPDE) is believed to be the ultimate carcinogen. In particular the (+) enantiomer of BPDE is the most potent carcinogen[98] and GSTs are implicated in its detoxification.[99–101] Studies in our laboratory on comparative catalytic efficiencies of the Alpha, Mu, and Pi classes of mouse GSTs indicate that their activity toward (+) BPDE are in the order Pi > Mu >> Alpha. Up to a certain threshold of B(a)P exposure, male mice, having higher constitutive levels of GST Pi, can more effectively detoxify BPDE. Beyond this threshold of B(a)P exposure, this differential carcinogenic effect of B(a)P is abolished and the male and female mice show similar susceptibility to B(a)P-induced carcinogenesis at high doses. The lack of protection against B(a)P-induced carcinogenesis by BHA in male

CD-1 mice as opposed to its protective effect in female CD-1 mice can also be explained in light of studies showing that BHA causes up to 10-fold induction of GST Pi in female CD-1 mouse livers. By contrast, induction of GST Pi in male livers is much less (1.5-fold).[70,83] Because the GST composition of tissues is significantly different in humans than in mice, it is difficult to extrapolate these findings to humans. However, human GST Pi is also highly efficient in conjugating (+) BPDE to GSH[101] and may contribute more toward the detoxification of (+) BPDE and PAH-derived carcinogens, particularly in human extrahepatic tissues where GST Pi is the predominant isozyme.[19,102] The existence of gender-related differences in humans in the expression of GST Pi, particularly in organs such as skin, colon, and erythrocytes[26,84,85] where GST Pi is the predominant enzyme, may suggest the possibility of differential carcinogenic insult of compounds such as PAH in males and females. However, extensive epidemiological as well as biochemical studies would have to be conducted to substantiate the idea of differential carcinogenic effects of environmental pollutants in male and female populations.

ACKNOWLEDGMENT

Research was supported in part by NIH grants EY04396 and ES012171.

REFERENCES

1. Kamisaka et al. Multiple forms of human glutathione S-transferase and their affinity for bilirubin. *Eur. J. Biochem.*, 60, 153–161, 1975.
2. Mannervik, B. and Danielson, U.H. Glutathione transferases — structure and catalytic activity. *CRC Crit. Rev. Biochem.*, 23(3), 283–337, 1988. Review.
3. Awasthi, Y.C., Sharma, R., and Singhal, S.S. Human glutathione S-transferases. *Int. J. Biochem.*, 26(3), 295–308, 1994. Review.
4. Hayes, J.D. and Pulford, D.J. The glutathione S-transferase supergene family: regulation of GST and the contribution of the isoenzymes to cancer chemoprotection and drug resistance. *Crit. Rev. Biochem. Mol. Biol.*, 30(6), 445–600, 1995. Review.
5. Hayes, J.D., Flanagan, J.U., and Jowsey, I.R. Glutathione transferases. *Annu. Rev. Pharmacol. Toxicol.*, 45, 51–88, 2005. Review.
6. Mannervik, B. et al. Nomenclature for human glutathione transferases. *Biochem. J.*, 15, 282, 305–306, 1992.
7. Awasthi, Y.C., Dao, D.D., and Saneto, R.P. Interrelationship between anionic and cationic forms of glutathione S-transferases of human liver. *Biochem. J.*, 191, 1–10, 1980.
8. Singh, S.V. et al. Different forms of human liver glutathione S-transferases arise from dimeric combinations of at least four immunologically and functionally distinct subunits. *Biochem. J.*, 232, 781–790, 1985.
9. Singh, S.V. et al. Purification and characterization of glutathione S-transferase of human kidney. *Biochem. J.*, 246, 179–186, 1987.
10. Corrigall, A.V. and Kirsch, R.E. Glutathione S-transferase distribution and concentration in human organs. *Biochem. Int.*, 16, 443–448, 1988.
11. Singh, S.V. et al. Characterization of glutathione S-transferases of human cornea. *Exp. Eye Res.*, 40, 431–437, 1985.

12. Peters, W.H.M. et al. Human intestinal glutathione S-transferases. *Biochem. J.*, 257, 471–476, 1989.
13. Sharma, R. et al. Independent segregation of glutathione S-transferase and fatty acid ethyl ester synthase from pancreas and other human tissues. *Biochem. J.*, 275, 507–513, 1991.
14. Singh, S.V., Srivastava, S.K., and Awasthi, Y.C. Purification and characterization of the two forms of glutathione S-transferase present in human lens. *Exp. Eye Res.*, 40, 201–208, 1985.
15. Peters, W.H.M., Wormskamp, N.G.M., and Thies, E. Expression of glutathione S-transferase in normal gastric mucosa and in gastric tumors. *Carcinogenesis*, 11, 1593–1596, 1990.
16. Ozer, N. et al. Resolution and kinetic characterization of glutathione S-transferases from human jejunal mucosa. *Biochem. Med. Metab. Biol.*, 44, 142–150, 1990.
17. Campbell, E. et al. A distinct human testis and brain mu class glutathione S-transferase: molecular cloning and characterization of a form present even in individuals lacking hepatic type mu isoenzymes. *J. Biol. Chem.*, 265, 9188–9193, 1990.
18. Di Ilio, C. et al. Purification and characterization of five forms of GST from human uterus. *Eur. J. Biochem.*, 171, 491–496, 1998.
19. Partridge, C.A., Dao, D.D., and Awasthi, Y.C. Glutathione S-transferases of lung: purification and characterization of human lung glutathione S-transferases. *Lung*, 162, 27–36, 1984.
20. Theodore, C. et al. Glutathione S-transferase of human brain: evidence for two immunologically distinct types of 26500-Mr subunits. *Biochem. J.*, 225, 375–382, 1985.
21. Singhal, S.S. et al. Glutathione S-transferases of human lung: characterization and evaluation of the protective role of the alpha class isozymes against lipid peroxidation. *Arch. Biochem. Biophys.*, 299, 232–241, 1992.
22. Zhao, T. et al. The role of human glutathione S-transferase hGSTA1-1 and hGSTA2-2 in protection against oxidative stress. *Arch. Biochem. Biophys.*, 367, 216–224, 1999.
23. Yang, Y. et al. Role of glutathione S-transferases in protection against lipid peroxidation. Overexpression of hGSTA2-2 in K562 cells protects against hydrogen peroxide-induced apoptosis and inhibits JNK and caspase 3 activation. *J. Biol. Chem.*, 276, 19220, 2001.
24. He, N.G. et al. Transfection of a 4-hydroxynonenal metabolizing glutathione S-transferase isozyme, mouse GSTA4-4, confers doxorubicin resistance to Chinese hamster ovary cells. *Arch. Biochem. Biophys.*, 333, 214, 1996.
25. Singh, S.V. et al. Purification and characterization of unique glutathione S-transferases from human muscle. *Arch. Biochem. Biophys.*, 264, 13–22, 1988.
26. Singhal, S.S. et al. Glutathione S-transferases of human skin: qualitative and quantitative differences in men and women. *Biochim. Biophys. Acta*, 1163, 266–272, 1993.
27. Del Boccio, G. et al. Identification of a novel glutathione transferase in human skin homologous with class alpha glutathione transferase 2-2 in rat. *Biochem. J.*, 244, 21–25, 1987.
28. Singh, S.P. et al. Membrane association of glutathione S-transferase mGSTA4-4, an enzyme that metabolizes lipid peroxidation products, *J. Biol. Chem.*, 227, 4232–4239, 2002.
29. Raza, H. et al. Multiple isoforms of mitochondrial glutathione S-transferases and their differential induction under oxidative stress, *Biochem. J.*, 366, 45–55, 2002.
30. Landi, S. Mammalian class theta GST and differential susceptibility to carcinogens: a review. *Mutat. Res.*, 463, 247–283, 2000.

31. Jowsey, I.R. et al. Mammalian class sigma glutathione S-transferases: catalytic properties and tissue-specific expression of human and rat GSH-dependent prostaglandin D2 synthases. *Biochem. J.,* 359, 507–516, 2001.

32. Bruin, W.C.C. et al. Expression of glutathione S-transferase θ class isoenzymes in human colorectal and gastric cancers. *Carcinogenesis,* 20, 1453–1457, 1999.

33. Hussey, A.J. and Hayes, J.D. Characterization of human class-theta glutathione S-transferase with activity towards 1-menaphthyl sulphate. *Biochem. J.,* 286, 929–935, 1992.

34. Tan, K.L. and Board, P.G. Purification and characterization of a recombinant human theta class glutathione S-transferase (GSTT2-2). *Biochem. J.,* 315, 727–732, 1996.

35. Webb, G. et al. Chromosomal localization of the gene for human theta class glutathione S-transferase (GSTT1). *Genomics,* 33, 929–935, 1996.

36. Tan, K.L. et al. Molecular cloning of cDNA and chromosomal localization of a human theta class glutathione S-transferase gene (GSTT2) to chromosome 22. *Genomics,* 25, 381–387, 1995.

37. Their, R. et al. Species differences in the glutathione transferase GSTT1-1 activity towards the model substrates methyl chloride and dichloromethane in liver and kidney. *Arch. Toxicol.,* 72, 622–629, 1998.

38. Whitington, A.T. et al. Characterization of cDNA and gene encoding and mouse theta class glutathione S-transferase mGSTT2 and its localization to chromosome 10B5-C1. *Genomics,* 33, 105–111, 1996.

39. Meyer, D.J. et al. Theta, a new class of glutathione transferases purified from rat and man. *Biochem. J.,* 274, 409–414, 1991.

40. Hiratsuka, A. et al. A new class of rat glutathione S-transferase Yrs-Yrs inactivating reactive sulfate esters as metabolites of carcinogenic arylmethanols. *J. Biol. Chem.,* 265, 11973–11981, 1990.

41. Hiratsuka, A. et al. Novel theta class glutathione S-transferases Yrs-Yrs′ and Yrs′-Yrs′ in rat liver cytosol: their potent activity toward 5-sulfoxymethylchrysene, a reactive metabolite of the carcinogen 5-hydroxymethylchrysene. *Biochem. Biophys. Res. Commun.,* 202, 278–284, 1994.

42. Buetler, T.M. and Eaton, D.L. Glutathione S-transferases: amino acid sequence comparisons, classification and phylogenetic relationship. *Environ. Carcinogen Ecotox. Revs.,* C10, 181–203, 1992.

43. Board, P.G. et al. Identification, characterization, and crystal structure of the omega class of glutathione transferases. *J. Biol. Chem.,* 275, 24798–24806, 2000.

44. Yin, Z.L. et al. Immunohistochemistry of omega class glutathione S-transferase in human tissues. *J. Biochem. Cytochem.,* 49, 983–987, 2001.

45. Kodym, R., Calkins, P., and Story, M. The cloning and characterization of a new stress response protein. *J. Biol. Chem.,* 274, 5131–5137, 1999.

46. Board, P.G. et al. Zeta, a novel class of glutathione transferases in a range of species from plants to humans. *Biochem. J.,* 328, 929–935, 1997.

47. Fernández-Cañón, J.M. and Penalva, M.A. Characterization of a fungal maleylacetoacetate isomerase gene and identification of its human homologue. *J. Biol. Chem.,* 273, 329–337, 1998.

48. Board. P.G. et al. Clarification of the role of key active site residues of glutathione transferase zeta/maleylacetoacetate isomerase by a new spectrophotometric technique. *Biochem. J.,* 374, 731–737, 2003.

49. Lim, C.E.L. et al. Mice deficient in glutathione transferase zeta/maleyacetoacetate isomerase exhibit a range of pathological changes and elevated expression of alpha-, mu-, and pi-class glutathione transferases. *Am. J. Pathol.,* 165, 679–693, 2004.

50. Fernández-Cañón, J.M. et al. Maleyleacetoacetate isomerase (MAAI/GSTZ)–deficient mice reveal a glutathione-dependent nonenzymatic bypass in tyrosine catabolism. *Mol. Cell. Biol.*, 22, 4943–4951, 2002.

51. Robinson, A. et al. Modeling and bioinformatics studies of the human kappa-class glutathione transferase predict a novel third glutathione transferase family with similarity to prokaryotic 2-hydroxychromene-2-carboxylate isomerases. *Biochem. J.*, 379, 541–552, 2004.

52. Carmichael, J.D. Glutathione and glutathione transferase levels in mouse granulocytes following cyclophosphamide administration. *Cancer Res.*, 46, 735–739, 1986.

53. Kitahara, A. et al. Changes in molecular forms of rat hepatic glutathione S-transferase during chemical carcinogenesis. *Cancer Res.*, 44, 2698–2703, 1984.

54. Moscow, J. et al. Elevation of π class glutathione S-transferase activity in human breast cancer cells by transfection of the GSTπ gene and its effect on sensitivity to toxins. *Mol. Pharmacol.*, 36, 22–28, 1989.

55. Horton, J.K. et al. Characterization of a chlorambucil-resistant human ovarian carcinoma cell line overexpressing glutathione S-transferase μ. *Biochem. Pharmacol.*, 58, 693–702, 1999.

56. Singhal, S.S. et al. A novel glutathione S-transferase isozyme similar to GST 8-8 of rat and mGSTA4-4 (GST 5.7) of mouse is selectively expressed in human tissues. *Biochim. Biophys. Acta*, 279–286, 1994.

57. Singhal, S.S. et al. Several closely related glutathione S-transferase isozymes catalyzing conjugation of 4-hydroxynonenal are differentially expressed in human tissues. *Arch. Biochem. Biophys.*, 311, 242–250, 1994.

58. Misra, P. et al. Glutathione S-transferase 8-8 is localized in smooth muscle cells of rat aorta and is induced in an experimental model of atherosclerosis. *Toxicol. Appl. Pharmacol.*, 133, 27–33, 1995.

59. Yang, Y. et al. Glutathione S-transferase A4-4 modulates oxidative stress in endothelium: possible role in human atherosclerosis. *Atherosclerosis,* 173, 211–221, 2004.

60. Cheng, J.Z. et al. Accelerated metabolism and exclusion of 4-hydroxynonenal through induction of RLIP76 and hGSTA5.8 is an early adaptive response of cells to heat and oxidative stress. *J. Biol. Chem.*, 276, 41213–41233, 2001.

61. Yang, Y. et al. Cells preconditioned with mild, transient UVA irradiation acquire resistance to oxidative stress and UVA-induced apoptosis: role of 4-hydroxynonenal in UVA-mediated signaling for apoptosis. *J. Biol. Chem.*, 278, 41380–41388, 2003.

62. Cheng, J. et al. Two distinct 4-hydroxynonenal metabolizing glutathione S-transferase isozymes are differently expressed in human tissues. *Biochem. Biophys. Res. Commun.*, 282, 1268–1274, 2001.

63. Pacifici, G.M. et al. Organ distribution of glutathione transferase isoenzymes in the human fetus: differences between liver and extrahepatic tissue. *Biochem. Pharmacol.*, 35, 1616–1619, 1986.

64. Faulder, C.G. et al. Studies on the development of basic, neutral, and acidic isoenzymes of glutathione S-transferase in human liver, adrenal, kidney and spleen. *Biochem. J.*, 241, 481–485, 1987.

65. Guthenberg, C. et al. Two distinct forms of glutathione transferase from human foetal liver. Purification and comparison with isoenzymes isolated from adult liver and placenta. *Biochem. J.*, 235, 741–745, 1986.

66. Khan, M.F. et al. Iron-induced lipid peroxidation in rat liver is accompanied by preferential induction of glutathione S-transferase 8-8 isozyme. *Toxicol. Appl. Pharmacol.*, 131, 63–72, 1995.

67. Partridge, C.A. et al., Rat lung glutathione S-transferases: subunit structure and the interrelationship with the liver enzymes. *Int. J. Biochem.*, 17, 331–340, 1985.

68. Singh, S.V., Srivastava, S.K., and Awasthi, Y.C. Binding of benzo(a)pyrene to rat lung glutathione S-transferases *in vivo*. *FEBS*, 179, 111–114, 1985.

69. Singh, S.V. et al. Glutathione S-transferase of mouse lung. *Biochem. J.*, 243, 351–358, 1987.

70. Chaubey, M. et al. Gender-related differences in expression of murine glutathione S-transferase and their induction by butylated hydroxyanisole. *Comp. Biochem. Physiol.*, 108C, 311–319, 1994.

71. Gupta, S. et al. Glutathione S-transferases of human lung evidence that two distinct mu-class subunits are expressed differentially in human population. *Clin. Chem. Enzym. Comms.*, 3, 115–124, 1990.

72. McLellan, L.I. and Hayes, J.D. Differential induction of class alpha glutathione S-transferases in mouse liver by the anticarcinogenic antioxidant butylated hydroxyanisole: purification and characterization of glutathione S-transferase $Ya_1 Ya_1$. *Biochem. J.*, 263, 393–402, 1989.

73. Raza, H. et al. Glutathione S-transferases in human and rodent skin: multiple forms and species-specific expression. *J. Invest. Dermatol.*, 96, 463–467, 1991.

74. Singh, S.V., Partridge, C.A., and Awasthi, Y.C. Rat lung glutathione S-transferases: evidence for two distinct types of 22,000 M_r subunits. *Biochem. J.*, 221, 609–615, 1984.

75. Singh, S.V. and Awasthi, Y.C. Two immunologically distinct Yc-type subunits are present in rat lung glutathione S-transferases. *Biochem. J.*, 224, 335–338, 1984.

76. Awasthi, Y.C. and Singh, S.V. Subunit structure of human and rat glutathione S-transferase. *Comp. Biochem. Physiol.*, 82B, 17–23, 1985.

77. Singh, S.V. et al. Comparative studies on the enzymes of glutathione S-transferase of rat brain and other tissues. *Comp. Biochem. Physiol.*, 86B, 73–81, 1987.

78. Awasthi, Y.C., Partridge, C.A., and Dao, D.D. Effect of butylated hydroxytoluene on glutathione S-transferase and glutathione peroxidase activities in rat liver. *Biochem. Pharmacol.*, 32, 1197–1200, 1983.

79. Sundseth, S.S. and Waxman, D.J. Sex-dependent expression and clofibrate inducibility of cytochrome P450 4A fatty acid ω-hydroxylase: male specificity of liver and kidney CYP4A2 mRNA and tissue-specific regulation by growth hormone and testosterone. *J. Biol. Chem.*, 267, 3915–3921, 1992.

80. Hatayama, I., Satoh, K., and Sato, K. Developmental and hormonal regulation of the major form of hepatic glutathione S-transferase in male mice. *Biochem. Biophys. Res. Commun.*, 140, 581–588, 1986.

81. Awasthi, S. et al. Purification and characterization of glutathione S-transferase of murine ovary and testis. *Arch. Biochem. Biophys.*, 301, 143–150, 1993.

82. Singhal, S.S. et al. Glutathione S-transferases of mouse liver: sex-related differences in the expression of various isozymes. *Biochim. Biophys. Acta*, 1116, 137–146, 1992.

83. Sharma, R. et al. Comparative studies on the effect of butylated hydroxyanisole on glutathione and glutathione S-transferases in tissues of male and female CD-1 mice. *Comp. Biochem. Physiol.*, 105C, 31–37, 1993.

84. Singhal, S.S. et al. Gender-related differences in the expression and characteristics of glutathione S-transferases of human colon. *Biochim. Biophys. Acta*, 1171, 19–26, 1992.

85. Srivastava, S.K. et al. Comparison of kinetic properties and expression of GSTπ (GSTP1-1) in tissues of men and women. *Biochem. Arch.*, 10, 155–160, 1994.

86. Sharma, R. et al. Differential carcinogenicity of benzo(a)pyrene in male and female CD-1 mouse lung. *J. Toxicol. Environ. Health*, 52, 45–62, 1997.

87. Wattenberg, L.W. Inhibition of chemical carcinogen–induced pulmonary neoplasia by butylated hydroxyanisole. *J. Natl. Cancer Inst.*, 50, 1541–1544, 1973.

88. Wattenberg, L.W. Effects of dietary constituents on the metabolism of chemical carcinogens. *Cancer Res.*, 35, 3326–3331, 1975.

89. Wattenberg, L.W. et al. Dietary constituents altering the response to chemical carcinogens. *Fed. Proc.*, 35, 1327–1331, 1976.

90. Wattenberg, L.W. Inhibition of chemical carcinogenesis. *J. Natl. Cancer. Inst.*, 60, 11–18, 1978.

91. Wattenberg, L.W. et al. Neoplastic effects of oral administration of (+/–) trans-7,8-dihydrobenzo(a)pyrene and their inhibition by butylated hydroxyanisole. *J. Natl. Cancer. Inst.*, 62, 1103–1106, 1979.

92. Wattenberg, L.W. Protective effect of 2 (3)-tert-butyl-4-hydroxyanisole on chemical carcinogenesis. *Food Chem. Toxicol.*, 24, 1099–1102, 1986.

93. Athar, M., Khan, W.A., and Mukhtar, H. Effects of dietary tannic acid on epidermal, lung, and forestomach polycyclic aromatic hydrocarbon metabolism and tumorigenicity in sencar mice. *Cancer Res.*, 49, 5784–5788, 1989.

94. Boogards, J.J. et al. Consumption of brussels sprouts results in elevated α-class glutathione S-transferase levels in human blood plasma. *Carcinogenesis*, 15, 1073–1075, 1994.

95. Spieir, J.L., Lam, L.K., and Wattenberg, L.W. Effects of administration to mice of butylated hydroxyanisole by oral intubation on benzo(a)pyrene-induced pulmonary adenoma formation and metabolism of benzo(a)pyrene. *J. Natl. Cancer Inst.*, 60, 605–609, 1978.

96. Awasthi, Y.C., Singhal, S.S., and Awasthi, S. Mechanisms of anti-carcinogenic effects of antioxidant nutrients. In *Nutrition and Cancer*, ed. Watson, R.R. and Mufti, S.I., 139–172, CRC Press, Boca Raton, 1995.

97. Glatt, H. and Oesch, F. Species differences in enzymes controlling reactive epoxides. *Arch. Toxicol.*, 10(Suppl.), 111–124, 1987.

98. Buening, M.K. et al. Tumorigenicity of the optical enantiomers of the diastereomeric benzo(a)pyrene 7,8 diol-9,10-epoxides in newborn mice. Exceptional activity of (+)-7β, 8α-dihydroxy-9α, 10α-epoxy-7,8,9,10-tetrahydrobenzo(a)pyrene. *Proc. Natl. Acad. Sci. USA*, 75, 5358–5361, 1978.

99. Robertson, I.G. et al. Glutathione S-transferases in rat lung: the presence of transferase 7-7, highly efficient in the conjugation of glutathione with the carcinogenic (+)-7-beta, 8-alpha-dihydroxy-9-alpha, 10-alpha-oxy-7,8,9,10-tetrahydobenzo(a)pyrene. *Carcinogenesis*, 7, 295–299, 1986.

100. Jernstrom, B., Martinez, M., and Dock, L. Glutathione transferase–catalyzed conjugation of benzo(a)pyrene diol epoxide with glutathione in rat hepatocytes. In *Glutathione S-Transferases and Drug Resistance*, ed. Hayes, J.D., Pickett, C.B., and Mantle, T.J., 111–120, Taylor & Francis, London, 1990.

101. Swedmark, S. et al. Studies on glutathione transferase belonging to class pi in cell lines with different capacities for conjugating (+)-7-beta, 8-alpha-dihydroxy-9-alpha, 10-alpha-oxy-7,8,9,10-tetrahydrobenzo(a)pyrene. *Carcinogenesis*, 12, 1719–1723, 1992.

102. Dao, D.D. et al. Human glutathione S-transferases: characterization of the anionic forms from lung and placenta. *Biochem. J.*, 221, 33–41, 1984.

103. Johansson, A.S. and Mannervik, B. Human glutathione transferase A3-3, a highly efficient catalyst of double-bond isomerization in the biosynthetic pathway of steroid hormones, *J. Biol. Chem.*, 276, 33061–33065, 2001.

104. Tsuchida, S., Maki, T., and Sato, K. Purification and characterization of glutathione transferases with an activity toward nitroglycerin from human aorta and heart. Multiplicity of the human class mu forms. *J. Biol. Chem.*, 265, 7150–7157, 1990.

105. Singh, S.V. et al. Immunohistochemical localization, purification, and characterization of human urinary bladder glutathione S-transferases. *Biochim. Biophys. Acta*, 1074, 363–370, 1991.

106. Di Ilio, C. et al. Glutathione transferase of human breast is closely related to transferase of human placenta and erythrocytes. *Biochem. Int.*, 13, 263–269, 1986.

107. Hirrell, P.A. et al. Studies on the developmental expression of glutathione S-transferase isoenzymes in human heart and diaphragm. *Biochim. Biophys. Acta*, 915, 371–377, 1991.

108. Marcus, C.J., Habig, W.H., and Jakoby, W.B. Glutathione transferase from human erythrocytes. Non-identity with the enzymes from liver. *Arch. Biochem. Biophys.*, 188, 287–293, 1978.

109. Singh, S.V. et al. Glutathione S-transferase of human fetal fibroblasts. *IRCS Med. Sci.*, 12, 1133–1134, 1984.

110. Singh, S.V. et al. Glutathione S-transferases of the human heart. *IRCS Med. Sci.*, 13, 973–974, 1985.

111. Di Ilio, C. et al. Selenium independent glutathione peroxidase activity associated with cationic forms of glutathione transferase in human heart. *J. Mol. Cell. Cardiol.*, 18, 983–991, 1986.

112. Koskelo, K. and Icen, A. Chromatofocusing of glutathione S-transferases from human kidney. *Scand. J. Clin. Lab. Invest.*, 44, 159–162, 1984.

113. Di Ilio, C. et al. Electrophoretic and immunological analysis of glutathione transferase isoenzymes of human kidney carcinoma. *Biochem. Pharmacol.*, 38, 1045–1051, 1989.

114. Polidoro, G. et al. Glutathione S-transferase activity from human lens. A single acidic form. *IRCS Med. Sci.*, 10, 962, 1982.

115. Warholm, M., Guthenberg, C., and Mannervik, B. Molecular and catalytic properties of glutathione transferase μ from human liver: an enzyme efficiently conjugating epoxides. *Biochemistry*, 22, 3610–3617, 1983.

116. VanderJagt D.L. et al. Isolation and characterization of the multiple glutathione S-transferases from human liver. Evidence for unique heme-binding site. *J. Biol. Chem.*, 260, 11603–11610, 1985.

117. Koskelo, K., Valmet, E., and Tenhunen, R. Purification and characterization of an acidic glutathione S-transferase from human lung. *Scand. J. Clin. Lab. Invest.*, 41, 683–689, 1981.

118. Singhal, S.S. et al. Purification and characterization of human muscle glutathione S-transferases: evidence that glutathione S-transferase ζ corresponds to a locus distinct from GST1, GST2, and GST3. *Arch. Biochem. Biophys.*, 285, 64–73, 1991.

119. Guthenberg, C., Akerfeldt, K., and Mannervik, B. Purification of glutathione S-transferase from human placenta. *Acta Chem. Scand.*, 33, 595–596, 1979.

120. Rogerson, K.S. et al. Studies on the glutathione S-transferase of human platelets. *Biochem. Biophys. Res. Commun.*, 122, 407–412, 1984.

121. Tew, K.D. et al. Glutathione S-transferases in human prostate. *Biochim. Biophys. Acta*, 926, 8–15, 1987.
122. Di Ilio, C. et al. Glutathione transferase isoenzymes from human prostate. *Biochem. J.*, 271, 481–485, 1990.
123. Del Boccio, G. et al. Purification and characterization of glutathione transferase of human thyroid. *Ital. J. Biochem.*, 36, 8–17, 1986.

16 Enzymology of Glutathione S-Transferases: Laboratory Methods

Sharad S. Singhal, Kenneth Drake, Sushma Yadav, Jyotsana Singhal, and Sanjay Awasthi

CONTENTS

16.1 INTRODUCTION

Glutathione S-transferases (GSTs, EC 2.5.1.18) are a large family of Phase II biotransformation enzymes that catalyze the formation of glutathione (GSH)-electrophile thioether conjugates (GS-E), the rate-limiting step for the elimination of electrophilic compounds through metabolism to mercapturic acids.[1-8] GSTs are among the most abundant proteins in some tissues, including the liver and kidneys, and have been referred to as intracellular albumin because of their ability to bind to structurally diverse endogenous and xenobiotic hydrophobic compounds. Indeed, before the description of the GSH S-transferase activity of GSTs, they were designated as ligandin, functioning to sequester certain toxins intracellularly.[9-16]

The identification of the enzymatic activity of GSTs toward a wide variety of substrates led to an understanding of the role of this GSH-transferase catalytic activity in defending cells from deleterious effects of endogenous electrophilic toxins generated from lipid peroxidation and exogenous toxins (xenobiotics), which are electrophilic by nature (i.e., alkylating agents) or are converted to genotoxic electrophilic intermediates by the catalytic action of cytochromes P450. Along with the recognition of the role of GSTs in electrophilic toxin defenses, studies aimed at purifying and fractionating GST activity into component enzymes have resulted in the definition of heterogeneity of GSTs, now known to consist of at least three major families (cytosolic, mitochondrial, and microsomal), and several immunologically and functionally distinguishable classes within each family.[4,5,8]

Different families and classes of GSTs display some remarkable differences in sequence, substrate specificity, tissue-specific expression, and patterns of induction in response to electrophilic stress.[4,5] These aspects of GST structure and functions are covered extensively in various chapters of this volume. Because GSTs are strongly implicated as a first line of defense against electrophilic stresses, the techniques for purifying and characterizing GSTs are still quite relevant in studies aimed at understanding an integrated picture of cellular electrophile/oxidant stress responses. Furthermore, uses of large quantities of tissue or recombinant GSTs are frequently necessary for purposes of enzymatic synthesis of GS-E. Recent development of GST-activated prodrugs has also resulted in the need for studies that require significant quantities of purified GSTs for pharmacological characterization of these drugs. This chapter focuses on laboratory methods routinely used to determine the activity and purification and characterization of these versatile enzymes.

16.2   GST ACTIVITY

Given striking differences in catalytic efficiency between the major classes and subclasses of GSTs for the known substrates, it is perhaps not surprising that the list of known substrates and assays is relatively lengthy. A comprehensive list of these substrates can be obtained from various review articles.[4,5,8] Because GSTs utilize such a wide array of structurally dissimilar chemical substrates for glutathione conjugation, the types of assays used for determining GST activity are also quite varied. Examples and applications of several methods are described below.

16.2.1   SPECTROPHOTOMETRIC METHODS

16.2.1.1   Transferase Activity

For monitoring the purifications of GSTs from most sources, spectrophotometric methods are well suited because of their relative simplicity. Both the transferases and the hydroperoxide oxidoreductase (glutathione peroxidase) activities of GSTs can be measured spectrophotometrically, the former by monitoring the appearance of product or disappearance of reactant, and the latter by monitoring the consumption of NADPH through a glutathione reductase (GR)–linked assay. Pioneering studies in Jakoby's laboratory provided a quick and convenient method to determine GST activity toward a number of substrates. In these studies, which stand as a major landmark in the field of GST research, Habig et al.[16] standardized methods for spectrophotometric determination of GST activity toward several substrates which are being still used to assay GST activity. These spectrophotometric methods have been extended to a variety of additional substrates listed in Table 16.1. Alternatively, GST activity is measured using HPLC, TLC, and fluorescence methods (Table 16.1)

TABLE 16.1
GST Assays

Assay Method	Substrates	References
Spectrophotometric	Styrene oxide	84
	4-hydroxydecenal	25
	Delta(5)-androstene-3,17-dione; delta(5)-pregnene-3,20-dione	85
	Acrolein; crotonaldehyde	86, 87
	Para-nitrophenyl acetate	79
	7-chloro-4-nitrobenzo-2-oxa-1,3-diazole	88
	Iodomethane	89
HPLC	Diol epoxides of chrysene; dibenz(a,h)-anthracene; benzo(a)pyrene; benzo(c)phenanthrene; benzo(c)chrysene; benzo(g)chrysene	90
	13-oxooctadeca-9,11-dienoic acid	91
Fluorescence	Monochlorobimane	92
Conjugation	15 deoxy-12,14-prostaglandin J2(15d-PGJ2)	8
	Aflatoxin B1	93

when spectrophotometric methods are unsuitable. GST substrates have limited aqueous solubility, thus kinetic studies at saturating concentrations of substrates are not possible for many. Absorbance maxima (λ_{max}) are generally in the UV range, making quartz cuvettes a requisite. The relatively small molar extinction coefficients, and relatively small differences in extinction coefficients at λ_{max} between the substrate electrophile and product thioether, may introduce significant systematic errors in determination of activities. Most GST isoenzymes display high activity toward 1-chloro-2,4-dintrobenzene (CDNB), thus it is the substrate used most commonly to monitor purifications. An exception is the Theta class, which neither binds well to GSH-affinity resins nor displays significant activity toward CDNB; for these enzymes the preferred substrate is p-nitrobenzyl chlorides (pNBC).[17] Hydroperoxides and ethacrynic acids (EA) are preferred substrates for cationic and anionic GST isoenzymes, respectively, while EPNP (1, 2-epoxy-p-nitro-phenoxy propane), LTA_4ME (leukotrienes A_4 methyl ester), and ESA (cis-9, 10-epoxystearic acid) are the preferred substrates for neutral GST isoenzymes.[18] 4-Hydroxynonenal (4-HNE) is the specific substrates for the mammalian GST isoenzymes rGST8-8, mGSTA4-4, hGSTA4-4, hGST5.8, and their drosophila counterparts.[19–28] Activity toward 4-HNE can also be quantified by spectrophotometrically monitoring at 224 nm.[19–28] Kinetics at saturating concentrations of 4-HNE is complicated because of the optical opacity of solutions at high 4-HNE concentrations. Similar problems hamper activity determinations on crude homogenates, particularly if overall activity is low.

In general, the protocols for determination of GST activity are simple and usually pose no technical problems. These methods can be used to measure GST activity in the purified preparation and also in crude tissue and cell extracts. The assays are carried out under optimal conditions and temperatures and the pH of the medium is preferred to be around 6.5 to minimize the nonenzymatic reaction between the electrophiles. The electrophilic substrates used in the assay are often hydrophobic and their precipitation in the cuvettes can lead to inconsistent results. A stock solution of the substrate can be made in ethanol and the final concentration of ethanol in the assay medium should not exceed 5%. GSTs are fairly stable enzymes and in our experience tissues stored at 4°C do not lose significant activity for up to 72 hours. This makes it easier for the experiments in which a large number of tissue samples are involved. Clear tissue extracts made in a buffer containing protease inhibitors can also be stored for up to 48 hours without significant loss in activity. It is important to add just enough enzymes to the medium. Higher amounts of enzymes will cause substrate exhaustion, resulting in a nonlinear rate. On the other hand, too few enzyme can lead to inaccurate measurements. Thus the conditions for assay need to be standardized in each laboratory and the linearity of the reaction with respect to time and protein concentration must be stringently established. Protocols for the determination of the transferase activity of GSTs toward some of the frequently used substrates are given below.

16.2.1.2   1-Chloro-2,4-Dinitro Benzene (CDNB)[16]

This is the most useful substrate for following GST activity during purification. The reaction mixture (1 ml) contains 830 μl potassium phosphate buffer (100 mM, pH

6.5), 100 μl GSH (stock 10 mM in 100 mM potassium phosphate buffer, pH 6.5), and appropriate amounts of GST (in 20 μl). Reaction is initiated by adding 50 μl CDNB (stock 20 mM in ethanol). The glutathionyl-thioether of CDNB (dinitrophenyl-S-glutathione, DnpSG) is formed through a substitution of the sulfur of sulfhydryl group on GSH for the chloride at the 1-position with formation of HCl and DnpSG. At room temperature, the nonenzymatic reaction is at least an order of magnitude slower at pH 6.5 than the enzyme-catalyzed reaction. Because the extinction coefficient of DnpSG at 340 nm ($\in$ 9.6 mM^{-1} cm^{-1}) is significantly greater than that of CDNB ($\in$ 0.6 mM^{-1} cm^{-1}), the reaction at room temperature can be followed by monitoring absorbance at 340 nm of the reaction mixture against a blank containing CDNB and GSH without enzyme spectrophotometrically. One unit of GST activity is defined as 1 μmol of DnpSG formed per minute at 25°C.

16.2.1.3   3,4-Dichloronitrobenzene (DCNB)[16]

For GST activity toward DCNB, the reaction mixture (1 ml) contains 830 μl of 100 mM potassium phosphate buffer, pH 7.5; 100 μl GSH (stock 50 mM in 100 mM potassium phosphate buffer, pH 7.5); and 20 μl GST. The reaction is initiated by adding 50 μl DCNB (stock 20 mM in ethanol). The blank contains all ingredients except GST. Activity is monitored spectrophotometrically at 345 nm at room temperature.

16.2.1.4   1,2-Epoxy-3-(P-Nitrophenoxy) Propane (EPNP)[16]

For GST activity toward EPNP, the reaction mixture (1 ml) contains 830 μl of 100 mM potassium phosphate buffer, pH 6.5; 100 μl GSH (stock 50 mM in 100 mM potassium phosphate buffer, pH 6.5); and 20 μl GST. The reaction is initiated by the addition of 50 μl EPNP (stock 100 mM in ethanol). The blank contains all ingredients except GST. Activity is monitored spectrophotometrically at 360 nm at room temperature.

16.2.1.5   Ethacrynic Acids (EA)[16]

For GST activity toward EA, the reaction mixture (1 ml) contains 830 μl of 100 mM potassium phosphate buffer, pH 6.5; 100 μl GSH (stock 2.5 mM in 100 mM potassium phosphate buffer, pH 6.5); and 20 μl GST. The reaction is initiated by adding 50 μl EA (stock 4 mM in ethanol). The blank contains all ingredients except the enzyme. Activity is measured spectrophotometrically at 270 nm at room temperature.

16.2.1.6   P-Nitrobenzyl Chloride (NBC)[16]

GST activity toward NBC is determined in a reaction mixture (1 ml) containing 830 μl of 100 mM potassium phosphate buffer, pH 6.5; 100 μl GSH (stock 50 mM in 100 mM potassium phosphate buffer, pH 6.5); and 20 μl GST. The reaction is initiated by adding 50 μl NBC (stock 20 mM in ethanol). The blank contains all ingredients except GST. Activity is measured spectrophotometrically at 310 nm at room temperature. One unit of enzyme utilizes 1 μmol of substrate per minute at 25°C.

16.2.1.7   4-Nitropyridine-N-Oxide (NPNO)[16]

GST activity toward NPNO is determined in a reaction mixture (1 ml) containing 830 µl of 100 mM potassium phosphate buffer, pH 7.0, 100 µl GSH (stock 50 mM in 100 mM potassium phosphate buffer, pH 7.0); and 20 µl GST. The reaction is initiated by adding 50 µl NPNO (stock 4 mM in ethanol). The blank contains all ingredients except GST. Activity is measured spectrophotometrically at 295 nm at room temperature.

16.2.1.8   Bromosulfophthalein (BSP)[16]

GST activity toward BSP is determined in a reaction mixture (1 ml) containing 830 µl of 100 mM potassium phosphate buffer, pH 7.5; 100 µl GSH (stock 50 mM in 100 mM potassium phosphate buffer, pH 7.5); and 20 µl GST. The reaction is initiated by adding 50 µl BSP (stock 0.6 mM in buffer). The blank contains all ingredients except GST. Activity is measured spectrophotometrically at 330 nm at room temperature.

16.2.1.9   Trans-4-Phenyl-3-Buten-2-One (tPBO)[16]

GST activity toward tPBO is determined in a reaction mixture (1 ml) containing 830 µl of 100 mM potassium phosphate buffer, pH 6.5; 100 µl GSH (stock 2.5 mM in 100 mM potassium phosphate buffer, pH 6.5); and 20 µl GST. The reaction is initiated by adding 50 µl tPBO (stock 1 mM in ethanol). The blank contains all ingredients except GST. Activity is measured spectrophotometrically at 290 nm at room temperature.

16.2.1.10   t-4-Hydroxy-2-Nonenal (4-HNE)[19]

GST activity toward 4-HNE is determined in 1 ml of a reaction mixture containing 780 µl of 100 mM potassium phosphate buffer, pH 6.5; 100 µl GSH (stock 5 mM in 100 mM potassium phosphate buffer, pH 6.5); and an appropriate amount of GST in 20 µl. The reaction is started by adding 100 µl of 4-HNE (stock 1 mM in 100 mM potassium phosphate buffer, pH 6.5). The utilization of 4-HNE is followed at 30°C at 224 nm ($\in$ 13.75 mM^{-1} cm^{-1}) in a spectrophotometer against a blank, which contains all the above reactants except GST. One unit of enzyme catalyzes the conjugation of 1 µmol of substrate per minute at 30°C.

16.2.1.11   Glutathione Peroxidase Activity

Spectrophotometric determination of the hydroperoxide-reductase activity of GSTs is performed through a glutathione reductase (GR)-linked assay. The assay buffer contains GR pre-equilibrated with GSH and NADPH. After adding GST and the hydroperoxide substrate, GSSG forms from GST-catalyzed coupled oxidation of GSH and reduction of the hydroperoxide to an alcohol. GSSG produced is reduced back to GSH by GR, coupled with oxidation of NADPH, the disappearance of which is monitored at 340 nm (λ_{340} for NADPH, 6.2 mM^{-1} cm^{-1}). A number of hydroperoxide substrates have been used in this assay, including cumene hydroperoxide, 9-hydroperoxylinoleic acid, 13-hydroperoxylinoleic acid, dilinoleoyl phosphatidylglycerol hydroperoxide, t-butyl hydroperoxide, dilinoleoyl phosphatidylcholine hydroperoxide, and dilinoleoyl phosphatidylethanolamine hydroperoxide.[18,24,29] Peroxidized phospholipids present in the biological membranes can also be used as substrates for this method.[30] Because these hydroperoxides are unstable, the exact concentration of each stock solution has to be determined just prior to assay using the iodometric method.[31] A similar approach is used for assaying glyceryl trinitrate (nitroglycerin)-reductase activity of GSTs.[32,33]

16.2.1.12   Lipid Hydroperoxides[18]

Glutathione-peroxidase activity using cumene hydroperoxide, 9-hydroperoxylinoleic acid, 13-hydroperoxylinoleic acid, dilinoleoyl phosphatidylglycerol hydroperoxide, dilinoleoyl phosphatidylcholine hydroperoxide, and dilinoleoyl phosphatidylethanolamine hydroperoxide as the substrates is determined in a 1-ml reaction mixture of 160 mM Tris-HCl, pH 7.0; 3.2 mM GSH; 0.32 mM NADPH; 1 unit GR; and 0.82 mM EDTA incubated with an appropriate amount of GST at 37°C for five minutes. The reaction is started by adding a hydroperoxide substrate (100 μM, prepared in methanol), and the utilization of NADPH (extinction coefficient for NADPH, 6.2 mM^{-1} cm^{-1}) is followed at 340 nm at 37°C in a spectrophotometer against a blank that contains all the above reactants except the hydroperoxide substrates, which are substituted by equivalent volumes of methanol. An additional blank is required to determine the nonenzymatic oxidation of GSH by hydroperoxides in a reaction mixture similar to that described above except that the enzyme is replaced by a buffer. The exact concentration of hydroperoxides in cumene hydroperoxide and phospholipid hydroperoxides to ensure their concentration in the reaction mixture should be determined.[31]

16.2.1.13   Nitroglycerin (Glyceryl Trinitrate, GTN)[33]

GST activity toward GTN is assayed by quantitating the formation of oxidized glutathione (GSSG). The 1-ml reaction mixture contains 50 mM potassium phosphate buffer, pH 7.0; 1 mM EDTA; 0.2 mM NADPH; 1 U of GR; 1 mM GSH; 1 mM GTN; and an appropriate amount of the enzyme. The formation of GSSG from GSH and GTN through the reaction catalyzed by GST is coupled with the conversion of NADPH to NADP in the presence of GR. Disappearance of NADPH is followed by measuring absorbance at 340 nm at 25°C.

16.2.2   THIN-LAYER CHROMATOGRAPHIC ASSAYS

Unlike the lipid hydroperoxide reductase activity, thioether formation from reactive lipids and other optically silent substrates is not readily amenable to monitoring by spectrophotometer. Instead, product formation or substrate consumption is monitored by chromatographic separation and quantitation of radio-labeled substrates. While the separation of the products and the substrate can be achieved by HPLC,[34,35] simplicity and economy often favor the use of thin-layer chromatography (TLC) where possible, particularly because of the usually marked differences in mobility between substrates and products due to formation of an anionic thioether. Where thin-layer chromatography is not feasible, high-performance liquid chromatography can be used to separate and quantify reaction products. The radio label may be present either on the electrophilic substrate or GSH, the latter being more convenient because it allows the use of a single radio-labeled reagent for assaying activity toward a number of substrates. Examples of substrates for which thin-layer chromatographic assay is necessary include LTA$_4$ME, ESA, and the alkylating agent chemotherapy drug chlorambucil (CLB). GST activity toward LTA$_4$ME,[36] ESA,[37] and CLB[38] can be determined by separating and quantifying the thioether conjugate of these electrophilic lipids by TLC. Protocols for some of these substrates are detailed below.

16.2.2.1   Leukotrienes A$_4$ Methyl Ester (LTA$_4$ME)[36]

GST activity using LTA$_4$ME as a substrate is measured in an incubation mixture (100 µl) that contains 25 mM potassium phosphate buffer, pH 7.0; 1 µg GST protein; 0.5 mM EDTA; 2 mM ^{3}H-GSH labeled on glycine (specific activity 0.25 µCi/µmol); and 20 µM LTA$_4$ME. The reaction mixture is preincubated at 30°C for two minutes. The reaction is then started by adding LTA$_4$ME and incubating for 30 minutes at 30°C. Reaction is terminated by adding 100 µl of ice-cold methanol. The reaction product, LTC$_4$ME, is separated using TLC. TLC plates are developed for four hours in a mixture of butanol:acetic acid:water (4:2:2) and visualized by 0.2% ninhydrin reagent. Formation of LTC$_4$ME is quantitated by scraping and measuring radioactivity of the ninhydrin positive spot, corresponding to the R_f value of authentic LTC$_4$ME. The control contains no enzyme and corresponding spot areas on the TLC plates are also scraped and radioactivity is counted. The rate of enzymatic conjugation is determined by subtracting the rate of the nonenzymatic reaction.

16.2.2.2   Cis-9, 10-Epoxystearic Acid (ESA)[37]

GST activity toward ESA is determined in a reaction mixture (100 µl) containing 2 mM ^{3}H-GSH (specific activity 0.25 µCi/µmol) in 100 mM potassium phosphate buffer, pH 7.4, and an appropriate amount of the enzyme. After five minutes of preincubation at 37°C, the reaction is initiated by adding 0.2 mM ESA (ESA solution in 100 mM potassium phosphate buffer, pH 7.4) and incubated for an additional 30 minutes. The reaction is terminated by adding 100 µl of ice-cold methanol and centrifuged at 3,000 rpm to remove precipitated protein. An aliquot from the supernatant is quantitatively applied on a TLC plate and developed in a presaturated chamber containing a mixture of n-butanol:acetic acid:water (4:2:2 v/v/v) as the mobile phase. The plates are visualized by spraying with 0.2% ninhydrin in ethanol and areas corresponding to authentic GS-ESA conjugate are scraped and radioactivity is quantified. The control contains no enzyme and corresponding spot areas on the TLC plates are also scraped and radioactivity is counted. The rate of enzymatic conjugation is determined by subtracting the rate of nonenzymatic conjugation.

16.2.2.3   Chlorambucil (CLB)[38]

GST activity toward CLB is determined in a reaction mixture (100 µl) containing 1 µg of enzyme protein; 2 mM ^{3}H-GSH labeled on glycine (specific activity 240 µCi/µmol); and 100 µM CLB in 100 mM potassium phosphate buffer, pH 7.4. The reaction mixture is preincubated for five minutes at 37°C before starting the reaction by adding CLB and is then incubated for an additional 30 minutes at 37°C. The reaction is terminated by adding 100 µl of ice-cold methanol and centrifuged to remove precipitated protein. The enzymatic reaction product (GS-CLB conjugate) is separated by TLC in a mixture of butanol:acetic acid:water (4:2:2) and visualized with 0.2% of ninhydrin reagent. The areas corresponding to authentic GS-CLB conjugate are scraped and the radioactivity is quantitated by a liquid scintillation counter (Beckman LS-6800). The control contains no enzyme and corresponding spot areas are also scraped and counted. The rate of enzymatic conjugation is determined by subtracting the rate of nonenzymatic reaction.

16.3   PURIFICATION

In earlier studies, the purification of GSTs was performed using ammonium sulfate precipitation, ion exchange chromatography, and gel filtration.[39–45] Purification of GSTs by GSH-affinity chromatography was first reported in Vander Jagt's laboratory[46] and is perhaps one of the most important advancements in GST research. This method[18,20–24,43,47–53] and affinity chromatography using S-hexylglutathione[54–57] have enormously contributed to studies on structure, function, and expression of GSTs.

16.3.1   GSH-Affinity Purification

GSH-Sepharose affinity ligand chromatography is the simplest and most effective method for purification of cytosolic GST from tissues and cells. GSH-affinity chromatography has been reported as highly useful to isolate GSTs from a wide variety of sources.[18,20–24,43,46] The method is relatively quick and yields a high purity of GST proteins. The specificity of GSTs for GSH-affinity chromatography and the ease of use of the purification method have led to widespread adoption of GST as an affinity tag, which enables identification or specific purification of GST-tagged recombinant peptides such as signaling proteins.[58–61]

In most cases, enzyme elution from the column can be accomplished using 5–10 mM of the counterligand GSH, or S-hexyl-GSH.[18,43,54] Typical dynamic capacity for these supports is approximately 0.2 mg of protein per ml of bed gel.[62] Bromosulfophthalein-GSH agarose affinity chromatography has also been used for the separation of the anionic and cationic GST isoenzymes from human erythrocytes, hearts, and lungs.[18,20,47,63] In our laboratory, GSH-affinity chromatography is a routine technique for isolating GSTs from humans as well as from rodents.[18,20–23,47–53] This chromatographic technique is highly selective and can be used as the initial chromatographic step, or in most cases it can serve as a final purification step.

16.3.2 Preparation of the GSH-Affinity Resin

GSH-Sepharose 6B[46] and S-hexylglutathione-Sepharose[54] affinity matrices are both well characterized and specific. Although preprepared GSH-affinity resins can be purchased alone or as part of a variety of kits, the preparation of GSH-affinity resins remains more economical for purifications at a larger scale. GSH-Sepharose 6B is prepared from epoxy-activated Sepharose 6B. In a typical protocol, the resin (5 g) is washed in water, equilibrated in a phosphate buffer at pH 7.0 under nitrogen, and coupled with 500 mg GSH overnight at 37°C. Unreacted GSH is washed off and unreacted epoxy groups are blocked by incubation with 1 M ethanolamine, pH 8.0, for four hours at 25°C. The resin is then sequentially washed with 0.5 M KCl in 0.1 M sodium acetate, pH 4.0, then 0.5 M KCl in 0.1 M sodium borate, pH 8.0, followed by equilibration in 22 mM potassium phosphate buffer, pH 7.0, containing 1.4 mM β-mercaptoethanol (β-ME). Approximately 3 ml affinity resin is obtained from 1 g of epoxy-activated Sepharose 6B. The resin can be stored in a sealed dark container in an affinity buffer for up to one month at 4°C. One ml bed volume of affinity resin binds up to 40 U GST activity in the homogenate.

In a typical protocol for the preparation of S-hexylglutathione-Sepharose 6B, S-hexylglutathione is synthesized by dissolving 1 g of GSH in 6 ml of 1 N NaOH and diluting the reaction mixture to 20 ml with 95% ethanol. Iodohexane (0.7 g) is added slowly with vigorous stirring, followed by incubation overnight at room temperature. S-hexylglutathione is precipitated out by adding 47% HI to adjust the pH to 3.5, collected and washed on a sintered glass filter, and vacuum dried. Epoxy-activated Sepharose 6B (5 g) washed with deionized water is mixed with 0.5 g of S-hexylglutathione in 20 ml of deionized water and the pH is adjusted to 12 with 2 N NaOH, then incubated at 30°C for 16 hours. After washing with water, the remaining active groups of the affinity resin are blocked by treating with 2 M ethanolamine, pH 9.0, for four hours at 30°C with gentle shaking. The affinity resin is then washed sequentially with 100 ml of 0.5 M NaCl, first in 0.1 M sodium acetate, pH 4.0, and then in 0.1 M Tris-HCl, pH 8.0. For storage, 0.02% sodium azide is added to prevent microbial growth.[54]

16.3.3 Purification of Total GST by GSH-Affinity Chromatography

16.3.3.1 Purification from Human Tissue

For the purification of GSTs from most sources including human tissues, a GSH-affinity procedure can serve as the initial step in the isolation of cytosolic GSTs from a crude mixture or it may be used as a final purification step after the several forms of GSTs are separated from one another.

For human tissue GST purification, appropriate IRB consent and universal precautions against infectious agents are requisite. Autopsy tissues should be collected as quickly as possible after death and stored at −20°C after removing the connective tissues and perfusing with PBS to remove erythrocytes (which contain GST-π). Most GST isoenzymes are sufficiently stable to be purified from frozen tissue several months later. Purification of GSTs from any source requires that all purification steps be performed at 4°C because of the known differential heat inactivation of GST isoenzymes.[48,52] Tissues are homogenized for ~ five minutes on ice using a rotary-blade homogenizer at pH 7.0, in a 10-mM potassium phosphate buffer with 1.4 mM β-ME (buffer A) and diluted to 10% (w/v) before centrifugation for 45 minutes at 28,000 g. The supernatant is dialyzed overnight against 100 volumes of buffer A, followed by repeat centrifugation for 30 minutes at 28,000 g. The supernatant is subjected to GSH-Sepharose 6B affinity column chromatography with the column preequilibrated with a 22 mM potassium phosphate buffer, pH 7.0, containing 1.4 mM β-ME (buffer B) at a flow rate of ~ 8 ml/hour. Subsequently, the column is washed with buffer B until the absorbance at 280 nm of the effluent is constant near zero. Thorough washing of the column to completely remove the unbound protein is essential for getting pure GSTs. Bound GSTs are eluted with 10 mM GSH in 50 mM Tris-HCl, pH 9.6, containing 1.4 mM β-ME. Fractions of the purified enzyme to be used for kinetic studies are dialyzed against buffer A and those used for structural studies (peptide fragmentation, SDS-PAGE, amino acid sequence analyses, etc.) can be dialyzed against 0.1% aqueous acetic acid. Purity is assessed by SDS-PAGE and by Western blotting against antibodies specific for each GST isoenzyme. The component isoenzymes in the total purified GST fraction can be separated by three different methods: DEAE cellulose (DE-52) anion exchange chromatography, isoelectric focusing (IEF) chromatography, or immuno-affinity chromatography. GST purifications using S-hexylglutathione-Sepharose 6B, using similar methods with minor differences, have been reported. In a typical protocol.[54] The 104,000 × g supernatant of homogenate is applied to S-hexylglu-tathione affinity resin, washed as described above, and the bound enzyme is eluted with a buffer containing 10 mM Tris-HCl, pH 7.8, 0.2 M NaCl, and 5 mM S-hexylglutathione. In this procedure, because of the presence of NaCl, desalting of the purified enzyme is required using a Sephadex G-25 column equilibrated with 10 mM sodium phosphate, pH 6.8, containing 1 mM EDTA. Both these procedures can be used for purification of GSTs from various tissues including from rats,[23,51] mice,[48,53] bovines,[64] rabbits,[33] and also insects.[65]

Purification of GSTs from blood presents the challenge of removing hemoglobin and contaminant GST isoforms from white blood cells. Purifications should be performed on fresh heparinized blood centrifuged at 2000 rpm for 10 minutes to remove plasma and the buffy coat (a white blood cell layer on the top). To minimize leukocyte contamination, erythrocytes should be filtered over glass wool. Packed red blood cells are hemolyzed by diluting in a hypotonic buffer (buffer A), hemoly-sate dialyzed against 100 volumes of buffer A for overnight, and centrifuged at 10,000 × g for one hour. The supernatant is collected and subjected to GSH-affinity chromatography. Extensive washing of the affinity resin is necessary to ensure adequate removal of hemoglobin. The GST content of blood is relatively low, with yields of activity near 0.1 to 0.2 U/ml blood, although GST-π is the major isoenzyme. GST-θ present in erythrocytes does not bind to GSH-affinity resin.

16.3.3.2 Cell Lines

In purification of GSTs from cultured eukaryotic cells, a yield of about 0.05 U/mil-lion cells may be expected.[66] However, this may vary for different cell lines. Large-scale cell cultures are necessary to purify sufficient GST proteins and to isolate and quantify individual isoenzymes. GSH-affinity resin can be used for the purification of GST isoenzymes from cultured cells. Cells used for GST purification must be harvested from culture and frozen at −80°C after washing with PBS until enough cells can be obtained for the experiments. The cells are thawed; dialyzed in a 10 mM potassium phosphate buffer, pH 7.0, containing 1.4 mM β-mercaptoethanol (buffer A); and sonicated (3 × 5 s) using a sonifier cells disrupter. The purification of total GSTs can be carried out according to the method described previously.[66] All purification steps should be performed at 4°C and enzyme activity during the purification can be monitored with CDNB as the substrate. The enzyme preparations used for kinetic studies should be dialyzed against buffer A and those used for structural studies should be dialyzed against 0.1% aqueous acetic acid.

10K relative molecular mass cut-off membranes amicons is useful to concentrate the enzyme prior to the next chromatographic step. To preserve enzymatic activity, each purification step should be optimized in relation to the type of buffer, pH, ionic strength, temperature, etc. Buffer systems based on phosphate (pH 6.5–7.4) or Tris-HCl (pH 7.4–9.6) have been most used. To prevent microbial contamination, sodium azide is often added to the buffer at 0.1 to 0.3% concentrations, which do not affect proteins. Many enzymes lose enzymatic activity upon oxidation, which is usually restored by adding thiols, such as β-mercaptoethanol or dithiothreitol (DTT). Because intracellular proteases are released by cell disruption, protease inhibitors such as phenylmethylsulfonyl fluoride (PMSF) must be added to the buffers to avoid protein degradation.

16.3.3.3 Bacterially Expressed Recombinant GST

GSH-affinity resin can also be used for purifying recombinant GST isoenzymes expressed in bacteria. Typically, the GST isoenzyme cDNA clone is expressed in competent *E. coli* (i.e., *E. coli* BL21[DE3]) using a prokaryotic vector (i.e., plasmid pET30a [+]), and expression of GST in large-scale cultures in Luria-Bertani (LB) medium is induced (i.e., by adding 0.4 mM isopropyl β-D-thiogalactoside) approximately eight hours before collecting the bacterial pellet. Homogenates prepared from bacterial lysates in a hypotonic phosphate buffer in the presence of 2 mM EDTA and 1.4 mM ME can be subjected to the GSH-affinity procedure described above.[24] Using this protocol, sufficient quantities of bacterially expressed mGSTA4-4, hGSTP1-1, hGSTA1-1, hGSTA2-2, and hGSTA4-4 have been obtained.[24–26,30,67–75]

16.3.4 SEPARATION OF INDIVIDUAL GST ISOENZYMES

Total purified cytosolic GST fractions, depending on the source, may contain several isoenzymes that are partially distinguishable in SDS-PAGE by differences in their M_r (Figure 16.1). Purification and identification of constituent isoenzymes from various sources have been the subject of numerous reports.[18,20–24,47–53,76] Separation of these isoenzymes has been achieved through various fractionation techniques, based on the differences in pI values of different classes of GSTs. One separation method based on pI differences is column chromatofocusing, in which proteins are

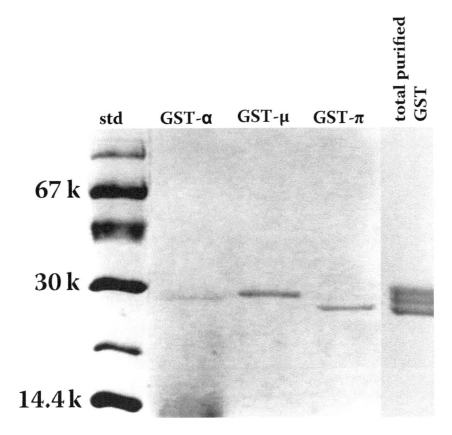

FIGURE 16.1 SDS/β-mercaptoethanol/polyacrylamide gel electrophoresis of purified GST α (25 K), GST μ (26.5 K), and GST π (22.5 K) isolated from human lungs using GSH-affinity column chromatography followed by an isoelectric focusing column.

bound to a solid-phase matrix functionalized with ion exchange groups, and elution is carried out by creating a pH gradient using Ampholines buffers. Among the various supports suitable for this technique, Mono P type columns from Pharmacia have been widely used to resolve the different isoforms of GSH-requiring enzymes.[63,77] Anion-exchange HPLC on a Synchrom AX 300 column also provides a rapid separation with excellent enzyme activity recovery (100%) and high purification fold.[63,76,77] Vander Jagt et al.[78] have separated 13 GST isoforms from the human liver, using affinity chromatography on GSH coupled to epoxy-activated Sepharose 4B resin and a two-step chromatofocusing design; the first used a PBE-118 column to focus protein in the 7.5–10 pH range and the second used a PBE-94 column to focus in the 4.5–7.5 pH range. This technique, using FPLC on a Mono P column, has also been reported by Alin et al.[79] Reversed phase HPLC is also an excellent tool for separating GST isoenzymes as well as quantifying them and assessing the purity of samples used for sequencing.[80] Ostlund Farrants et al.[81] reported a method using C_{18} columns, which separates GST μ and π from rat tissues.

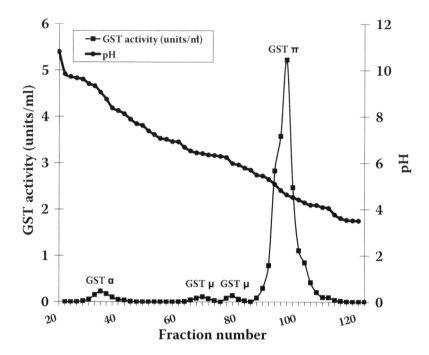

FIGURE 16.2 Isoelectric focusing profile of human lung GST isoenzymes obtained by GSH-affinity chromatography. Ampholines in the 3.5–10.0 pH range were used. (Square), GST activity toward CDNB; (circle), pH gradient. GSH-affinity purified total GST, 54 units (equivalent to 20 g human lung tissue) was used for IEF.

16.3.4.1 Liquid Column Electrofocusing

For separating GST isoenzymes of humans and other species, our laboratory has extensively utilized a liquid column isoelectric focusing approach, the principal advantage of which is that it allowed for the application of relatively large amounts of protein. In a water-jacketed column at 4°C IEF, a sucrose-density gradient column is created with the cathode solution at the bottom and the anode at the top. Application of an electrical field results in a sharply defined pH gradient due to the presence of Ampholines in the solutions. The focused proteins are eluted into 1-ml or smaller fractions from the column by gravity though a valve at the bottom of the column. The isoenzyme peaks obtained during IEF can be pooled separately and subjected to immunological and kinetic characterization. We have used this technique to resolve the major cytosolic GST isoenzymes from many tissues[18,53] (Figure 16.2).

16.3.5 IMMUNO-AFFINITY PURIFICATION

Though effective, column IEF and most ion exchange–based approaches are more tedious than immuno-affinity purification using class-specific anti-GST antibodies. Methods for preparing and purifying GST-class specific rabbit–anti-human polyclonal antibodies have been described by us previously.[43,50,82] In brief, a highly purified preparation of isoenzymes belonging to the specific class of GSTs is required for antibody preparation. Most of the GST antibodies specific to various classes are available commercially. Alternatively, antibodies against the human GSTs α, μ, and π can be raised in rabbits as described by us previously.[43,82] The IgG fractions from these antibodies can be purified and their immunological cross-reactivities exclusively to their respective classes should be established before their use for immunoaffinity studies.[43,82] The IgG fractions obtained from the antibodies can be bound to CNBr-activated Sepharose 4B resin by the procedure of Porath et al.[83] with slight modifications as described previously.[50] The GSH-affinity purified total GST can be incubated with the anti-GST α bound to Sepharose 4B beads with gentle shaking for 12 hours at 4°C. The unabsorbed enzyme can be collected after centrifugation at 3000 × g and the immuno-affinity beads should be thoroughly washed with the 10 mM K-PO$_4$ buffer, pH 7.0. The procedure can be repeated if the enzyme in the unabsorbed fraction indicated cross-reaction with anti-GST α antibodies in Western blot analyses. The enzyme absorbed on immuno-affinity beads can be extracted with 4 M KCNS and immediately subjected to dialysis against 10 mM K-PO$_4$ buffer, pH 7.0, containing 1.4 mM β-mercaptoethanol to remove KCNS. The portion of the enzyme not bound by the anti-GST α immuno-affinity beads can then be treated sequentially with the antibodies raised against the μ-class GST bound to CNBr-activated Sepharose 4B followed by the antibodies raised against the π-class GST bound to CNBr-activated Sepharose 4B in a manner similar to that described above for GST α antibodies.

While studies of the mechanisms for the expression of GST gene family, structures, and physiologic/pharmacologic functions, and mechanisms for the regulation of tissue specific expression of GST isozymes have immensely contributed to our current understanding of this complex enzyme system, the methods for their purification and determination of activities towards various substrates are still central to GST research. Protocols compiled in this chapter should be useful to biochemists, toxicologists, and clinicians. Depending on the constraints of the model system, the nature of the questions asked, and resources in investigators' laboratories, these protocols can be modified to serve specific purposes.

ACKNOWLEDGMENT

Research for this chapter was supported in part by USPHS grants CA 77495 and CA 104661.

REFERENCES

1. Booth, J., Boyland, E., and Sims, P., An enzyme from rat liver catalyzing conjugations with glutathione, *Biochem. J.,* 79, 516, 1961.
2. Jakoby, W.B., The glutathione S-transferases: a group of multifunctional detoxification proteins, *Adv. Enzymol. Relat. Areas Mol. Biol.,* 46, 383, 1978.
3. Chasseaud, L.F., The role of glutathione and glutathione S-transferases in the metabolism of chemical carcinogens and other electrophilic agents. *Adv. Cancer Res.,* 29, 175, 1979.

4. Mannervik, B. and Danielson, U.H., Glutathione S-transferases structure and catalytic activity, *CRC Crit. Rev. Biochem.*, 23, 283, 1988.
5. Hayes, J.D. and Pulford, D.J., The glutathione S-transferase supergene family: regulation of GST and the contribution of the isoenzymes to cancer chemoprotection and drug resistance, *Crit. Rev. Biochem. Mol. Biol.*, 30, 445, 1995.
6. Awasthi, Y.C. et al., Regulation of 4-hydroxynonenal–mediated signaling by glutathione S-transferases, *Free Radic. Biol. Med.*, 37, 607, 2004.
7. Awasthi, Y.C., Sharma, R., and Singhal, S.S., Human glutathione S-transferases, *Int. J. Biochem.*, 26, 295, 1994.
8. Hayes, J.D., Flanagan, J.U., and Jowsey, I.R., Glutathione transferases, *Annu. Rev. Pharmacol. Toxicol.*, 45, 51, 2005.
9. Kamisaka, K. et al., Multiple forms of human glutathione S-transferase and their affinity for bilirubin, *Eur. J. Biochem.*, 60, 153, 1975.
10. Listowsky, I. et al., Intracellular binding and transport of hormones and xenobiotics by glutathione S-transferases, *Drug Metab. Rev.*, 19, 305, 1988.
11. Oakley, A.J. et al., The ligandin (non-substrate) binding site of human pi class glutathione transferase is located in the electrophile binding site (H-site), *J. Mol. Biol.*, 291, 913, 1999.
12. Waxman, D.J., Glutathione S-transferases: role in alkylating agent resistance and possible target for modulation chemotherapy — a review, *Cancer Res.*, 50, 6449, 1990.
13. Morrow, C.S. and Cowan, K.H., Glutathione S-transferases and drug resistance, *Cancer Cells*, 2, 15, 1990.
14. Townsend, D.M and Tew, K.D., The role of glutathione S-transferase in anti-cancer drug resistance, *Oncogene*, 22, 7369, 2003.
15. Pickett, C.B. and Lu, A.Y., Glutathione S-transferases: gene structure, regulation, and biological function, *Annu. Rev. Biochem.*, 58, 743, 1989.
16. Habig, W.H., Pabst, M.J., and Jakoby, W.B., Glutathione S-transferases. The first enzymatic step in mercapturic acid formation, *J. Biol. Chem.*, 249, 7130, 1974.
17. Meyer, D.J. et al., Theta, a new class of glutathione transferases purified from rat and man, *Biochem. J.*, 274, 409, 1991.
18. Singhal, S.S. et al., Glutathione S-transferases of human lung: characterization and evaluation of the protective role of the α-class isozyme against lipid peroxidation, *Arch. Biochem. Biophys.*, 299, 232, 1992.
19. Alin, P., Danielson, U.H., and Mannervik, B., 4-Hydroxyalk-2-enals are substrates for glutathione transferase, *FEBS Lett.*, 179, 267, 1985.
20. Singhal, S.S. et al., Several closely related glutathione S-transferase isoenzymes catalyzing conjugation of 4-hydroxynonenal are differentially expressed in human tissues, *Arch. Biochem. Biophys.*, 311, 242, 1994.
21. Singhal, S.S. et al., A novel glutathione S-transferase isozyme similar to GST 8-8 of rat and mGSTA4-4 (GST 5.7) of mouse is selectively expressed in human tissues, *Biochim. Biophys. Acta*, 1204, 279, 1994.
22. Singhal, S.S. et al., Glutathione S-transferases of human ocular tissues: characterization of several new isozymes catalyzing the conjugation of 4-hydroxynonenal to glutathione, *Invest. Ophth. Vis. Sci.*, 36, 142, 1995.
23. Khan, M.F. et al., Iron-induced lipid peroxidation in rat liver is accompanied with preferential induction of glutathione S-transferase 8-8, *Toxicol. Appl. Pharmacol.*, 131, 63, 1995.
24. Zimniak, P. et al., Estimation of genomic complexity, heterologous expression, and enzymatic characterization of mouse glutathione S-transferase mGSTA4-4 (GST 5.7), *J. Biol. Chem.*, 269, 992, 1994.

25. Hubatsch, I., Ridderstrom, M., and Mannervik, B., Human glutathione transferase A4-4: an alpha class enzyme with high catalytic efficiency in the conjugation of 4-hydroxynonenal and other genotoxic products of lipid peroxidation, *Biochem. J.*, 330, 175, 1998.

26. Patrick, B. et al., Depletion of 4-hydroxynonenal in hGSTA4-transfected HLE B-3 cells results in profound changes in gene expression, *Biochem. Biophys. Res. Commun.*, 334, 425, 2005.

27. Singh, S.P. et al., Catalytic function of *Drosophila melanogaster* glutathione S-transferase DmGSTS1-1 (GST-2) in conjugation of lipid peroxidation end products, *Eur. J. Biochem.*, 268, 2912, 2001.

28. Sawicki, R. et al., Cloning, expression, and biochemical characterization of one epsilon-class (GST-3) and ten delta-class (GST-1) glutathione S-transferases from *Drosophila melanogaster*, and identification of additional nine members of the epsilon class, *Biochem. J.*, 370, 661, 2003.

29. Awasthi, Y.C., Beutler, E., and Srivastava, S.K., Purification and properties of human erythrocyte glutathione peroxidase, *J. Biol. Chem.*, 250, 5144, 1975.

30. Yang, Y. et al., Role of glutathione S-transferases in protection against lipid peroxidation. Overexpression of hGSTA2-2 in K562 cells protects against hydrogen peroxide–induced apoptosis and inhibits JNK and caspase 3 activation, *J. Biol. Chem.*, 276, 19220, 2001.

31. Kokatnur, V.R. and Jelling, M., Iodometric determination of peroxygen in organic compounds, *J. Am. Chem. Soc.*, 63, 1432, 1941.

32. Tsuchida, S., Maki, T., and Sato, K., Purification and characterization of glutathione transferase with an activity toward nitroglycerin from human aorta and heart. Multiplicity of the human class mu forms, *J. Biol. Chem.*, 265, 7150, 1990.

33. Singhal, S.S. et al., Rabbit aorta glutathione S-transferases and their role in bioactivation of trinitroglycerine, *Toxicol. Appl. Pharmacol.*, 140, 378, 1996.

34. Chang, M. et al., Isozyme specificity of rat liver glutathione S-transferases in the formation of PGF2 alpha and PGE2 from PGH2, *Arch. Biochem. Biophys.*, 259, 548, 1987.

35. Arttamangkul, S. et al., 5-(Pentafluorobenzoylamino)fluorescein: a selective substrate for the determination of glutathione concentration and glutathione S-transferase activity, *Anal. Biochem.*, 269, 410, 1999.

36. Soderstrom, M. et al., Leukotriene C4 formation catalyzed by three distinct forms of human cytosolic glutathione transferase, *Biochem. Biophys. Res. Commun.*, 128, 265, 1985.

37. Sharma, R. et al., Glutathione S-transferase catalysed conjugation of 9,10-epoxystearic acid with glutathione, *J. Biochem. Toxicol.*, 6, 147, 1991.

38. Horton, J.K. et al., Characterization of a chlorambucil-resistant human ovarian carcinoma cell line over-expressing glutathione S-transferase μ, *Biochem. Pharmacol.*, 58, 693, 1999.

39. Chang, S.H., High-performance liquid chromatography of proteins, *J. Chromatogr.*, 125, 103, 1976.

40. Heinitz, M.L. et al., Chromatography of proteins on hydrophobic interaction and ion-exchange chromatographic matrices: mobile phase contributions to selectivity, *J. Chromatogr.*, 443, 173, 1988.

41. Clark, A.G. et al., The purification by affinity chromatography of a glutathione S-transferase from larvae of *Galleria mellonella*, *Life Sci.*, 20, 141, 1977.

42. Fausnaugh, J.L., Kennedy, L.A., and Regnier, F.E., Comparison of hydrophobic-interaction and reversed-phase chromatography of proteins, *J. Chromatogr.*, 317, 141, 1984.

43. Singhal, S.S. et al., Purification and characterization of human muscle glutathione S-transferases: evidence that glutathione S-transferase ζ correspond to a locus distinct from GST1, GST2, and GST3, *Arch. Biochem. Biophys.*, 285, 64, 1991.

44. McLellan, L.I., Wolf, C.R., and Hayes, J.D., Human microsomal glutathione S-transferase. Its involvement in the conjugation of hexachlorobuta-1,3-diene with glutathione, *Biochem. J.*, 258, 87, 1989.

45. Maddipati, K.R. and Marnett, L.J., Characterization of the major hydroperoxide-reducing activity of human plasma. Purification and properties of a selenium-dependent glutathione peroxidase, *J. Biol. Chem.*, 262, 17398, 1987.

46. Simons, P.C. and Vander Jagt, D.L., Purification of glutathione S-transferases from human liver by glutathione affinity chromatography, *Anal. Biochem.*, 82, 334, 1977.

47. Awasthi, Y.C. and Singh, S.V., Purification and characterization of a new form of glutathione S-transferase from human erythrocytes, *Biochem. Biophys. Res. Commun.*, 125, 1053, 1984.

48. Singhal, S.S. et al., Glutathione S-transferases of mouse liver: sex-related differences in the expression of various isozymes, *Biochim. Biophys. Acta*, 1116, 137, 1992.

49. Singhal, S.S. et al., Glutathione S-transferases of human skin: qualitative and quantitative differences in men and women, *Biochim. Biophys. Acta*, 1163, 266, 1993.

50. Singhal, S.S. et al., Characterization of a novel α-class anionic glutathione S-transferase isozyme from human liver, *Arch. Biochem. Biophys.*, 279, 45, 1990.

51. Singhal, S.S. et al., Purification and characterization of glutathione S-transferases from rat pancreas, *Biochim. Biophys. Acta*, 1079, 285, 1991.

52. Singhal, S.S. et al., Gender-related differences in the expression and characteristics of glutathione S-transferases of human colon, *Biochim. Biophys. Acta*, 1171, 19, 1992.

53. Awasthi, S. et al., Purification and characterization of glutathione S-transferase of murine ovary and testis, *Arch. Biochem. Biophys.*, 301, 143, 1993.

54. Guthenberg, C. and Mannervik, B., Purification of glutathione S-transferases from rat lung by affinity chromatography. Evidence for an enzyme form absent in rat liver, *Biochem. Biophys. Res. Commun.*, 86, 1304, 1979.

55. Guthenberg, C. et al., Two distinct forms of glutathione transferase from human foetal liver. Purification and comparison with isoenzymes isolated from adult liver and placenta, *Biochem. J.*, 235, 741, 1986.

56. Stenberg, G. et al., Effects of directed mutagenesis on conserved arginine residues in a human class alpha glutathione transferase, *Biochem. J.*, 274, 549, 1991.

57. Didderstrom, M. et al., Mutagenesis of residue 157 in the active site of human glyoxalase I, *Biochem. J.*, 328, 231, 1997.

58. Miki, H. et al., Association of Ash/Grb-2 with dynamin through the Src homology 3 domain, *J. Biol. Chem.*, 269, 5489, 1994.

59. Pearce, S.F.A., Wu, J., and Silverstein, R.L., Recombinant GST/CD36 fusion proteins define a thrombospondin binding domain evidence for a single calcium-dependent binding site on Cd36, *J. Biol. Chem.*, 270, 2981, 1995.

60. Odai, H. et al., Purification and molecular cloning of SH2- and SH3-containing inositol polyphosphate-5-phosphatase, which is involved in the signaling pathway of granulocyte-macrophage colony-stimulating factor, erythropoietin, and Bcr-Abl, *Blood*, 89, 2745, 1997.

61. Kim, K. et al., Effect of Fgd1 on cortactin in Arp2/3 complex-mediated actin assembly, *Biochemistry*, 43, 2422, 2004.

62. Simons, P.C. and Vander Jagt, D.L., Purification of glutathione S-transferases by glutathione-affinity chromatography, *Methods Enzymol.*, 77, 235, 1981.

63. Singh, S.V., Ansari, G.A.S., and Awasthi, Y.C., Anion-exchange high-performance liquid chromatography of glutathione S-transferases. Separation of the minor isoenzymes of human erythrocyte, heart, and lung, *J. Chromatogr.*, 361, 337, 1986.

64. Srivastava, S.K. et al., A group of novel glutathione S-transferase isozymes showing high activity towards 4-hydroxy-2-nonenal are present in bovine ocular tissues, *Exp. Eye Res.*, 59, 151, 1994.

65. Tewari, N.K. et al., Purification and characterization of glutathione S-transferases of *E. Loftini* (Dyar) and *Diaterea Saccharalis* (F), *J. Econ. Entomol.*, 84, 1424, 1991.

66. Singhal, S.S. et al., Comparison of glutathione S-transferase isoenzymes in human leukemia K562, HL60, and U-937 cells, *Biochem. Arch.*, 15, 163, 1999.

67. Singh, S.V. et al., Catalytic differences between alpha class human glutathione transferases hGSTA1-1 and hGSTA2-2 for glutathione conjugation of environmental carcinogen benzo(a)pyrene-7,8-diol, *Biochemistry*, 43, 9708, 2004.

68. Zhao, T. et al., The role of glutathione S-transferases hGSTA1-1 and hGSTA2-2 in protection against oxidative stress, *Arch. Biochem. Biophys.*, 367, 216, 1999.

69. Gu, Y. et al., Crystal structure of human glutathione S-transferase A3-3 and mechanistic implications for its high steroid isomerase activity, *Biochemistry*, 43, 15673, 2004.

70. Sharma, R. et al., Transfection with 4-hydroxynonenal-metabolizing glutathione S-transferase isozymes leads to phenotypic transformation and immortalization of adherent cells, *Eur. J. Biochem.*, 271, 1690, 2004.

71. Yang, Y. et al., Glutathione-S-transferase A4-4 modulates oxidative stress in endothelium: possible role in human atherosclerosis, *Atherosclerosis*, 173, 211, 2004.

72. Gardner, J.L. and Gallagher, E.P., Development of a peptide antibody specific to human glutathione S-transferase alpha 4-4 (hGSTA4-4) reveals preferential localization in human liver mitochondria, *Arch. Biochem. Biophys.*, 390, 19, 2001.

73. Desmots, F. et al., Genomic organization, 5′-flanking region and chromosomal localization of the human glutathione transferase A4 gene, *Biochem. J.*, 336, 437, 1998.

74. Pandya, U. et al., Activity of allelic variants of pi class human glutathione S-transferase toward chlorambucil, *Biochem. Biophys. Res. Commun.*, 278, 258, 2000.

75. Reinemer, P. et al., Three-dimensional structure of class pi glutathione S-transferase from human placenta in complex with S-hexylglutathione at 2.8 A resolution, *J. Mol. Biol.*, 227, 214, 1992.

76. Radulovic, L.L. and Kulkarni, A.P., A rapid, novel high performance liquid chromatography method for the purification of glutathione S-transferase: an application to the human placental enzyme, *Biochem. Biophys. Res. Commun.*, 128, 75, 1985.

77. Radulovic, L.L. and Kulkarni, A.P., H.p.l.c. separation and study of the charge isomers of human placental glutathione transferase, *Biochem. J.*, 239, 53, 1986.

78. Vander Jagt, D.L. et al., Isolation and characterization of the multiple glutathione S-transferases from human liver. Evidence for unique heme-binding sites, *J. Biol. Chem.*, 260, 11603, 1985.

79. Alin, P. et al., Purification of major basic glutathione transferase isoenzymes from rat liver by use of affinity chromatography and fast protein liquid chromatofocusing, *Anal. Biochem.*, 146, 313, 1985.

80. Esworthy, R.S. et al., Characterization and partial amino acid sequence of human plasma glutathione peroxidase, *Arch. Biochem. Biophys.*, 286, 330, 1991.

81. Ostlund Farrants, A.K. et al., The separation of glutathione transferase subunits by using reverse-phase high-pressure liquid chromatography, *Biochem. J.*, 245, 423, 1987.

82. Singhal, S.S. et al., Polyclonal antibodies specific to human glutathione S-transferase 5.8 (hGST 5.8), *Biochem. Arch.*, 11, 189, 1995.

83. Porath, J., Axen, R., and Ernback, S., Chemical coupling of proteins to agarose, *Nature*, 215, 1491, 1967.

84. Pacifici, G.M. et al., Detoxification of styrene oxide by human liver glutathione transferase, *Hum. Toxicol.*, 6, 483, 1987.

85. Johansson, A.S. and Mannervik, B., Human glutathione transferase A3-3, a highly efficient catalyst of double-bonded isomerization in the biosynthetic pathway of steroid hormones, *J. Biol. Chem.*, 276, 33061, 2001.

86. Stenberg, G. et al., Cloning and heterologous expression of cDNA encoding class alpha rat glutathione transferase 8-8, an enzyme with high catalytic activity towards genotoxic, α,β-unsaturated carbonyl compounds, *Biochem. J.*, 284, 313, 1992.

87. Bjornestedt, R., Tardioli, S., and Mannervik, B., The high activity of rat glutathione transferase 8-8 with alkene substrates is dependent on a glycine residue in the active site, *J. Biol. Chem.*, 270, 29705, 1995.

88. Board P.G., Identification of cDNAs encoding two human Alpha class glutathione transferases (GSTA3 and GSTA4) and the heterologous expression of GSTA4-4, *Biochem. J.*, 330, 827, 1998.

89. Habig, W.H., Pabst, M.J., and Jakoby, W.B., Glutathione S-transferase AA from rat liver, *Arch. Biochem. Biophys.*, 175, 710, 1976.

90. Jernstrom, B. et al., Glutathione S-transferase A1-1-catalysed conjugation of bay and fjord region of diol epoxides or polycyclic aromatic hydrocarbons with glutathione, *Carcinogenesis*, 17, 1491, 1996.

91. Bull, A.W. et al., Conjugation of the linoleic acid oxidation product, 13-oxooctadeca-9,11-dienoic acid, a bioactive endogenous substrate for mammalian glutathione transferase, *Biochim. Biophys. Acta*, 157, 77, 2002.

92. Eklund, B.I. et al., Screening for recombinant glutathione transferases active with monochlorobimane, *Anal. Biochem.*, 309, 102, 2002.

93. Guengerich, F.P. et al., Activation and detoxification of aflatoxin B1, *Mutat. Res.*, 402, 121, 1998.

Index

T - #0348 - 071024 - C8 - 234/156/18 - PB - 9780367390532 - Gloss Lamination